PLANET EARTH
AND THE NEW GEOSCIENCE

by
Victor A. Schmidt
Department of Geology and Planetary Science
University of Pittsburgh

through the
University External Studies Program
University of Pittsburgh

in coordination with the television series
PLANET EARTH
Produced by WQED/Pittsburgh
in association with
The National Academy of Sciences

Major Funding by The Annenberg/CPB Project
Corporate Funding by IBM

Kendall/Hunt
Publishing Company
Dubuque, Iowa

This edition has been printed directly from the author's manuscript copy.

Copyright © 1986 by Metropolitan Pittsburgh Public Broadcasting, Inc.

Library of Congress Catalog Card Number: 85–81502

ISBN 0–8403–3809–0

Printed in the United States of America

B   403809   02

NATIONAL ACADEMY OF SCIENCES
GEOPHYSICS FILM COMMITTEE

Roger R. Revelle (Chairman)
University of California, San Diego

G. Arthur Barber
Deep Observation and Sampling
of the Earth's Continental
Crust, Inc. (DOSECC)

Charles L. Drake
Dartmouth College

Herbert Friedman
National Academy of Sciences

Laurence M. Gould
University of Arizona

Thomas F. Malone
St. Joseph College

Stanley Ruttenberg
University Center for
Atmospheric Research

John P. Schaefer
The Research Corporation

Alan H. Shapley
NOAA/NESDIS

Eugene M. Shoemaker
U.S. Geological Survey

Walter S. Sullivan
The New York Times

Verner E. Suomi
University of Wisconsin

James A. Van Allen
University of Iowa

J. Tuzo Wilson
Ontario Science Centre

James H. Zumberge
University of Southern California

Richard W. Birnie (Consultant)
Dartmouth College

ADVISORY PANELS

The National Academy of Sciences appointed the Geophysics Film committee that named seven advisory panels to work with WQED/Pittsburgh and the University of Pittsburgh. Each of these panels is composed of scientists whose expertise is appropriate to the subject matter of the individual programs; they have checked the authenticity, accuracy and scientific integrity of each of the seven programs in the series, as well as the content of **Planet Earth and the New Geoscience** by Victor A. Schmidt.

THE LIVING MACHINE:
Plate Tectonics and Continental Tectonics and the Earth's Interior

Charles L. Drake (Chairman)
Dartmouth College

Bruce A. Bolt
University of California, Berkeley

Allan V. Cox
Stanford University

V. R. Murthy
University of Minnesota

A. R. Palmer
Geological Society of America

Eugene M. Shoemaker
U.S. Geological Survey

J. Tuzo Wilson
Ontario Science Centre

THE BLUE PLANET:
Physical and Chemical Makeup of the Oceans and Dynamics of the Oceans

Arthur E. Maxwell (Chairman)
University of Texas, Austin

D. James Baker
Joint Oceanographic Institutions, Inc.

Charles D. Hollister
Woods Hole Oceanographic Institution

James J. McCarthy
Harvard University

Pearn P. Niiler
Scripps Institution of Oceanography

William B. F. Ryan
Lamont-Doherty Geological Observatory

Karl K. Turekian
Yale University

THE CLIMATE PUZZLE:
The Atmosphere and Climates of Earth

Warren M. Washington (Chairman)
National Center for Atmospheric Research

John Imbrie
Brown University

Judith Totman Parrish
U.S. Geological Survey

Charles R. Bentley
University of Wisconsin

John E. Kutzbach
University of Wisconsin

Stephen H. Schneider
National Center for Atmospheric Research

James Hansen
Goddard Institute for Space Studies

TALES FROM OTHER WORLDS:
The Solar Family and Origins

Michael H. Carr (Chairman)
U.S. Geological Survey

Bruce C. Murray
California Institute of Technology

Carle M. Pieters
Brown University

James R. Arnold
(Chairman June 1984 - March 1985)
University of California, San Diego

Tobias Owen
State University of New York
Stony Brook

David M. Raup
University of Chicago

Harold Masursky
U.S. Geological Survey

GIFTS FROM THE EARTH:
Mineral Resources and Energy Resources

G. Arthur Barber (Co-Chairman)
Deep Observation and Sampling
of the Earth's Continental
Crust, Inc. (DOSECC)

Lawrence W. Funkhouser
Chevron Corporation

Charles W. Meyer
University of Arizona

Charles J. Mankin (Co-Chairman)
University of Oklahoma

Harold J. Gluskoter
Exxon Production Research Company

Siegfried Muessig
Getty Mining Company

Wendell A. Duffield
U.S. Geological Survey

Richard W. Hutchinson
Colorado School of Mines

THE SOLAR SEA:
The Sun and Interactions Between Sun and Earth

Raymond G. Roble (Chairman)
National Center for Atmospheric Research

Louis A. Frank
University of Iowa

George Reid
National Oceanic and Atmospheric
Administration

John A. Eddy
National Center for Atmospheric Research

Louis J. Lanzerotti
Bell Laboratories

FATE OF THE EARTH:
The Balance of Nature and The Impact of Man

F. Kenneth Hare (Chairman)
University of Toronto/Trinity College

David W. Deamer
University of California, Davis

Peter H. Raven
Missouri Botanical Gardens

Stanley Awramik
University of California,
Santa Barbara

Michael B. McElroy
Harvard University

Stephen H. Schneider
National Center for Atmospheric Research

CREDITS FOR PLANET EARTH

Thomas Skinner, Project Director and Executive Producer, WQED
Frank Press, President, The National Academy of Sciences
Marc Pollock, Project Manager, WQED
Barbara Valentino, Project Administrator for The National Academy of Sciences

Television Series

Gregory Andorfer, Series Producer
Amy Barraclough, Production Assistant
Robin Bates, Producer
Patricia Colvig, Production Assistant
Debbie Glovin, Associate Producer/Researcher
Georgann Kane, Coordinating Producer/Producer
Wayne Morris, Production Manager/Post-production Supervisor
Deane Rink, Associate Producer/Researcher
Michelle Rodriguez, Production Assistant/Secretary
Ted Thomas, Producer
Jim Toll, Unit Manager
Bonnie Waltch, Production Assistant
William Walter, Associate Producer/Researcher

Textbook

Frank W. Benacquista, Graphic Artist and Research Assistant
Fred G. Bost, Project Coordinator
M. Bernardine Gasior, Computer Systems Management
Barbara Pace, Technical Editor
Erin E. Peszko, Text Design and Layout
Michael B. Spring, Director, University External Studies Program

ACKNOWLEDGMENTS

We would like to thank the following people for their assistance in the preparation of Planet Earth and the New Geoscience:

The Annenberg/CPB Project: Hyman Field
The National Academy of Sciences: Barbara Valentino
Metropolitan Pittsburgh Public Broadcasting, Inc: Marc Pollock
Xerox Corporation: Kenneth Holes, William H. Moses, John Robison, and Gill Winkler
NAS Content Consultant: Richard W. Birnie, Dartmouth College

A NOTE ON DESIGN

Planet Earth and the New Geoscience was produced via version 5.0 of the Xerox Integrated Composition System (XICS) output to a XEROX 8700 MOD V GHO laser printer. Selected graphics were composed on a XEROX Star workstation and all graphics and images were scanned using a XEROX 150 Graphic Input Station. The text is composed of more than 150 segments which are concatenated and composed by XICS on demand from individual faculty. Final copies of the textbook use the Bodoni font family.

DEDICATION

My efforts in writing this text are dedicated to
Marji, Wendy, and Ben, who understood and helped.

Victor A. Schmidt

TABLE OF CONTENTS

LIST OF FIGURES

UNIT I A SENSE OF TIME

THE GRAND CANYON

A. INTRODUCTION

1. Overview

Any understanding of modern Earth science must begin with a comprehension of the vast stretches of geologic time. The essential processes that have shaped the Earth are slow-acting, yet all parts of the Earth's surface show evidence of profound and numerous changes. The Earth must be very ancient to accommodate these changes. We shall first consider the history of attempts to determine the age of the Earth.

Today, two broad methods are used for determining the age of rocks: <u>relative</u> methods that determine only the order or sequence in which different layers of rocks formed, and <u>absolute</u> methods, which determine the age of individual rocks in years.

Radioisotope age determination is the principal method for the latter purpose, and we shall study it in some detail. Because many people regard the great antiquity of the Earth as inconsistent with their religious beliefs, a discussion of the method of science is necessary so that you can understand how scientists choose between conflicting theories. We will continue this discussion throughout the telecourse.

Finally, you will be invited to stretch your mind to accommodate the vastness of geologic time. This will provide you with the necessary framework upon which to develop the concepts of this telecourse.

2. Objectives

Upon completion of this unit you should be able to:

1. recognize evidence which indicates the antiquity of features on the Earth's surface
2. discuss the history of geologists' attempts to determine the age of the Earth
3. explain the difference between the catastrophist and uniformitarian approaches to geological explanation
4. explain how geologists can place different rock layers in sequence according to their ages
5. use the law of radioactive decay to determine the ages of certain rocks and minerals
6. describe the scientific method as an interplay between theory and observations
7. discuss the nature of scientific truth
8. identify the eras and periods of geological time, place them in chronological order, and relate them to ages determined radioisotopically.

3. Key Terms and Concepts

erosion

strata

Principle of Original Horizontality

Principle of Superposition

lithification

uniformitarianism

catastrophism

fossils

paleontology

extinctions

natural radioactivity

absolute ages

relative ages

radioisotope

parent atoms

daughter atoms

radioisotope age dating

half-life

Law of Radioisotope Decay

creationists

scientific method

scientific truth

Eras: Precambrian, Paleozoic, Mesozoic, Cenozoic

Periods: Cambrian, Ordovician, Silurian, Devonian, Carboniferous, Permian, Triassic, Jurassic, Cretaceous, Tertiary, Quaternary.

B. "...AS OLD AS THE HILLS"

The familiar expression recognizes that the features of the landscape are old compared to the lifetimes of humans. But how old? A thousand lifetimes? A million? A billion?

1. How Old is the Earth?

When monuments have been constructed that are designed for permanence, the material of choice has usually been that of the hills: stone. Yet few of man's ancient monuments have withstood the passage of time without change. The etchings of wind, sand, water, and ice have taken their toll on the once-proud features of ancient stone kings, and even the modern colossus carved from granite at Mt. Rushmore requires annual maintenance. The hills may be old, but they are changing every day.

That the graven images of Egyptian kings have survived at all, shows that the processes of weathering and erosion are generally slow. Yet year by year, the angular becomes more rounded, cracks are enlarged and blocks are loosened by the alternate freezing and thawing of water, and the surfaces of all exposed rocks gradually weather and crumble to form gravel, sand, soil, and clay.

Some hills are clearly older than others, and it does not take too much imagination to guess that the low, rounded, and stable shapes of the Appalachians or the mountains of Germany's Black Forest are older than the sharp upthrust spires of the Alps, Himalayas, and Rockies. Our original question becomes more complicated. How old are the hills? Some older than others, it would seem, but we are still left with a failure of sense of time.

The hills must be far older than the Egyptian kings, for the statues have changed little, while the hills have changed much. Even the youngest mountains have seen numerous massive rockfalls and had valleys cut into their flanks. Perhaps they are vastly older than the most ancient monuments.

Such were the musings of James Hutton in the eighteenth century. A Scottish physician and geologist, his observations had led him to contrast the relative effects of weathering on the mountains and hills of his homeland and on Hadrian's Wall, a stone edifice built by the Romans in 136 AD in the north of England to defend their northern borders.

Following our same line of reasoning, Hutton concluded that the hills had to be very ancient. Hadrian's Wall was then over 1,600 years old but its rocks had changed only slightly. On the other hand, the mountains he had studied had been eroded to such an extent that in places many layers had been stripped away and their hearts had been exposed. Hutton concluded the time available to the processes that shape the Earth must be very great.

Perhaps his most famous observation was made at Siccar Point in Scotland. At this site, the North Sea had laid bare a succession of rock layers (called strata in the plural, or stratum in the singular). The configuration is diagramed in Figure I-1. You will visit the actual site in the film. All the rocks at Siccar Point are of a type called sedimentary rocks, which are formed by the depositing of layers of sediments such as sand, mud, or clay followed by compaction and chemical cementing of the grains into rock. This process is called lithification.

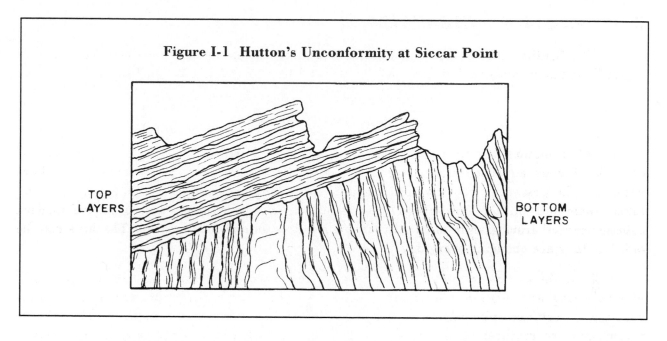

Figure I-1 Hutton's Unconformity at Siccar Point

TOP
LAYERS

BOTTOM
LAYERS

In order to interpret how this sequence of strata came to be, Hutton drew on geological principles that had been established by earlier workers. Two of these are the Principle of Original Horizontality and the Principle of Superposition. The former states that sedimentary rocks generally form with each layer flat and approximately horizontal. Any folding or tilting of the strata must have occurred at a later time.

The latter principle states that in a pile of layers or strata, the topmost formed latest, and the oldest are on the bottom of the pile. The situation is analogous to the pile of pancakes produced on a platter beside the stove by a cook. When the platter is passed around, whoever gets the topmost pancake gets the freshest one. These two simple principles allowed Hutton to deduce the history of the Siccar Point rocks. He constructed a mental model of this history that is shown in Figure I-2.

Hutton's theoretical model began with the deposition of layers of sediment. The deposition had to take place underwater, because that is where that type of sediment is normally found being deposited today. After the sediment was lithified into sedimentary rock, it was compressed and folded, somewhat in the manner that a large rug will rumple and fold if one end of it is pushed.

Next, erosion planed off the rock to a flat surface, a process that cannot occur underwater, but must occur at or above sea level. Then fresh sediment was deposited on top of the folds. This sediment in turn became lithified and the whole body of rock was tilted, perhaps as part of another episode of folding. Finally, erosion shaped the rocks to the form that Hutton saw in the eighteenth century.

In the film you will see the same rocks as they appear today, a bit more than 200 years later. They have not changed much in that time interval.

Hutton was overwhelmed by the implications of his model. Each of the many steps clearly required enormous amounts of time, and the model contained two complete cycles of

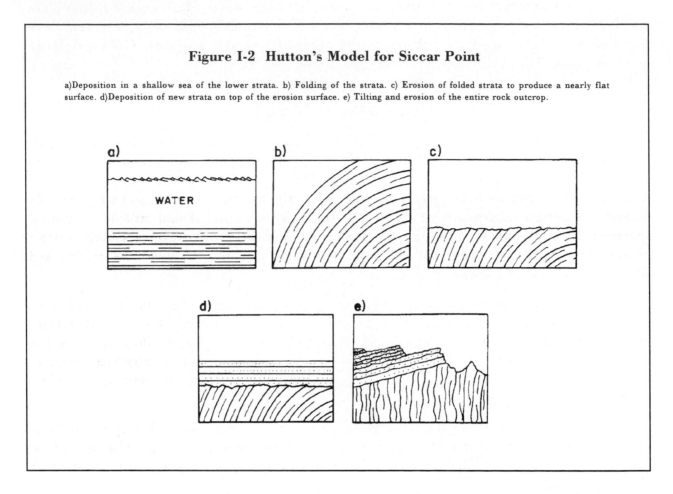

Figure I-2 Hutton's Model for Siccar Point

a)Deposition in a shallow sea of the lower strata. b) Folding of the strata. c) Erosion of folded strata to produce a nearly flat surface. d)Deposition of new strata on top of the erosion surface. e) Tilting and erosion of the entire rock outcrop.

deposition, lithification, folding, and erosion. The rocks that told this story were only a small fraction of the many strata he had examined. He saw a "succession of former worlds" in which mountains and landmasses were built and destroyed in endless cycles, with "no vestige of a beginning, no prospect of an end."

Hutton's view of the extent of geological time was at sharp variance with the prevailing opinions of his age. It had been recognized for some time that the written record of human history encompassed only some 5,000 years. An anthropocentric, or human-centered, view of the world needed to add but little to that span. Almost 200 years before Hutton, Shakespeare in <u>As You Like It</u> had Rosalind say, "The poor world is almost six thousand years old...."

Only fifty years later, Shakespeare's surmise was to receive strong support when in 1654 the Anglican Archbishop James Ussher counted back through the "begats" in the Book of Genesis, adding up the years of each generation back to Adam and Eve. His work was duplicated and refined a few years later by the Cambridge scholar Dr. John Lightfoot, who wrote:

> Heaven and Earth, center and circumference, were made in the same instance of time, and clouds full of water, and man was created by the Trinity on the 26th of October 4004 BC at 9 o'clock in the morning.

Today's scientists can only envy the precision of that dating, but other religions place the date of Creation at other times ranging back to thousands of millions of years. Since Christianity was the dominant religion of Western Europe, the less than 6,000-year span ascribed to the Bible would become a dogma difficult to dislodge.

2. Uniformitarianism and Catastrophism

This brief span provided geologists with considerable difficulty. One of Hutton's principal accomplishments was to insist that no ancient geological processes may be invoked that could not be observed to be in operation in historical times. This became known as the "uniformitarian" approach to geology, in the sense that natural processes operate uniformly through the ages, and processes that operated in the ancient past should still be observable operating in the present. With the uniformitarian approach, 6,000 years clearly doesn't provide enough time for any but the most minor of changes, let alone the raising and destruction of mountain ranges!

An alternative approach was already available. If great changes have to be accomplished in a short time, then they must be catastrophic when they occur. Observing that some geological processes, such as volcanic eruptions, earthquakes, and severe flooding can effect substantial changes in the landscape in a short time, some geologists of the nineteenth century argued that all geologic change could be explained by invoking horrendous catastrophes, orders of magnitude greater than any observed in historic times.

The Bible provided a ready example in Noah's Flood, dated by Ussher as occurring in the year 2348 BC. Here was a catastrophe of heroic proportions, perhaps great enough to carve out mountains and valleys in a matter of days, and to deposit sea shells near the summits of mountain peaks.

Throughout the eighteenth century, the distribution of fossils in rocks provided some interesting arguments. Studies of these fossils showed very clearly that the remains of present life forms were not found in all strata. In fact, it was one of the singular accomplishments of early geology to recognize that the appearances and disappearances of fossil life forms are so well organized within the different strata that it is possible to correlate the strata from one place to another solely on the basis of the characteristic life forms found within them.

As an example, the trilobite (Figure I-3) was a hard-shelled creature that lived in shallow seas. All over the world its fossils are found only in a certain range of strata that are called Paleozoic, meaning "ancient life." Rocks above (younger than) these strata do not contain their fossils, nor do rocks below (older than) them. Trilobites appeared at the beginning of the Paleozoic Era and at a much later time became extinct, marking the end of the Paleozoic Era. There are no trilobites (nor even any close relatives) living today.

Thousands of other fossils may similarly be used to correlate particular strata around the world. Dinosaur bones are found only in rocks that lie atop the Paleozoic strata. These have been termed Mesozoic strata, meaning "middle life." Above these, still younger rocks contain abundant mammal fossils, but no dinosaur bones. Only in the most recent (topmost) layers, do the bones of primates and men make an appearance, in the era of the Cenozoic, meaning "recent life."

Figure I-3 Sketch of a Trilobite Fossil

Part O. Arthropoda 1: Fig. 245, 1, p. 0333, "From Treatise on Invertebrate Paleontology," courtesy of the Geological Society of America and University of Kansas.

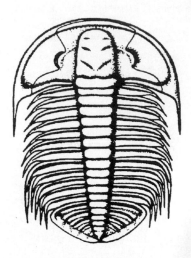

Geologists who study fossils are called <u>paleontologists</u>, and the science of paleontology even today provides the most precise techniques for determining the sequence of events throughout geologic time. The comings and goings of the thousands of different fossil lifeforms mark the passage of geologic time, and make it possible to subdivide the eras into shorter periods, as diagramed in Figure I-4. The geological time scale was constructed by the end of the nineteenth century, using the basic Principles of Superposition, Original Horizontality, and others that aided in solving the gigantic jigsaw puzzle of worldwide rock strata.

Paleontology, together with these principles makes it possible to determine the <u>sequence</u> of geologic time, but not the actual number of years that have elapsed during each era or period. They provide only <u>relative ages</u>, not <u>absolute ages</u>. Largely because of this, both uniformitarians and catastrophists used the observations of paleontology to support their views.

The catastrophists noted that at various points in geologic time (especially at the end of the Paleozoic and Mesozoic Eras) large numbers of species appeared to become extinct at the same time. Surely, they argued, these great extinctions must mark worldwide cataclysms far beyond anything experienced in historic times.

On the other hand, the uniformitarians pointed out that the fossil record seemed to require no less than 27 distinct catastrophes, while the Bible mentioned only one: Noah's Flood. In addition, the absence of human bones from all but the very topmost strata argued that geologic time encompassed very many times the span of human history.

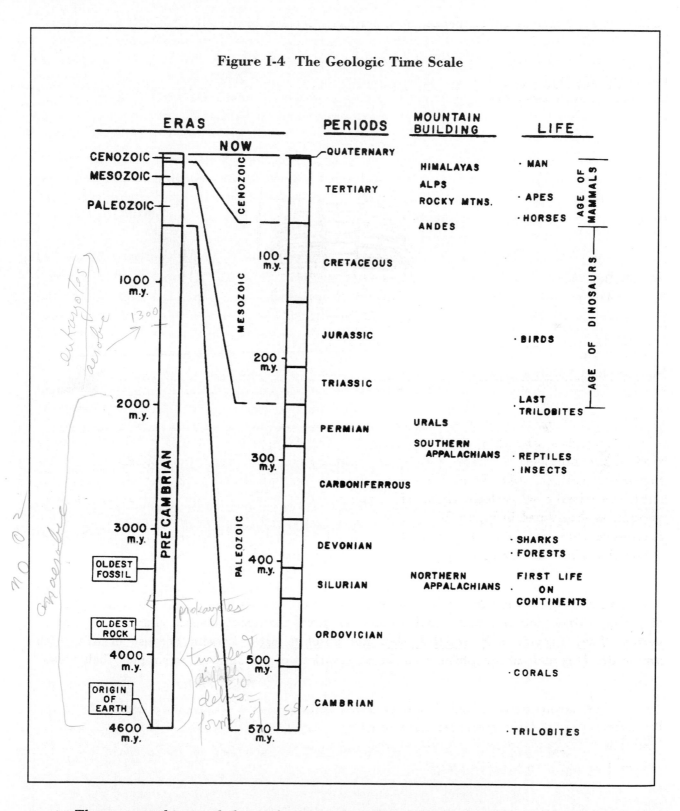

Figure I-4 The Geologic Time Scale

The catastrophists and the uniformitarians waged verbal war with one another through-out the nineteenth century, with the uniformitarians gradually gaining the upper hand by the turn of the century. Instrumental in this development was the work of two contemporary

Englishmen who strongly influenced one another: Charles Lyell and Charles Darwin. Lyell wrote the first comprehensive textbook on geology and stressed uniformitarian principles. Darwin's theory of evolution used the succession of fossils worked out by paleontologists to show how life had evolved from simple primitive forms in the earlier strata to ever-increasing complexity and more highly-developed creatures in the later strata. Darwin's "survival of the fittest" provided the mechanisms to explain extinctions and appearances of new species.

Darwin's evolution was based on slow change from one generation to the next, and so required vast amounts of time. It meshed well with uniformitarian principles, and by the late 1900's, the Darwinists and the uniformitarians had concluded Hutton was literally correct in proclaiming "no vestige of a beginning." They regarded the Earth as incomprehensibly old.

Other nineteenth century scientists, less concerned with the fossil record, attempted to measure the age of the Earth using independent methods. In 1715, Edmund Halley, of comet fame, tried to use the saltiness of the oceans to accomplish the task. He surmised that the oceans were salty because the rivers of the world carried down into the ocean small amounts of salt dissolved from the continental rocks and soil. The water would eventually evaporate and recycle back through the rivers as rain, but the salt unable to evaporate, would be left behind and accumulate. Halley did not carry out the actual calculations, but two hundred years later, the Irish scientist John Joly did. He measured the average salt content of river water, estimated the total flow of all the rivers of the world, and calculated how long it would take to bring the total volume of ocean water to its present level of saltiness. His answer: 80 to 89 million years.

Lord Kelvin, one of the greatest physicists of all time, took a different tack. Assuming the Earth to have begun as a ball of incandescent molten rock, he calculated how long it would take to cool to its present state of heat loss, which had recently been determined by measuring how the temperature increases with depth in deep mines. His answer: 100 million years.

Though in fairly close agreement with one another, these ages shocked everyone. Too long for the catastrophists and too short for the uniformitarians, they did little to settle the controversy.

STUDY QUESTIONS

I-1. When sediments are deposited in a shallow sea, why are they likely to produce flat, horizontal layers?

I-2. In what kind of environment might you find each step of the process in Figure I-2 occurring today?

I-3. Which of the following hypothetical events would be classed as catastrophist rather than uniformitarian in nature: a) the eruption of a volcano, b) the destruction of a city by an earthquake, c) the collision of an asteroid with the Earth, d) the carving of the Grand Canyon over a million years by erosion, e) Noah's Flood?

I-4. Which came first, dinosaurs or trilobites? Base your answer on the observation that in localities where the undisturbed fossils of both are found, the layers containing dinosaur fossils always lie above the layers containing trilobite fossils.

C. RADIOISOTOPE AGE DATING

A momentous discovery in 1896 changed the whole situation. In that year, the French scientist Henri Becquerel discovered natural radioactivity. Natural radioactivity is the process by which the nucleus of an unstable atom spontaneously changes (decays) to a different atom, releasing energy (and sometimes other particles) as it does so. If the new nucleus just produced is also radioactive, the process will repeat until the end product is a stable (non-radioactive) atom. The original radioactive atom is called a parent, and the final stable atom is called a daughter. It was soon discovered that the rate at which parent changed to daughter was governed by a simple relation that was unaffected by temperature, pressure, or chemical environment. At last, a reliable clock had been found for measuring geologic time.

Because it is so important, let us examine the process in some detail. Particular atomic nuclei are called isotopes, and the method is called radioisotope age dating. In the processes that we will consider, each time a radioactive parent atom decays it is replaced by a stable daughter atom which is no longer radioactive. The total number of atoms remains constant. Consider what happens if we start with a number of parent atoms and no daughter atoms in a rock. As time goes on, more and more of the parents will decay, and the parent population will decrease while the daughter population will increase. If we know the rate of decay, then we can determine the age of the rock by measuring the relative proportions of parent and daughter at the present time. The greater the proportion of daughter present, the older the rock.

TABLE I-1: SOME RADIOISOTOPES USED IN AGE DATING

Decay scheme	Use	Half-life
Carbon-14 to Nitrogen-14	Archaeology	5,730 years
Potassium-40 to Argon-40	Geology	12,500 million years
Rubidium-87 to Strontium-87	Geology	48,800 million years
Uranium-235 to Lead-207	Geology	704 million years
Uranium-238 to Lead-206	Geology	4,470 million years
Thorium-232 to Lead-208	Geology	14,010 million years

Figure I-5 shows how the proportions change with time. Time is measured along the horizontal axis in half-lives, where one half-life is a particular length of time that is characteristic to each radioisotope and determines its rate of decay. Table I-1 lists the length of a half-life for several radioisotopes used in age dating. Each isotope is identified by an element

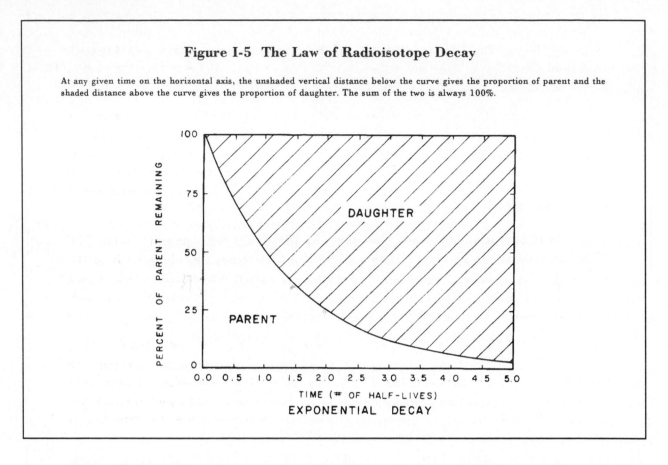

Figure I-5 The Law of Radioisotope Decay

At any given time on the horizontal axis, the unshaded vertical distance below the curve gives the proportion of parent and the shaded distance above the curve gives the proportion of daughter. The sum of the two is always 100%.

name followed by a number that specifies the weight of the particular isotope. The weight is given as the sum of the number of protons plus the number of neutrons in the nucleus. The kind of element and its weight and the proportions of each isotope can be determined using a sensitive instrument called a mass spectrometer. It is important to recognize that even though each radioisotope has a different half-life, all radioisotopes are observed to decay according to the law given by Figure I-5.

The decay curve in Figure I-5 has some interesting properties. At the beginning of the process we start with 100% parent and no daughter. After one half-life of time has elapsed the proportion of parent has dropped to one-half (hence the name). After another half-life, the percentage of parent has decreased by another factor of one-half, to one quarter (25%), and so on. Of the <u>parent</u> atoms that exist at any particular time, just half will decay during the next half-life of time.

The Law of Radioisotope Decay is a statistical law. It works according to the following principle: in any fixed interval of time, each radioactive atom has a certain fixed probability of decaying. Mathematicians refer to this as "exponential decay."

Try the following exercise to see how it works. When a coin is flipped, there is a 50% probability of it coming up "heads." We can take one flip of a coin as analogous to the half-life of a radioactive atom.

EXERCISE

Gather as many coins as you can find (at least 20 and preferably 60-100) and place them in a pile. Each coin in this pile represents a radioactive atom. Count the coins and record the number.

Go through the pile one-by-one, flipping each coin in turn. Place the flipped coin in one of two new piles, depending on whether it comes up heads or tails. The "heads" pile represents radioisotopes that have survived the round without decaying, and the "tails" pile represents atoms that have decayed to daughter atoms. After you have flipped each coin, count and record the number of coins in the "heads" (parent) pile.

Now repeat the process, this time flipping only the "parent" coins (the "daughter" coins are no longer radioactive). If a coin comes up tails, add it to the daughter pile. If it comes up heads, place it in a new parent pile. This will insure that each parent coin is flipped only once in each round. At the end of the round, count and record the number of coins in the parent pile.

Repeat the process three or four times and then plot your results on a graph similar to that in Figure I-5. Don't bother figuring percentages, just plot the number of "parent" coins on the vertical axis. Plot the total number of coins that you started with on the vertical axis line (at zero half-lives), then plot the number of "parent" coins remaining after the first round at one half-life, the number of "parent" coins after the second round at two half-lives, etc.

Connect the points on your graph with a smooth curved line. Your graph should now resemble in shape the curve in Figure I-5. The more coins you use, the better will be the resemblance. The reason for this is that statistics do not apply well to small numbers. An individual gambler can never predict how he might do on a given try at a slot machine, but averaged over many thousands of gamblers, the casino owners are assured of a steady flow of profits. In the same way, we can never predict just when an individual radioactive atom will decay, but when we are dealing with billions of atoms in rocks (which is always the case), we can be assured that the Law of Radioactive Decay is followed very closely.

Now let's see how the age-dating scheme can use the law of radioactive decay. Let us say that we begin with 16 billion Uranium-235 atoms in a rock and no Lead-207 atoms (see Table I-1 and Figure I-6).

After 700 million years have elapsed (that is, one half-life for Uranium-235 to Lead-207), half (8) of the original uranium atoms have decayed to lead. After another half-life has elapsed, bringing the total to 1,400 million years, half of the remaining 8 uranium atoms have decayed. Only 25% of the original parent atoms remain (4 out of the original 16).

If we find a rock today which contains 4 billion Uranium-235 atoms and 12 billion Lead-207 atoms, then we can determine the age of the rock by using the simple formula:

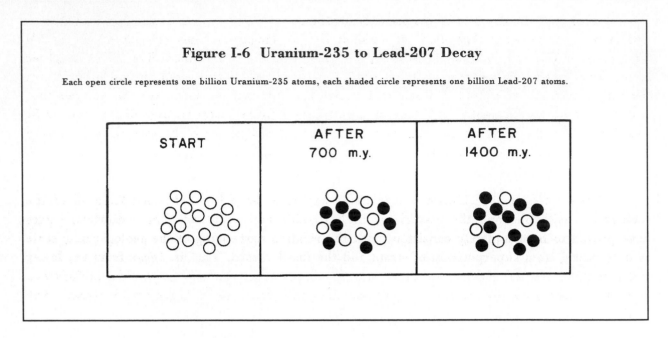

Figure I-6 Uranium-235 to Lead-207 Decay

Each open circle represents one billion Uranium-235 atoms, each shaded circle represents one billion Lead-207 atoms.

$$\% \text{ Parent remaining} = \frac{\#\text{Parent}}{\#\text{Parent} + \#\text{Daughter}} \times 100$$

where "#Parent" and "#Daughter" mean the number of parent and daughter atoms observed in the rock today. The percentage of parent remaining (determined from the formula) can then be used with the curve in Figure I-5 to find the number of half-lives, and hence the age of the rock.

Try it for each case shown in Figure I-6 and see if you can determine the correct age for the three cases shown. Simply count the number of parent and daughter atoms shown in each case, substitute these numbers into the formula above, and use the decay curve in Figure I-5 to determine the number of half-lives. In this case, each half-life lasts 700 million years. Do it now, and be sure you understand how it works.

Critical to this process is the requirement that no daughter atoms be present at the start. In fact, this turns out to be the essential definition of when the process "starts." How can we be sure that the condition was met when a rock that we are dating formed so long ago?

In the case of uranium-lead decay, nature has very kindly supplied us with the necessary assurance in the mineral <u>zircon</u>. When zircon forms by solidifying from molten magma, it readily incorporates uranium into its crystal structure as an impurity, but rigorously excludes lead. This is directly observable in newly-formed zircons, and the way in which it now happens assures that it happened this way in the distant past as well.

The solidification of the zircon crystal "starts" the uranium-lead clock ticking because that is when no daughter lead atoms are present. Lead atoms formed by the decay of uranium atoms are trapped in the solid crystal structure of the zircon, however, and remain there as long as the zircon is not heated excessively.

In each of the decay schemes used for geologic age dating, either nature has provided us with a way of knowing that there are essentially no daughter atoms present at the rock's formation, or else we are able to use some logical stratagem to determine how many daughter atoms were present at the start of the process. If we know the latter, then we can subtract out the initial number of daughter atoms and proceed as before. We must also be assured that neither parent nor daughter are lost in appreciable numbers during the life of the rock. Satisfying all of these demands means that only a limited number of rocks and minerals qualify for reliable radioisotope age dating.

Nonetheless, many thousands of reliable age dates have been obtained from all of the rock strata available, and the results are both striking and profound. The radioisotope ages have proved to be completely consistent with the relative sequencing of the geologic time scale as determined from superposition of strata and the fossil record. That is, rocks from the lower strata consistently give older dates than rocks from the higher strata, and the time periods and eras shown in Figure I-4 are shown in correct order by either the relative or the radioisotope methods.

STUDY QUESTIONS

I-5. Cite several examples of the Principles of Original Horizontality and Superposition demonstrated by sediments in your neighborhood. Hint: mud and gravel are common sediments in residential areas.

I-6. Why does the sum of parent and daughter atoms in a zircon remain constant as time goes on?

I-7. A zircon is found which contains 100 billion Uranium-235 atoms and 300 billion Lead-207 atoms. a) How many Lead-207 atoms were present when the zircon formed? b) How many Uranium-235 atoms were present when the zircon formed? c) How old is the zircon?

I-8. A piece of wood is dated using the Carbon-14 method. It is found that of the original Carbon-14 in the wood when it was alive, only 30% remains today. How old is the wood?

D. HOW OLD IS THE EARTH?

The numerical age dates assigned to the boundaries between the geologic periods in Figure I-4 were determined from radioisotope data. The Precambrian Era is by far the longest, taking up more than 4,000 of the 4,600 million years of Earth history. Precambrian rocks contain relatively few fossils and have often been so changed, or metamorphosed, that it is difficult to determine exactly what conditions were like in the Precambrian world.

The Paleozoic, the era of ancient life, began some 570 million years ago (m.y.) and the Mesozoic, the Age of the Dinosaurs, extended from 245 m.y. to 65 m.y. The oldest fossil evidence for life dates back to about 3,300 m.y., while the oldest Earth rocks found so far have been recovered from the coast of Greenland and yield radioisotope ages of about 3,800 m.y. Recently, Australian minerals have yielded a date of 4,200 m.y.

Actually, the oldest rocks found on the Earth are not from the Earth. They are meteorites, which are rock and metal fragments that have fallen to Earth from space, and the oldest date back to 4,600 m.y. For reasons that will become clear in **The Solar Family** unit, this is generally taken to be the age of the Earth and of the Solar System as a whole.

There are also complex and somewhat indirect radioisotope methods that use not individual rocks or minerals, but treat the entire Earth as though it were a single sample for age dating. These methods also yield ages of about 4,600 m.y.

Does this resolve the controversy between the uniformitarians and the catastrophists? In terms of the antiquity of the Earth, most scientists think so. The evidence in favor of great age is simply overwhelming at this point in time.

There remains, however, a small group of scientists who prefer the Biblical span of 6,000 years and the mechanisms of catastrophism to an antiquity reaching back 4,600 million years and the slow processess of uniformitarianism and evolution. Terming themselves scientific creationists, they are generally allied with fundamentalist religions which regard every word in the Book of Genesis as literal Truth.

Where does the truth lie? Before we can proceed with this study of Earth science, it is quite necessary for us to examine the nature of scientific truth.

E. SCIENTIFIC METHOD AND SCIENTIFIC TRUTH

How can we tell when one scientific theory is "right" and another "wrong"? We already have an excellent example before us. We have just learned that radioisotope dating tells us that the Earth is 4,600 million years old. Yet, we have seen that two of the last century's greatest scientists, Halley and Kelvin, concluded from different evidence that the Earth was no more than 100 million years old. Which result do we believe?

As it turns out, we now know that Halley and Kelvin both made important errors of omission in the assumptions that went into their models. Halley and Joly assumed that the oceans are still becoming more salty every year. They were wrong. The discovery of vast salt deposits beneath the sea floor in a number of places shows that large quantities of salt have been removed from the ocean waters at various times. The oceans long ago appear to have reached their maximum saltiness. Joly may have been right in maintaining that it took about 90 million years for the oceans to reach their present saltiness, but that may have occurred in the first few hundred million years of Earth history, more than 4,000 million years ago.

Kelvin could not possibly be blamed for his error in calculating the cooling history of the Earth. In making his calculations in 1862, he did not know about the large quantities of heat produced by natural radioactivity, which is now suspected as being the dominant source of heat flow from the Earth's interior. After all, natural radioactivity was not discovered until 1896.

If Kelvin could not know that he was wrong in 1862, then how can we be certain that some new discovery a dozen or a hundred years hence won't invalidate the assumptions of radioisotope dating? The answer may surprise you: we can't.

Most scientists are confident that the 4,600 m.y. age is likely to prove more accurate than the 6,000-year age. Why? Because the older age is supported by far more observational evidence than the younger age. This, in fact, is the one essential criterion that is used in making choices between differing scientific explanations. The method of science can be loosely stated in a single sentence: That theory or model is best which best explains all available observations. There are provisos, of course. To insure that the observations are reliable and not fraudulent, they must be repeatable by anyone who has the necessary equipment and training. But that's really all you need to know in order to understand the workings of science. This statement of scientific method seems almost self-evident, but in fact has far-reaching implications. In deciding among competing theories, they must be tested against all the available observational evidence, not just those that support one pet theory.

If this is the objective backbone of science, then why do scientists ever disagree with one another? First of all, scientists are human, and if one of the theories happens to be yours, it can be very hard to be truly objective. In addition, scientists nowadays are specialists and usually are more familiar with one area of observational evidence than with others. This can skew the process of testing theories against all the available observations. Most of us try very hard to be objective, but just as two different juries might come to different decisions on the same case, different scientists may also differ in interpreting the same evidence. The analogy with a jury is not a bad one: both jury and the scientific community as a whole examine all the available evidence and then try to come up with a logical and consistent explanation for that evidence. There is no guarantee, however, that either one will always come up with the correct verdict.

This may impress you as a very unsatisfactory state of affairs. Isn't science supposed to be able to lead us to the Truth, with a capital T? The answer to this one has to be hedged: Maybe science can lead us toward the Truth about how the universe works, but the trouble is we can never be certain that we are all the way there. If the statement of the scientific method is meaningful, then as long as there are still new observations to be made, the nature of scientific truth is always subject to change. If the scientific method is to be followed, then absolute or dogmatic Truth cannot be part of science.

This is the heart of the debate between the creationists and the scientific majority. The creationists are unwilling to abandon the dogmatic Truths which they ascribe to the Book of Genesis, while the scientists maintain that dogmatic beliefs are properly a part of religion, but not of science. Science does not lay claim to absolute Truth, but rather to a kind of scientific truth-in-progress. With that understanding, we can ask which age of the Earth best explains all the presently available observations: 6,000 years or 4,600 million years? It should not be a close race, since the two ages differ from one another by a factor of nearly one million.

The race is not close. For every observation supporting creationism, there are hundreds supporting the great antiquity of the Earth. In addition, many of the former are subject to more than one interpretation. It is not an anti-religion bias, but the weight of evidence written in the rocks, that has convinced most Earth scientists of the great antiquity of our planet.

F. THE SPAN OF GEOLOGIC TIME

Accepting the antiquity of the Earth is one thing, but grasping it is another. So many geological and geophysical processes are slow-acting: erosion, mountain-building, drifting of the continents, the formation of sedimentary rocks and mineral and energy resources. They are all but imperceptible in the span of a single human life. Only placed in the context of immense stretches of time do their effects accumulate to the point where they are capable of reshaping the face of the Earth.

This is not to say that catastrophists have been totally shut out of modern geology. Evidence is growing that our planet has seen a number of catastrophic events, some perhaps caused by collisions with extraterrestrial bodies (see the **Origins** unit). But during the past 600 million years these events appear to have been the exception and not the rule. The principal shapers of Earth's present features have been the slow-acting, long-term processes.

Geology teachers have devised many analogies to help their students place the span of geologic time in proper perspective. One of the best was used by the eminent science writer, Nigel Calder.[1] He likened Mother Earth to a woman of 46, where each of her "years" is equivalent to 100 million years of geologic time. Her early history is poorly recollected, and the mementos of her early childhood are lost, scrapped and recycled by geological processes.

Our first glimpse of her is at age eight, but the snapshot is faded and only dim outlines are visible. The earliest record of primitive life in her oceans appears at age 13, but development is so slow that abundant fossils of complex hard-shelled creatures like trilobites did not appear until she was 40. Her continents were bare of all life until she reached 42, just four of her "years" ago. Calder points out that she flowered, literally, in her middle age.

Her past four years have been busy ones. By the time she was 43, sharks had appeared in her oceans, insects in her air, and forests and reptiles on her continents. At 44, only two of her "years" ago, the dinosaurs held sway on her continents and the trilobites had already become extinct in her oceans.

The dinosaurs died out only eight months ago to be replaced by the mammals, and only two weeks ago ape-like hominids made their first appearance. Last week the glaciers of the Ice Ages advanced for the first time and the last of them did not withdraw until 50 minutes ago.

Calder eloquently concludes:

> Just over four hours have elapsed since a new species calling itself <u>Homo sapiens</u> started chasing the other animals and in the last hour it has invented agriculture and settled down. A quarter of an hour ago, Moses led his people to safety across a crack in the Earth's shell, and about five minutes later Jesus was preaching on a hill farther along the fault-line. Just one minute has passed, out of

---

[1]Nigel Calder, <u>The Restless Earth: A Report on the New Geology</u>, Penguin Books, 1972.

Mother Earth's 46 'years', since man began his industrial revolution, three human lifetimes ago. During that minute he has multiplied his numbers and his skills prodigiously and ransacked the planet for metal and fuel.

STUDY QUESTIONS

I-9. You should try to memorize the order of the geologic eras and periods and know which periods belong to which eras. Here is a mnemonic to help you. Both start at older periods and progress to younger periods.

Precambrian:

Paleozoic:	Cambrian	Come
	Ordovician	Over
	Silurian	Soon;
	Devonian	Drinking and
	Carboniferous	Card
	Permian	Playing.
Mesozoic:	Triassic	This
	Jurassic	Judgement
	Cretaceous	Creates
Cenozoic:	Tertiary	Three
	Quaternary	Quandaries.

The underlined letters are the keys to remembering the period names, especially those with the same first letters. Note that "tertiary" means "third."

Now, cover the above and place the following into order from youngest to oldest: Jurassic, Tertiary, Carboniferous, Ordovician, Triassic.

I-10. If the same five periods were represented at one locality by five strata, list how the strata would appear, from top to bottom.

RECOMMENDED READING

Frank Press, Earth, Third Edition, W.H. Freeman and Co. (1982). This is an excellent general introduction to geology and geophysics for the non-scientist.

David G. Smith, ed. The Cambridge Encyclopedia of Earth Sciences, Cambridge University Press (1981). A comprehensive general reference with illustrations.

W. Lee Stokes, <u>Essentials of Earth History</u>, Fourth Edition, Prentice-Hall, Inc. (1982). A well-written historical geology text, outlining the principal events that have shaped Earth's surface. The first few chapters are especially relevant to the discussions in Unit I.

Jonathan Wiener, <u>Planet Earth</u>, Bantam Books, 1985. Because of the process used in the production of this textbook, we have not been able to include color graphics, relying instead on your viewing of the television programs to fill this need. This tradebook for popular consumption has been written to accompany the television series, however, and it contains numerous fine illustrations.

UNIT II GEOGRAPHICAL REVIEW

EARTH FROM SPACE

A. INTRODUCTION

1. Overview

Do you know where the Ural Mountains are? The Atacama Desert? The East Pacific Rise? Your tour of **Planet Earth** will take you to many of its lesser-known reaches, and you will need a guide to help find your way. This unit may serve as your guide, and you should develop the habit of referring to it frequently throughout the course. It begins with a description of the major surface features of Earth: the continents and ocean basins. It then describes some of the rock types that you will encounter again and again in later units. Finally, a map of the world is keyed to every place and region name used throughout the telecourse. A handy table enables you to look up any place name and find out where it is.

---

2. Objectives

Upon completion of this unit you should be able to:

1. describe the difference between continents and ocean basins in terms of the elevations of their surfaces

2. describe the three main type of rocks: igneous, sedimentary, and metamorphic

3. classify igneous rocks on the basis of their chemical composition and texture

4. locate major surface features of the Earth

5. describe and use a map resource of the world in order to locate and reference sites mentioned in the text and video.

B. THE FACE OF THE EARTH

The logo for this telecourse shows the familiar view of our planet from space: the deep blue of the oceans contrasting with the tan of the continents, spangled with white clouds and softened by the haze of the atmosphere. We are a very different kind of planet from the others in our solar system, and the strong contrast between continents and oceans is one of our most striking features.

The contrast is not only one of rock and water, though the visible presence of liquid water is unique in itself. The waters are gathered into oceans because these are the low points

Figure II-1 Hypsometric Curve

The curve shows the proportion of Earth's rocky surface that stands at each elevation. Adapted by permission of Smithsonian Institution Press from Continental Drift: The Evolution of a Concept, by Ursula B. Marvin. Figure 25, page 69. (c) Smithsonian Institution, Washington, D.C. 1973.

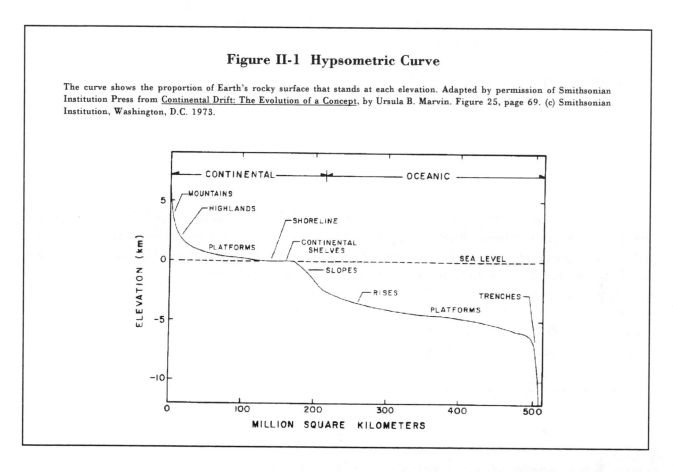

of the Earth's surface. Even if there were no water at all on Earth, the ocean basins would be remarkable features. In fact, the elevations of the rocky surface are dominated by two elevation ranges: that of sea level to 1,000 meters (3,300 feet) above it and that of 4,000 to 5,000 meters (13,000 to 16,400 feet) below sea level. Figure II-1 shows graphically the distribution of elevations on Earth. The curve starts at the left at the highest elevations (8848 m or 29,030 ft at Mt. Everest) and proceeds to the lowest (11,000 m or 36,100 ft below sea level in the Mariana Trench.) Each elevation is shown in direct proportion to the area of the Earth that stands at that elevation. Note how the two elevation ranges just cited mark the continental and oceanic platforms, which together account for the lion's share of the surface. The continental slope bridges these two elevations, while the extremes -- mountains and oceanic trenches -- take up only a tiny fraction of the available acreage.

Earth is indeed a split-level planet. The liquid oceans are where they are because that is where there is room for their waters. In places, the oceanic waters lap onto the edges of the continents, forming shallow seas floored by continental, not oceanic, rocks. These are the continental shelves which in places like the Grand Banks off Newfoundland or the Falkland Banks near southern Argentina can extend hundred of kilometers beyond the present-day shorelines.

As it happens, the rocks flooring the oceans and continents differ from one another, as we shall see in the **Plate Tectonics** unit. There, also, we shall investigate the reasons for this dichotomy of elevations. But before we can do that, we need to learn something about the rocky materials of the Earth's surface.

C. ROCK TYPES

Rocks are divided into three varieties, based on their mode of origin: igneous, sedimentary, and metamorphic. Igneous rocks are produced by freezing of molten rock, which is called magma within the crust of the Earth, or lava when it erupts onto the surface.

Sedimentary rocks are formed when grains of sediment (mud, clay, sand, gravel, skeletal remains of plants and animals) are deposited by water, wind, or ice. All sediment is derived from previously existing rocks. With time, these sediment grains may become lithified (made into rock) by burial, compaction, and cementation, often by precipitation of minerals that bind together the sediment grains into solid rock. Sandstones are formed from sandy deposits, shales from mud, and limestones from calcium-rich sediments -- often made up largely of the shells and skeletal remains of marine creatures large and small. Sedimentary rocks form a relatively thin veneer covering some three-quarters of Earth's surface but making up only about one-twentieth of the crust by volume.

Metamorphic rocks are any rocks that have been heated and/or squeezed to such an extent that their chemical and physical makeup are altered.

Igneous rocks are the primary rocks within the Earth. To be able to deal with them, it will be necessary for you to understand a brief and simplified classification scheme, given in Figure II-2. This scheme is based on the chemical composition of the rock (represented by the three vertical columns) and the texture of the rock (represented by the two horizontal rows).

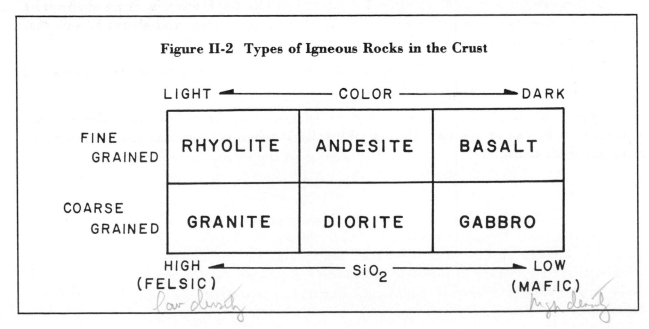

Figure II-2 Types of Igneous Rocks in the Crust

The chemical classification is based on the abundance of certain minerals within the rock, especially silica (silicon dioxide, or SiO_2). Rocks with a high silica content are termed felsic; those with a low silica content are termed mafic. Felsic rocks tend to be light in color and have a low density; mafic rocks are usually dark in color and have a higher density.

There is another chemical class of igneous rocks -- the ultramafic rocks. These make up much of the deeper part of the Earth, called the mantle, and in our classification scheme would be found to the right of basalt and gabbro. They are dark in color and even denser than the mafic rocks.

The physical classification is based on grain size within the rock. When a mass of magma cools slowly deep within the Earth, its atoms have time to aggregate into large grains of individual minerals: the rock becomes coarse-grained. On the other hand, lavas on the surface generally cool quickly, resulting in fine-grained rocks.

Granites are familiar rocks, often used for decorative or memorial purposes. Its large grains give it a speckled appearance, and its overall color is often a light gray. Basalt, on the other hand, is a dark, nearly black rock that looks almost uniform in color. Only close inspection with a lens will reveal the small grains of the mostly dark minerals that constitute it. Andesite is a fine-grained rock of intermediate chemical composition. Rhyolite is a fine-grained rock whose chemical composition is very similar to that of granite, but differs from it in grain size.

Study Figure II-2 and become familiar with the rocks shown in it. We shall refer especially to basalts, granites, andesites, and rhyolites in later units.

D. WHERE IS IT?

Figure II-3 is a map of the world for your reference use. Marked on it are the features and localities mentioned throughout this text. Refer back to it often, and always be sure that you know where you are during our world travels.

Table II-1 is the key to find places on the map. The table lists all the features and localities alphabetically and by number.

Two other important resources are provided for you in the glossary of geological terms and index at the end of this text. Use both liberally and you will find that you are getting far more out of the course.

TABLE II-1: WORLD INDEX

Map Number	Place Name	Location
1	Adirondack Mountains, N.Y.	E-2
2	Afar Triangle	K-3
3	Africa	J-3
4	African Plate	J-4
5	Alaska	B-1
6	Alberta, Canada	D-2
7	Alcatraz Island	C-2
8	Aleutian Islands	B-2
9	Alexandria, Egypt	J-2
10	Allende, Mexico	D-3
11	Alps	J-2
12	Amazon Basin	E-4
13	Anatolian Fault	K-2
14	Anchorage, Alaska	C-1
15	Andes Mountains	E-4
16	Antarctic Circle	D-6
17	Antarctic Plate	L-5
18	Antarctica	M-6
19	Appalacians	E-2
20	Arabian Peninsula	K-3
21	Arctic Circle	M-1
22	Arctic Ocean	B-1
23	Argentina	E-5
24	Arizona	D-2
25	Asia	L-2
26	Aswan High Dam	K-3
27	Atacama Desert	E-5
28	Atlantic Ocean	F-2
29	Australia	N-4
30	Australian Plate	M-4
31	Azores	F-2
32	Baja California	D-3
33	Baltic Sea	J-2
34	Bangladesh	L-3
24	Barringer Crater, Arizona	D-2
35	Beaufort Sea	C-1
36	Benue Trough	J-3
37	Bering Strait	B-1
38	Bermuda	F-2
39	Birmingham, England	H-2

40	Black Forest, Germany	J-2
41	Black Sea	K-2
42	Bolivia	E-4
43	Boulder, Colorado	D-2
44	Brazil	F-4
45	Brisbane, Australia	N-5
46	Bushveld Igneous Complex, S.A.	J-4
47	Cairo, Egypt	K-2
48	Caledonian Mountains	H-2
49	Calgary, Alberta	D-2
7	California	C-2
50	Cambridge, England	J-2
51	Camp Century, Greenland	F-1
52	Campo del Cielo, Argentina	E-4
53	Canada	D-2
54	Canadian Shield	D-2
55	Cape Canaveral	E-3
56	Cape Hatteras	E-2
57	Cape Mendocino	C-2
58	Cape of Good Hope	J-5
59	Caribbean Sea	E-3
60	Caspian Sea	L-2
61	Central America	E-3
62	Charleston, South Carolina	E-2
63	Chesapeake Bay	E-2
27	Chile	E-5
64	China	M-2
65	Cincinnati, Ohio	E-2
66	Cocos Plate	D-3
67	Colombia River Plateau	D-2
43	Colorado	D-2
68	Congo River Basin	J-4
69	Cuba	E-3
70	Cyprus	K-2
71	Dead Sea	K-2
72	Death Valley	C-2
73	Deccan Traps, India	L-3
43	Denver, Colorado	D-2
50	Dover, England	J-2
74	Dresden, Germany	J-2
76	Dvina River	K-1
77	East African Rift	K-3
78	East Indies	M-4
79	East Pacific Rise	D-4
80	Equador	E-4
26	Egypt	K-3

RECOMMENDED READING

For discussions of rock types and distinctions between continents and ocean basins, see Frank Press, <u>Earth</u>, Third Edition, W.H. Freeman and Co., (1982).

Every student should have access to a world atlas. One of the best of moderate cost is put out by the National Geographic Society, Washington, D.C.

UNIT III THE LIVING MACHINE:

PLATE TECTONICS

THE PLATES OF PLANET EARTH

A. INTRODUCTION

1. Overview

By the turn of the century, geologists had come to regard the major surface features of the Earth, especially the continents and ocean basins, as fixed and permanent. Today we find that the entire surface of the Earth is mobile, with continents moving about, breaking up and reforming, and ocean floor being created and destroyed. We now know that the ocean floor is far younger than most of the continents. The interactions between moving blocks of the Earth's surface, called plates, accounts for many of the dominant geological processes, such as earthquakes, volcanoes, and mountain building. This new view of the Earth is described by the theory of plate tectonics, which we shall explore in this unit.

2. Objectives

Upon completion of this unit you should be able to:

1. understand Wegener's theory of continental drift

2. describe the broad structure and properties of lithospheric plates

3. describe the three major types of plate boundaries: spreading oceanic ridges and rifts, subduction zones, and transform faults

4. describe the behavior of hot spots

5. predict the general motions of plates and the geological activity around their perimeters by inspecting a map showing types of plate boundaries which separate adjacent plates

6. predict geographical locations of earthquakes and volcanoes in relation to the three types of plate boundaries

7. explain chemical fractionation of silicate rocks and apply it to the creation of oceanic and continental crust

8. recognize that magnetic information recorded in rocks permits us to determine the history of continental drifting

9. apply magnetic-field reversals to sea-floor spreading in order to explain the striped magnetic anomaly pattern on the sea floor

10. relate the broad history of continental drifting during the past 600 million years to the origin of mountain chains

11. place the major discoveries leading to the concept of plate tectonics in historical perspective

12. relate the development of plate tectonic theory to the scientific method.

3. Key Terms and Concepts

plate tectonics	oceanic spreading ridge
continental drift	transform fault
Pangea	oceanic hot spots
lithosphere	continental hot spots
asthenosphere	volcanoes
isostasy	andesitic volcanism
Mohorovicic discontinuity (Moho)	basaltic volcanism
mantle	earthquakes
crust	mountain building
sea-floor spreading	magnetic polarity reversals
subduction	striped magnetic pattern on ocean floor

4. Corresponding Video

This program lays the foundation for the discovery of Plate Tectonics by exploring geologic time, paleomagnetism, and other early discoveries leading to the confirmation of this revolutionary new theory about the Earth. Graphic modeling illustrates the dynamics of the theory including subduction, collision, transform faults, and sea-floor spreading. Scientists searching for clues to the still unsolved mystery of what drives the plates demonstrate tomography and geochemistry as tools for studying the Earth's interior.

B. CONTINENTAL DRIFT

In 1915, Alfred Wegener published a book entitled **The Origin of Continents and Oceans** that shocked the scientific world. In it, he advanced the startling hypothesis that the continents of the world were actually moving and that at one time the Atlantic Ocean did not exist (Figure III-1).

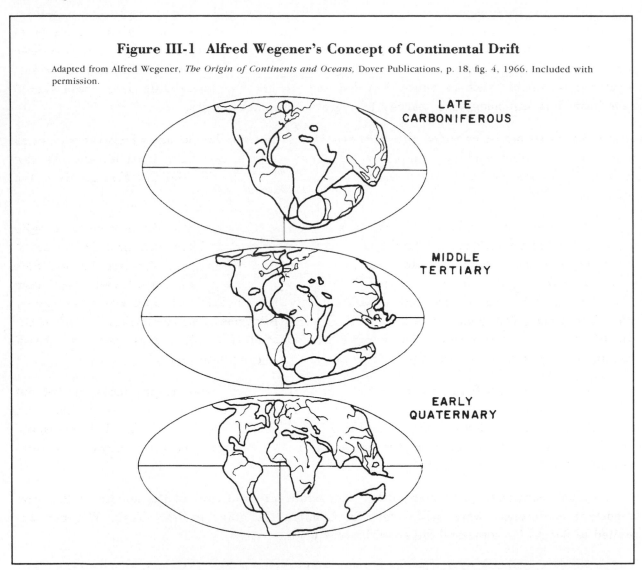

Figure III-1 Alfred Wegener's Concept of Continental Drift

Adapted from Alfred Wegener, *The Origin of Continents and Oceans*, Dover Publications, p. 18, fig. 4, 1966. Included with permission.

LATE CARBONIFEROUS

MIDDLE TERTIARY

EARLY QUATERNARY

Wegener's hypothesis brought down a storm of controversy because it challenged established scientific wisdom. The prevailing view of the Earth was based on continuity of geological processes and on the permanence of major surface features such as continents and ocean basins. Wegener's concept of drifting continents was a radical departure in thought for the geological community.

In spite of this, the idea was not simply dismissed as a crackpot notion. Though not a geologist, Wegener was an established scientist, a German meteorologist and polar explorer.

Furthermore, he had done his homework. He marshaled evidence in support of his idea from a wide variety of sources, the bulk of which was drawn from publications of respected geologists.

The story of the development of his idea during the next half century is one of the most fascinating chapters in the history of science. Far from ignoring Wegener, geologists of the world hotly debated the notion of continental drift for the next eleven years.

That the debate continued at all shows us that Wegener had many supporters as well as detractors among geologists. Paleontologists (geologists who study the fossilized remains of ancient organisms), found much to admire in Wegener's handiwork. They had long wondered why so many fossils of land or shallow-water organisms were found in identical form in widely separated continents such as South America and Africa. The intervening deep ocean should have formed an insurmountable barrier to migration for such species.

As an example, consider the relatives of the monkey-like lemur which can be found living in trees in Madagascar and India. The lemur cannot swim any great distance so how could its ancestors have migrated thousands of kilometers from the island of Madagascar to the Asian continent?

Wegener had a ready answer: in the Carboniferous period 300 million years ago, India was attached, along with Madagascar, to the east coast of Africa. They were part and parcel of a huge supercontinent that, he claimed, existed in the Carboniferous and Permian Periods. This supercontinent has since been given the name Pangea, meaning "all lands." Thus, the lemur could be found in widely separated localities because its original homeland was subsequently split. One piece (Madagascar) remained close to its original position near Africa, but the other (India) traveled to the north at an average rate of some 5 cm (2 inches) per year. Fossil associations between Africa and South America were explained similarly.

Other scientists found much to doubt in this scenario. Some geophysicists pointed out that continents are made up of low density rocks like granite, while the ocean floors are composed of a harder, denser rock called basalt. So how could the softer rocks of the continent plow through the hard rocks of the ocean floor without the continent breaking up and becoming deformed in the process?

There continued to be many other arguments, pro and con. In the middle 1920's, two important conferences were held -- one in London, the other in New York. Wegener was invited to defend his ideas and did so with great persuasion.

Even so, at the end of the London conference, two of his critics expressed what seemed to be the majority view:

> In examining ideas so novel as those of Wegener, it is not easy to avoid bias. A moving continent is as strange to us as a moving Earth was to our ancestors, and we may be as prejudiced as they were. On the other hand, if continents have moved many former difficulties disappear, and we may be tempted to forget the difficulties of the theory itself and the imperfection of the evidence...

and

We are discussing his hypothesis seriously because we should like him to be right, and yet I am afraid we have to conclude... that in essential points he is wrong. But the underlying idea may yet bear better fruit.

After the American conference, which went even more poorly for Wegener, the idea of drifting continents passed into an oblivion that was to last for over 25 years.

The story of how the hypothesis of continental drift was revived and reworked into a new theory of plate tectonics is a long one -- too long for us to treat in detail here. However, the story is related in three fine books: Walter Sullivan's **Continents in Motion**,[2] Ursula Marvin's **Continental Drift: The Evolution of a Concept**,[3] and William Glen's **The Road to Jaramillo**[4]. While interesting as history, these books also serve as fascinating guides to the workings of science.

Our approach will take a different route. First, we shall examine the essential concepts of plate tectonic theory as it currently is understood. Then, at the end of this unit, we shall return briefly to the historical perspective and try to answer the question: Did the scientific community blunder in rejecting Wegener's hypothesis?

STUDY QUESTIONS

III-1. According to Wegener's theory, how did the world of the Late Carboniferous differ from today's world?

III-2. Fossil bones of the Mesosaur, an extinct reptile that lived in swamps and could not swim the open oceans, have been found by paleontologists in Africa, South America, and Antarctica. How could the theory of continental drift explain these occurrences?

C. MOVING PLATES

What is meant by the name of the theory: plate tectonics? The word tectonics shares a common root with the word architecture -- it relates to processes of construction. In geology, the word has a further connotation implying motion of huge blocks of rock. The name, then, simply states that the surface of the Earth is constructed of a series of moving plates.

We may make an analogy to a piece of an eggshell, in that it is rigid and brittle, and its thickness is small compared to its breadth. Because the Earth is spherical, a plate that made up part of its surface would have to share its spherical shape.

Let us take the latter analogy yet a step further. Consider a hard-boiled egg that has been dropped onto the floor. Its shell has cracked into a number of pieces. The cracks surround each plate of eggshell and define its boundaries.

---

[2]Walter Sullivan, Continents in Motion, New York: McGraw Hill, 1974.
[3]Ursula Marvin, Continental Drift: The Evolution of a Concept, Smithsonian Institution Press, Washington, D.C., 1973.
[4]William Glen, The Road to Jaramillo, Stanford University Press, 1982.

Each eggshell plate is still as rigid as the original whole, but now the existence of the cracks allows each piece to move slightly in relation to its neighbor. Hence, the eggshell has gained a degree of mobility that it did not have before it cracked. In the egg, each fragment of shell can only move a fraction of a millimeter before it is stopped by its neighbors. If we want to move a piece of eggshell a significant distance, it would be necessary to overlap it with other eggshells in some places and to leave gaps between pieces in other places.

Earth plates move in a similar manner, and as a result they can move very long distances. Notice that so far we have referred only to the motion of <u>plates</u> on the Earth rather than to the motion of <u>continents</u>. This is the critical difference between modern plate tectonics and Wegener's continental drift. The plates may contain continents, or ocean floor, or both.

The most intense geological activity tends to occur around the perimeter of the plates, at the boundaries where one plate interacts with its neighbors. Where the edges of the plates grind together, tear apart, or slip past one another, they produce earthquakes.

On a map of the world the locations of earthquakes during a recent seven-year period (Figure III-2) outline the plate boundaries for us.

Figure III-3 diagrams the major plates of the world. Compare the two maps and note how the boundaries are clearly shown by the earthquakes in the oceans, but are often broad or indistinct on the continents.

Now consider the North American Plate. It contains not only the North American continent, but also the western half of the North Atlantic Ocean floor. The western boundary of the plate is the San Andreas fault in California; the eastern boundary is the crest of the Mid-Atlantic Ridge, a huge undersea mountain chain that bisects the Atlantic from north to south.

The shift in emphasis from continents to plates is an important one. It resolves one of the principal difficulties of Wegener's hypothesis. In the plate tectonics model, the continents do not move <u>through</u> a stationary ocean floor: the ocean floors are moving also.

The way these plates move, and how they interact with one another at their boundaries, will provide us with a model for understanding many of the geological processes that occur on the Earth's surface. This is a theme that will run throughout much of the rest of this course.

Plate tectonics has proved to be a unifying principle for the geological sciences: a kind of megatheory that philosophers call a <u>paradigm</u>. With its introduction, we have gained a conceptual tool of immense power and a whole new way of looking at Planet Earth.

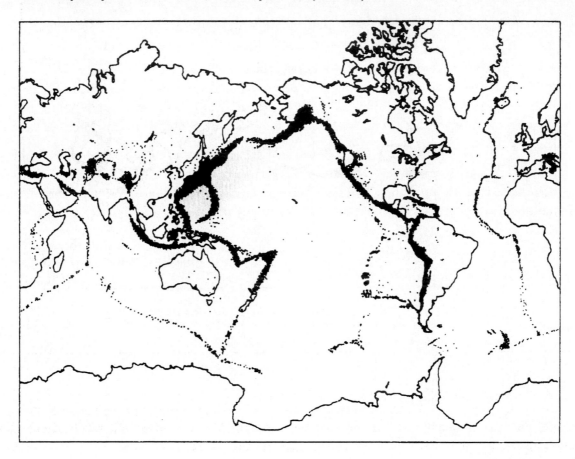

Figure III-2 World Seismicity

Recent earthquake epicenters, shown as dots, outline the plates as they exist today.

STUDY QUESTIONS

III-3. What does the term <u>plate tectonics</u> imply about the construction of the surface of the Earth?

III-4. The boundaries of each plate are best outlined by what type of geological activity?

III-5. What is the principal difference between Wegener's theory of continental drift and the theory of plate tectonics?

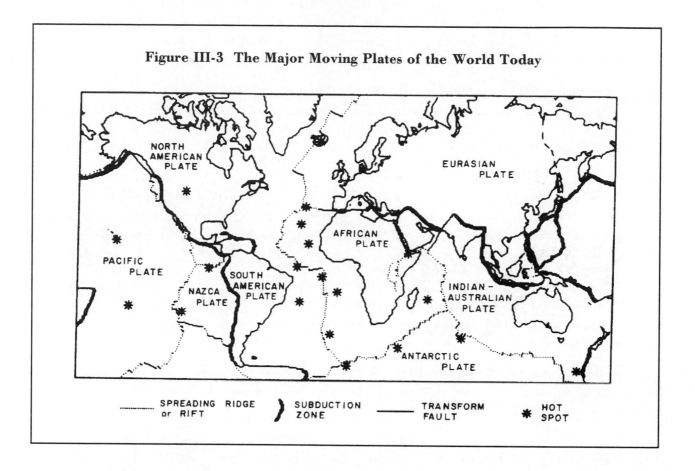

Figure III-3 The Major Moving Plates of the World Today

NORTH AMERICAN PLATE

EURASIAN PLATE

AFRICAN PLATE

PACIFIC PLATE

NAZCA PLATE

SOUTH AMERICAN PLATE

INDIAN - AUSTRALIAN PLATE

ANTARCTIC PLATE

SPREADING RIDGE or RIFT   SUBDUCTION ZONE   TRANSFORM FAULT   HOT SPOT

D. THE STRUCTURE OF THE PLATES

How is it possible for the plates to move at all? They are, after all, made of solid rock. Let us examine in some detail the uppermost parts of the Earth.

To begin with, earthquakes make waves -- vibrations similar to sound waves that can travel through the deep parts of the Earth. The speed with which these waves travel through different rock layers can tell us much about the density and rigidity of each layer. The study of these waves is called seismology.

Seismologists have noted that in the depth range of 100 to about 300 km (60 to 190 mi) beneath the surface, earthquake waves tend to travel more slowly than at either shallower or greater depths and they also lose their strength more quickly in that zone. For this reason it is called the low velocity zone. The physical properties of the rock in this zone indicate that it is solid but soft -- rather like asphalt on a hot day.

Many materials have this property of becoming soft at elevated temperatures, even though technically they remain classified as solids. The effect is particularly noticeable at temperatures that approach, but do not actually reach, the material's melting temperature. At cold temperatures, the same material can be quite rigid and brittle. A stick of butter that has been in the freezer is very hard, while one that has been sitting at room temperature for several hours can be spread easily with a knife. Rocks behave in a somewhat similar manner. It is believed that rocks in the low velocity zone are very near their melting temperature and that

45

this accounts for their soft or weak character. The soft zone probably extends somewhat deeper than the low velocity zone, and this expanded region is referred to by the name asthenosphere, (derived from the Greek for "weak zone").

The upper 100 km of the Earth behaves in a much more rigid manner because there the temperature is far below the melting point. This zone is called the lithosphere. The name is also derived from the Greek and refers to the more familiar rock-like behavior of this region. It is the lithosphere that makes up our rigid plates, and we shall refer to them as lithospheric plates.

Lithospheric plates can move as coherent pieces because they can slide about on the mushy underlying asthenosphere. Taking the egg analogy another step further, we can liken the asthenosphere to the slippery layer below the shell that enables us to peel it from the hard-boiled white.

The proportions of the eggshell are similar to those of the lithosphere. The radius of the Earth is 6371 km (3959 mi), and so the lithosphere extends less than one-sixtieth of the way to the center of the Earth. Lithosphere and asthenosphere together constitute only the top ten percent of the Earth.

The division of the upper parts of the Earth into lithosphere and asthenosphere is based on the physical property of rigidity. The same kind of rock could occur in both regions and yet, because of differences in temperature, be more rigid in one than in the other.

The lithospheric plates are themselves layered; this time based not on temperature differences but on different types of rock materials. On this basis, we can divide the lithosphere into crust and a portion of what is known as the mantle (see Figure III-4).

The crust is made up of all the surface rocks familiar to geologists: granite, basalt, limestone, and so on. The mantle is composed of ultramafic rocky materials that are denser than the crust. It is separated from the crust by a sharp boundary called the Mohorovicic discontinuity, or Moho, for short. Look at Figure III-4 and try to recognize that the division into crust and mantle and into lithosphere and asthenosphere are two very different ways of describing the uppermost layers of the Earth. The mantle differs from the crust in having a higher density (about 3.3 grams per cubic centimeter -- g/cc). The average density of the crust is about 2.7 g/cc. Compare this to the density of water -- 1.0 g/cc.

The lithosphere includes both crust and the uppermost mantle (in the region where it is cool and quite rigid), while the asthenosphere includes that portion of the mantle in the depth range of roughly 100 km (60 mi) to 640 km (400 mi) where the mantle is near its melting point and rather soft or mushy. The boundary between lithosphere and asthenosphere is not at all sharp. Indeed, the asthenosphere is hard to detect, and may not even exist in some regions. It often is identified most clearly beneath the oceans.

Figure III-4 also shows that the crust varies greatly in thickness. Beneath the oceans, the crust is only about 5 km (3 mi) thick, and composed mostly of basalt. Beneath the continents, the crustal thickness averages about 35 km (22 mi), increasing to 70 or 80 km (44 or 50 mi) beneath the largest mountain ranges, such as the Himalayas. The shape of the Moho is an inverted and exaggerated version of the surface topography.

Figure III-4 Division of the Earth's Upper Layers

The upper layers of the Earth may be divided into <u>crust</u> and <u>mantle</u> on the basis of rock type, or into <u>lithosphere</u> and <u>asthenosphere</u> on the basis of rigidity.

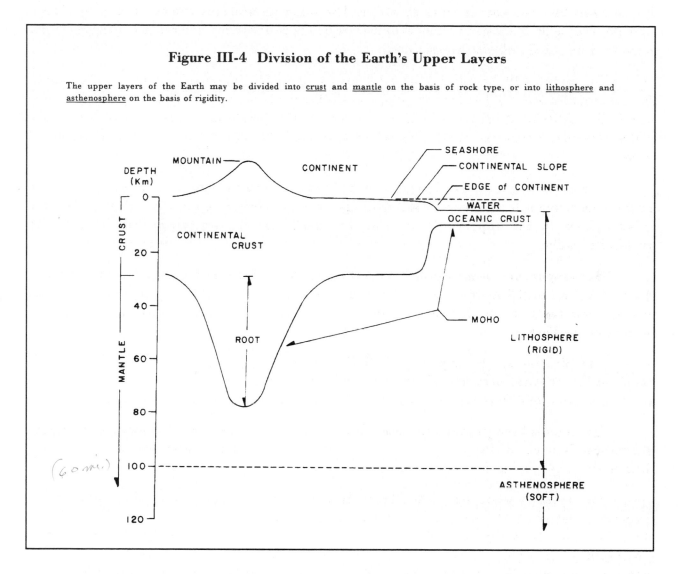

These large changes in the depth of the Moho are readily understood once we recognize that the crust is less dense than the mantle. As a result, the crust has buoyancy, and literally "floats" on the denser mantle.

Figure III-5 shows how this works using as an example blocks of ice floating in water. An ice cube floats in tap water such that 10% of its mass protrudes above the water surface and 90% of its mass is submerged. As a result, the bottom of a thick block will float at a greater depth than that of a thin block. Also, the top of a thick block will stand at a higher elevation than the top of a thin block.

If this still seems obscure, you can try a simple experiment by floating different size cubes or chunks of ice in water. The tallest chunks will stick out of the water the most, and they will also reach to the greatest depth. For the case of a mountain, the buoyant "root" beneath it is approximately 4-1/2 times the height of the mountain itself, as shown in Figure III-4.

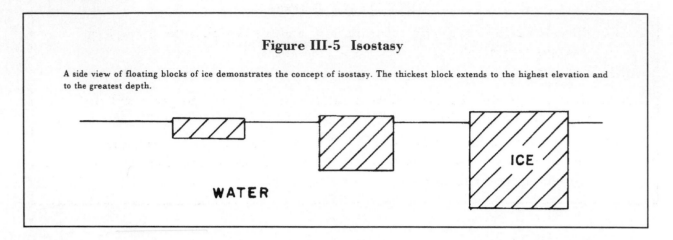

Figure III-5 Isostasy

A side view of floating blocks of ice demonstrates the concept of isostasy. The thickest block extends to the highest elevation and to the greatest depth.

WATER

ICE

The concept of a solid block floating in a liquid is a familiar one; less familiar is the situation where a light solid "floats" in a denser solid. Such is the case for the crust and mantle. In order for this to work, the mantle must be able to flow somewhat in order to accomodate vertical motions of crustal blocks. Because of the high temperatures and pressures found in the mantle, it is reasonable to expect this kind of behavior provided that it takes place slowly, over a long period of time.

Density of the blocks is also a factor. The denser the block, the lower it will float in the mantle, just as ice, having a higher density than cork, will float lower in water than an equal size block of cork.

The concept of low density crust floating on high density mantle material is called isostasy. At one stroke it explains both the changing depth of the Moho and the major changes in elevation of the Earth's surface: the Earth's topography.

We can now make some progress toward answering a very important question: Why is the surface of the Earth divided into continents and ocean basins? We can see from Figure III-4 that continents exist where the crust is thick, and that ocean basins exist where the crust is very thin. Actually, we have only rephrased the question: Why is the crust thick in some places and thin in others? Nonetheless, we have taken an important step toward understanding the origin of continents and oceans. **Whenever we produce a thick crust composed of a light rock like granite, then we will have produced a continent.**

According to this definition, the edge of a continent is where the crust changes from thick to thin. This is not necessarily the location of the seashore. In many cases, the water filling the ocean basins covers a part of the edge of the continent, called the continental shelf. The seashore is then substantially inland from the true edge of the continent.

This rather long preamble has been necessary to define the essential structure of the Earth's crust and lithospheric plates. We shall make frequent use of these concepts later on.

STUDY QUESTIONS

III-6. What is the most important characteristic that distinguishes the lithosphere from the asthenosphere?

III-7. What distinguishes the crust from the mantle?

III-8. How is it possible for the mantle to include both the asthenosphere and the lower portion of the lithosphere?

III-9. Mt. Everest reaches to a height of 8,800 meters above sea level. According to the theory of isostasy, what is the vertical extent of the buoyant root supporting the mountain?

E. PLATE BOUNDARIES AND HOW THEY WORK

The plates are all interlocking, so when a plate moves, something has to give. The simple case shown in Figure III-6 can help us discover how this works.

In this illustration, we consider only two flat plates. Plate A is a square cut from surrounding plate B. If we move it to the left in a rigid manner, that is, without distorting its shape, then we produce three distinctly different types of boundaries. We produce a gap on its right margin, an overlap on its left margin, and a side-by-side sliding motion on its top and bottom margins, where there is neither gap nor overlap.

If the concept is not clear, trace the top half of Figure III-6 on a piece of paper, cut out plate A so it is loose from plate B, and slide it to the position shown in the bottom half of the figure. Watch what happens on all four boundaries.

Nature tends to behave in the same way. Plate margins may be classified according to three main types corresponding to the kinds of motions demonstrated by our simple model. They are spreading ridges or rifts, subduction zones, and transform faults. Spreading ridges occupy boundaries where two plates are pulling away or diverging from one another; subduction zones are found where two plates are converging with one plate diving beneath the other; and transform faults are characterized by a mostly horizontal side-by-side sliding motion where neither gap nor overlap are produced. In addition, transform faults connect to other plate boundaries at either end where the sliding motion abruptly changes to divergence at a spreading ridge or convergence at a subduction zone. It is the transformation of spreading motion to side-by-side motion at either end of a transform fault that gives this feature its name.

Our flat paper model is obviously an over-simplification of how plates move. Figure III-7 is a cross section or profile view of the lithosphere and shows how spreading ridges and subduction zones work in the real world. To fully understand it, let us look more closely at each kind of plate boundary.

Figure III-6 Plate Boundaries at Work

A demonstration of the kinds of motions characteritic of the three types of plate boundaries. In this simple case, plate A is completely surrounded by plate B. In the real world, each plate is surrounded by two or more other plates, but the same three types of boundaries are found there as well.

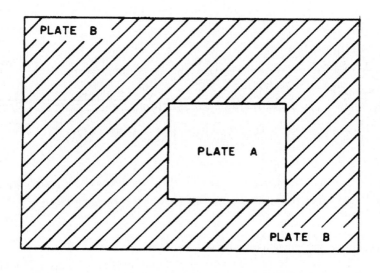

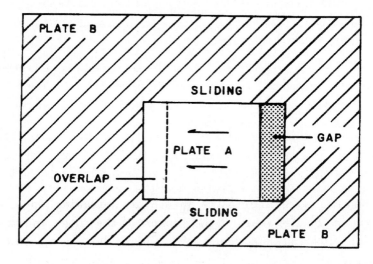

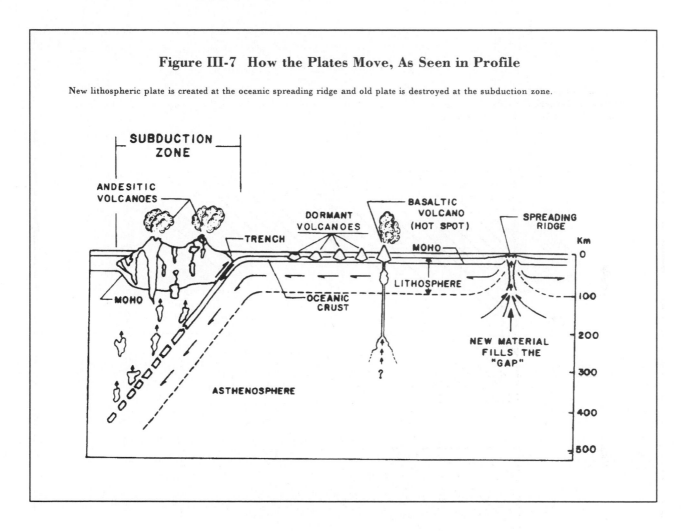

Figure III-7 How the Plates Move, As Seen in Profile

New lithospheric plate is created at the oceanic spreading ridge and old plate is destroyed at the subduction zone.

1. Spreading Ridges or Rifts

Spreading ridges or rifts occur where two plates are moving apart. The resultant gap is constantly filled with newly created crustal material. As it happens, the gap is always filled with oceanic crust, never with continental crust. We shall see why a little later on.

Figure III-8 diagrams what happens at an oceanic spreading ridge. As the two plates move apart, cracks form in the rift valley at the summit of the ridge, and the magma from the hot, molten region below fills and seals the cracks. This is how new ocean crust is added to each plate. This process, called sea-floor spreading is repeated indefinitely, with new cracks forming and more lava intruding and solidifying, ever widening the ocean floor.

Spreading oceanic ridges are found in all the major ocean basins of the world. These ridges are huge structures whose summits often stand two to three kilometers (7,000 to 10,000 feet) above the surrounding ocean floor. They are nearly all interconnected into one vast ridge system that stretches for some 65,000 km (40,000 mi) (see the spreading ridges in the world's oceans in Figure III-3). Far broader than most mountain ranges, individual spreading ridges are typically 1,000 to 4,000 km (600 to 2,500 mi) wide at their base. Indeed the suboceanic ridge system constitutes the greatest mountain range on Earth.

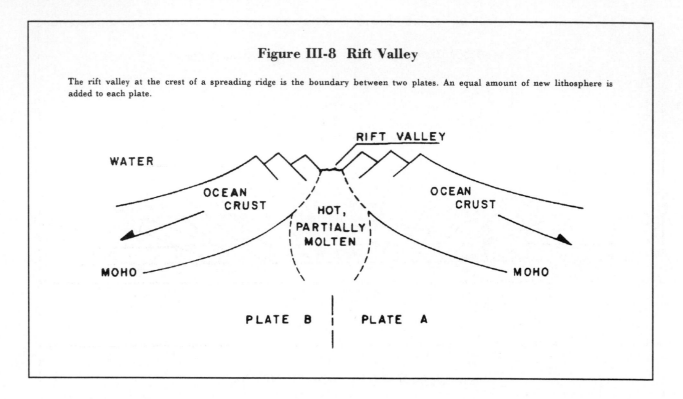

Figure III-8 Rift Valley

The rift valley at the crest of a spreading ridge is the boundary between two plates. An equal amount of new lithosphere is added to each plate.

RIFT VALLEY

WATER

OCEAN CRUST

OCEAN CRUST

HOT, PARTIALLY MOLTEN

MOHO

MOHO

PLATE B   PLATE A

The rift valley seen at the summit of the ridge in Figure III-8 is often the site of volcanic activity. In recent decades, undersea technology has provided researchers the means to study this phenomenon. During the early 1970's, a joint French-American project investigated a section of the Mid-Atlantic Ridge south of the Azores. Dubbed FAMOUS , for French-American Mid-Ocean Undersea Study, the project sent manned research submersibles 3,000 meters (10,000 ft) below the sea's surface to visit the rift valley.

Look for some of the bizarre features observed by the team of scientists as you watch the corresponding television program. These include the creation of pillow basalts and buds by the extrusion of molten lava directly into cold ocean water.

A rift valley might seem like an unusual feature for a ridgetop -- most continental mountains don't have them. But the oceanic ridges are not ordinary mountains. The rift valleys develop because the ridges are spreading and creating new oceanic crust.

A simple model demonstrates how the process works (Figure III-9). A number of books are placed in the usual way on a shelf, but without bookends. The end books will tend to flop over, and if we try to insert an additional book into the center of the group, the configuration shown in the diagram will develop. Our model would be more realistic if the books were added from below rather than from above, but it does demonstrate a number of features shared by actual spreading ridges. Consider what happens when books tip to one side (Figure III-10). As they tip over, they also slide against one another as shown by the pair of arrows.

If we replace our books with blocks of rock, then the sliding action is called <u>faulting</u> and the plane of contact between the two blocks is called a <u>fault plane</u>. It occurs in a spreading ridge as shown in Figure III-10. The faults cut the ridge crest into a series of blocks arranged somewhat like our books, except that the amount of tipping is much smaller than is the case for

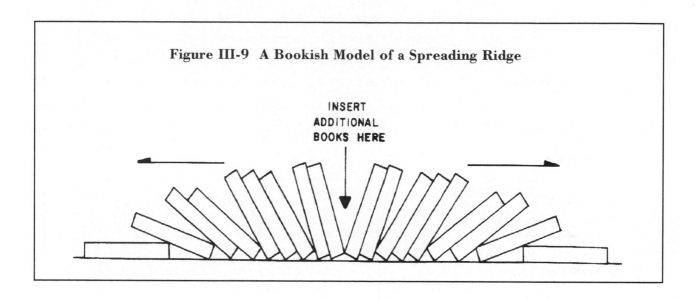

Figure III-9 A Bookish Model of a Spreading Ridge

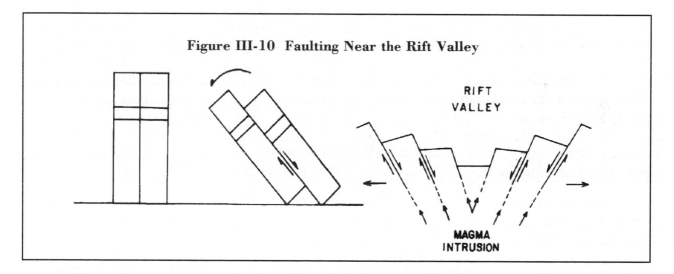

Figure III-10 Faulting Near the Rift Valley

the books. Because the oceanic crust continues for great distances on either side of the ridge, the crustal blocks never have room to flop over onto their sides.

Instead of inserting whole blocks, however, the crust is extended by the intrusion of magma into the faults from below. The magma cools in the widening faults and forms sheet-like layers of basalt called <u>dikes</u>. Because there are many faults and cracks over the source of magma, half of the newly-formed dikes spread out to the left and half to the right, adding equally to each plate.

Slippage along the faults usually is not smooth and steady, but tends to proceed in a series of jumps with intervening periods of quiet. This is due to the friction between the blocks of rock in the fault plane. Energy stored during a quiet period is released suddenly, creating an <u>earthquake</u>.

The map in Figure III-2 shows that the earthquakes associated with the oceanic ridges form a very narrow band about the ridge crest. The normal faults are actively slipping only while they are very close to the rift valley, and they become dormant once they have traveled some distance from it.

The continuous pattern of shallow earthquakes along the oceanic ridges indicates that all of these ridges are actively spreading at the present time, constantly renewing the floor of the world's oceans.

The structure of the oceanic crust has been studied through seismology, actual drilling of the oceanic crust by the Deep-Sea Drilling Program, and the study of suites of rocks which are believed to have formed as oceanic crust and later have been pushed onto the edge of a continent at a subduction zone. From this evidence, the top of the oceanic crust is believed to consist of a layer (about 1 km thick) of pillow basalts atop a thicker layer of closely spaced sheets of solidified magma called dikes oriented similarly to the faults in Figure III-10c. In fact, the crustal blocks themselves are made up of the sheeted dikes, so new faults and dikes intrude older dikes that had formed earlier by the same process.

In addition to the rift valleys, the ocean ridges show another striking difference from normal mountains. Continental mountain ranges have isostatic roots -- the crust becomes thicker beneath them, providing the buoyancy necessary to support their weight (see Figure III-4). Beneath the oceanic ridges, however, the crust becomes thinner (Figure III-8). The height of the ridge is supported by the buoyancy not of the crust but of the hot mantle material below it. Hot rock expands and becomes more buoyant and molten rock is lighter yet.

2. Subduction Zones

Subduction zones are the second type of plate boundary. They occur where two plates are coming together and overlapping. Figure III-7 shows the structure of a subduction zone where two lithospheric plates are converging. In this case, the plate on the left is overlapping the plate on the right, which is being pulled or forced down (subducted) into the asthenosphere.

Let us consider the case in which the subducted plate is oceanic. This process is very different from that at the spreading ridges. Here we are destroying, not creating, plate. Also, the spreading process is a very symmetric one, with new material being added equally to each plate. When oceanic plate is being subducted, the process is asymmetric, with only the oceanic plate being destroyed. Subduction zones are characterized by earthquakes within the subducting slab, down to depths as great as 600 km (370 mi). This contrasts with the case of spreading ridges, which are characterized by shallow earthquakes confined to depths of less than 70 km (44 mi).

Two important geological features result from subduction of ocean floor and are shown in Figure III-7 on page 51: trenches and volcanoes.

Oceanic trenches are significant features of the ocean floor. The deepest is the Mariana Trench in the southwestern Pacific, which reaches to 11,000 m (36,100 ft) below sea level. The trench floor is some 5,500 m (18,000 ft) farther below the surrounding ocean floor. The trenches result from the tendency of the descending plate to drag the edge of the overriding plate downward.

A line of volcanoes is found parallel to the subsea trench, directly over the downgoing lithospheric slab. These volcanoes tend to erupt andesitic lava. If the topmost plate is also oceanic, then a volcanic island arc -- an arcuate chain of islands -- is formed in the ocean. If the topmost plate is continental, then a linear belt of volcanic mountains is formed at the continent's edge, as in the Andes of South America.

This volcanism is different from that associated with the spreading ridges. As a consequence of greater viscosity and a higher content of dissolved gases in andesitic magma, subduction zone volcanoes tend to erupt much more explosively.

The violent eruptions of Krakatoa in Indonesia in 1883 and of Mount St. Helens in the state of Washington in 1980 were examples of subduction zone volcanism. So was the eruption of Mt. Pelee on the Caribbean island of Martinique in 1902 which totally destroyed the town of St. Pierre along with all 30,000 of its inhabitants, as well as the eruption of Vesuvius that wiped out Pompeii and Herculaneum in 79 AD.

Contrast this with the 1973 eruption of Kirkjufell on the outskirts of the Icelandic fishing town of Vestmannaeyjar. Not only were the inhabitants spared, many remained in the town and valiantly defended their homes by sweeping heavy layers of volcanic ash from their roofs. They even tried to stem the flow of lava toward the heart of the village by spraying water onto advancing flows with fire hoses. In the end, much of the village escaped destruction.

Iceland sits astride the Atlantic spreading ridge, and its volcanism, like that of all oceanic ridges, is basaltic in nature. Basaltic magma is relatively fluid and low in volatiles and results in effusive lava fountains and wide-ranging lava flows, but seldom in the kind of cataclysm in which the whole top of a volcano might be blown away.

The structure of a subduction zone shown in Figure III-7 indicates other features of interest. At the trench, the plate that is being subducted makes a quite abrupt bend and plunges downward into the asthenosphere. The downgoing oceanic crust rubs against the adjoining plate as shown by the coupled arrows in the diagram.

Slippage along the subduction zone and breakage of the subducting slab can give rise to violent earthquakes. In addition, other earthquakes may be generated where the brittle lithospheric slab is being bent near the trench and where it is breaking up as it descends into the asthenosphere. Because rock is a very poor conductor of heat, it takes a long time for the descending lithosphere to heat up, lose its relative rigidity, and be absorbed into the mantle. Once it becomes hot and soft, earthquakes can no longer occur within it. Instead of rupturing, it simply deforms.

Figure III-11 diagrams the occurrence of earthquakes (shown as dots) beneath the Japanese island arc. The earthquakes clearly indicate the downgoing slab, while the asthenosphere, shown as a shaded zone, is seismically quiet. Earthquakes occur only in rigid and brittle rock and cannot occur in the asthenosphere because of its soft and malleable nature. Earthquakes at a depth of more than 100 km are found only in subduction zones.

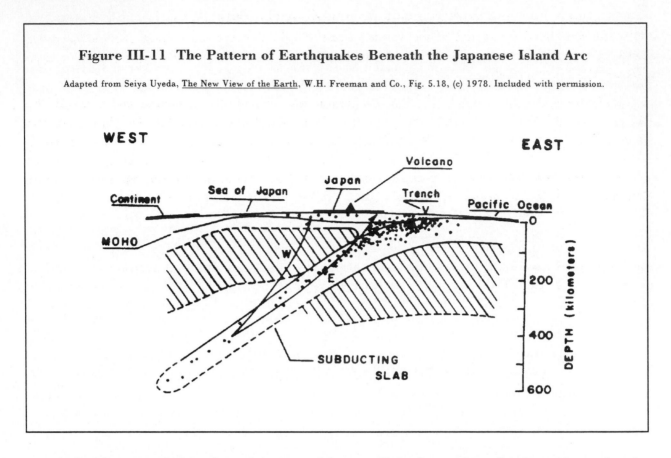

Figure III-11 The Pattern of Earthquakes Beneath the Japanese Island Arc

Adapted from Seiya Uyeda, <u>The New View of the Earth</u>, W.H. Freeman and Co., Fig. 5.18, (c) 1978. Included with permission.

Further proof that the subducting slab is still fairly rigid at depth is shown by the effects of an earthquake occurring at a depth of 400 km (250 mi) within the slab. Earthquake waves travel via path W in Figure III-11 to reach the west coast of Japan while waves travel via path E to reach the east coast. Even though path W is much shorter than path E, the earthquake was felt much more strongly on the east coast of Japan. The reason is that path E lies entirely within the rigid subducting slab, which conducts the waves with little loss of strength, while path W must traverse the asthenosphere above the slab, in which the seismic waves tend to lose strength rapidly due to the mushy nature of the asthenosphere.

3. Transform faults

Transform faults are long and relatively continuous where they cut across continents, but tend to appear as short discontinuous segments offsetting sections of spreading ridges on the ocean floor (see Figure III-3 on page 45.) While subduction zones are often arcuate in shape, the combination of oceanic ridges and transform fault segments tends to form a rectilinear zigzag. It is still not clear just how this zigzag pattern is formed initially.

The transform fault is in many respects an astonishing feature of the Earth. It is simply a fault connecting two other kinds of active plate boundaries, but that is a deceptively simple definition. In the simple model of Figure III-6 on page 50, the transform faults connect the spreading ridge to the subduction zone. In the real world, however, the most common type offsets two spreading ridges. Let us examine this type in some detail.

Figure III-12 is a working model of a transform fault that you can construct and study. Take the time now to cut and fold it and try some simple experiments with it. Notice that your model is very similar to portions of the Mid-Atlantic Ridge between South America and Africa (Figure III-3 on page 45).

Begin with the model in its closed position. The two points marked A and B should be juxtaposed, and the gaps at the two spreading ridges should be closed tightly. Now open the model. Note how new plate is "created" at each spreading ridge and an equal area is added to each of plate A (the shaded plate) and plate B (unshaded). Also, notice that after spreading has occurred, the two ridges are no farther apart than before spreading. Verify this by measuring distance DD[1] with the model closed and distance EE[1] with the model open.

The model illustrates the fact that the shape of an oceanic ridge does not change with time. In Figure III-3 on page 45, the shape of the Mid-Atlantic Ridge between South America and Africa today remains an exact replica of the rift that initially tore the two continents asunder. The ridge has not altered its shape in over 150 million years of sea-floor spreading.

Now turn your attention to the fault itself. With the model closed, imagine yourself standing on point A, looking across the fault to the juxtaposed point B. Slowly open the model, watching how point B moves as seen from point A. An observer at point A sees point B move to the right.

Turn the model upside down and repeat the process. An observer standing at point B also sees point A moving to the right. Because of this independence of where you happen to stand, this particular transform fault is said to display right-lateral motion. That is, the motion is side-to-side and the other side of the fault always moves to your right.

Transform faults may also be left-lateral, if the ridges are offset in the opposite sense. To see this, turn the model page over to the other side and trace the positions of points A and B on the reverse side of the paper. Holding the closed model so that you are looking at its reverse side, open it and note the motions of points A and B. If you are standing at point A, you will see point B move to your left.

The combination of ridge segments and transform faults forms a rectilinear zigzag pattern for oceanic plate boundaries that may be seen clearly in Figure III-3 on page 45. It is still not clear just how this zigzag pattern is formed initially, but because the pattern does not generally change shape with time, it must have come into existence at about the same time as the ridges themselves. The process by which this happens is still not fully understood.

Where transform faults cut across continents, however, they tend to be long and relatively continuous, with few, if any, spreading segments. The best known and most studied example of a continental transform fault is California's San Andreas. In Figure III-13, land and ocean floor to the west of the San Andreas Fault is part of the Pacific Plate and is moving to the northwest, parallel to the fault. To the east of the fault is the North American Plate. (You may also want to look at Figure III-3 on page 45.) Where the San Andreas Fault crosses the North American continent it is long and unbroken, but where it goes out to sea, it is cut into shorter segments separated by spreading ridges. Some of the ridge segments themselves are quite short, as in the Gulf of California.

Note that the ridge and fault geometry is similar to that of your paper model, for which right-lateral motion is expected along the fault. This is in fact what is observed along the length of the San Andreas.

Along the fault, the rocky edges of the plates grind against one another. In a few places, the slippage occurs smoothly. Here, any structure such as a fence or road that crosses the fault is offset at a rate of up to six centimeters (2-1/2 inches) per year. But in other places, the fault is jammed and does not move steadily. As the plates continue their inexorable motion, the forces exerted on the pinned fault build up with each passing year. Finally the rock can stand no more and it breaks, unleashing the pent-up energy as strong vibrations of the ground: an earthquake.

Figure III-13 The San Andreas Fault as a Plate Boundary

The San Andreas is a transform fault, where the Pacific Plate is sliding past the North American Plate, carrying Los Angeles and Baja California along with it.

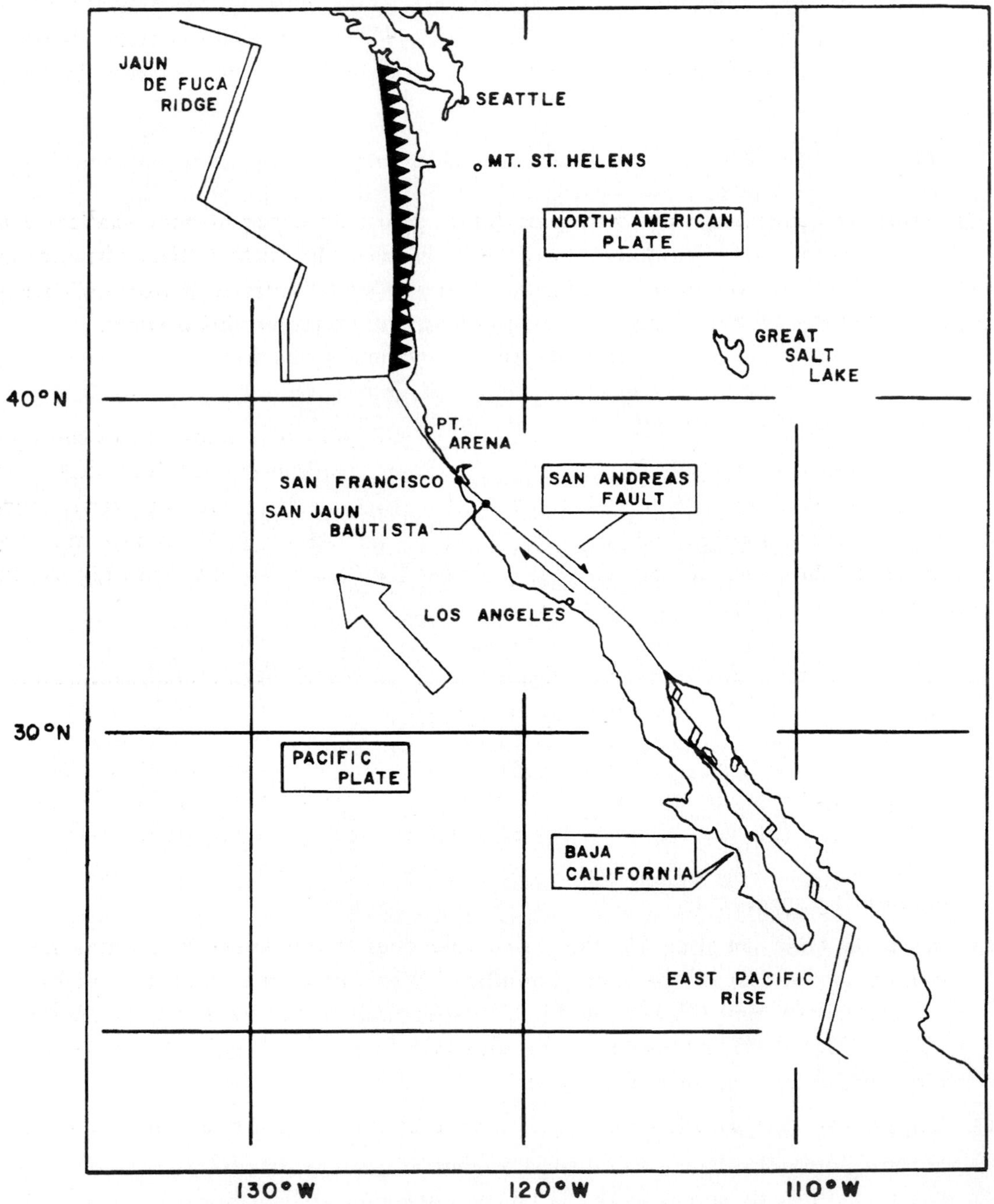

Those vibrations are capable of shaking buildings to the ground, as San Franciscans discovered at 5:12 AM on April 18, 1906. The great earthquake, measuring 8.3 on the Richter scale, lasted less than two minutes. At the end of those two minutes, much of the city was rubble and in the next hour much of the rest was in flames. Seven hundred lives were lost.

The fault had ruptured along 430 km (270 mi) of its length, extending from San Juan Bautista in the south to Point Arena in the north. At places along the rupture, it was found that the ground west of the fault had lurched 4.3 meters (14 feet) northwestward in relation to the ground east of the fault.

Earthquakes associated with transform faults generally occur at very shallow depths -- less than 20 km (13 mi) deep. Because they occur so close to the surface, they often cause more damage then a deep earthquake of similar magnitude. The Anatolian Fault in Turkey is a transform fault that has claimed many thousands of lives during recorded history.

Perhaps the greatest impact of plate tectonic theory for most people has been upon our view of the occurrence of earthquakes. Long regarded as random acts of God that might not recur, many earthquakes are now seen to be recurrent events along plate boundaries. Indeed, a long respite from earthquake activity at a plate boundary is not a sign of safety -- it is a sign of danger. The longer the period of inactivity, the larger the forces that are building up, and the greater the earthquake that inevitably must come.

STUDY QUESTIONS

III-10. The age of the ocean floor has been directly determined from samples brought up by the Deep Sea Drilling Program. It has been found that the age of the ocean floor increases with distance from the spreading ridge crest on both sides of the ridge. Why is this regarded as proof of the concept of sea-flooring spreading?

III-11. In Figure III-3, on page 45, the active volcanoes of the Andes Mountains are located to the east of the plate boundary. Why are they located east of the trench, on the South American Plate, rather than to the west, on the Nazca Plate? What is the general rule that determines on which side of the trench are subduction zone volcanoes to be found?

III-12. Why are oceanic trenches found only in association with subduction zones?

III-13. Trace the outline of the Baja California peninsula along with the portion of the California coast west of the San Andreas Fault from Figure III-13. Can you slide this sliver of land along the transform faults back to where it originally came from? Where is it going to in the future?

F. HOT SPOTS

There is another part of plate tectonic theory that is perhaps the most curious of all: the <u>hot spots</u>. They may be found on the plate boundaries or in the interiors of the plates, and consist of a line of dormant volcanic features that ends abruptly in a single spot of active volcanism .

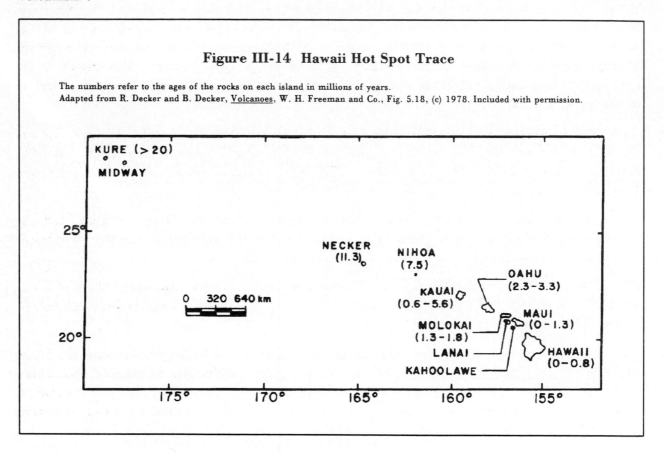

Figure III-14 Hawaii Hot Spot Trace

The numbers refer to the ages of the rocks on each island in millions of years.
Adapted from R. Decker and B. Decker, <u>Volcanoes</u>, W. H. Freeman and Co., Fig. 5.18, (c) 1978. Included with permission.

The Hawaiian island chain presents the clearest example of a hot spot trace (see Figure III-14). The eight main Hawaiian Islands are but the most southeastern members of a linear chain of islands that stretch for 3500 km (2200 mi) northwest to Midway Island. Most of the islets are tiny and uninhabited, but in fact they sit on large undersea volcanic mountains whose tops have been planed off by the erosive action of the sea. Coral reefs have built up on these submerged platforms, keeping the last visible trace of the ancient volcanoes from disappearing altogether.

As we travel southeast, we find that the islands become larger and younger. We do not encounter recent volcanism until the last two islands of the chain: Maui and the big island of Hawaii. Historic activity is largely confined to the two most southeastern volcanoes on the island of Hawaii. These are the Mauna Loa and Kilauea volcanoes. As this is being written, both are in active eruption, spewing basaltic lava fountains into the air and sending flows down their slopes toward the sea.

Fifteen kilometers southeast of the island of Hawaii, an undersea volcano has been detected. Its name is Loihi, and so far it has built up about 5 km (3 mi) above the ocean floor and now reaches to within 960 m (3,150 ft) of the ocean surface. If we continue southeast from Loihi, we find only relatively featureless ocean floor.

The concept of a hot spot envisions a small source of heat fixed deep in the Earth, below the lithospheric plates (Figure III-7 on page 51). Molten rock from the heat source rises rapidly, melting its way through the overlying lithospheric plate and emerging on the surface to form a volcano. But because the plate is moving, the volcano is soon carried away from the point over the heat source, and becomes dormant. The hot spot burns its way through the lithosphere directly above it and begins to construct a new volcano. The result is an ever-lengthening line of volcanoes whose abrupt beginning at an active vent marks the location of the hot spot itself.

As an analogy, consider the action of a sewing machine. The needle can move up and down, but not sideways, repeatedly piercing the cloth which moves under the needle. The result is a line of stitches starting at the needle and proceeding away from it in the direction that the cloth is moving.

In the same way, the Hawaiian Islands are created by the fixed hot spot, and the moving Pacific Plate carries the finished products away to the northwest. The line of volcanic islands defines the direction toward which the plate moves.

Figure III-3 on page 45 shows the locations of some two dozen hot spots. Many of them, such as Hawaii, are located far from plate boundaries, and others, such as Iceland and the Azores, are located on spreading ridges.

Oceanic hot spot volcanoes erupt basaltic lava that is rich in the mantle mineral olivine. Continental hot spots must melt their way through many kilometers of granitic continental crust, and so the volcanoes associated with them range from basaltic to rhyolitic in composition. Rhyolite has the same composition as granite, but is fine-grained and results from rapid cooling of granitic magma once it is erupted on the surface as lava.

The origin of hot spots is still unknown. While Figure III-7 on page 51 shows a hot spot source extending down into the asthenosphere, that only represents how far down they have been traced to date. That they may have their origin very deep in the mantle is suggested by studies of the motions of hot spots.

Once the motions of all the major plates were worked out for the past few million years, it was found that the volcanic traces of all the hot spots could be correctly predicted by a model in which the hot spots are fixed in position in relation to one another. Recent research indicates that over long periods of geological time, the hot spots do move, but this motion is much slower than that of the lithospheric plates.

This is unlikely to be the case if their source is in the highly mobile asthenosphere. The deeper mantle is more rigid than the asthenosphere, and it has been suggested that the hot spots have their source in this less mobile region.

If so, the nature of the hot spots is truly remarkable. A source of heat, perhaps thousands of kilometers deep within the Earth, produces superhot material that gathers into a thin rising column. Almost like a laser beam it burns its way upward to the surface to produce volcanic activity like that of Hawaii and Yellowstone.

STUDY QUESTIONS

III-14. Using Figure III-14 on page 64, where was the Hawaiian hot spot located three million years ago? Which islands did not yet exist at that time?

III-15. In what direction is the Pacific Plate moving with respect to the Hawaiian hot spot?

G. THE ORIGIN OF CONTINENTS AND OCEANS

Now at last we can answer a question posed earlier: <u>Why are there continents and ocean basins on Earth?</u> We got as far as realizing that continents exist where the crust is thick and is composed of low density rocks such as granite. Ocean basins exist where the crust is thin and basaltic in composition. The process of isostasy then determines the relative elevations of the continents and ocean floors.

Plate tectonic theory allows us to go another step toward an explanation. Basaltic ocean crust is created at spreading ridges, and granitic and andesitic continental crust is created beneath the volcanoes associated with subduction zones. Volcanic island arcs are bits of newly created continental crust that eventually coalesce into small continents, like the Japanese Islands, or are plastered onto the edges of larger continents.

Oceanic crust, on the other hand, is very ephemeral. Created at the oceanic ridges and destroyed at the subduction zones, the floor of the ocean basins can never reach a truly ancient age. The age of the oldest ocean floor does not exceed 200 million years. This can be compared with ages for continental rocks that range back to 3,800 million years. All the ocean floor that exists today has been created since Pangea began to break up.

We need to answer one final question: <u>Why do spreading ridges produce only basaltic crust, and why do subduction zone processes generate andesitic and granitic crust?</u>

The process begins at the oceanic ridges, where hot mantle material wells up beneath the ridge crest and partially melts. The mantle is composed of a rich mix of many different minerals. Dominant are the dense silicate minerals olivine, garnet, and pyroxene. But also mixed in are smaller proportions of the lighter silicates that make up granite and basalt.

The partial melting beneath the ridge results in a kind of distillation or fractionation process that selectively incorporates the lighter minerals of the mantle into the oceanic crust. The process may be compard with the distillation of fermented grain, in which the more volatile alcohol is concentrated in the finished product. At the oceanic ridges, the end product is a basaltic rock, dark in color and moderately dense, though far less dense than the parent mantle rock. This becomes the thin crust of the ocean floor and is carried away rapidly by the plate tectonic conveyor belt.

At a subduction zone, the basaltic ocean crust is carried down below the trench and cooked by the Earth's internal heat along with frictional heat produced where the plates rub together. The subducted crust has been exposed to the oceanic environment for many millions of years, during which time it has been altered by chemical reactions with seawater and has accumulated a thick covering of sediment. Some of this sediment is scraped off the oceanic crust when it is subducted, but recent evidence indicates that some of it is also subducted. In addition, large quantities of water contained in the crust are also subducted.

Partial remelting of the altered oceanic crust along with subducted ocean floor sediments results in a second stage of fractionation, further concentrating the lighter minerals and producing molten rock of andesitic and granitic composition. Eventually this material is added to the continental crust by the andesitic volcanoes.

The question of whether the total volume of all the continents is now increasing with time is still a subject of debate. While new continental material is added by subduction zone volcanics, continental material is also removed by erosion, deposited on the ocean floor, and eventually subducted once again. We cannot as yet determine which, if either, of these two processes is dominant at the present time.

STUDY QUESTIONS

III-16. What is meant by he term _fractionation_? Can you think of any commercial application of fractionating or distilling processes for the purpose of concentrating desired substances from extremely heterogeneous mixtures?

III-17. Why do you think that newly-formed oceanic crust tends to be thin, while "continental" andesitic and granitic crust built up by subduction zone volcanoes tends to be much thicker? Look carefully at Figure III-7 on page 51 and Figure III-8 on page 52 before deciding on an answer.

H. PLATE TECTONICS AND MOUNTAIN BUILDING

Turn again to Figure III-3 on page 45. Because new plate is created at the ridges and old plate is destroyed at subduction zones, plates naturally move from the ridges toward the trenches.

Consider the Pacific Plate. It is gaining new crust at the East Pacific Rise on its southeastern edge and it is being destroyed in the trenches at the north and west sides of the Pacific Ocean. As seen from its neighboring plates, its motion is from the southeast toward the northwest.

The Nazca Plate moves in an easterly direction from the East Pacific Rise toward the Peru-Chile Trench.

The motion of North America seems more complex, but follows the same general rules. As seen from Europe, the North American Plate moves westward, away from the Mid-Atlantic Ridge. At the San Andreas Fault, the North American Plate encounters the north-westerly moving Pacific Plate. Because both plates are moving to the west, the result is the slipping motion of a transform fault rather than subduction.

These motions can become very complex, since the direction of motion of a plate, as it appears to you, depends on where you happen to be standing. It is rather like the case of a ferry slowly moving away from its dock. To someone on the dock, the ferry is moving out to sea. But to someone on the ferry, it can appear as though the dock is moving away from the ferry. In the same way, someone standing on South America sees Africa moving away to the east, while an observer on Africa sees South America moving away to the west.

As a final case, consider the Indo-Australian Plate. It is moving to the north, away from the spreading ridge in the Indian Ocean. On its north-central boundary, there is indeed a subduction zone marked by the Java Trench. But its northwestern edge falls along the northern boundary of India -- the Himalayas.

Here we see two continents in the act of colliding. India, in its long northward trek from its original home near Madagascar, has rammed into the underbelly of Asia, and the result is the highest mountain range on Earth. Much the same is happening in the Alps, where the northern motion of Africa is forcing Italy against the European continent.

As though they were two automobiles in a collision, the continents display the equivalent of crumpled fenders and twisted frames in the tortured and distorted strata of rock found in these mountain-building collision zones (see Figure III-15).

Figure III-15 Folded Strata in the Swiss Alps

(Cross section view) Adapted from <u>Cambridge Encyclopedia of Earth Sciences</u>, Cambridge University Press, 1981. Included with permission.

Geologists have found evidence for just such a history in the Himalayas and Alps, which are still active today, as well as in the long-dormant Urals of Russia and the Appalachians of the eastern United States.

Africa is moving to the north, driving the north of Italy into the heart of Europe to form the young Alps, just as India is forcing its way into Asia. The Urals would appear to mark an ancient collision between Europe and Asia. But what of the Appalachians?

The bend of the mountain ridges as they pass through Pennsylvania seems to hint that the bulk of the colliding mass came from the southeast. Could it have been Africa that smashed into the east coast of North America at the end of the Paleozoic, some 300 million years ago? If so, the existence of the Appalachians testifies to an event that was part of the assembly of Pangea.

In all of this we sense a mobility through the ages that seems to transcend anything that Wegener envisioned. <u>Now that we know how the plates are moving today, is it possible to reconstruct the motions of the plates throughout the history of Planet Earth?</u> The answer to this question comes from an unexpected source: the study of the Earth's magnetic field.

STUDY QUESTIONS

III-18. As seen from the North American Plate, which way is the Eurasian Plate moving? Which way is the African Plate moving?

III-19. Collisions between which continents formed the Appalachians? Alps? Urals?

I. PALEOMAGNETISM

Magnetic compasses have been in use as navigational aids for at least 700 years. A magnetized needle is suspended so that it can rotate freely about a vertical axis. When it comes to rest, we find that it points approximately toward the North Magnetic Pole.

When certain kinds of volcanic rocks cool and crystallize from the molten state, they "freeze in" a memory of the direction of the magnetic field in which they cooled. In a sense the magnetic atoms in the rock act rather like fossilized compasses that can tell us which way was north at the time the rock formed.

As it turns out, rock magnetism can tell us even more about the ancient magnetic field than an actual compass could. Figure III-16 shows the shape of the present day Earth's magnetic field.

<u>Lines of force</u> of the magnetic field emerge from the southern hemisphere, loop through space, and reenter the Earth in the northern hemisphere. Notice that the lines of force point down into the ground in the northern hemisphere and up out of the ground in the southern hemisphere.

Magnetic lines of force are a convenient way of representing the direction that a freely suspended compass needle would point at any location within a magnetic field. They do not really exist as lines, and the magnetic field itself is invisible. A magnetic field is capable of exerting a force on a magnetized needle, turning it so that it points in a preferred direction. We say that the magnetic field points in this direction.

At the South Magnetic Pole the magnetic field points straight up out of the ground; at the magnetic equator it is horizontal (parallel to the ground surface), and at the North

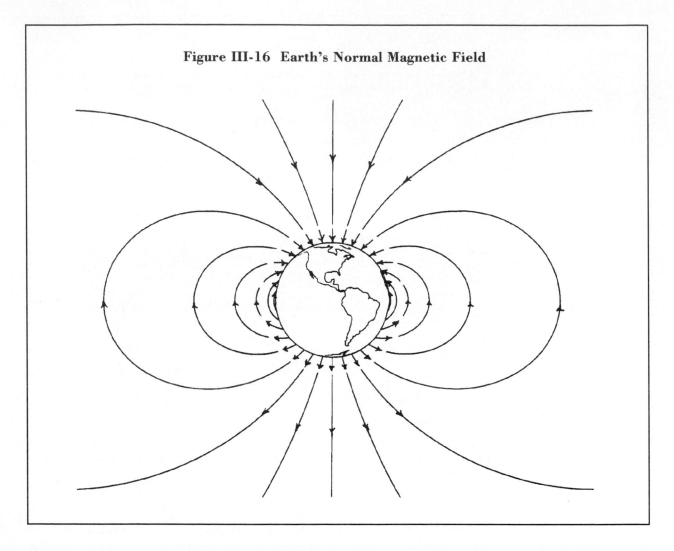

Figure III-16 Earth's Normal Magnetic Field

Magnetic Pole it points straight down into the ground. At intermediate magnetic latitudes the angle that the magnetic field makes with the Earth's surface varies smoothly as shown in the graph in Figure III-18. In other words, if we know the angle that the magnetic field makes with the horizontal and whether it is pointing up or down, then we can determine our magnetic latitude.

For example, if you were to find yourself imprisoned in a room with no windows, you might learn your approximate latitude by measuring the direction of the magnetic field in the room. If, say, you found that the magnetic field pointed down into the ground making an angle of 40⁰ (down) on the vertical axis of the graph, draw a horizontal line from it to the curve, then drop straight down to the horizontal axis to find that you are somewhere around 23° North latitude. On the other hand, if you were to find that the magnetic field pointed (up) out of the ground at an angle of 60°, then you could conclude that you must be somewhere around 41° South latitude.

There is a problem here because at the present time the magnetic axis is tilted some 11 degrees from the geographic axis of the Earth. What we would really like to know is the geographic, not the magnetic, latitude.

Figure III-17 Earth's Reversed Magnetic Field

Note that in the normal configuration of the Earth's magnetic field, the North Magnetic Pole is near the North Geographic Pole while in the reversed configuration of the magnetic field, the North Magnetic Pole is near the South Geographic Pole. Both diagrams show the average positon of the field.

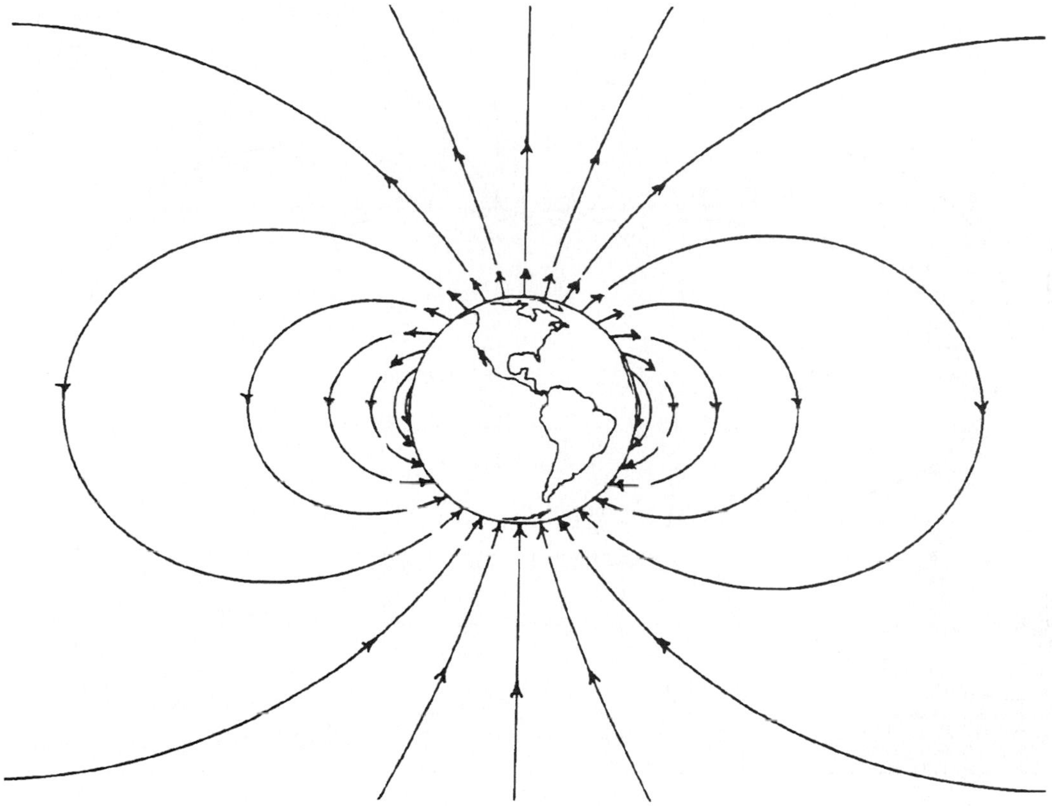

Studies of rocks which are no more than a few million years old show that at any given time the magnetic and geographic poles may differ by as much as 15 degrees, but on the average, the magnetic poles are in fact centered on the geographic poles. The magnetic pole moves with time, shifting about the geographic pole in a more-or-less random manner. Over a period of a few thousand years, however, its average position is found to coincide fairly closely to that of the geographic pole. This has apparently been the case throughout most of geologic time. In paleomagnetic investigations, large numbers of samples spanning a time interval of more than a few thousand years must be examined in order to ensure that the ancient geographic latitude has been determined.

An ordinary compass cannot give us enough information to do this, but frozen in place in a cooled volcanic rock, the magnetic atoms can. Using sensitive instruments called magneto-meters, geophysicists coax from cooperative rocks information about the direction of the recorded magnetic field, enabling them to determine which way was north when the rock

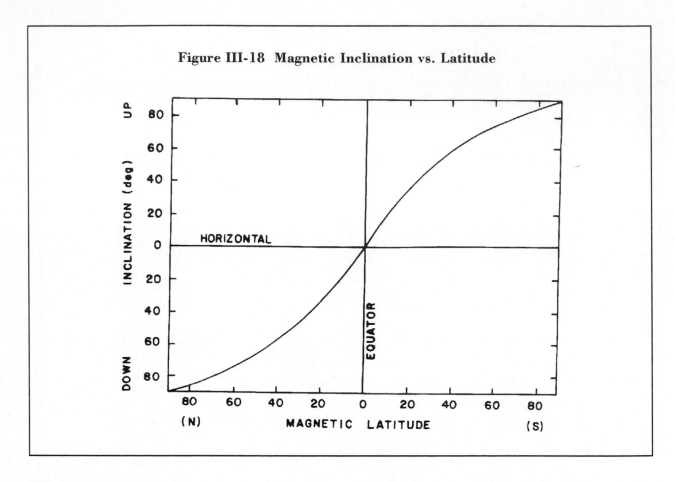

Figure III-18 Magnetic Inclination vs. Latitude

crystallized; and at what latitude did it form. Fortunately, some sedimentary rocks are also capable of recording the direction of the magnetic field when they form.

Here is a tool of tremendous utility. The study of <u>paleomagnetism</u> (the word literally means "ancient magnetism") provides us with much of the information we need to retrace the wanderings of the continents during the past several hundred million years.

Nonetheless, paleomagnetism does not give us all that we would wish. For a given continent at a given age, we can determine the ancient direction of north (that is, the orientation of the continent) and the ancient latitude. It tells us nothing about the ancient longitude.

But that is enough to tell us a great deal about the motion of the continents. Rocks of various ages in North America have been examined and yield the information shown in Figure III-19.

We can see that in the Early Paleozoic, North America was located on the equator and was tipped far over to the right. As time went on, the continent changed its orientation rapidly in a counterclockwise direction and then shifted in latitude by moving to the north.

We can do the same thing for all the continents, but the absence of longitude control makes it hard to know just how they should be placed in relation to one another.

Figure III-19 How North America Has Moved

From paleomagnetic evidence alone, we can determine a continent's ancient latitude and orientation, but not its longitude. North America's latitude and orientation are shown for each of five geological periods.

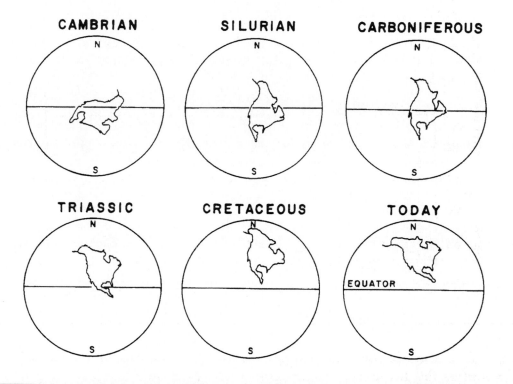

STUDY QUESTIONS

III-20. What inclination angle does the present Earth's magnetic field make with the ground surface at New Orleans (latitude N 30°)? Oslo, Norway (N 60°)? Porto Alegre, Brazil (S 30°)?

III-21. At the present time, where would you be on Earth if you measured the magnetic field direction and found it to be pointing north and horizontal to the ground surface? Pointing north and up out of the ground?

III-22. What two pieces of information about the ancient positioning of the continents can we determine from the magnetic field recorded in rocks? What important piece of information <u>can't</u> we determine in this way?

J. MAGNETIC POLARITY REVERSALS

Yet another feature of the Earth's magnetic field comes to our assistance by providing the missing information, at least for the past 200 million years. This feature is the propensity of the Earth's field to suddenly interchange its North and South Magnetic Poles. These events are termed <u>magnetic polarity reversals</u>.

After a polarity reversal occurs, the magnetic field at every point on Earth points in the opposite direction to that which it had prior to the reversal. (Compare Figure III-16 with Figure III-17).

Polarity reversals happen very quickly, at least from a geological point of view. It may take no more than a few thousand years for the field to flip to its new opposite configuration. During the past 50 million years, polarity reversals have occurred at an average rate of a few per million years, though exactly when they occur seems to be randomly determined.

The history of polarity flips during the last five million years is shown in Figure III-20. It was determined by collecting lava flow samples of different ages. Each sample was age dated by radioisotope methods and also was measured in a magnetometer. If the magnetization in the rock pointed north, the sample was classified as being magnetized in the <u>normal</u> sense. If it pointed south, the sample was classified as being magnetized in the <u>reversed</u> sense.

From the diagram, you can see that the magnetic field has been normal for the past 730,000 years, was reversed from 730,000 years ago until 900,000 years ago, etc. This pattern of polarity reversals has been verified by many thousands of additional samples measured since the original determination.

At first sight, the presence of polarity reversals seems to be an enormous complication in applying paleomagnetism to continental drift. In fact, the average time of polarity intervals (less than a million years) is so short compared to the time needed for significant continental motion (tens to hundreds of millions of years) that separating the two effects is usually possible.

For instance, if two rocks differ little in age but contain magnetizations that point in opposite directions, then it is clear that we are dealing with a polarity reversal and not with some sudden jump in the continent's position.

The Earth's magnetic field arises from large electric currents circulating in the liquid iron core. These currents in turn arise from convective motions of the fluid and a complex interaction between them, the electric currents, the magnetic field, and the Earth's rotation. The Earth acts like an electric generator, or <u>dynamo</u>, using its rotation and internal heat energy to produce electric and magnetic fields. When the polarity of the field changes, all that is needed to accomplish this is for the electric currents to reverse their direction as well. The physical motions of the Earth are not likely to be affected in this process.

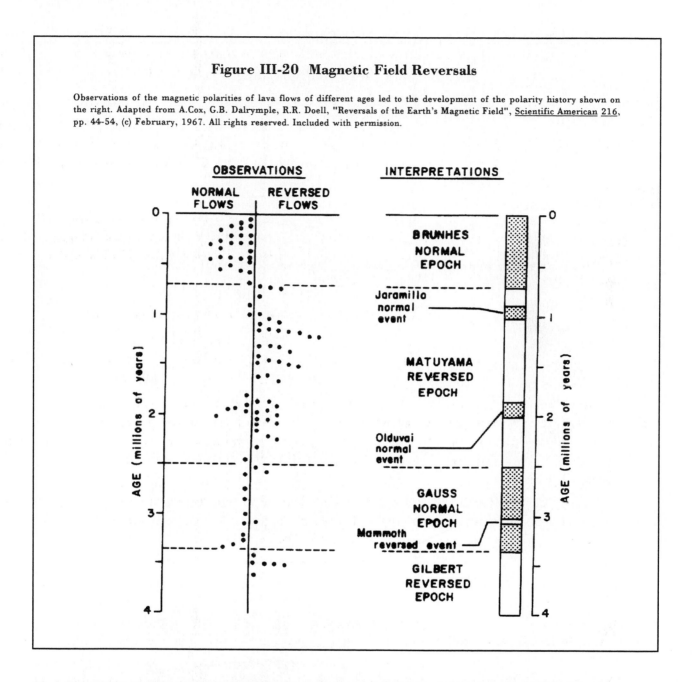

Figure III-20 Magnetic Field Reversals

K. SEA-FLOOR SPREADING AND THE MAGNETISM OF THE OCEAN FLOOR

The polarity reversals provide us with an extraordinary tool for measuring the motions of the oceanic portions of the plates. When new oceanic crust is created at the spreading ridges, the pattern of magnetic field reversals is imprinted in the rocks of the ocean floor.

The process is shown in Figure III-21. The animated version of this diagram on television is especially helpful in understanding how it works. Basaltic lava cools beneath the crest of the spreading ridge and records the direction of the magnetic field, freezing it into the

Figure III-21 Sea-Floor Spreading

As the sea floor spreads, the magnetic polarity is recorded in the ocean crust at the ridge axis. This record is then carried away from the ridge, so older polarity zones are found farther from the ridge crest.

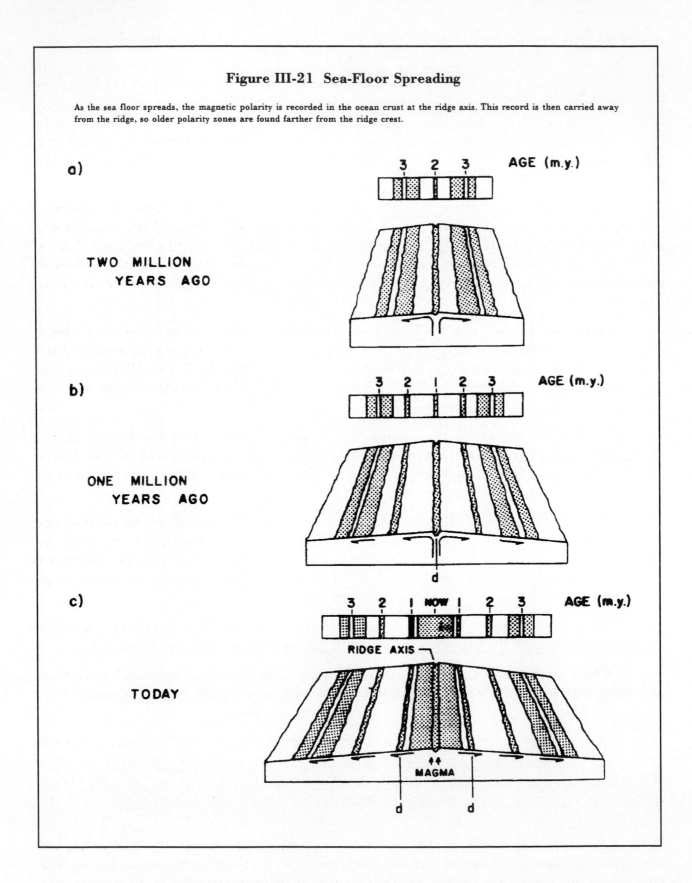

atomic structure of the rock. Half of this newly formed crust is added to the plate on either side of the ridge. (You may want to review Figure III-8 at this point.)

When the magnetic field is in the normal direction, then the new oceanic crust becomes normally magnetized; when the field is reversed, the new crust becomes reversely magnetized. The newly formed crust is then carried away from the ridge and forms striped patterns of alternately normally and reversely magnetized ocean floor. If sea-floor spreading proceeds at a smooth and steady rate, then the striped magnetic pattern on the ocean floor should match the pattern of polarity reversals in Figure III-20.

The magnetic pattern on the ocean floor can be detected and mapped by towing a sensitive magnetometer in a criss-cross path over the ridge. The polarity of the crustal rocks may be determined because over normally-magnetized crust, the Earth's magnetic field is slightly intensified, while over reversely-magnetized crust the field is slightly weakened. Worldwide results from the vicinity of the spreading ridges show that the patterns of stripes and of magnetic polarity reversals in fact do match.

At one stroke, our interpretation of the sea-floor patterns not only provides evidence that sea-floor spreading is actually happening, but it also allows us to trace the motion of the plates on either side of the ridge with precision. For instance in Figure III-21c, we can reconstruct the relative positions of the two plates on either side of the ridge as they were one million years ago by the following procedure.

On the polarity time scale bar at the top of the diagram, find the normal (black) polarity that occurred approximatly one million years ago. The stripe formed at that time is labeled (d) on each plate. Everything between the two stripes labeled (d) has been created since then.

The two stripes (d) must both have been located at the crest of the ocean ridge one million years ago, since that is where all ocean crust is formed. So we need only cut along the (d) stripes with scissors, lift out the younger crust between them, and slide the two plates together to close the gap. The result is seen in Figure III-21b. We can repeat the process to go back in time to two million years ago (Figure III-21a).

What we are doing here is, in effect, running the sea-floor spreading animation backward in time from the present, until we reach the time that we wish to reconstruct.

Where continents are part of the plates on either side of a spreading ridge, we can trace the relative motions of those continents with considerable accuracy. There is an exercise at the end of this section that will help you to understand how this works.

The striped magnetic pattern on the ocean floor allows us to reconstruct continental positions far more accurately than by using paleomagnetic data alone. In combination, the two methods complement one another -- paleomagnetism gives us the ancient latitude and orientation of each continent, while the sea-floor spreading pattern gives us the separations between the continents.

These analytical tools should allow us to reconstruct the ancient visage of our planet throughout geologic time. That we can do so only partially at present is due to two factors. The first is that there is no ocean floor (and hence striped magnetic pattern) older than 200 million years, and the second is that paleomagnetic data are not yet available for all continental blocks for very ancient times.

EXERCISE

In this exercise you will reconstruct the relative positions of Africa and South America at intervals of 5, 38, and 65 million years ago. The map in Figure III-22 shows the age of the ocean floor separating these two continents as determined from the stripe magnetic pattern in the South Atlantic Ocean.

Take a piece of tracing or onionskin paper and trace onto it the outlines of South America and Africa as they appear on the map.

The arrows at the bottom of the diagram show which portions of the ocean floor were created within the last 5, 38, and 65 million years. The spreading ridge is today contained within the central black area and in the past five million years has created the floor shown in black. The age of the ocean floor becomes progressively older as we move away from the ridge to either side. To reconstruct the position of Africa and South America as they stood with respect to one another five million years ago, you need only cut out the black area around the ridge so that the two plates are separated completely. Then fit the plates back together so as to close the gap caused by the missing five million years of ocean floor. South America and Africa should now be in their correct relative position. This process can be repeated to reconstruct the continental positions for 38 and 65 million years ago.

After each step of the exercise, reposition your tracing paper so that the traced and printed images of South America exactly coincide. Then without moving the paper, trace the new position of Africa at intervals of 5, 38, and 65 million years, and label the age of each position for Africa. Your tracings should now represent a partial history of the opening of the South Atlantic Ocean.

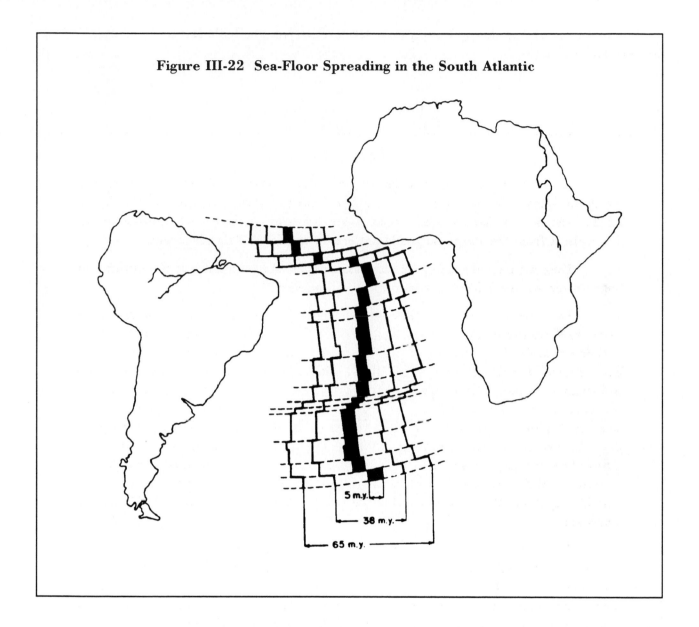

Figure III-22 Sea-Floor Spreading in the South Atlantic

STUDY QUESTIONS

III-23. How does the direction of the magnetic field at a particular point on Earth differ before and after a polarity reversal takes place? Refer to Figure III-16 and Figure III-17.

III-24. In the exercise using Figure III-22 you used the age of the ocean floor as determined from the striped magnetic pattern to reconstruct the ancient positions of South America and Africa with respect to one another. Can you reconstruct the ancient latitudes of South America and Africa from the ocean floor age data? If so, what are they? If not, what additional data would you need?

L. EARTH HISTORY FROM A PLATE TECTONIC PERSPECTIVE

Now let us put together all the available evidence and attempt to reconstruct the ancient positions of the continents. The principal evidence is derived from the paleomagnetic data, giving us latitudes and orientations of the continents. For the past 200 million years, we have the additional control provided by the striped magnetic patterns on the ocean floor.

There is also additional input from many other sources of information: ancient suture zones, which indicate continental collisions; paleoclimatological data, buttressing the magnetic data in distinguishing polar from tropical zones; and paleontological data, indicating which continental fragments shared common fauna and were likely to be attached or in close proximity to one another.

When we have done all this and uncertainties remain, then we apply the principle called <u>Ockham's Razor</u>: when we have to choose among several possible models and the evidence is not sufficient to make the choice unambiguous, then we choose the simplest. For the most ancient times, continents are placed where they will be in the best position to partake in known future events such as continental collisions. We cannot be certain that the result is correct, but it should show the essential relationships among the continents, and in any case, it is the best we can do at present.

Figure III-23 through Figure III-32[5] outline how the continents and pieces of continents have moved about for the past 540 million years, according to a recent synthesis. Some aspects of these reconstructons remain speculative in nature. For this reason, the description that follows should be regarded as a tentative scenario based on these interpretive maps.

Take a few minutes at this point to study the maps. Start with today's familiar world in Figure III-32 and work backwards through each map in turn, tracing the motions of the various continental blocks.

---

[5]Figure III-23 through Figure III-32 were prepared by Christopher R. Scotese, University of Texas, Institute for Geophysics.

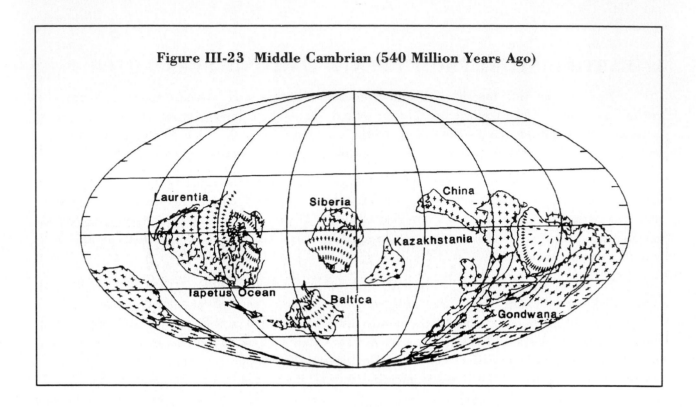

Figure III-23 Middle Cambrian (540 Million Years Ago)

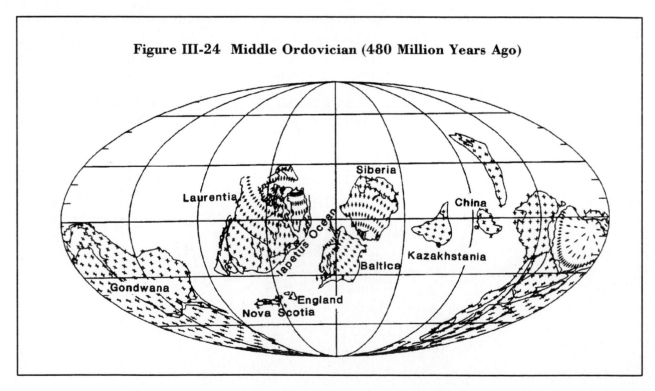

Figure III-24 Middle Ordovician (480 Million Years Ago)

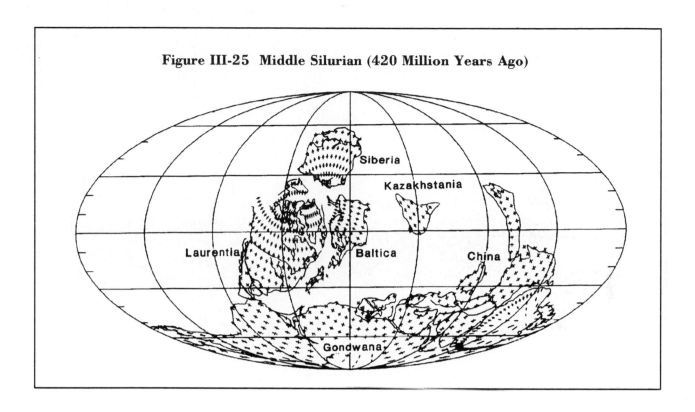

Figure III-25 Middle Silurian (420 Million Years Ago)

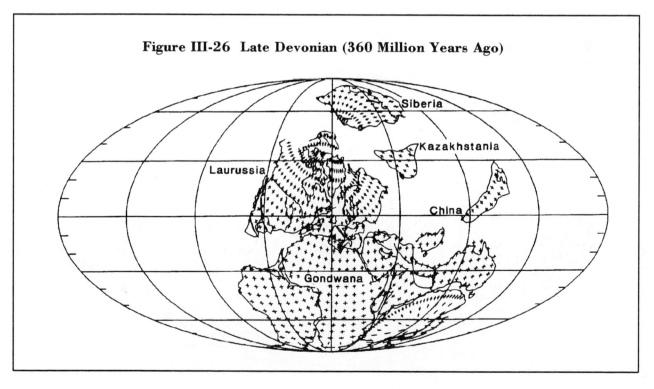

Figure III-26 Late Devonian (360 Million Years Ago)

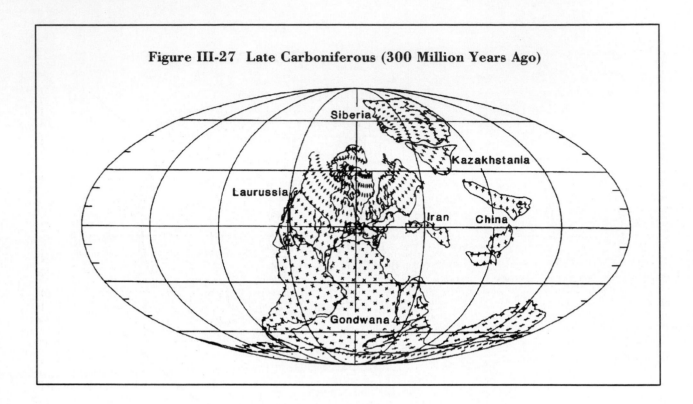

Figure III-27 Late Carboniferous (300 Million Years Ago)

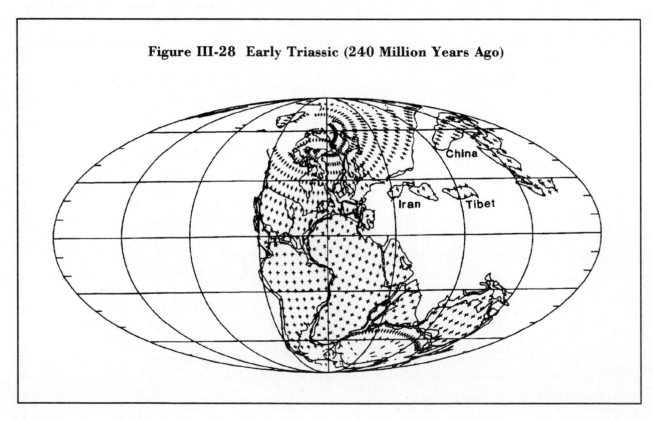

Figure III-28 Early Triassic (240 Million Years Ago)

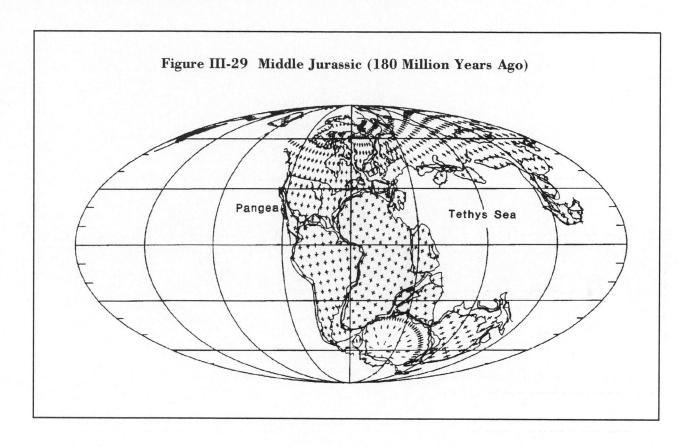

Figure III-29 Middle Jurassic (180 Million Years Ago)

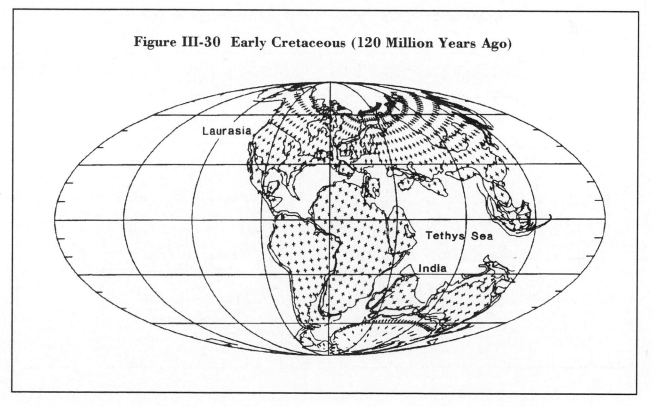

Figure III-30 Early Cretaceous (120 Million Years Ago)

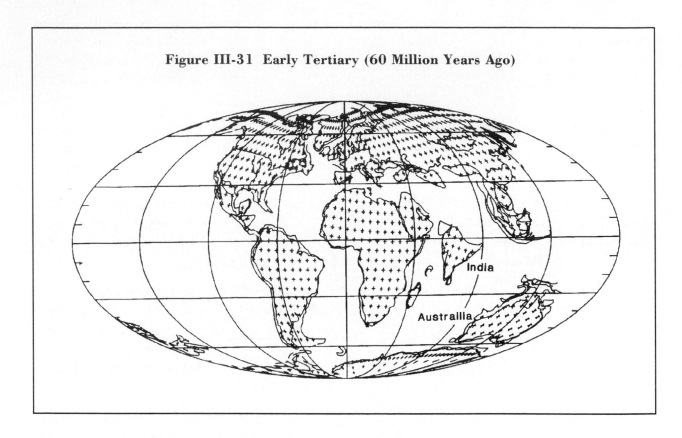

Figure III-31 Early Tertiary (60 Million Years Ago)

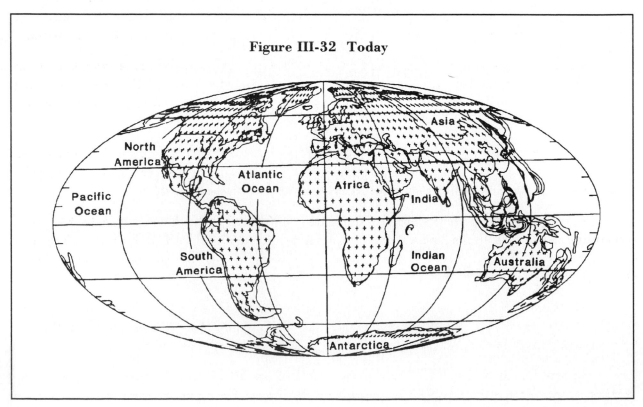

Figure III-32 Today

Note that unfamiliar names appear on some of the earlier maps for assemblages of different blocks: <u>Laurussia</u> for North America, Greenland, northern Europe, and eastern Russia; <u>Laurentia</u>, for North America, Greenland, and a bit of northern Europe, including Scotland; and <u>Gondwana</u>, for South America, Africa, Antarctica, Australia, New Zealand, southern Europe, India, and much of the Middle East.

Gondwana appears to have existed intact as a supercontinent throughout the entire Paleozoic and much of the Mesozoic. Then, in the Cretaceous, it broke into four major fragments, one of which, India, began moving northwards toward Asia. Today, Africa continues to fragment as the Arabian peninsula splits off from it, with a newly-formed spreading ridge in the Red Sea widening the gap between them. A fresh crack is forming in east Africa - the East African Rift - perhaps foreshadowing the creation of a new subcontinent made up of parts of Somalia, Ethiopia, Kenya, Tanganyika, and Mozambique (see Figure III-3). Or perhaps the rift will become dormant, leaving only an inactive trough called an <u>aulacogen</u>.

At the beginning of the Paleozoic (Figure III-23) Gondwana shared the world with five major island continents: Laurentia, Baltica (northern Europe), Siberia, China, and Kazakhstania (a triangular slice of south-central Russia and northeastern China). Laurentia was separated from Baltica by the Iapetus Ocean -- a previous incarnation of the North Atlantic. Whimsical scientists derived the name from mythology, in which Iapetus was the father of Atlantis.

By Mid-Ordovician time (Figure III-24) a subduction zone formed off the east coast of North America. It swallowed up part of the Iapetus Ocean floor, and Baltica collided with the northern part of Laurentia, raising the Caledonian mountains on the east coast of Greenland, in Scotland, and on the west coast of Norway (Figure III-25) forming the continent of Laurussia.

Meanwhile, portions of New England, Nova Scotia, and Newfoundland may have been sheared northeasterly along the east coast of North America (Figure III-26), continuing the construction of the Appalachians that had begun with the onset of subduction in the Iapetus Ocean.

Gondwana had moved to the north (Figure III-26 and Figure III-27) and collided with Laurussia, bringing the Appalachians to their climax. By late Carboniferous time, the Appalachians may well have been the most magnificent mountain range on Earth.

Throughout the Late Paleozoic, and into the Triassic, the blocks of Asia had been assembling, and around the end of that era a series of collisions were raising the Urals and other mountains of Asia, wresting prominence away from the now-dead and eroding Appalachians.(Figure III-28)

The construction of Pangea was now complete (Figure III-29). What a strange, lopsided world it was, with all the landmasses splashed across one hemisphere in a continuous supercontinent, while the other hemisphere was given over entirely to ocean.

Pangea began to break up in the Jurassic (Figure III-29 and Figure III-30) with the northern continents rotating clockwise to destroy the Tethys Sea separating India from Asia. In the process, the North Atlantic began to open. Ocean floor that exists today had finally made an appearance at this late date in Earth history.

Intrusion of basaltic magma accompanying the first rifting of the Atlantic Ocean created the Palisades of New York and New Jersey, and the subduction of ocean floor beneath the west coasts of North and South America continued the construction of what would become the Andes and the Sierra Nevada of California.

With the fragmentation of Gondwana (Figure III-30), the South Atlantic began to open and India began its northward trek and the South Atlantic began to open. (Figure III-31.)

By the beginning of the Tertiary (Figure III-31) Australia had broken loose from Antarctica and the northward motion of Africa and India had all but destroyed the Tethys Sea floor.

Today, (Figure III-32), Africa continues to move northward, subducting Mediterranean ocean floor to form the volcanoes of Italy while the northern part of Italy is forced into the heart of Europe to form the Alps. And to the east, the ongoing collision between India and Asia has raised the Himalayas to the roof of the world.

The hypothetical history we have presented here goes back only some 540 million years, out of the 4,500 million years of total Earth history. Evidence is now being accumulated that may eventually allow us to construct world maps going far back into the Precambrian. Were there other accumulations of all the continents into supercontinents in the Precambrian? What did they look like? Did plate tectonics operate then in the same manner as it does today? We will have to await further evidence before we can answer these questions.

STUDY QUESTIONS

III-25. What present-day continents made up the supercontinent of Gondwana, and when did it exist?

III-26. At the time of Pangea's existence (Figure III-29), the Tethys Sea was a major arm of the world ocean. Why can't you find the Tethys Sea on any present world map?

M. HISTORICAL PERSPECTIVE

If you compare Figure III-29 to Figure III-32 with Figure III-1 (page 40), you will see that Wegener certainly had the right general idea in his continental reconstructions. So why was he repudiated by the scientific community in the 1920's?

In the first place, Wegener was unable to provide a convincing mechanism for moving the plates. It is important to recognize that Wegener did not come up with an early version of plate tectonics -- his theory was concerned only with the drifting of the continents. For him the ocean floor was a stationary impediment rather than the most mobile part of the whole scheme. In Wegener's day, the floor of the ocean was virtually unknown to scientists.

Recall that in science, that theory is best which best explains all available observations. The emergence of plate tectonic theory was based largely on observations in paleomagnetism, oceanography and seismology that were not available to Wegener.

Most of Wegener's arguments were based on paleontology and paleoclimatology -- two of the better developed sciences of his time. His detractors, however, were able to propose with alternative theories that did not violate the entrenched belief in the permanence of the oceans and continents. Temporary land bridges that have since sunk beneath the waves, or natural rafts of floating vegetation, had allowed creatures to migrate from one continent to another across the vast oceans, they said. And everyone knew that past climates were very different from those of today. Wegener's genius lay in his ability to draw correct inferences from circumstantial and often scanty evidence. But the scientific community was under no obligation to accept his inferences when the weight of the available evidence was not decisively in his favor.

There is also a natural reluctance to discard familiar and long-established ideas that have proved useful in one's career. Scientists are human, too.

Nonetheless, once the necessary observations became available, the scientific community made an abrupt about-face, revived Wegener's idea of drifting continents and quickly transformed it into the theory of plate tectonics. Three independent developments in the fifteen years following the Second World War set the stage for the revival. The U.S. Navy, idled by the end of the war but impressed by the capabilities of submarine warfare, supported the monumental task of mapping the ocean floor throughout the world. Seismologists were amassing data on the distribution of earthquakes and the behavior of earthquake waves in the upper mantle, aided by fledgling new devices called computers. And a few scientists in England, Japan, France, and the United States were pursuing studies aimed at working out the history of the Earth's magnetic field.

In its early days, paleomagnetism must have seemed a very esoteric science. With no application in sight, scientists were simply pursuing their own curiosity about the origin and behavior of the magnetic field.

Oceanography had become a burgeoning science. By the mid 1950's the continuity of the worldwide oceanic ridge system had been established. By 1961 ocean floor maps had become available and data showing the striped magnetic patterns associated with the oceanic ridges saw publication. A correct explanation for the magnetic stripes actually appeared in print as early as 1963, but was largely ignored. The evidence for both sea-floor spreading and magnetic reversals was considered very inconclusive by most geologists at that time. But the evidence for continental drift and sea-floor spreading mounted.

By 1966, the worldwide occurrence of magnetic field reversals had been confirmed and a computer-drawn reassembly of the continents bordering the Atlantic into a Pangea configuration showed that the continental shapes really did fit together very well. The concept of the transform fault was introduced. Using seismic evidence, subduction was demonstrated in the Tonga Trench in the South Pacific.

Now the pace accelerated and, like a large and difficult jigsaw puzzle that mostly has gone together slowly and painfully, the last few pieces went together in a rush. Three of the key pieces were the discoveries of sea-floor spreading and subduction, and the recognition of the distinction between the lithosphere and the asthenosphere based on seismic studies. By 1968, the essentials of plate tectonic theory had been worked out in a series of papers by two dozen or so scientists of half a dozen nationalities.

Unlike the theory of continental drift, which was clearly Wegener's, there is no one name associated with plate tectonics. It was a truly international and interdisciplinary effort. There were many outstanding scientists involved, and the television series will introduce you to some of them.

In order to concentrate on the concepts themselves, we have not listed any of these scientists here, but instead refer you to the three books by Sullivan, Marvin, and Glen, cited early in this unit. The complete story is interesting, exciting, and very human, and we earnestly recommend it to you.

STUDY QUESTIONS

III-27. Why was Wegener's theory of continental drift rejected in the 1920's?

III-28. Physical oceanography, paleomagnetism, and seismology were once considered separate disciplines within geophysics. Do you think it likely that plate tectonic theory would have emerged from any one of these disciplines alone, without communication and interdisciplinary work among them?

III-29. The development of plate tectonic theory was very international in terms of the scientists involved. What kinds of international cooperation are necessary to encourage these kinds of multinational efforts?

N. THE HAZARDS OF EARTHQUAKES

We have seen that earthquakes result from the tremendous mobility of the Earth's crust and are ubiquitous on the surface of our planet. Most of them are associated with plate boundaries, but quite a few are found in diffuse zones in the continental interiors.

The locations of earthquakes have been mapped in detail, and on this historic evidence, maps have been drawn up showing those areas at greatest risk. Compare Figure III-33 with the United States portion of Figure III-2.

California, of course, carries the greatest risk, but large historic earthquakes in South Carolina (1886) and southeastern Missouri (1811-12) show that it is not alone in seismic hazard. Large cracks may still be seen in antebellum houses in the Carolinas, and the New Madrid earthquakes of 1811 and 1812 were great enough to shift the Mississippi River into a new course.

Figure III-33 Seismic Risk Map

Zone 0 = No damage expected from seismic activity. Zone 1 = Minor damage; distant earthquakes may cause damage to structures; corresponds to V and VI of Modified Mercalli Scale. Zone 2 = Moderate damage; corresponds to VII of Modified Mercalli Scale. Zone 3 = Major damage; corresponds to VII and higher of Modified Mercalli Scale.

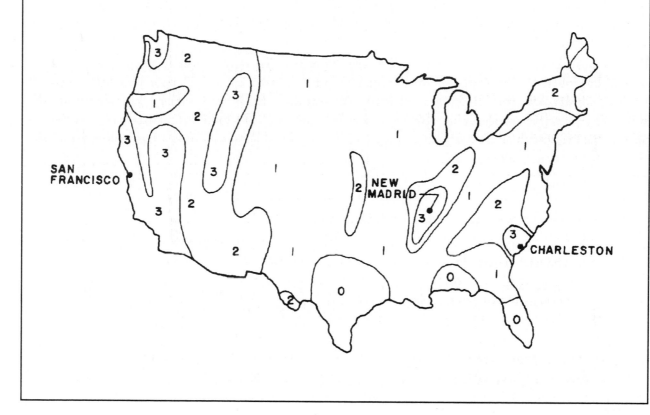

The inter-mountain seismic belt stretching from Utah to Montana reflects continued activity in the Great Basin and the Rocky Mountains, and a line of earthquake activity from Erie, Pennsylvania, along the St. Lawrence River shows that the straight course of that river follows a still-active fault.

The strength of an earthquake can be measured by two very different scales. The first is the <u>Richter Magnitude Scale</u>, which is based on the amplitude of ground shaking at a fixed distance from the earthquake. This is an open-ended scale, on which the weakest felt earthquake registers a magnitude of about 2 and the largest recorded to date fall in the range 8.8 - 8.9.

The Richter scale is sometimes used as a very rough indicator of the total energy released in an earthquake. Because the scale is logarithmic, the energy increases geometrically with the magnitude. For example, a magnitude 3 earthquake releases roughly 30 times the energy of a magnitude 2 earthquake; and magnitude 4 releases approximately 30 times that of a magnitude 3 earthquake. A great earthquake of magnitude 8 releases roughly:

$$30 \times 30 \times 30 \times 30 \times 30 \times 30 = 729,000,000$$

times the energy of a barely-felt magnitude 2 earthquake. An earthquake of magnitude 8 is energetically equivalent to the simultaneous explosion of approximately 30 one-megaton hydrogen bombs.

Another scale that is often used is the Modified Mercalli Intensity Scale. The Mercalli scale is based on observations of observers, and ranges from i - xii. Roman numerals serve to distinguish Mercalli intensities from Richter magnitudes. Each step carries a description (see box). After a major earthquake, surveys are conducted and Mercalli intensities are assigned to each area. The result is an isoseismal map (see Figure III-34) which shows how widespread the effects of the earthquakes were.

THE MODIFIED MERCALLI INTENSITY SCALE

i.   Not felt except by a very few under especially favorable circumstances.

ii.   Felt only by a few persons at rest, especially on upper floors of buildings. Delicately suspended objects may swing.

iii.   Felt quite noticeably indoors, especially on upper floors of buildings, but many people do not recognize it as an earthquake. Standing motor cars may rock slightly. Vibration like passing truck. Duration estimated.

iv.   During the day felt indoors by many, outdoors by few. At night some awakened. Dishes, windows, doors disturbed; walls make creaking sound. Sensation like heavy truck striking building. Standing motor cars rocked noticeably.

v.   Felt by nearly everyone; many awakened. Some dishes, windows, etc., broken; a few instances of cracked plaster; unstable objects overturned. Disturbances of trees, poles, and other tall objects sometimes noticed. Pendulum clocks may stop.

vi.   Felt by all; many frightened and run outdoors. Some heavy furniture moved; a few instances of fallen plaster or damaged chimneys. Damage slight.

vii.   Everybody runs outdoors. Damage negligible in buildings of good design and construction; slight to moderate in well-built ordinary structures; considerable in poorly built or badly designed structures; some chimneys broken. Noticed by persons driving motor cars.

viii.   Damage slight in specially designed structures; considerable in ordinary substantial buildings, with partial collapse; great in poorly built structures. Panel walls thrown out of frame structures. Fall of chimneys, factory stacks, columns, monuments, walls. Heavy furniture overturned. Sand and mud ejected in small amounts. Changes in well water. Disturbs persons driving motor cars.

ix.   Damage considerable in specially designed structures; well-designed frame structures thrown out of plumb; great in substantial buildings, with partial collapse. Buildings shifted off foundations. Ground cracked conspicuously. Underground pipes broken.

---

x. Some well-built wooden structures destroyed; most masonry and frame structures destroyed with foundations; ground badly cracked. Rails bent. Landslides considerable from river banks and steep slopes. Shifted sand and mud. Water splashed over banks.

xi. Few, if any, (masonry) structures remain standing. Bridges destroyed. Broad fissures in ground. Underground pipelines completely out of service. Earth slumps and land slips in soft ground. Rails bent greatly. Damage total. Waves seen on ground surfaces. Lines of sight and level distorted. Objects thrown upward into the air.

Figure III-34 Isoseismal Map Kern County Earthquake, July 21, 1952

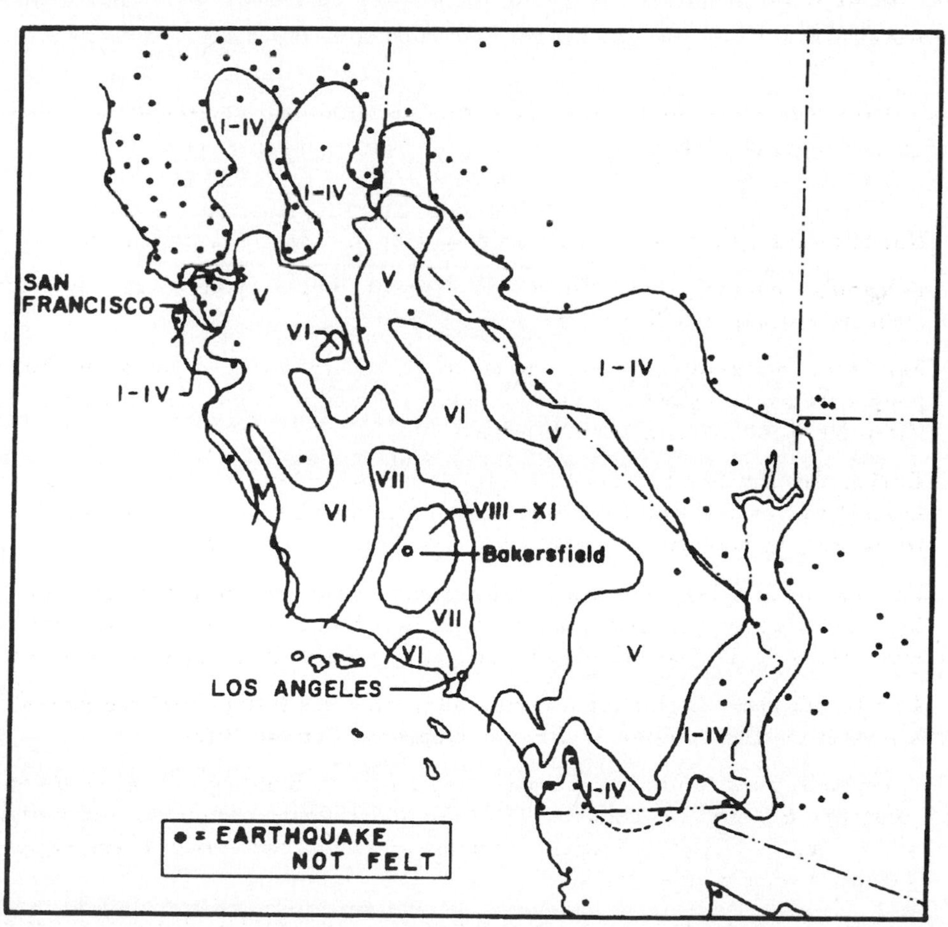

The duration of an earthquake seldom exceeds one minute, though it undoubtedly seems much longer to a terrified participant. The ground along a ruptured fault may suddenly lurch by as much as ten or more feet, and the shaking may induce strong sympathetic vibrations which are sufficient to collapse substantial buildings.

Most vulnerable is unreinforced masonry construction, which cracks and crumbles under the large forces exerted on it. Steel frame and steel-reinforced concrete fare better, and best of all is ground-hugging wood frame construction, which absorbs shearing forces well and is also flexible and resilient.

Fire is another hazard that often follows a great earthquake in an urban area. Gas pipes in the ground and in buildings are likely to be ruptured, and sparks or open flames can start a holocaust. More of San Francisco was destroyed in 1906 by the ensuing fire than by the direct shaking of the earthquake. Since water mains are likely to be broken in many places, the fire can rage unchecked. It is probably difficult to overestimate the horror of a truly great earthquake in a congested urban setting.

Secondary natural effects can also take a high toll in lives and property. Where there are steep slopes, landslides may be triggered by seismic shaking. A tragic example occurred in Peru in 1970, when an earthquake shook loose a huge chunk of Mt. Huascaran, one of the more impressive peaks in the Andes. Pulverized rock and ice came sweeping down from the mountainside and buried whole towns, leaving 20,000 people dead. In 1964, houses in parts of Anchorage, Alaska, received severe damage due to the slumping of ground beneath them when a magnitude 8.4 earthquake struck.

Loose soil that is saturated with water can literally turn to quicksand due to seismic shaking. In 1964 in Niigata, Japan, apartment buildings tipped crazily onto their sides even though they suffered relatively little structural damage. Their foundations had sunk into the fluidized soil.

Dams may be vulnerable to damage from earthquakes, not only from the ground shaking, but also from the waves, called seiches, that may be produced in reservoirs by earthquakes. These waves travel back and forth and repeatedly slam into the dam, possibly weakening it.

Seismic sea waves can pose substantial hazards to people living on low-lying coastal areas, especially around the Pacific Ocean. The misleading term, tidal wave, has been applied to them, but the attraction of the moon has nothing to do with these earthquake-generated waves. The Japanese name tsunami is also frequently used.

Sudden motions along seabed faults set tsunamis into motion. Traveling at great speed, they can cross the whole of the Pacific Ocean in one day. While at sea, they are all but unnoticeable, but as they approach a shoreline, they can build up into waves of huge proportions.

Sometimes ten meters (30 ft) or more in height, these surges can also be extremely broad, with many minutes elapsing between successive crests. The waves can continue for hours. It is the tremendous volume of water in each of these surges that is responsible for their destructive power and their danger. Large boats are sometimes carried far inland, houses swept off their foundations, and people unfortunate enough to be caught by one may be drowned in the powerful currents.

The Pacific is plagued by tsunamis because it is ringed by subduction zones where large earthquakes are spawned that can shift the sea floor, starting the seismic sea wave on its path.

Through international cooperation, a tsunami warning system has been set up for the Pacific, saving lives by issuing warnings in time for evacuation of threatened coastal areas.

If earthquakes are inevitable in places like California, then what can we do about them? It is obvious that San Francisco, Los Angeles, and Tokyo will not be abandoned through fear of a possibly devastating earthquake. However, the possibility of a severely damaging earthquake cannot be ignored. Increased attention has been given in recent years to earthquake-resistant construction of buildings and disposition of existing buildings that are most likely to collapse. Still, the economic costs of earthquake anticipatory measures can be very high, and there is a natural human tendency to ignore a danger that is not clearly imminent.

As an example, consider the case for the removal of parapets on San Francisco houses. Parapets, balconies, and other forms of overhanging stonework have long been a charming feature of San Franciscan architecture. In an earthquake, however, they are all too likely to come loose and crash to the sidewalk below, at great risk to pedestrians. For some time, the city has offered homeowners incentives to remove these features in the interests of safety. Nonetheless, the homeowner is faced with the prospect of expensive renovation that may deface much of the house's beauty and possibly result in lower resale value, all in hopes that this might save someone's life if there is an earthquake during the building's remaining life. If you owned such a building, what would you do?

Once the decision has been made to reside in earthquake country, the remaining decisions can be fraught with uncertainty. Where to build or buy one's home? Houses built on bedrock are more apt to survive than those built on deep fill. Steep slopes, especially those oversteepened by earthmoving, should be avoided. A visit to the State Geologist's office may be wise before such a home purchase decision is made. There you can inspect maps showing areas of maximum risk.

Everyone living in an earthquake-prone area should be aware of what to do in the event of an earthquake. In general, there will be little or no warning, and once begun, the shaking may be over in thirty seconds. This does not leave much time for meaningful action. The most important thing is to avoid panic and to move as quickly as possible to an area that offers some protection from falling objects. If outside, move away from buildings, light poles and power lines; an open space is your safest harbor in an earthquake.

If you are indoors at home, leave the house, avoiding chimneys and other masonry objects. But if you are indoors in an urban area, stay inside. The narrow streets of a city may become deathtraps of falling glass, bricks, and stone. Avoid the centers of rooms, as ceiling collapse is most likely there. Move to a structurally strong place: a doorway, a closet, or even a corner of a room can offer some protection.

Families should prearrange to gather at some definite place; after an earthquake, telephone communication may be impossible for some time. Know where the main gas valve is in your home, and keep a tool handy that can be used to turn it off.

Much of the hazard to life would be mitigated if it were possible to predict with accuracy just when and where large earthquakes are going to occur. Unfortunately, this is not yet possible with any degree of reliability. It has been done in a very few instances, most notably in 1975 in Haicheng, China. Both long- and short-range predictions proved to be accurate, and, equally important, government and citizens cooperated with the scientists in heeding the warnings and in moving the populace to the safest possible locations. When the magnitude 7.3 earthquake hit, 50% of all homes were destroyed or badly damaged, but only some 300 people were killed out of a population of 100,000.

As encouraging as this success may be, our hopes for reliable earthquake prediction should be tempered by the realization that only one year later the Tangshan area of China was unexpectedly shaken by a magnitude 8.2 earthquake. This proved to be one of the most tragic earthquakes in modern history. Occuring without any warning it claimed half a million lives. A television segment shows damage done by the Tangshan earthquake, attesting to its tremendous destructive power.

Current research in earthquake prediction is following several different tracks. Long-range forecasts are aimed at identifying those areas of major faults that are at highest risk during the coming decade or so. These forecasts are often based on the periodicity of previous earthquakes and on the buildup of strain around the fault.

Areas identified by these forecasts are then studied intensively with the intent of identifying earthquake precursors that might provide a warning of an imminent earthquake. Changes in the pattern of microearthquakes, bulging of the ground surface in a fault zone, increased radon gas concentrations in well water, changes in the speed with which seismic waves travel through a fault zone, and even changes in the local magnetic field, have all appeared at one time or another to presage earthquakes. But none of them has proved to be totally reliable.

More research into earthquake prediction is clearly needed. But even if this research were reasonably successful, there are staggering social problems that would have to be faced. For example: if you were the mayor of Los Angeles and a delegation of distinguished scientists informed you that in their opinion there was a 50% chance of a devastating earthquake during the next twelve months, what would you do?

STUDY QUESTIONS

III-30. What five areas of the United States have the highest seismic risk?

III-31. The 1906 San Francisco earthquake measured 8.3 on the Richter scale. In terms of energy release, how much stronger was that earthquake than one that measured 6.3 on the Richter scale?

RECOMMENDED READING

Walter Sullivan, Continents in Motion, McGraw Hill (1974). A good history and analysis of the development of continental drift and plate tectonic theory.

Ursula Marvin, Continental Drift: The Evolution of a Concept, Smithsonian Institution Press (1973). More than a history of continental drift, this fascinating book also delves into the early development of geological thought and early ideas on the permanence of Earth's continents and oceans.

William Glen, <u>The Road to Jaramillo</u>, Stanford University Press (1982). A well-written and recent narrative describing the development of plate tectonic theory with insights into the personalities of the scientists involved.

Nigel Calder, <u>The Restless Earth</u>, Penguin Books (1972). A nicely-written and illustrated exploration of plate tectonic theory for popular audiences. Now a bit dated in some respects, but still worthwhile.

P.J. Wyllie, <u>The Way the Earth Works</u>, John Wiley and Sons (1976). A good introductory textbook for non-scientists covering much of solid-Earth geophysics.

UNIT IV   THE LIVING MACHINE:
CONTINENTAL TECTONICS AND THE EARTH'S INTERIOR

MOUNT SAINT HELENS

A. INTRODUCTION

1. Overview

In this unit you will examine a number of topics in continental tectonics that have been of recent interest. Some are explainable within the framework of plate tectonic theory and others do not appear to be at present. Then, in the second part of the unit, you will look into the deep interior of the Earth, where the driving mechanisms for the plates must surely lie.

---

2. Objectives

Upon completion of this unit you should be able to:

1. describe how the study of earthquake (seismic) waves allows us to deduce the internal structure of the earth

2. identify the various kinds of seismic waves and how they travel through the earth

3. use travel-time curves to locate earthquake epicenters

4. recognize that heat is being produced within the Earth by radioactive decay and other processes

5. relate the pattern of heat flow at the Earth's surface to plate tectonic principles

6. describe different models of mantle convection

7. relate models of mantle convection to motions of the lithospheric plates.

---

3. Key Terms and Concepts

continental tectonics	P wave
continental hot spot	S wave
magma chamber	seismic ray path
geysers	travel-time curve
hot springs	crust
caldera	mantle
fumarole	inner core
rhyolite	outer core
basaltic lava	ultramafic minerals
aulacogen	conduction
terrane	convection
accreted terrane	solid-state creep
seismology	seismic tomography

4. Corresponding Video

Plate Tectonic theory is not only a powerful tool for understanding the oceans and mountain ranges of the Earth, but it is also generating exciting new theories about the formation of continents: hot spots and ancient fault zones may be tearing the continents apart; and "microplate tectonics" and "accreted terranes" may be shaping the continents of the future.

B. CONTINENTAL TECTONICS

1. Seismicity Not Associated with Plate Boundaries

Plate tectonics has been remarkably successful in explaining tectonic activity such as earthquakes and volcanoes associated with plate boundaries, but not all geological activity is confined to boundary zones. Earthquakes may be found far from the nearest plate boundaries, as may be seen in Figure III-2 on page 44.

Diffuse patterns of earthquakes are found on several continents, notably in China, eastern Africa, and the United States. Seismicity in Tibet and southwestern China is probably related to the ongoing collision between India and Asia, but what of the rest?

In recent years the attention of geologists and geophysicists has turned back toward the continents, seeking an answer to the question of whether this intraplate activity is related to plate motion or whether new and independent mechanisms are at work.

The most widely-felt earthquakes in the United States occurred in the winter of 1811-12 at New Madrid, Missouri, near the junction of the Ohio and Mississippi Rivers. Chimneys fell as far away as Nashville and Cincinnati; church bells rang and pendulum clocks stopped in Charleston and Washington, D.C. The area is still prone to numerous small shocks. In 1886 a major shock hit Charleston, South Carolina, and again was felt throughout a wide region. It would seem that major earthquakes occur less often in the eastern United States, but when they do, they have the potential for much more widespread damage.

Why is it that current plate tectonic theory does not provide a ready explanation for these earthquakes? The fact that they occur at all indicates that at least some geological structures within the plates are not fixed and static, but are changing and active. In many cases it would appear that earthquakes occur in regions of the plates that have pre-exisiting weaknesses that are exploited by large-scale forces acting on the plates. But what is the source of the forces? In its present form, plate tectonics provides an extremely good description for the kinematics of the plates -- that is, how are the plates moving, at what speed and in which direction? The theory has been much less successful in answering questions about the dynamics of the plates -- what are the forces acting within the plates, and what is making them move? Problems of continental tectonics often seem to be related to this latter class of questions, and the possibility of a significant advance in our understanding has many scientists excited.

STUDY QUESTION

IV-1. Turn to Figure III-2 on page 44 and Figure III-3 on page 45. Are earthquakes in the interiors of plates (far from plate boundaries) found more often on continents or ocean floors?

---

2. Continental Hot Spots

Yellowstone National Park in the western United States is the scene of active volcanism both now and in the very recent geological past. Geysers, hot springs, and recent lava flows are found in abundance, creating a geological spectacle that draws millions of visitors each year. These features are caused by a large body of hot, low density rock only a few kilometers beneath the Yellowstone caldera, a 70 km (44 mi) wide bowl-shaped depression that includes Old Faithful geyser and much of Yellowstone Lake in the south-central part of the park.

The source of heat may be a magma chamber -- a large room or chamber that has been melted out of the solid rock and completely filled with liquid magma. Crustal fractures give ground water access to this extremely hot region, and the resulting superheated steam powers the geysers, hot springs, and bubbling paint pots that fascinate tourists and scientists alike.

Figure IV-1 Track of the Yellowstone Hot Spot

The arrow shows the direction that the North American Plate is moving with respect to the Yellowstone hot spot. Adapted from R.B. Smith and R.L. Christiansen, "Yellowstone Park as a Window on the Earth's Interior", <u>Scientific American</u>, (c) February 1980, (Captioned "Map of Yellowstone Areas".) All rights reserved. Included with permission.

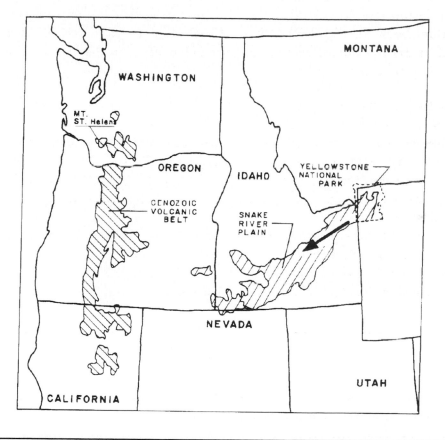

Figure IV-1 maps the location of Yellowstone Park in relation to the Snake River plain, a swath of basaltic lava extending to the southwest across Idaho. The Snake River plain leads directly into the Yellowstone caldera. To the northeast, beyond the intense activity of Yellowstone, we find nothing. The pattern is similar to that of Hawaii, and Yellowstone is generally regarded as a manifestation of a hot spot.

The arrow in Figure IV-1 shows the direction that the North American Plate is moving in relation to all the major hot spots of the world. We can see, then, that the Snake River plain is the trace of former locations of the Yellowstone hot spot as the North American Plate moves over it to the southwest. It took the Yellowstone hot spot some 15 million years to move from the Oregon border to its present location. Or perhaps we should say that it took 15 million years for the North American Plate to move the distance from the Oregon border to Yellowstone Park. Over the next few million years, we can expect to see the center of volcanism move to the northeast out of the present park boundaries.

Unlike the Hawaiian hot spot, the magma supplying the Yellowstone hot spot must melt its way through thick continental crust before it reaches the surface. The result is a mixture of volcanic rocks ranging in composition from rhyolite to basalt. The brilliant colors of the rocks in the Grand Canyon of the Yellowstone River within the park result from the presence of rhyolite which has been chemically altered by the activity of fumaroles and hot springs.

Fumaroles are vents for hot gases escaping from the ground, while hot springs are conduits for ground water that has been heated by steam or contact with hot rock far below the surface. Hot water or steam in contact with rocks can chemically alter them in a process known as hydrothermal alteration. Reactions of this kind are of importance in volcanic areas everywhere: at hot spots, subduction zone volcanoes, and at the spreading ridges of the ocean floor. We shall see in the **Mineral Resources** unit, that these reactions are often important to the accumulation of mineral deposits of economic value.

STUDY QUESTIONS

IV-2. Oceanic hot spots usually erupt lava that has a basaltic composition, while continental hot spots yield lavas with a broad range of composition. Why is this?

IV-3. If a hot spot is active beneath a plate that is moving to the north, it will produce a volcanic trace on the plate that is oriented north-south. Would you expect to find presently active volcanism at the north or south end of the trace?

---

3. Explosive Volcanism

The Yellowstone caldera had its origin 600,000 years ago in a cataclysmic eruption that must have affected nearly the entire western half of the country. For about a half-million years previous to the eruption, a rhyolitic magma chamber below the caldera had given rise to viscous lava flows extruded through ring-like fractures that developed over the periphery of the chamber.

Magma often has a considerable content of dissolved gases, mostly water vapor. Several kilometers beneath the Earth's surface, the gases remain dissolved in the magma: the pressure is too great to permit the existence of gas bubbles. But if something should occur to lower the ambient pressure below some critical value, then the volatiles begin to come out of solution to form bubbles.

You can observe this effect every time you open a bottle of soda pop. With the cap in place, the pressure inside the bottle is sufficient to keep most of the carbon dioxide dissolved in the liquid. When the cap is removed, the pressure drops, allowing the carbon dioxide gas to come out of solution in a froth of gas bubbles.

In a magma chamber, the weight of the overburden of rock supplies the pressure that normally prevents gas bubbles from forming. Occasionally, something triggers a pressure drop in the chamber, perhaps a fracture or failure of the chamber roof. The magma froths and expands and begins venting to the surface through any opening that it can find or make. In the same way that a hastily-opened bottle of champagne may empty itself of much of its contents, the magma chamber erupts massively, expelling huge quantities of chilled and shattered rock fragments into the atmosphere. With the magma chamber partially emptied, the roof may collapse, forming a giant caldera.

The eruption that formed the Yellowstone caldera vented more than 1,000 cubic kilometers (240 cubic miles) of dust and ash that were deposited in a centimeters-thick layer that extended from the Pacific Ocean nearly to the Mississippi River. The ecological effects of such an eruption must be devastating. Total destruction would extend over an area of up to 30,000 square kilometers (12,000 square miles) about the vent, while additional millions of square kilometers would be covered by a blanket of heavy ash. Worldwide climate would probably be affected for sometime by the dust injected into the atmosphere. The total effect on national and global agriculture should such an event occur today would be difficult to assess, but it would likely be severe.

Fortunately, explosive volcanism of this magnitude is rare. Contrast the Yellowstone eruption with that of Mt. St. Helens in 1980, in which less than one cubic kilometer of ash was expelled. Nonetheless, Mt. St. Helens (Figure IV-1) killed 72 people and spread choking clouds of ash eastwards for hundreds of kilometers. Though on a much smaller scale, the explosive eruption of Mt. St. Helens derived its power from the same mechanism of sudden pressure release as the giant caldera events. An earthquake triggered a massive landslide on the north flank of the mountain (Figure IV-2), effectively uncovering the magma chamber. The dissolved volatiles flashed out of solution, unleashing a superheated lateral blast called a pyroclastic surge that destroyed everything to the north of the mountain to a distance of 20 km (12 mi). A pyroclastic flow is so hot that rock fragments and particles can weld together once again after they have settled to the ground.

Another giant volcanic eruption occurred in Long Valley (Figure IV-3) just to the east of the Sierra Nevada Mountains of California some 700,000 years ago. Approximately 170 cubic kilometers of ash were deposited over nine states extending as far east as Kansas and Nebraska.

There have been recent indications that magma is once again on the move beneath the Long Valley caldera. Uplift of the ground surface and a type of seismicity usually associated with magma migration have raised concerns for the population of the area. The town of Mammoth Lakes plays host to tens of thousands of skiers each winter, and substantial efforts have been undertaken to increase monitoring activity so that evacuation can be carried out if significant volcanic activity seems to be imminent. Fortunately, minor eruptions are much more likely than major caldera-forming events.

Mt. St. Helens is a volcano associated with the subduction of a small oceanic plate just off the coast of Oregon and Washington (see the top of Figure III-13), and Yellowstone is

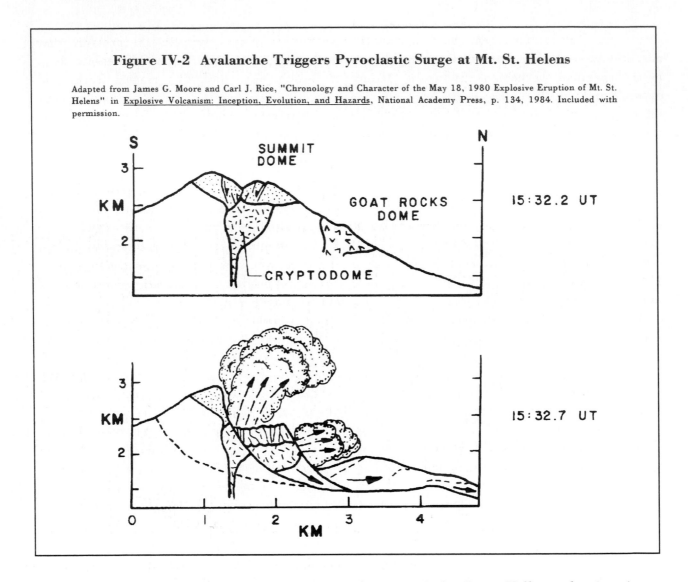

Figure IV-2 Avalanche Triggers Pyroclastic Surge at Mt. St. Helens

Adapted from James G. Moore and Carl J. Rice, "Chronology and Character of the May 18, 1980 Explosive Eruption of Mt. St. Helens" in Explosive Volcanism: Inception, Evolution, and Hazards, National Academy Press, p. 134, 1984. Included with permission.

caused by a continental hot spot. The essential cause of the Long Valley volcanism is an accumulation of magma below the caldera, but why is magma being introduced into this region?

Figure IV-3 maps recent volcanism in the western United States. There has been substantial volcanic activity in eastern California, Utah, Arizona, and New Mexico for which plate tectonic theory does not provide a ready explanation. Many of these regions, such as the Rio Grande Rift running north-south in New Mexico, are areas of apparent crustal extension or rifting where the continental crust is being pulled apart and thinned, like a piece of taffy that is being stretched. The extended and thinned crust probably encourages the formation of magma. Why crustal extension is taking place in these regions is not well understood.

Figure IV-3 Western U.S. Volcanism of the Past Five Million Years

Black pattern indicates basaltic and andesitic volcanism, white spots within black are rhyolitic and dacitic volcanism. The grid patterns serve to emphasize the different volcanic zones. Adapted from R.L. Smith and R.G. Luedke, "Potentially Active Volcanic Liniaments and Loci in the Western Coterminous United States" in Explosive Volcanism: Inception, Evolution, and Hazards, National Academy Press, p. 50, 1984. Included with permission.

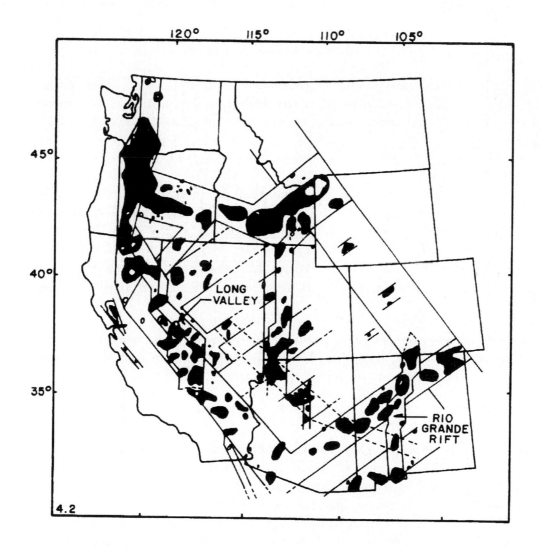

---

4. Continental Rifting

Crustal extension, of course, is precisely the process that takes place at oceanic spreading ridges. Could continental rift zones mark the beginning of continental breakup? Africa's Afar Triangle has seen intensive study by scientists interested in the process of continental rifting. The Afar Triangle is a region of recent volcanism at the intersection of three great rifts: the Red Sea, the Gulf of Aden, and the East African Rift.

The Red Sea and Gulf of Aden are underlain by recently formed ocean floor, born in the splitting away from Africa of the Arabian peninsula. The East African Rift extends north from Lake Nayasa and Lake Tanganyika through Ethiopia to the Afar Triangle, with several branchings along the way. The rift zones show evidence of extension and crustal thinning, and with the introduction of plate tectonics, it was assumed that the Horn of Africa was in the early stages of splitting away from the rest of the continent, a continuation of the breakup of Gondwana.

This notion was soon challenged when it was recognized that in some cases where a triple junction of rifts forms, only two of them continue to form open ocean, while the third eventually dies, leaving a relict rift valley called an aulacogen. These low-lying valleys eventually fill with sediment and, because of their access to the ocean, frequently become conduits for major rivers.

There in fact are triple junctions in which all three arms are active spreading ridges. The Azores, at the juncture of the American, Eurasian, and African Plates in the North Atlantic, is one and the Galapagos Islands lie near another in the east Pacific, where the Cocos, Nazca, and Pacific Plates meet. An important clue to their origin lies in the fact that an active hot spot is often found at or near a triple junction. This is true for the Azores and Galapagos, and in all likelihood, for Afar as well. It would appear that the oceanic triple ridge junctions were initiated and maintained by the hot spots.

In the case of the initial splitting of a continent, it may be that a hot spot or some other localized source of heat uplifts and stretches the overlying plate, causing it to fracture with a three-armed crack spreading out from the center of the uplift. A similar effect can sometimes be seen in a baking pie crust when the pie filling expands and overly stretches the crust above it.

Several triple rifts of this sort within a continental mass might link up and split the continent apart (Figure IV-4). In general, only two of the three rift arms are needed to form a

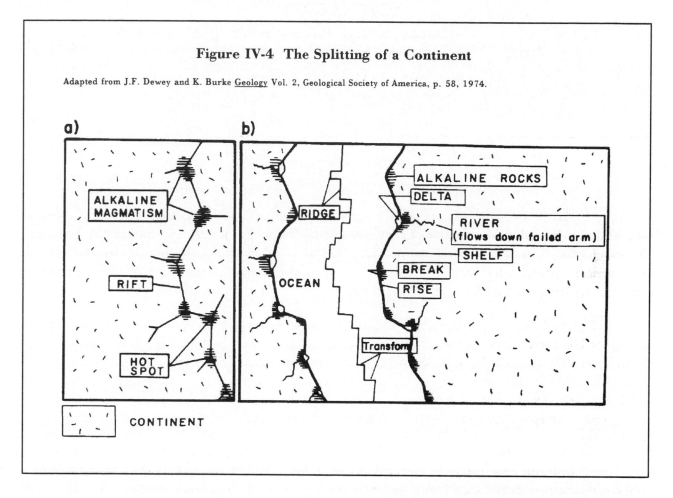

Figure IV-4 The Splitting of a Continent

Adapted from J.F. Dewey and K. Burke Geology Vol. 2, Geological Society of America, p. 58, 1974.

continuous line linking the hot spots in what will eventually become a widening ocean. The unneeded third arm becomes dormant and forms aulacogens.

This appears to have happened in the case of the Benue Trough, at the mouth of the Niger River in Nigeria. The sudden bend in the African coastline at this point would be a natural consequence of the continued widening of two of the original three rifts.

The same may hold true for the Afar Triangle as well. Nonetheless, the East African Rift is both seismically and volcanically active today. Will it continue to widen and form a new plate boundary or will it expire and become an aulacogen? No one knows for certain.

Let us return for a moment to the New Madrid earthquakes of 1811-12. The area remains seismically active today, and concern over a possible recurrence of a major earthquake has prompted intense study by teams of scientists. Sophisticated computer displays of seismic activity have helped reveal something of the mechanisms that are causing the earthquakes.

Beneath New Madrid are the remains of an ancient continental rift, dating back to the Precambrian. Long dormant, the rift is nonetheless a zone of weakness and recently has been reactivated not in an extensional sense, but by compressive forces acting on the whole midsection of the continent. Recognition of these structures and the forces acting upon them will help in evaluating seismic risk in the area and in determining those regions where sensitive structures such as nuclear reactors should not be built.

A search for the source of those compressive forces, however, raises more questions than answers. Do they arise from the motion of the whole North American Plate? Or are their origins more local in nature? Questions like these quicken the pulse of a geoscientist. They lead us onward, into new and exciting areas of research that are yet unexplored.

STUDY QUESTIONS

IV-6. Are aulacogens associated with regions of extension or compression at the time that they form?

IV-7. Continental rifts often take the form of valleys. Why?

---

5. Accreted Terranes

Another stop on the television tour of continental tectonics brings us to the Marin Headlands, just across the Golden Gate Bridge from San Francisco. The Marin Headlands are remarkable in that they are part of a sliver of land that appears to have originated as part of the ocean floor. Nearby tracts such as Alcatraz Island, however, appear to have origins that would place them in widely different parts of the Earth at the time that they formed.

Each block that contains land of similar geological characteristics is called a terrane. What makes these terranes so exotic is their juxtaposition as parallel slivers that appear to have been plastered onto the edge of the continent. Because they are additions to the continent, they are often referred to as accreted terranes.

How can geologists distinguish between different terranes? Perhaps the most obvious way is to observe sudden changes in rock type and age as we cross a boundary from one terrane to another. Sedimentary rocks of marine origin often contain microfossils such as radiolaria and fusulinids that may give clues as to where the terrane formed.

These tiny shelled creatures come in a number of varieties, some of which have adapted to particular environmental or geographical settings. The radiolaria, for instance, are sensitive to ocean temperature and so can tell us whether a rock orginated in tropical or subpolar latitudes, while fusilinids found in some of the west coast terranes belong not to the type common in the North American mid-continent but to a type common to Asia. Their presence on the west coast indicates that their host rocks originated far across the Pacific and were subsequently accreted onto North America.

A final technique for determining displaced terranes is already familiar to you. Paleomagnetism allows us to find the original latitude and orientation of the terrane provided it contains rocks that are suitable magnetic recorders.

Figure IV-5 shows some of the larger accreted terranes that have been identified in western North America. These microplates must have been carried in and plastered onto the continental edge at a time when active subduction was taking place along the west coast. Some

Figure IV-5 Accreted Terranes in Western North America

Some of the larger accreted terranes are shown as outlined areas. The Laramide deformation was a major episode of mountain building in the Rocky Mountains and other parts of western North America that culminated roughly 50 million years ago. It has been suggested that the arrival of the accreted microplates was one of several causes for the Laramide deformation. Adapted from D.L. Jones, et. al. "Growth of North America," <u>Scientific American</u>, (c) May 1982. All rights reserved. Included with permission.

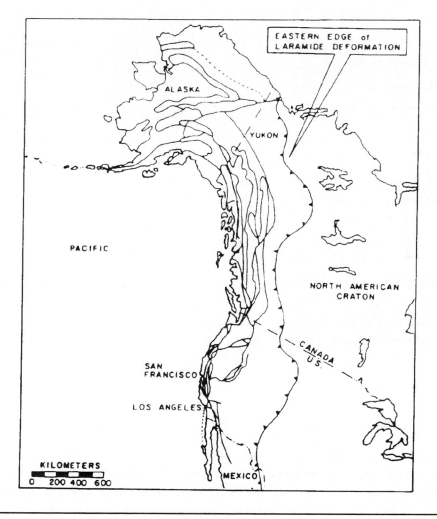

of these terranes are oceanic in origin -- volcanic edifices or rises in the ocean floor that were thick and buoyant enough to have survived the subduction process. Others are continental in origin and appear to have originated far to the south.

This process is still going on in modified form. In Figure IV-5 and Figure III-13 on page 62, notice how the Mexican peninsula of Baja California and everything west of the San Andreas Fault is being sheared from the edge of the continent and transported on the Pacific Plate to the northwest. In fifty million years or so, the Los Angeles Microplate will probably be accreted onto the southern Alaska coast. Since this sliver of land already contains some displaced terranes, those terranes are being displaced for (at least!) a second time.

It is becoming clear that the wanderings and accretion of microplates are important geological processes. Figure III-24 on page 83 shows a hypothetical stage in the accretion of a sliver consisting of northern New England, Nova Scotia, Newfoundland, Ireland, England, Spain, and part of France (the Armorican Microplate). By Devonian or Carboniferous time, this plate had moved to the northeast and docked with New England and the Maritime Provinces of Canada. Scotland was added to England (or was England added to Scotland?) and, at a much later time, the European portions split off as part of the Eurasian Plate on the other side of the newly-opened Atlantic Ocean. If you look closely at Figure III-24 to Figure III-31 beginning on page 83, you can trace this particular bit of microplate give-and-take.

The concept of microplates is a quite new one, and the identification of microplates in all parts of the world and the working out of their individual origins and histories has only begun.

STUDY QUESTIONS

IV-8. Why must the accretion of terranes along the west coast of the United States have taken place at a time when subduction was active there?

IV-9. Most of the United States west coast accreted terranes became attached to North America within the last 100 million years. Why have none arrived on the east coast of North America in the same time span? Note that the Armorican Plate arrived some 300 million years ago.

C. JOURNEY TO THE CENTER OF THE EARTH

The surface of Mars is much more accessible to direct observation than the rock layers only 20 km (12 mi) beneath our feet. We can observe Mars telescopically. We have sent spacecraft there to photograph the planet from close range, and robot landing craft have sampled and analyzed its surface. On the other hand, the deepest well drilled on Earth reaches only 12 km (7.5 mi) below the surface, compared to the 6371 km (3,960 mi) distance to the center of the Earth. How, then, can we learn the structure and composition of the deep interior of the Earth?

In a number of cases, the present surface of the Earth consists of rocks that at one time were buried perhaps to depths of tens of kilometers. Erosion and uplift have removed the rock that once covered them and exposed them to our scrutiny. Even so, only a tiny portion of the Earth's interior may be studied directly. For the most part, we have to rely on indirect observations in order to study the truly deep parts of the Earth.

1. Seismology

Seismology, the study of earthquake waves, provides the most detailed information available to us about Earth's deep interior. Like a bat in flight in a dark cave or a submarine moving through the dark recesses of the ocean, we can use seismic waves as a kind of sonar to probe the inside of the Earth and to determine its structure.

When the vibrations of an earthquake are released by the sudden slipping of blocks of rock along a fault, waves of vibration spread out from the source of the earthquake in all directions, similar to the manner in which ripples spread through a quiet pond after a rock has been dropped into it. Before long, these waves fill the whole Earth resulting in a very complex pattern that is constantly changing and is very hard to describe in detail.

For this reason, seismic waves are often described in terms of <u>rays</u>, just as rays of light are described as emanating from a light bulb and traveling outward to illuminate objects at some distance. Shadows are easy to understand using the concept of rays -- a shadow occurs in the region behind an object that blocks the direct passage of the rays.

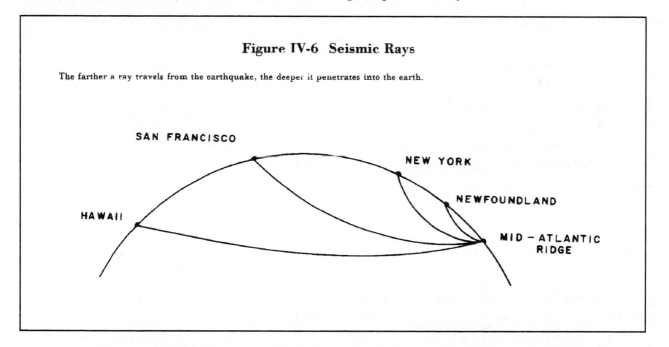

Figure IV-6 Seismic Rays

The farther a ray travels from the earthquake, the deeper it penetrates into the earth.

The spherical shape of the Earth helps to make it possible to use seismic rays to probe different regions at depth. Figure IV-6 shows how seismic rays travel from an earthquake on the Mid-Atlantic Ridge to receivers located in Newfoundland, New York, San Francisco, and Hawaii. Note that the rays tend to curve up toward the surface. This results from the fact that in the region shown, the speed of travel for seismic waves generally increases with depth.

As the wave travels along the ray path, it sees a lower speed region above it and a higher speed region below. The result is similar to what happens to an automobile whose right wheels are off the pavement into mud while the left wheels remain on the road. The tendency of the right wheels to be slowed down by the mud, drags the car around toward the right. The same tendency bends the rays in the diagram upward, toward the lower speed region.

Partly because of this effect and partly due to the spherical shape of the Earth, the farther that a ray travels, the deeper it penetrates below the surface. Seismic waves from nearby earthquakes gather information about the uppermost layers, while waves from distant earthquakes add information gleaned from greater depths.

One way to interpret this information is to look for shadows cast by large obstructions to seismic rays. The Earth's core was discovered in this manner in 1906. Figure IV-7 shows how the presence of the core deflects the passage of seismic waves, causing shadow zones in the hemisphere opposite to that in which the earthquake occurs. Note that the rays entering the core are bent strongly into it. This is caused by the same effect that bends the rays in the mantle upward, except in this case the waves in the outer core travel more slowly than at the bottom of the mantle. Once again, the rays are bent toward the region with the lowest speed of travel.

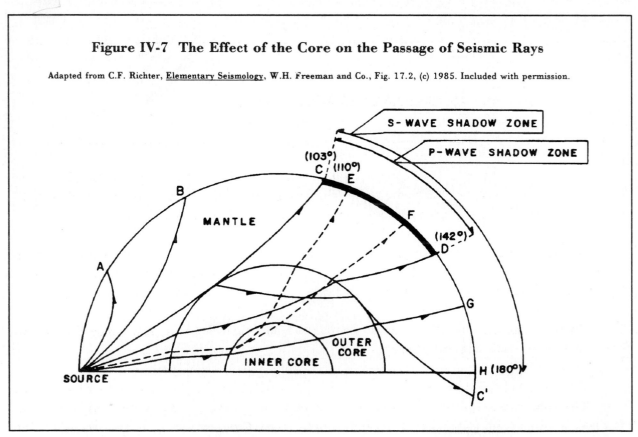

Figure IV-7 The Effect of the Core on the Passage of Seismic Rays

Adapted from C.F. Richter, Elementary Seismology, W.H. Freeman and Co., Fig. 17.2, (c) 1985. Included with permission.

There are several different types of seismic waves, and we may use their different properties and behaviors to probe the internal constitution of the Earth. Waves that can travel through the deep body of the Earth come in two types: P (for primary) and S (for secondary). Each type of wave distorts the rock through which it passes in a distinct manner, as shown in Figure IV-8.

Figure IV-8 The Difference Between P and S Waves in Solids

Reprinted from Bruce Bolt, <u>Inside the Earth</u>, W.H. Freeman and Co., p. 30, (c) 1982. Included with permission.

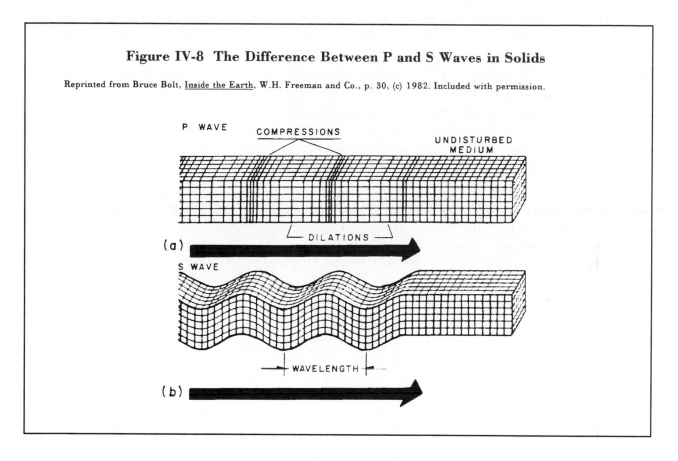

P waves are identical in action to sound waves in air in that they consist of alternate stretchings and squeezings that travel as waves through the medium. S waves are similar to the snake-like waves that travel along a rope when one end of it is whipped sideways.

P waves may travel through solids, liquids, and gases, since they depend on the medium's resistance to compression. S waves, however, depend on the medium's resistance to sideways deformation (shear) and as a result they cannot travel in liquids and gases.

In the same medium, P waves always travel faster than S waves, and hence they are the first to arrive at a seismograph station.

Look once again at Figure IV-7. All the rays shown in the figure are P waves, and the P-wave shadow zone is produced because rays entering the core are deflected as they cross the core-mantle boundary. The P-wave shadow zone is not totally dark, however. In 1936 Inge Lehmann proposed the existence of an <u>inner core</u> to explain the presence of waves in the shadow zone. If the P-wave velocity is higher in the inner core than in the outer, then some of the rays entering the inner core can be deflected into the shadow zone.

On the other hand, the shadow zone for direct S waves is much larger and more complete (Figure IV-7), indicating that S waves cannot penetrate the core-mantle boundary at all. This is strong evidence that the outer core is liquid.

When rays traveling downward encounter an interface between two regions of differing seismic velocity, they can split into two rays. One crosses the boundary and is deflected either up or down, depending on whether the wave travels with a higher or lower velocity in the bottom layer. This bending of the ray is called <u>refraction</u> and is similar to the bending of a light ray as it enters a glass or water surface. The other ray is <u>reflected</u> mirror-like back into the upper region. Again, light behaves in a similar fashion. When you look into a store window, you not only see light rays that have passed through the glass from behind it, but you also can see your own reflection -- light rays that emanated from you and reflected back to your eyes from the glass surface.

Figure IV-9 Seismic Ray Paths Within the Earth

Explanation of symbols: Focus=location of the earthquake, P=P wave in the mantle, S=S wave in the mantle, K=P wave in the outer core, I=P wave in the inner core, J=S wave in the inner core, c=reflection from the core-mantle boundary, I=reflection from the inner-outer core boundary. Reprinted from Bruce Bolt, <u>Inside the Earth</u>, W.H. Freeman and Co., p. 65, (c) 1982. Included with permission.

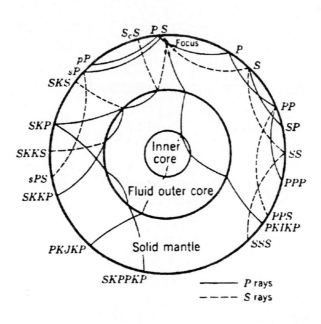

In the Earth, the behavior of reflections and refractions becomes even more complex because P waves can generate S waves, and vice-versa, at these boundaries. After a large earthquake or nuclear explosion, seismic waves spread throughout the entire earth, with reflections and refractions occurring at both the core-mantle and the inner-outer core boundaries. The surface of the Earth is a very effective reflector of waves as well. Figure IV-9 shows how complex the ray paths within the Earth can get before they are recorded at various points on the surface by seismographs.

Figure IV-10 Travel-Time Curves for Earthquake Waves

Angular distance is the great circle distance in degrees spearating the earthquake and the seismograph station. An angular distance of one degree is equivalent to 111.2 km (69.1 mi). Reprinted from Bruce A. Bolt, <u>Inside the Earth</u>, W.H. Freeman and Co., p. 65, (c) 1982. Included with permission.

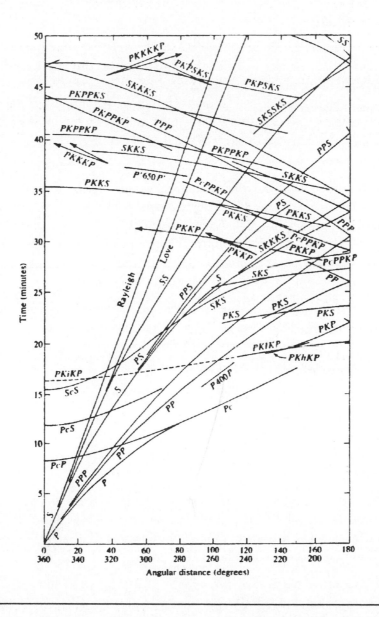

Ever since the birth of modern seismology around the turn of the century, sensitive recording seismographs have allowed seismologists to determine how long it takes these various waves to travel the distance measured along the surface from the earthquake to seismograph stations. From this information, it is possible to construct a profile of the Earth's interior showing how fast seismic waves travel at every depth from the surface to the center of the Earth. The velocity profile contains a wealth of information from which most of our knowledge of the Earth's deep interior has been derived.

Travel times for seismic waves have been graphed as shown in Figure IV-10. As an example, let us say that the earthquake occurred 6,672 km (60°) distant from the seismograph. Place a straight edge vertically on the 60° distance marks at top and bottom of the graph and read off the times of arrivals of the various waves: 10 min. for the P wave, 12 min. for the PP, 18 min. for the S wave, 32 min. for the PKKP, etc.

With only two exceptions, all the lines in the travel-time diagram are curved because the waves take short cuts through the body of the Earth, while the great circle distances on the horizontal axis are measured along the curve of the Earth's surface. The two exceptions are straight lines labeled Rayleigh and Love, named after two prominent British mathematician-physicists. These are a type of wave different from P and S waves in that they cannot travel through the deep parts of the Earth but are confined to traveling at and just below the surface at more or less constant speed.

Because both Rayleigh and Love surface waves gradually diminish in strength with depth, they have proved extremely useful to seismologists who wish to probe the structure of the top few hundred kilometers of the Earth (to which they do penetrate) without the interference from complexities at greater depth. They are, however, the bane of earthquake country dwellers, since surface waves generated by strong shallow earthquakes produce the longest duration of strong ground shaking and hence do much damage.

Travel-time curves serve two major applications: locating earthquake epicenters and modeling the structure of the Earth's interior. Let us look at each in turn.

a. Locating Earthquake Epicenters

Figure IV-11 reproduces the travel-time curves for P and S waves only. Note how they diverge from one another as the distance from the earthquake increases. Because P waves travel at the highest speed, they always arrive first. The S waves arrive some time later. What is important to note is that the time interval between the arrivals of the P and S waves increases with increasing distance to the earthquake.

The same effect may be noted in foot races. In a 100-yard dash, the runners may all cross the finish line within a second or two of one another, while in a 10 km run, they may finish over an interval of several minutes. This spreading out of the waves with distance is the essence of the method for locating earthquakes.

As an example, suppose we record the arrival of a P wave at our seismograph station at 2:57:40 PM (40 seconds after 2:57 PM). The S wave arrives 4 minutes and 20 seconds (or 260 seconds) later at 3:02:00 PM. There is only one distance from the earthquake at which the P and S waves arrive just 260 seconds apart and that is at 2,900 km. That is the distance from the seismograph to the earthquake.

Verify this result using Figure IV-11. Place the edge of a piece of paper on the vertical time axis. Make two marks on the edge of the paper, one at time zero and the other at 260 seconds. Then move the paper to the right until you find the only place where you can match

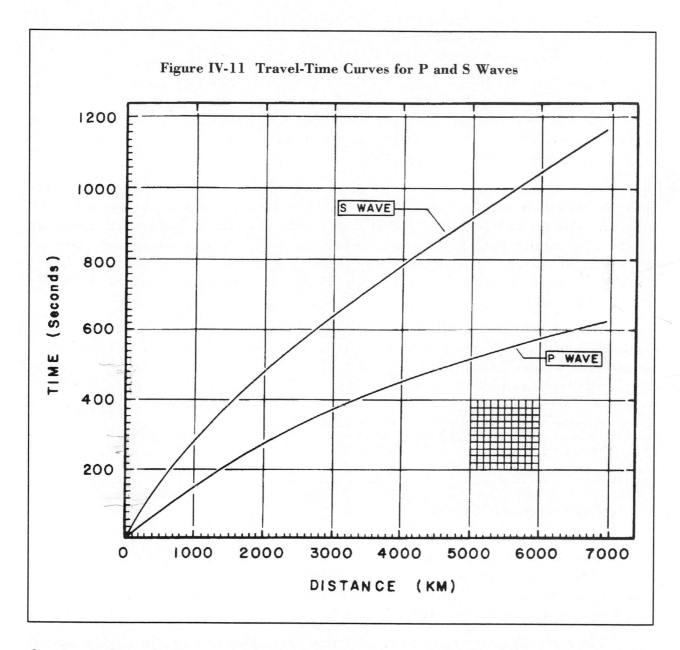

Figure IV-11 Travel-Time Curves for P and S Waves

the one mark to the P curve and the other to the S curve. You must keep the paper edge vertical. When you have done this correctly, your paper edge should be aligned with the distance of approximately 2,900 km on the horizontal axis.

Since this procedure gives us only the distance to the earthquake, we will need at least three stations to pinpoint the actual epicenter. In the exercise that follows, you will use data from four stations. In this way, if you make an error, only three of the circles that you draw will intersect at the same point and the disagreeing circle is the one that is in error.

EXERCISE

Two earthquakes are recorded in the morning and evening of the same day at four seismograph stations. Table IV-I gives the times of arrival for the P and S waves for each earthquake at the four stations.

Find the time interval in seconds between the P- and S-wave arrivals for each earthquake and then use Figure IV-11 to determine the distance to the epicenter from each station. Note that you must use the time difference between the P- and S-wave arrivals, and don't forget that there are 60 seconds in a minute and 60 minutes in an hour. Check your procedure on the morning earthquake recorded in Honolulu. You should find that the time interval between P- and S-wave arrivals was 7 minutes 50 seconds (470 seconds) and that this yields a distance to the earthquake of 5,900 km.

Once you have determined all eight distances, turn to Figure IV-12. You will need a compass to complete the exercise. For each distance that you have determined, place the point of the compass on the zero of the map scale below the map and adjust the spread of the compass so the pencil point is on the correct distance. Without changing the compass setting, move the point of the compass to the location of the seismograph station for which that distance was determined and draw a circle about the station. The earthquake epicenter should lie somewhere on that circle.

Repeat the procedure for the other three stations for the morning earthquake, using the appropriate distances that you determined from Table IV-I. Your four circles should all approximately intersect at one point, marking the location of the epicenter of the morning earthquake.

Now repeat the whole procedure for the afternoon earthquake. You will find that even if you do the exercise with great care, your circles will not meet precisely at one point. There are several reasons for this, but the main one is that you are doing the exercise on a flat map rather than on a spherical globe. Any map projection introduces distortions in the distances between different points on Earth. Nonetheless, the effect here is not too severe, and your circles should all intersect to within about a half centimeter (one-quarter inch) or less.

TABLE IV-1: EARTHQUAKE ARRIVAL DATA

Station	AM Earthquake		PM Earthquake	
	P wave	S wave	P wave	S wave
Honolulu (Hawaii)	4:56:30	5:04:20	7:29:35	7:37:30
Tokyo (Japan)	4:52:20	4:56:20	7:30:20	7:39:10
Manila (Phillippines)	4:52:40	4:57:00	7:30:00	7:38:20
Brisbane (Australia)	4:55:40	5:02:20	7:24:00	7:27:00

b. Modeling The Earth's Interior

There is obviously a lot of information in the complete travel-time curves of Figure IV-10. Can we use these curves to construct a model of how fast seismic waves travel in each region of the Earth's interior? Each curve tells us how long it takes for a seismic wave traveling along a particular path to complete its journey from the earthquake source to the recording seismograph, but not how fast each wave travels in each particular region of the Earth's interior.

The answer is: yes, with difficulty and a small amount of uncertainty. Actually, our task would be much easier if we could turn the question around and ask instead: If we have an accurate model showing how fast seismic waves travel in each region of the Earth, then can we use the model to reproduce the travel-time curves in Figure IV-10? The answer to this question is an unambiguous yes, and the task is much less difficult (though still formidable).

This is a frustrating but very common situation in geophysics. The direct problem (given a well-defined model, predict what the observations should be) is relatively straight-forward to solve and provides results which have no uncertainty. On the other hand, the inverse problem (deriving a model from the observations) is difficult and may yield more than one model that fully explains or predicts all the observations. In spite of this difficulty, the inverse problem for seismic wave velocities can be solved and the uncertainties in the model turn out to be fairly minor, largely because of the huge amount of information available in the travel-time curves.

Figure IV-13 shows a recent model based on seismic observations. Sudden changes in the velocities of P and S waves clearly outline the major divisions of the Earth.

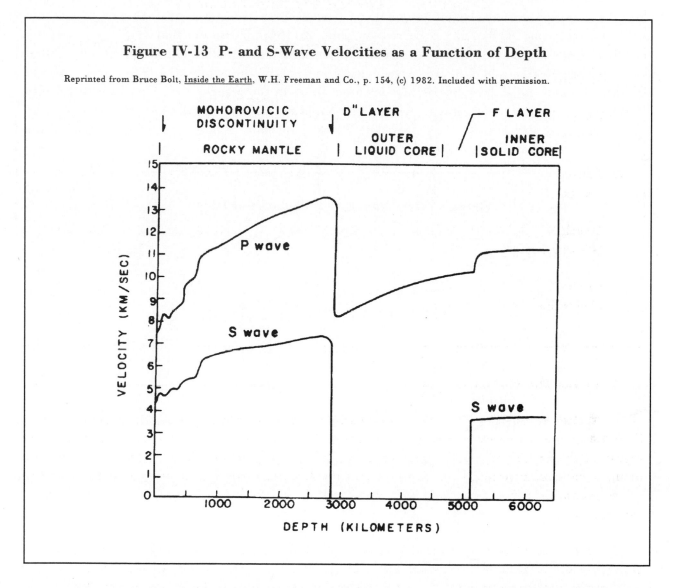

Figure IV-13 P- and S-Wave Velocities as a Function of Depth

Reprinted from Bruce Bolt, <u>Inside the Earth</u>, W.H. Freeman and Co., p. 154, (c) 1982. Included with permission.

The construction of a comprehensive model of the interior structure and composition of the Earth must draw on information from a wide variety of disciplines. Principal among these is seismology via the wave velocity profile in Figure IV-13. Measurements of the Earth's gravity field allow the determination of the mass of the Earth. Astronomical measurements of the motions of the Earth in its orbit place strong contraints on how that mass can be distributed within the Earth.

Great earthquakes set the entire planet to ringing like a bell, and the characteristic frequencies of this slow ringing (one to a few vibrations per hour) place further constraints on the physical properties of the Earth. The pitch of a bell is determined by its size, shape, thickness of its walls, and the material of which it is made. In the same way, the characteristic frequencies of Earth's free oscillations are determined by the distribution of mass within the Earth and by the physical properties of the materials of which it is made.

The presence of these vibrations had been predicted on theoretical grounds, with early work on the response of elastic spheres extending back to the eighteenth century. It was not until a devastating earthquake occurred in Chile in 1960, however, that they were actually observed. Seismographs designed to respond to these extremely slow vibrations of the ground showed them clearly and allowed seismologists to analyze the vibrations. They were able to separate the individual frequencies of vibration, much in the manner that a musical chord may be separated into its constituent notes. The analysis of free oscillations of the Earth in subsequent large earthquakes has proved to be an important tool in the determination of the properties of the Earth's interior.

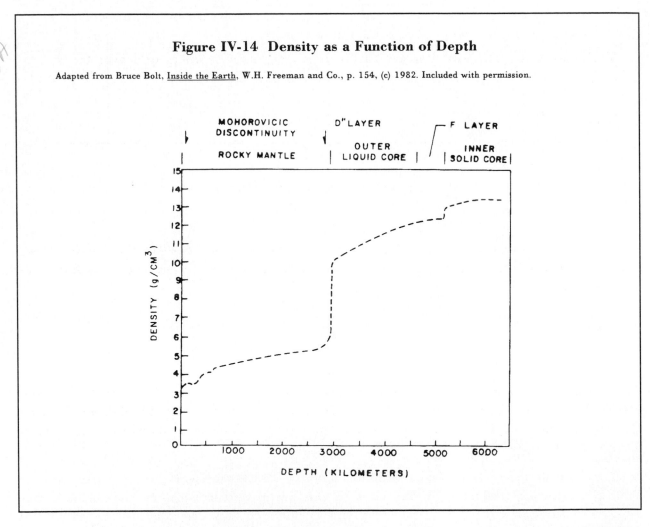

Figure IV-14 Density as a Function of Depth

Adapted from Bruce Bolt, Inside the Earth, W.H. Freeman and Co., p. 154, (c) 1982. Included with permission.

Putting all of this information together allows the construction of a reasonably reliable model showing how three physical properties vary with depth: density, rigidity, and compressibility. The construction of the model, however, entails substantial uncertainties. Nonetheless, the different models advocated by various workers agree with one another to within 10% and so we may have some confidence that at least the larger features of the Earth's interior are reasonably portrayed. Figure IV-14 shows how the density changes in one such model.

So far our model-making has been concerned with physical properties: wave velocities, density, and so on. Can we extend our model-making to tell us what each region is made of?

This is much more difficult to do, and there is not yet general agreement on the exact compositions of each layer within the Earth, but something can be said about the kinds of materials within each layer. In essence, we can ask what kinds of materials have the appropriate physical properties for each region. We shall do this in the next section.

STUDY QUESTIONS

IV-10. Follow ray C^1 in Figure IV-7. Note that it bends sharply as it passes from the mantle to the outer core and again as it pases from the outer core to the mantle. Explain why it bends the way it does.

IV-11. Ray G in the same diagram hardly bends at all as it enters or leaves the inner and outer core. Why does it behave so differently from ray C^1?

IV-12. Why is the outer core believed to be liquid?

---

2. The Major Divisions of the Earth

We can now summarize our knowledge of the internal structure of the Earth. If we could take a pie-cut slice of the planet from surface to center, we would expect to find it as diagramed in Figure IV-15.

a. Crust

We have already described the thin oceanic crust in the section **The Structure of the Plates**. The much thicker continental crust is composed of a thin veneer of sediments atop an igneous and metamorphic substrate called the "basement", which on the average is granitic to dioritic in composition.

Since most drilled wells are less than 10 km in depth and average continental crustal thickness is 30 km, the lower crust is nearly as inaccessable to us as the mantle. On the basis of the seismic evidence available, many geophysicists today think the lower crust is a highly metamorphosed rock in the range of composition of diorite (intermediate composition equivalent to andesite).

The Moho appears to be a boundary between two very different mixes of minerals: the dioritic crust above and the very dense mantle minerals below. The crust is solid except in a few very localized magma chambers in volcanic areas.

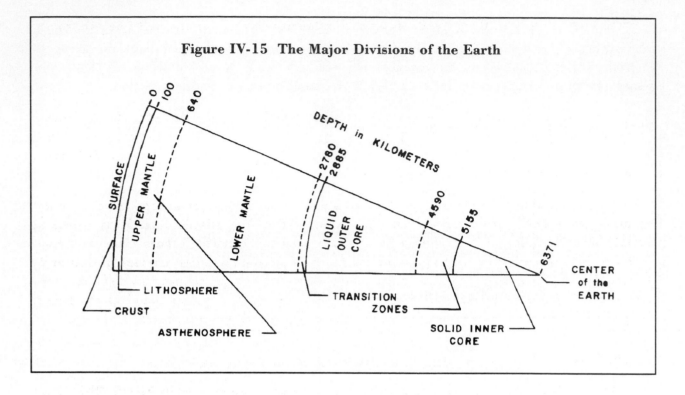

Figure IV-15 The Major Divisions of the Earth

b. Mantle

On the basis of a jump in seismic wave velocities and density at 640 km (400 mi), the mantle is usually divided into upper and lower layers.

The very uppermost portion of the mantle, at depths of less than 100 km, is part of the rigid lithosphere along with the crust. The upper mantle also contains the asthenosphere, extending approximately from 100-640 km, which is the region where the mantle material is hot and soft and can be penetrated by the cool and rigid subducting slabs. You should realize that the asthenosphere probably does not differ significantly in chemical composition from the portion of the mantle above 100 km -- that is, the part of the mantle included in the lithosphere. The difference between lithosphere and asthenosphere is purely one of rigidity, caused by increasing temperature with depth.

P- and S-wave velocities and the density of the mantle are substantially higher than in the crust, strongly indicating that the mantle is made up of rocks that are not common in the crust. These physical properties have been measured in the laboratory for most of the known minerals, and the only significant class that falls in the appropriate range are the ultramafic minerals: olivine, garnet, pyroxenes and spinel. These dense, dark-colored minerals are the main candidates to make up the mantle. Various mixes of them have been proposed, but there is no general agreement yet on exactly which mix is best for each part of the mantle.

In addition to the jump in seismic velocities at 640 km that divides the upper from the lower mantle, there is another at about 400 km (250 mi). Each change could signify either a chemical boundary (between different mixes of minerals) or a physical boundary (between two different crystalline forms of the same mineral).

The mantle is solid throughout, though the asthenosphere is soft and many contain some partial melting.

c. Outer Core

The core-mantle boundary marks a very fundamental division within the Earth. Below it, P-wave velocity drops, S waves disappear entirely, and the density jumps to nearly twice the value it had in the mantle. The absence of S waves in the outer core proves that it is liquid, and the only liquids with such high density are molten metals.

Iron is an abundant element in the Earth and in the solar system, and its density at core pressures is only a little higher than that expected from the seismic data. Many workers presently favor an alloy of iron and sulfur, another common element; others favor iron and silicon or perhaps a mixture of iron, oxygen, sulfur, and silicon. A molten mixture of iron and sulfur, for example, with sulfur comprising about 12% of the total weight, not only has the correct density but also transmits P waves at a speed comparable to that found in the outer core, provided that the immense pressures prevailing there are taken into account.

d. Inner Core

The small increase in P-wave velocity as we enter the inner core in Figure IV-13 has been interpreted as requiring that the inner core be solid. If this is the case, then it must support an S-wave velocity as shown in Figure IV-13, since S waves can travel in solids.

The solidity of the inner core could be verified directly if the wave labeled PKJKP in Figure IV-9 could be detected with certainty. So far it has not been, probably because it is expected to be extremely weak. Traveling as a P wave in the liquid outer core, it converts to an S wave at the inner-outer core boundary, then converts back to a P wave when it leaves the inner core. Seismologists continue to search for this elusive messenger bearing information from the center of the Earth.

Either pure iron or an iron-nickel alloy are possible candidates for the composition of the inner core.

e. Summary

The emerging picture of the Earth's interior is an altogether remarkable one. We might recall our earlier analogy to a boiled egg, in which the eggshell corresponded to the lithosphere and the slippery layer below it to the asthenosphere. Now we may take the white as an analog to the mantle, and the yolk as an analog to the core.

Even though the proportions are not too bad, our egg-model somehow seems too mundane to compare to the majesty of the real Earth. Below the thin crust with all its familiar rocks, we enter the mantle, a realm of dense minerals such as olivine, garnet, pyroxenes, and spinel; minerals that when found on the surface in pure form are sometimes used in jewelry as dark semiprecious stones. The mantle is immense, comprising some 83% of the total volume of the Earth.

Almost halfway to the center of the Earth, we encounter the core. It is a white-hot, turbulent mixture of molten iron and perhaps sulfur and in its center, with no solid connection to the rest of the Earth, floats a huge iron cannonball, 2,400 km (1,500 mi) across.

STUDY QUESTIONS

IV-13. What is the principal difference between the core and the mantle in terms of their compositions?

IV-14. What parts of the Earth are liquid and what parts are solid?

3. Temperatures within the Earth

Gold miners working the deep mines of Africa are well aware that temperature increases with depth within the Earth. As measured in mines and deep boreholes, the temperature is found to increase at an average rate of 30°C per kilometer of depth (54°F per thousand feet). If this rate were to continue unchanged, the mantle would be completely molten. Since the mantle is solid, the rate of temperature increase must slow deeper down.

In this unit we have discussed a number of models constructed by geophysicists to describe the interior of the Earth. All of them are based on observations made at the Earth's surface. The P- and S-wave velocity model (Figure IV-13) is probably fairly accurate, since it is based directly on the wealth of data obtained from the arrival times of earthquakes. The density model (Figure IV-14) is somewhat further removed from the observational data, and the uncertainties grow correspondingly.

This is in keeping with the dictates of the scientific method: that model is best that best explains all available observations. If the observations are few in number or are linked to the model by relations that are fraught with uncertainty, then it may be hard to decide which model really is the best.

Such is the case with internal temperature profiles for the Earth. Data are very few: we know that the mantle is soft and may be near melting in the asthenosphere, but is solid otherwise. Clearly the temperature must be above the melting point of the liquid outer core but below that of the solid inner core, and we have a rough idea of what kind of material each layer is made. Unfortunately, melting temperatures are often strongly dependent on the exact composition of the material, and the data allow considerable variation, perhaps by as much as 1500°C (2700°F) at the core-mantle boundary.

Figure IV-16 shows the range of possible temperatures within the Earth, along with values for the melting temperature of the material in each region.

The temperature behavior within the mantle is significant in terms of the structure of the plates. Note that within the lithosphere (0 - 100 km depth), the temperature is far below melting, resulting in rigid behavior. Within the asthenosphere just below it, however, temperatures are very close to melting, resulting in the soft, plastic behavior characteristic of that region. While the temperature continues to rise with depth, the increasing pressure forces the

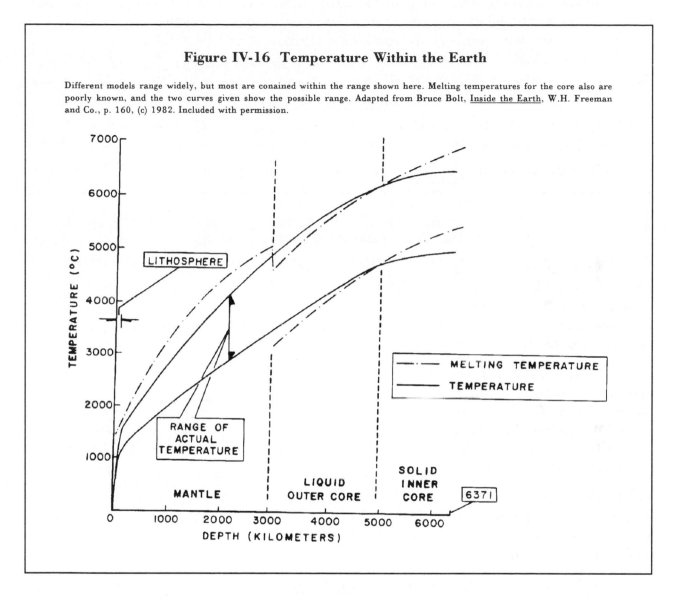

Figure IV-16 Temperature Within the Earth

Different models range widely, but most are conained within the range shown here. Melting temperatures for the core also are poorly known, and the two curves given show the possible range. Adapted from Bruce Bolt, Inside the Earth, W.H. Freeman and Co., p. 160, (c) 1982. Included with permission.

melting temperature to rise even faster, with the result that mantle at greater depth is actually farther from melting than mantle at shallower depths. Thus the lower mantle becomes more rigid than the mantle of the asthenosphere. You can probably see from the diagram why the lower boundary of the asthenosphere is not nearly as well defined as the upper boundary. The actual and melting temperatures only gradually diverge, without any relatively sharp changes such as we see at the base of the lithosphere. The increase of melting temperature with increasing pressure also explains why the hotter inner core is solid while the cooler outer core is liquid.

Why is the inside of the Earth so hot? There are two likely sources for this heat -- natural radioactivity from elements such as uranium, thorium, and potassium, and heat left over from the formation of the Earth 4,600 million years ago. If the solid inner core were growing by gradual solidification of the outer core, this would release additional heat into the base of the mantle.

Available heat travels to the surface of the Earth where it is radiated into space and lost. Within the Earth, it can travel by two different mechanisms: conduction and convection.

Conduction is the process by which heat travels through stationary materials. When an empty pot is placed on a fire, the inside of the pot soon becomes hot. The heat has traveled through the solid bottom of the pot by conduction.

On the other hand, heat can be very efficiently transported by convection, in which the heat is carried from one place to another by a moving material. The hot air rising from a radiator is carrying heat away from the radiator by convection.

If we fill the pot with water, the liquid at the bottom of the pot will heat up, expand and become less dense than the water above it, and rise toward the surface. There it cools, losing heat to the colder air above. As a result, it becomes more dense again, and sinks to the bottom to complete the cycle. The process is diagramed in Figure IV-17.

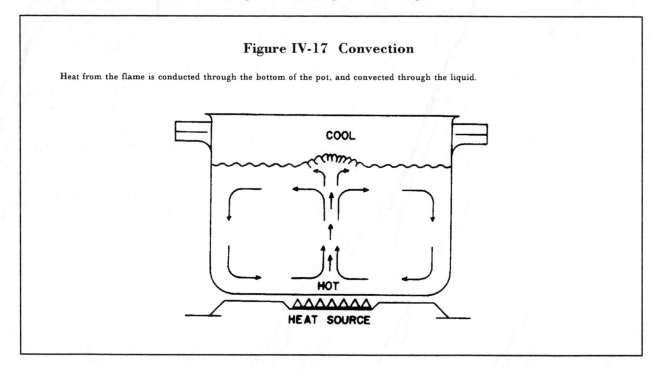

Figure IV-17 Convection

Heat from the flame is conducted through the bottom of the pot, and convected through the liquid.

The temperature profile for the Earth in Figure IV-16 was constructed using the assumption that convection is occurring in the outer core and in the mantle. Without this assumption, the temperature in the model would rise too quickly with depth, and a solid mantle and inner core would not be explainable.

Convection in the liquid outer core seems reasonable enough, but how can the mantle convect if it is solid?

STUDY QUESTIONS

IV-15. How hot might we expect the crustal rocks to be at a depth of 10 km?

IV-16. Are the following effects examples of heat transport by <u>conduction</u> or <u>convection</u>? a) A cool breeze blows off the ocean onto the shore; b) In wintertime, outside walls of a house feel colder than inside walls; c) Heat generated by electricity in a hot iron is transferred to the clothes being pressed.

---

4. Mantle Convection

On page 46 we discussed the tendency of solids to become soft when the temperature approaches the melting point of the solid. This "softness" at high temperatures takes the form of a <u>solid-state creep</u> in which any force exerted on the material will eventually produce a permanent deformation of it.

A familiar example of softness at high temperature is asphalt used in paving roads. On a winter day it feels quite hard and brittle, but on a hot summer day it is resilient and can yield and deform when you step on it. Other examples are found in the slow motion of a glacier flowing down a valley, or in the distortion of some kinds of plastic utensils when run through a hot dishwasher cycle.

Because so much of the mantle appears to be at temperatures that are not too far from its melting point (Figure IV-16), the solid mantle is probably capable of slow convection. If so, then we can ask: Do the convective motions in the mantle provide the forces that move the plates about on the Earth's surface? Is all of the mantle convecting, or only part of it? Do rising plumes of convecting mantle determine the location of hot spots, oceanic ridges, and continental rifts? At the present time none of these questions can be answered with certainty.

Figure IV-18 shows two types of competing models for mantle convection.

In the layered model, strong convection taking place in the upper mantle is effectively decoupled or isolated from slow convection in the more rigid lower mantle. This means that there would be little or no mixing of material between the upper and lower mantle.

In the whole-mantle convection model, we have a situation more analogous to our pot of convecting fluid in Figure IV-17. Here a single cycle of convection reaches from the core-mantle boundary to the base of the lithosphere.

Which of the two types of models is most likely depends to a considerable extent on whether the upper-lower mantle boundary at 640 km is a boundary between two different mineral mixes or is caused by a change in the crystal structure of a single mantle material. If the former, then the mantle is sorted into two compositional layers, and convection that crosses the boundary would mix the two layers. Whole-mantle convection is likely only in the latter

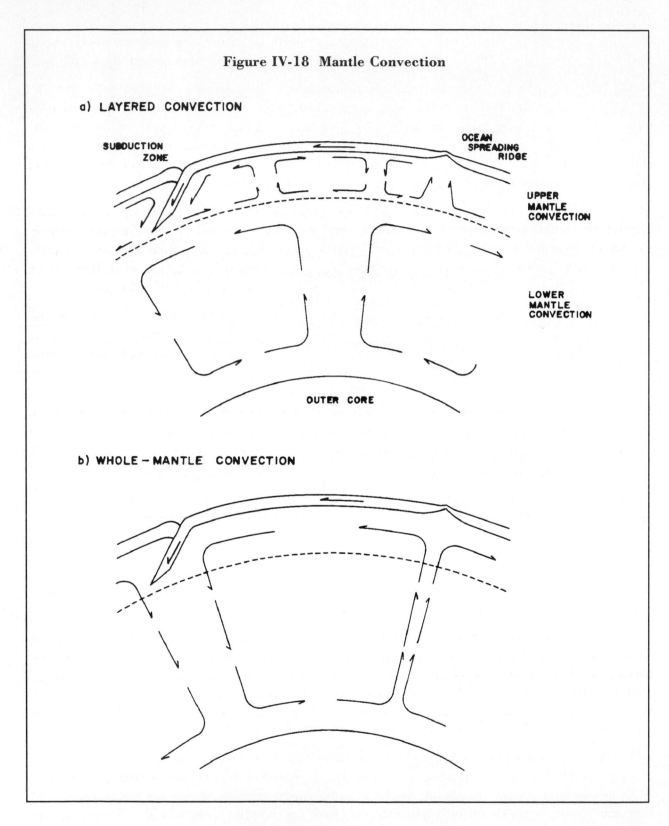

Figure IV-18 Mantle Convection

a) LAYERED CONVECTION

SUBDUCTION ZONE

OCEAN SPREADING RIDGE

UPPER MANTLE CONVECTION

LOWER MANTLE CONVECTION

OUTER CORE

b) WHOLE–MANTLE CONVECTION

case, in which the boundary could be maintained because minerals crossing it would respond to the pressure by taking the crystal form appropriate to each depth.

Each of these models has several variants, reflecting the scarcity of direct observations and the tenuous link between the models and the seismic data. At present it does not seem possible to directly observe the motions of material beneath the lithosphere.

For some time seismologists have been aware that the simple picture presented in Figure IV-13, in which seismic velocity in the mantle varied only with depth and not location, needed modification. S-wave velocities appeared to be higher beneath continents than beneath oceans, perhaps to depths as great as 500 km (310 mi). This observation casts doubt on the idea that the moving plates are exactly the same as the rigid lithosphere.

Perhaps, it was suggested, the continental portions of the plates have some kind of deep appendage in the asthenosphere that follows the continents around in some manner and so should be regarded as integral parts of the plates. These appendages might consist of mantle material from which the basaltic components have been removed, resulting in differences in chemical composition between the mantle underlying oceanic and continental lithosphere.

If this hypothesis turns out to be true, then some fundamental notions concerning the workings of plate tectonics may need to be revised. With so many conflicting models to consider, the time is ripe for new data that can provide the ability to discriminate between them.

STUDY QUESTION

IV-17. Why would we expect mantle convection currents usually to be rising beneath the oceanic ridges? Compare Figure IV-17 and Figure III-7.

---

5. Seismic Tomography

An exciting new observational tool called seismic tomography promises to improve in a dramatic way our view of the mantle and its detailed structure.

It operates in a manner similar to the CAT-scan X-ray machines that have been developed in recent years for probing the details of structures within a living human. In that device, the X-ray beam is rotated around the body of the subject, building up a large number of images taken from different angles. These images are processed by a computer and a detailed three-dimensional image is built of the subject's internal organs. The result is a great improvement over the usual two-dimensional X-ray photograph.

The process works somewhat like the following: You see the shadow cast on a curtain from the head of someone standing behind it. If the person stands facing the curtain, you can make out the hair style, head shape, and perhaps the ears, but nothing of the facial features. If the person slowly turns around, you will see successive shadow images that will include the profile and oblique views. From this information your mind can construct a three-dimensional image of the person's appearance that may be good enough to allow you to identify that person if it turns out to be a friend.

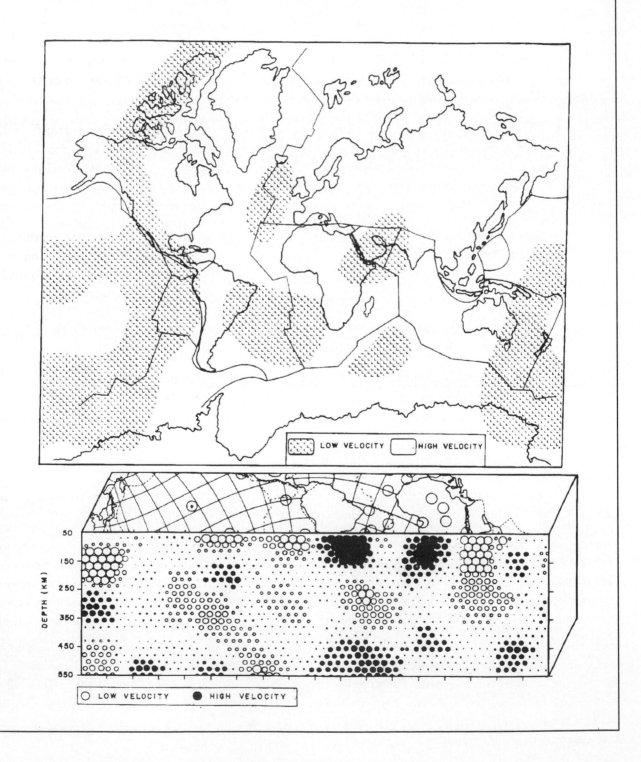

Figure IV-19 Tomographic Views of the Mantle

(Top) S-wave velocities at a depth of 250 km. Shaded areas show regions of slow velocity. (Bottom) S-wave velocities in a vertical slice through the upper mantle. Circles on the map in the lower diagram represent possible hot spots. Adapted from Don Anderson, California Institute of Technology. Included with permission.

Seismic tomography works in a similar manner. Seismic rays passing at many different angles through a region of the mantle are analyzed by computer to reconstruct the three-dimensional pattern of high and low velocities that occur throughout that region. The three-dimensional model in the computer can be viewed in plan-view or as cross-sectional slices.

Two of these are shown in Figure IV-19. The top map shows a plan view of mantle S-wave velocities at a depth of 250 km (155 mi). Shaded areas show slow velocities which might be indicative of hot, rising mantle material. Note that most of the oceanic ridges lie over slow regions, which is what we would expect, but the ridge separating the Antarctic and Australian Plates seems to lie atop a relatively fast and cool region.

The lower diagram displays a slice taken vertically along a line passing through India, Africa, South America, and the Mid-Pacific Ocean. The slice extends from the top of the mantle to a depth of 550 km (340 mi), including nearly the whole of the upper mantle. Note how cool, fast regions lie under South America and Africa, while in this slice, the hot region beneath the Mid-Atlantic Ridge does not extend to a very great depth.

Interpretation of these complex patterns has only now begun. Tomographic pictures of the entire mantle are becoming available, and it should soon be possible to examine models of mantle convection in the light of seismic observations. Major advances in the Earth sciences have often come about as a result of the development of new kinds of observations.

STUDY QUESTION

IV-18. In what ways is the use of X-ray photography of human internal organs analogous to the use of seismic rays in probing the Earth's interior?

---

6. What Drives the Plates?

It seems ironic that the question that gave Alfred Wegener so much trouble in the 1920's should remain essentially unanswered so late in the development of plate tectonic theory. Nonetheless, that is the case today.

Three main kinds of forces acting on the moving plates have been identified. Since the lithosphere is denser and heavier than the asthenosphere, the subducting slabs may exert a pull on the plate toward the trenches. In addition, the lithosphere may tend to slide down and away from the oceanic ridges under the influence of gravity. Finally, motion of the asthenosphere beneath the plates may tend to carry them along or slow them down, depending on how the asthenosphere is moving. At the present time, the first of these forces seems to be stronger than the second, but the role of mantle convection in the third force is still too poorly understood to evaluate its contribution.

Plate tectonic theory has advanced our understanding of the workings of Planet Earth greatly, but it is clear that there is much exciting work yet to be done.

STUDY QUESTION

IV-19. Figure IV-18b are the convection currents helping to move the plates from ridge toward subduction? What about in Figure IV-18a?

---

7. Summary: The Three-Dimensional Structure of the Earth [6]

We have seen already how much our knowledge of the structure of the deep interior of the Earth depends on the location of earthquakes and on measurements of earthquake waves by earthquake observatories around the world (Unit III). For this reason, more than most sciences, seismology is international. Since the first earthquake measurements from individual observatories were established around 1900, it has been clear that earthquake measurements from individual observatories have to be combined in order to be sensibly interpreted. A single earthquake observatory, working in splendid isolation, can make only a limited contribution to the resolution of the constitution of the Earth. Knowledge of the deep parts of the globe has drawn on readings from seismographs located on all continents and many oceanic islands and has involved scientists from many countries. Growth of this knowledge continues to depend on cooperation and exchange of earthquake data. At the present time, there are some 2,000 seismographic observatories involved in data exchange.

We have also seen (see Figure III-2) that the epicenters of earthquakes are widely distributed on continents and oceans around the world. Clearly, the distribution is not uniform geographically. Most earthquakes are generated at the plate edges where the geological structures are complicated rather than under the deep oceans and continental shields. Nevertheless, by appropriate geographical selection of earthquake sources and seismographic stations, it is possible to obtain sets of measurements of seismic waves whose paths have sampled many different parts of the Earth's interior. By careful statistical methods, we may then look for significant differences in the measurements along different paths.

It has been found, for example, that the average travel times shown in Figure IV-10 differ slightly for different regions of the Earth. These variations, which deviate usually by only a few percent from the average, are evidence for small but significant differences in the physical structure of the deep interior from purely symmetrical conditions such as assumed in Figure IV-15. One such case is the difference in average physical conditions in the mantle under spreading ridges and subduction zones.

Work to resolve fine structure and lateral changes in the deep Earth's interior has been going on for many years, but there are experimental problems in that the earthquake waves become scattered by structural irregularities just as light is scattered by a rough reflecting surface. It is necessary to disentangle seismic waves that have traveled along direct rays, such as in Figure IV-7, from indirect scattered waves. In addition, changes in geology near the Earth's surface mask the deeper irregularities. These problems present research challenges which are now being explored by methods such as seismic tomography.

---

[6]This section was written solely for use in this text by Dr. Bruce A. Bolt of the Seismographic Station at the University of California, Berkeley.

RECOMMENDED READING

National Academy of Sciences, <u>Continental Tectonics</u>, National Academy Press (1980).

National Academy of Sciences, <u>Explosive Volcanism</u>, National Academy Press (1984). Fairly technical in nature but good and recent coverageof subjects discussed in this unit.

David L. Jones, Allan Cox, Peter Coney, and Myrl Beck, <u>The Growth of Western North America</u>, Scientific American, November 1982 issue. A good illustrated discussion of the concept of accreted terranes.

Bruce A. Bolt, <u>Inside the Earth</u>, W.H. Freeman and Co. (1982). A very readable and interesting discussion of the exploration of Earth's interior via seismology - highly recommended.

Don L. Anderson and Adam M. Dziewonski, <u>Seismic Tomography</u>, Scientific American, October 1984 issue. A recent review of attempts to discern convective flow within the mantle.

UNIT V THE BLUE PLANET:
PHYSICAL AND CHEMICAL MAKEUP OF THE OCEANS

THE GLOMAR CHALLENGER

A. INTRODUCTION

1. Overview

Early explorers knew only the surface of the seas and their maps showed them as featureless blanks. Once the oceans and continents had been mapped, the observations of seafarers provided the first information on the currents and winds of the oceans. The laying and maintenance of the transatlantic cables provided impetus to obtaining knowledge of the deep ocean environment. Though the oceans are thin compared to their breadth, the greatest volume of the oceans remains inaccessible to the investigator using only SCUBA gear. Modern oceanography uses a wide array of tools to probe the oceanic depths, including depth sounders, physical and chemical measurements of seawater, and cores taken from the sea floor. Plate tectonic processes determine the geography of the ocean floor, while physical, chemical, and biological processes contribute the blanket of sediment that coats the sea floor. Explaining the chemistry of seawater requires interaction between the water and the mantle at the spreading ridges, leading to the conclusion that world geological and geochemical processes are more interrelated than we had previously thought.

---

2. Objectives

Upon completion of this unit you should be able to:

1. visualize the shape and scale of the ocean basins

2. describe the history of oceanic exploration

3. understand that oceanographic studies go far beyond studies using aqualungs near the water-air interface

4. describe types of measurements being made in modern oceanography

5. describe the structure and dynamics of the ocean floor

6. relate small changes in sea level observed by satellites to mapping of the ocean floor

7. compare and contrast sedimentation on the sea floor with that on the continents

8. compare old and new explanations for the saltiness of seawater, and recognize the interrelated behavior of the solid, liquid, and gaseous parts of the Earth

9. describe the major zones of the oceans according to depth.

---

3. Key Terms and Concepts

hypsometric diagram

Franklin's map of the Gulf Stream

voyage of H.M.S. Challenger

sounding line

acoustic echo sounder

bathymetric maps

Deep Sea Drilling Project

submersible research vessels

seismic studies at sea

pinger

piston corer

grab sampler

instrumented buoys and tripods

research submersible Alvin

spreading rates and ocean basin volume

topography of ocean floor

recycling of sediments in subduction zones

sediment thickness and distance from ridge

continental shelf

continental slope

continental rise

continental margin

terrigenous sediments

biogenic sediments

ooze

turbidity currents

abyssal plains

benthic storms

side-scanning sonar

Seasat oceanographic satellite

using gravity to map the ocean floor

unusual properties of water

ions

salts and salinity

nutrients

trace elements

hydrothermal circulation

factors affecting salinity

seawater density

surface waters

thermocline

deep or bottom waters

4. Corresponding Video

In this program, you will explore the oceans, the last great frontier on Earth. Dramatic footage will show that the oceans contain at least as much "weather" as the atmosphere. Historical ocean mapping and recent satellite data on the sea floor will reveal the ocean's geography. Animation of the entire world circulation will depict the Gulf Stream and its eddies.

B. AN OCEANIC PERSPECTIVE

1. The Ocean Basins

The ocean waters cover 71% of the surface of the Earth and are the most distinctive feature of our planet. Take out a map of the world (or look at Figure II-3 on page 34) and consider the dimensions of the oceans. The scale of most flat maps is accurate only at the equator, so measure the widths in kilometers of the Atlantic and Pacific Oceans along the equator in order to get an idea of their size.

How deep is the ocean bottom? From the earlier units you already know that the depth to the ocean floor varies considerably, being shallow on the continental shelves and extending to great depths in the oceanic trenches. You can turn to the hypsometric diagram on page 23 for an average depth to the abyssal plains -- approximately 4.5 km (15,000 ft). Now compare this depth to the width that you obtained for the Pacific Ocean. The proportion is such that if you were to construct a true scale model of the water in the Pacific Basin, its thickness would be very nearly that of the paper on which your map is printed. Even the deepest point in the oceans (the Mariana Trench) at 11,033 m (36,198 ft), would be represented on your model as an inconspicuous doubling of its thickness in the very small area occupied by the trench. The waters of the oceans, then, constitute a very thin layer covering the mud-covered ocean floor.

In spite of its relative thinness, however, a column of water 5 kilometers in thickness is still extremely heavy, exerting a pressure of some 520 kilograms per square centimeter. This is over 500 times atmospheric pressure at sea level, and is a formidable adversary to explorers who wish to venture to such depths. For most of our history our explorations of the oceans have been confined to the top few meters of the waters and the vast expanse of the oceanic surface.

2. Early Explorations

Before oceanography emerged as a science, knowledge of the oceans was being compiled by sailors. As early as 600 BC, Phoenician explorers had circumnavigated Africa and maps of the world began to expand outwards from the Mediterranean. From the first millenium BC on, Indian, Chinese, and Arabian sailors explored the Indian Ocean, and from the ninth to fifteenth centuries, Arabian and Persian pilots compiled navigational instructions that included information on winds and currents in addition to the mapping of coasts, islands, and ports. Knowledge of the geography of the world grew with oceanic exploration, and reliable world maps began to emerge. By 330 BC Aristotle had deduced the fact that the world was spherical by observing the shape of Earth's shadow on the moon during a lunar eclipse and in 250 BC Eratosthenes, a librarian in Alexandria, measured the radius of the Earth to remarkable precision. He also published a world map that included Europe and northern Africa and extended as far east as India. His map represented the land as being surrounded by an all-encompassing ocean.

In 140 AD Ptolemy published a map that included China and a bit more of Africa, but he represented the Atlantic and Indian Oceans as enclosed seas, like the Mediterranean. In addition, he used a value for the radius of the Earth that was substantially smaller than that of

Eratosthenes. By the time that Columbus set out in search of a new route to the spice islands of the East Indies, more complete maps of the eastern hemisphere continents existed. Along with most other educated people of his age, Columbus knew full well that the world was spherical in shape. In constructing his own charts of the world, however, he made two extremely interesting errors that conspired to influence the course of history. The first was that Columbus chose to follow Ptolemy's lead in using far too small a radius for the Earth. On this shrunken world, he wrapped a Eurasian continent that was significantly too large. Since the measurement of longitude requires accurate timepieces, he could really only guess the distance eastward from Europe to Southeast Asia from travel times of boats and overland expeditions.

In any case, the combination of these two errors -- wrapping a too-large Eurasian continent around a too-small globe -- conspired to convince Columbus that he could set out to the <u>west</u> and, after traveling for less than five thousand kilometers, arrive in Japan -- a vast saving in time and expense over the more traditional route that first rounded the Cape of Good Hope on the southern tip of Africa, and then crossed the Indian Ocean in an eastward direction.

The rest is ironic history, and explains why Native Americans came to be called "Indians". Columbus sighted land just about where he had expected to, and was convinced that he had, in fact, reached the fabled islands of the Indies. To his dying day, he never did realize that he had discovered a whole New World.

But Columbus did more than just sail in search of wealth (which he never found). Along the way, he made scientific observations that established the fact that the declination of the Earth's magnetic field -- the angle between true north and magnetic north -- has different values at different places on Earth. This practice of taking observations during the course of long cruises would prove to be the seed from which modern oceanography sprouted.

Almost all early knowledge of the oceans was derived from the experiences of seafarers. But a truly clever scientist could still make important discoveries without even setting out to sea. Such was the case with Benjamin Franklin's discovery of the Gulf Stream. Franklin was Postmaster General for the American Colonies from 1764 to 1775 and wondered why mail packets sailing from Falmouth, England to New York took weeks longer than heavier laden merchant ships traveling from London to Rhode Island. In <u>Maritime Observations</u>, published by Franklin in 1786, he says:

> About the year 1769 or 70, there was an application made by the board of customs at Boston, to the lords of the treasury in London, complaining that the packets between Falmouth and New York, were generally a fortnight longer in their passages, than merchant ships from London to Rhode Island, and proposing that for the future they should be ordered to Rhode Island instead of New York.

> ...it appearing strange to me that there should be such a difference between two places, scarce a day's run asunder, especially when the merchant ships are generally deeper laden, and more weakly manned than the packets, and had from London the whole length of the river and channel to run before they left the land of England, while the packets had only to go from Falmouth, I could not but think

the face misunderstood or misrepresented. There happened then to be in London, a Nantucket sea-captain of my acquaintance, to whom I communicated the affair. He told me he believed the fact might be true; but the difference was owing to this, that the Rhode Island captains were acquainted with the gulf stream, which those of the English packets were not. We are well acquainted with that stream, says he, because in our pursuit of whales, which keep near the sides of it, but are not to be met with in it, we run down along the sides, and frequently cross it to change our side: and in crossing it have sometimes met and spoke with those packets, who were in the middle of it, and stemming a current, that was against them to the value of three miles an hour; and advised them to cross it and get out of it; but they were too wise to be counselled by simple American fisherman. When the winds are but light, he added, they are carried back by the current more than they are forwarded by the wind: and if the wind be good, the subtraction of 70 miles a day from their course is of some importance. I then observed that it was a pity no notice was taken of this current upon the charts, and requested him to mark it out for me, which he readily complied with, adding directions for avoiding it in sailing from Europe to North America. I procured it to be engraved by order from the general post office ... and copies were sent down to Falmouth for the captains of the packets, who slighted it however ...

Observations of sea captains continued to be an important source of oceanographic information, but as time went on, expeditions whose goals were purely scientific became more numerous and influential. Significant impetus for these studies was provided by the desire to know more about the deep ocean environment on the part of the managers responsible for the maintenance of the transatlantic cables that were laid starting in 1858.

Perhaps the most famous and successful of the early scientific voyages was that made by H.M.S. Challenger which set out from England in 1872. The trip lasted 3-1/2 years, covered 68,890 nautical miles, and took physical, chemical, and biological observations at hundreds of places in the Atlantic, Pacific, and Indian Oceans. The scientific results took two decades to analyze and eventually filled 29,500 pages in 50 volumes. It ranks as one of the great scientific achievements of all time.

Throughout the nineteenth century, only the vaguest outlines began to emerge of the shape of the ocean floor. The only measurement method available was the sounding line, in which a light line with a weight on its end was lowered over the ship's side until the weight came to rest on the ocean floor. The person lowering the line might then notice the change in the apparent weight of the line and could mark the point at which this occurred. The depth to the ocean floor would then be just the length of the line that had been played out.

The difficulty with this method is that once four kilometers of line have been played out, the weight of the line is more than that of the weight on the bottom, and the heaving of the ship in waves and the fact that it is impossible to bring the ship to a total halt makes it very hard to judge just when the weight has struck bottom. It was not until the development of the acoustic echo sounder in the twentieth century that a detailed view of the ocean floors became possible. It works on the principle of sending out a pulse of sound from the ship and measuring the time needed for the pulse to travel down to the bottom and for its echo to return

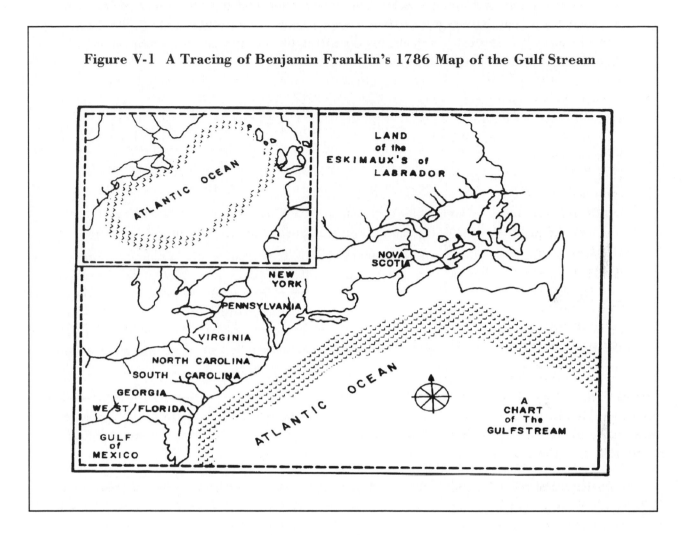

Figure V-1 A Tracing of Benjamin Franklin's 1786 Map of the Gulf Stream

to the ship. Now, by assuming a constant velocity of sound in seawater, it was possible to make continuous soundings of the ocean floor while the ship was moving.

A notable oceanographic expedition was that of the ship Meteor, which set out in 1925 for 25 months and crisscrossed the South Atlantic Ocean, gathering more than 70,000 soundings of the ocean floor. For the first time it became possible to at least partially fill in the blank blue spaces on world globes. The exploration of the remaining two-thirds of the Earth had begun.

In this period oceanography grew as a discipline, with the establishment of major new oceanographic institutes such as Scripps Institution of Oceanography in La Jolla, California, and Woods Hole Oceanographic Institution in Woods Hole, Massachusetts in the United States.

Submarine warfare during World War II generated interest on the part of naval authorities for more detailed knowledge of the ocean floor, and with the end of the war, mapping of the deep sea continued with renewed vigor, eventually resulting in the highly detailed bathymetric (ocean floor) maps that are now available.

Within the span of a single decade, the heights and depths of the Earth were visited by humans. In 1953, Mt. Everest was climbed for the first time, and in 1960 the bathyscaph Trieste slowly settled into the depths of the Mariana Trench at its lowest point, named Challenger Deep after H.M.S. Challenger.

Perhaps the most ambitious deep ocean project since the war, and one of the most successful to date, was the Deep Sea Drilling Project. A large ship, aptly named the Glomar Challenger, was constructed with a hole in the middle and a drilling rig mounted astride it. Six engines and propellors could keep the ship nearly motionless even in heavy seas while drilling bits were lowered through kilometers of water to core the ocean bottom.

In this process, a hole is bored in the ocean floor using a bit that has a hole in its center. Into this hole, a cylinder of sediment and rock is extruded into a chamber and can be brought back up to the surface for detailed analysis. These ocean cores have been taken at over 500 sites around the world and constitute an invaluable record of the upper layers of the ocean floor. Dating of the sediments has provided direct confirmation of sea-floor spreading by showing how the ocean floor becomes older as you proceed away from the ridge crests, and in addition has provided evidence on topics as diverse as studies of magnetic field behavior, fluctuations in world climate, and a history of circulation patterns.

3. The Third Dimension

When the word "oceanographer" is spoken, most people probably envision someone in scuba gear, investigating coral reefs or sunken wrecks. If this were the complete picture, however, oceanography would be a two-dimensional science and oceanographers would be confined to a tiny fraction of the total volume of the oceans -- the water-air interface. Free-swimming divers are restricted to only the top hundred meters or so of the oceans, but modern oceanographers are determined to explore the third dimension of the oceans as well, taking their investigations right down to the ocean floor.

Until recently oceanographers had to make their measurements from ships on the two-dimensional water-air interface, using acoustic sounders and instrument packages lowered into the depths to extend their view into the third dimension. Now, three important developments are extending that view. One is the small submersible research vessel that can take the oceanographer down into the great depths of the ocean for first-hand observations. Another is the use of robotics for unmanned precise sampling and mapping. The third is the satellite or manned orbiting vehicle. In 1978, Seasat, the first satellite devoted entirely to oceanographic measurements, was launched, providing ocean scientists with powerful new tools for viewing their domain. In 1984 a space shuttle soared into orbit carrying the first oceanographer to make his observations from space. It seemed very fitting that the shuttle bore a name distinguished in the annals of the science: Challenger.

<u>STUDY QUESTIONS</u>

V-1. What width did you measure for the Atlantic Ocean at the equator?

V-2. How did Aristotle prove that the Earth was spherical?

V-3. How far did Columbus think he had to travel in order to reach Japan?

V-4. Why did mail boats traveling from England to New York take longer than merchant ships traveling from England to Rhode Island in Franklin's day?

V-5. What method was used to determine the depth to the ocean floor in the nineteenth century? In the twentieth?

V-6. What was the principal task of the Deep Sea Drilling Project?

C. OBSERVATIONAL TECHNIQUES

Let us briefly review the kinds of measurements that oceanographers make and the equipment that they use. An oceanographic research vessel must not only provide living facilities for the scientists and crew, but must also have laboratory space, a winch for lowering instrument packages into the depths of the sea, and often an on-board computer as well, for analysis of the data while the scientists are still at sea.

A wide variety of instrumentation is used on a research vessel. We have already mentioned the acoustic depth sounder, which sends out sound pulses and measures the distance to the bottom by measuring the time needed for the sound to go down and back to the ship. By the way, the term "sounding" predates by far the use of sound to measure depth -- the weighted line previously used was called a "sounding line" -- and provides us with a highly appropriate coincidence. The acoustic sounder can sometimes provide us with a view of more than just the topography of the ocean floor. Because some sediments resting there are relatively transparent to sound waves, it is often possible, by recording echoes from deeper layers, to get a picture of the shape of the sediment layers from the sounder as well.

Seismology allows us to examine the interior of the earth, and its use at sea provides a detailed look at the structure of the sea floor and the mantle beneath it. It is not practical, however, for a ship to sit at one station for a long enough period to wait for natural earthquakes to provide the necessary seismic waves, and so explosive charges or other artificial sources are used instead.

Different techniques have been employed for gathering seismic data at sea. One uses two ships, one moving and producing seismic waves at intervals, while the other is stationary and records them using special low-frequency microphones, called hydrophones, that are placed in water. Another arrangement uses a single ship that produces the seismic wave, usually with an air gun that fires about once per second. The seismic wave travels to the bottom, penetrates the sea floor, and is reflected from the various layers of rock beneath the sea floor to depths of a few kilometers. The echoes from these layers return to a string of hydrophones being towed by the ship. In this way, a continuous profile of sea-floor structure may be obtained as the ship moves along its course. These seismic techniques work similarly to those used to explore the continental crust for minerals and oil.

Measurements at depth are often carried out by instruments lowered at the end of a cable to whatever depth desired by the scientist. A "pinger" in the instrument package allows the determination of just how far above the ocean floor it is. A pinger works similarly to the acoustic depth sounder in that it emits a sound pulse that is received by a microphone on the surface. The time required for the sound wave to travel directly from the pinger is measured, as is the time required for the sound to bounce off the ocean floor and return to the ship. The greater the difference between the arrival times of the direct and the echoed sound waves, the farther the pinger is from the ocean floor. When the two waves arrive at exactly the same time, the pinger has arrived on the bottom.

Many different kinds of instruments may be lowered on cables. Among these are cameras, both film and television, and the lights necessary for them to record anything in the total darkness of the deep ocean. Light from the Sun fails to penetrate any farther down than about 100 m (330 ft). Water samples may be taken from various depths by bottles that are opened automatically when they arrive at the desired level. Electrode systems measure the electrical resistance of ocean water and determine its salinity, since dissolved salts make water more conductive to electricity.

An important device in the study of the soft sediments that blanket the ocean floor is the piston corer, illustrated in Figure V-2. The device consists of a cylindrical coring tube with a movable piston inside. As it is lowered to the ocean floor, a smaller tube, the "trigger core", takes a sample of the topmost layers, which are the ones most disturbed in the main tube. In the process, the main core tube is released and falls, thrusting itself into the sediment. The piston is held at the level of the sediment surface while the core tube continues to penetrate the sediment. The piston acts much like the plunger in a hypodermic needle used to withdraw blood, sucking the sediment up into the core tube with a minimum of disturbance to the sediment layers. When the corer is pulled back up to the surface, the sediment in the core tube goes with it and can be removed on shipboard for study. Cores as long as thirty meters have been recovered using piston corers.

In addition to corers, dredges and grab samplers may be used to recover samples from the ocean floor. A grab sampler operates much like the clamshell bucket used in excavation machinery to bite off chunks of dirt in construction projects.

When measurements over an extended period of time are required, instrumented buoys may be used. One type uses a floating buoy that is anchored to the sea bottom with a cable. Instruments may ride the buoy or the cable at specified depths, and can record data concerning weather or ocean conditions for long periods of time. Free-floating buoys may be used to track oceanic currents for long distances, and more recently, subsurface buoys whose buoyancy has

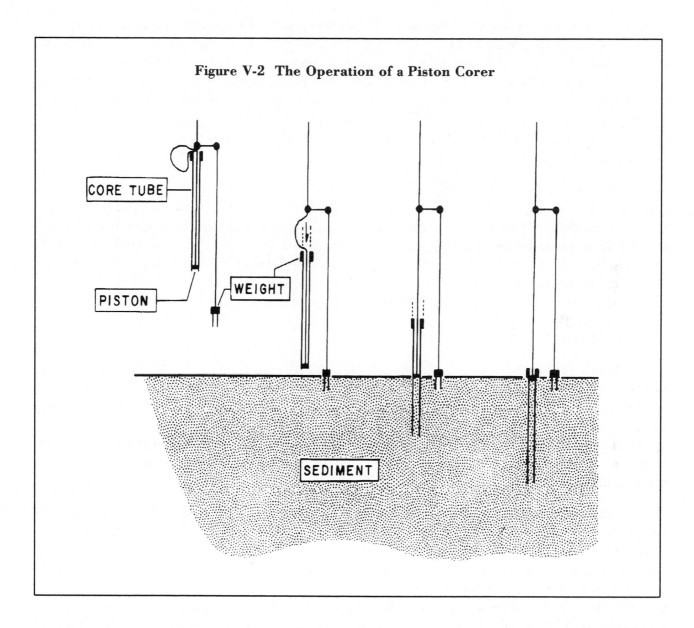

Figure V-2 The Operation of a Piston Corer

been carefully adjusted to float at a specified depth are being employed to chart the wanderings of the hidden layers of the ocean.

In addition, instrument-bearing tripods have been lowered to the ocean floor for the purpose of observing sea-bottom phenomena. Powered by batteries for periods of six months to a year, the package may contain sophisticated microcomputers that can sense the environment and turn on particular instruments when something of interest occurs. These undersea robots can be made to perform chemical, physical, and biological experiments on the spot, recording data in a digital form for later retrieval. When it is time to recover the instruments, an acoustic signal from a ship on the surface triggers a release mechanism that separates the instrument package from weights that hold it to the ocean floor and flotation chambers provide sufficient buoyancy to allow the package to float to the surface where it can be picked up and reused.

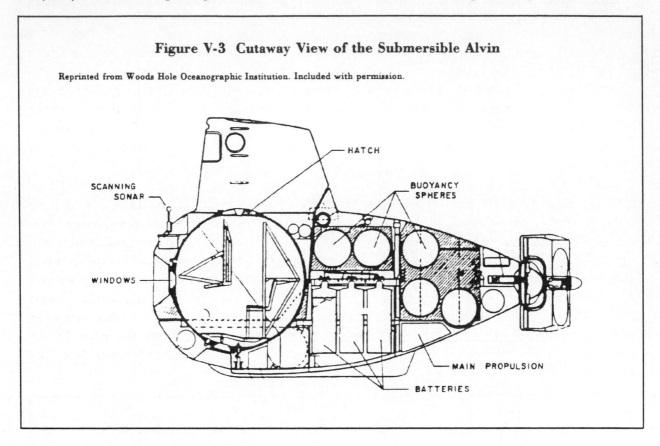

Figure V-3 Cutaway View of the Submersible Alvin

Reprinted from Woods Hole Oceanographic Institution. Included with permission.

Submersible research vehicles have evolved to a point of considerable sophistication. Figure V-3 shows a cutaway view of the submersible vessel Alvin, owned by the U. S. Office of Naval Research. In order to provide protection from the crushing pressures of the deep ocean environment, the passengers are confined to a thick-walled spherical chamber only two meters (six feet) in diameter. Nevertheless, they are able to observe through thick glass portholes, and a mechanical arm allows them to pick up objects and manipulate machinery on the ocean floor. Other countries that operate research submersibles are France and Japan.

STUDY QUESTIONS

V-7. What kind of material is sampled by a piston corer?

V-8. Why must oceanographic research vessels carry powerful winch systems?

D. GEOPHYSICS OF THE OCEAN FLOOR

The geography of the ocean floor is dominated by the ocean basin floor and the ocean ridge system. The ocean basin floors are those parts of the oceanic plates that are between the spreading ridges and trenches. Their elevation stands at an intermediate depth within the range of three to five kilometers. The spreading ridges owe their height to the buoyant force of the low-density hot or molten rock beneath them (see Figure III-8 on page 52). This low density mass of rock below the ridge plays the role of an isostatic "root" and supports the weight of the ridge.

The ridge is also spreading, creating new plate on either side of it. As it does so, the lithospheric plates recede from the source of heat at the ridge and slowly cool and become more dense, sinking in the process. As long as the ocean floor remains in isostatic equilibrium, the extent to which the ocean floor has sunk is a function primarily of the age and temperature of that segment of ocean floor. On the other hand, the width of the ridge will be a strong function of how fast the ridge is spreading. Figure V-4 shows this effect. Both ridges are the same height, but the fast-spreading ridge is much wider than the slow-spreading one because new, hot, and therefore high-elevation ridge is being carried away rapidly from the crest in the former case. The result is that the fast-spreading ridge has a much larger volume than the slow-spreading ridge.

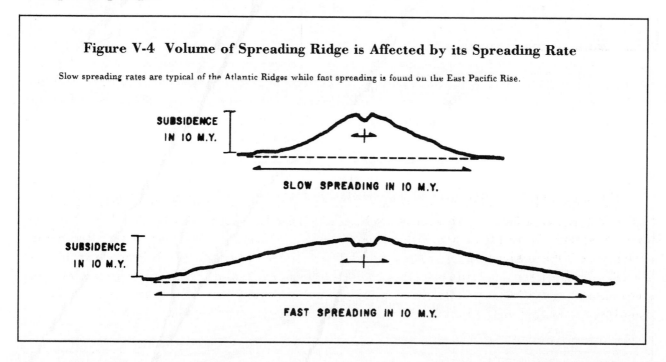

Figure V-4 Volume of Spreading Ridge is Affected by its Spreading Rate

Slow spreading rates are typical of the Atlantic Ridges while fast spreading is found on the East Pacific Rise.

SUBSIDENCE
IN 10 M.Y.

SLOW SPREADING IN 10 M.Y.

SUBSIDENCE
IN 10 M.Y.

FAST SPREADING IN 10 M.Y.

The more volume there is in the oceanic ridges, the more water is displaced by them, with the result that worldwide sea levels will stand higher on the continents during times of fast average sea-floor spreading rates than during times of slower spreading rates. For this reason, it has been proposed that those periods of geological time during which the sea covered substantial portions of the continents, such as during the Cretaceous period, were times of rapid sea-floor spreading.

The transform faults that offset oceanic spreading ridges also contribute to the shape or topography of the ocean floor (see Subsection III-E-3). These are often the sites of very rugged terrain, caused by the sliding of the two plates past one another for millions of years, the results of countless earthquakes. These fracture zones, as they are called, often extend for thousands of kilometers along the ocean floor.

The trenches, of course, are the result of the subduction of the ocean floor. Here one plate is being forced down and under the other, and in the process, both are buckled downwards (see Figure III-7 on page 51). Into these great depths move the ocean floor and its load of sediments. It was initially thought that all the sediment would accumulate in the trench or be added to the plate that was not being subducted, but recent seismic studies show that substantial amounts of oceanic sediment are being subducted along with the oceanic crust. For the most part, these are very wet sediments, and so in this way a part of the ocean waters is recycled back into the mantle. Some of it at least will be erupted through the mouths of the andesitic volcanoes associated with the subduction zone and returned to the atmospheric/oceanic environment. As we shall see, this is not the only case in which ocean waters are brought into close interaction with mantle rocks.

1. Sediments and Sedimentation

Most of the ocean floor is blanketed in soft sediments of one kind or another. Only young surfaces, found on the spreading ridges or on active volcanic undersea mountains, have large areas of bare rock exposed. There is a continuous rain of sediment down through the ocean waters, adding layer after layer, year after year. The result is that the sediment blanket becomes thicker with increasing age of the ocean floor. Near the ocean ridges, the layer is thin or absent, but it increases steadily as we go farther from the ridge (Figure V-5).

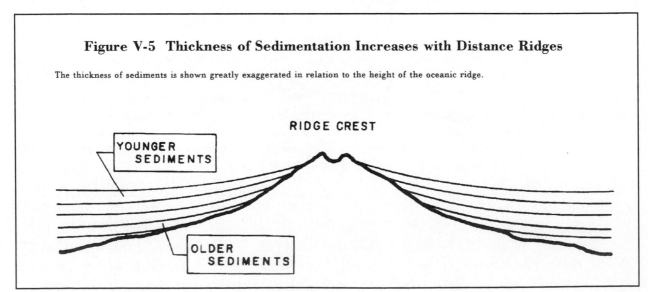

Figure V-5 Thickness of Sedimentation Increases with Distance Ridges

The thickness of sediments is shown greatly exaggerated in relation to the height of the oceanic ridge.

Oceanic sediments fall into two main categories -- terrigenous and biogenic.

Terrigenous sediments originate on the continents, as their name implies. The sediment load of rivers is a major contributor, but wind can carry terrigenous sediments large distances offshore. Most of these sediments enter the ocean at the shoreline, but soon disperse in a variety of ways throughout the continental margin, which consists of the continental shelf, slope, and rise (see Figure II-1 on page 23).

The continental shelves are relatively flat and end fairly abruptly at the continental slope. Here the sea bottom slopes downward gradually at an average slope of about 4°. The continental rise is a still more gradual slope (generally less than half a degree) extending out to the deep sea floor. The greatest accumulations of sediment are found on the continental rise, having migrated there under the influence of gravity. In fact, in the North Atlantic Ocean, sediments have accumulated on the continental rise since the opening of the Atlantic in the Jurassic, sometimes reaching a thickness of 10 km (6 mi). This is somewhat thicker than the average depth of the oceans.

Coarse terrigenous sediments are found in the deep ocean at high latitudes, where melting icebergs can dump large quantities of terrigenous sediment far from land.

Biogenic sediments owe their origin to life in the oceans. Most of it comes from the skeletal remains of microscopic plants and animals that live in the biota-rich environment of the uppermost water layer. In the shallow-water environment of the continental shelves, reefs also contribute their mass to biogenic sediments. Ocean-bottom sediments rich in material of biological origin are referred to as ooze.

Inorganic clays form much of the deep-ocean sediment layer. In many places, calcium carbonate-rich skeletal remains dissolve in their long trip from the surface waters to the cold depths of the ocean floor. Carbon dioxide from the atmosphere is dissolved in seawater, producing carbonic acid. This is the same mechanism by which carbon dioxide gas is put into carbonated soft drinks to give them their fizz. The result is an acidity that attacks the calcium carbonate of the skeletal remains and dissolves them. Soda pop holds more carbonation when it is cold than when it is warm, and the same is true for seawater. Cold bottom water tends to be more acidic as a result, and if the supply of biogenic sediment is not so high, all the skeletal material may dissove in the acidic water before it reaches the bottom. In warmer waters, or in areas of higher biotic production, calcium-rich oozes can be the dominant sediment.

Some sediments are chemically precipitated directly from seawater. Most notable among these are phosphorites and manganese nodules. Phosphorites are phosphorus-rich deposits that tend to form on the sea floor near the continents and manganese nodules are rounded concretions that contain concentrations of iron, manganese, copper, and other metals. The nodules, ranging in size from 1 - 20 cm across, are mostly found in the deep ocean where sedimentation rates are low. They form by precipitation from seawater, coating some object such as a shark's tooth, growing at extremely slow rates on the order of one millimeter per million years. As you will see in the **Mineral Resources** Unit, mangenese nodules may prove to have economic value as a source of metals for the future.

Volcanic dust may be distributed very far over the oceans by major eruptions. The dust settles out of the atmosphere and slowly rains down onto the ocean floor to join other sediments.

The rate of sediment accumulation varies greatly depending on the type and place. In general, sedimentation is rapid for terrigenous sediments near the continents, with accumulation rates of around 20 cm per thousand years being common. On the other hand, deep ocean sediments accumulate much more slowly at rates that average only 2 mm per thousand years. It is this slow but steady rate of accumulation that makes sediment cores from the deep ocean floor so valuable as repositories of scientific information spanning vast stretches of time.

The great difference between accumulation rates for the continental margins and the deep ocean floor means that sediment thickness near the land can become very great. The sheer weight of this load of sediment has an isostatic effect, depressing the continental shelves. As a result, the edge of the continent often sinks somewhat, and this appears as a rising of sea level in that area. The east coast of the United States is a case in point, with the drowned river valleys of the Maine coast and the Chesapeake Bay as evidence of the effect. In these places, the sea reaches up the valleys of former rivers, creating long brackish bays and tidewaters.

Still another effect of the sediment buildup near the continents is the occurrence of turbidity currents. The continental margin is in many places cut by undersea canyons, often leading from the mouths of rivers out to the deep ocean floor. Sediments accumulate in the heads of these canyons until the pile becomes unstable, at which point the sediments can begin to slide down the canyon, gathering speed and picking up more sediment along the way. Traveling at substantial speed, the turbidity current can travel for long distances -- sometimes right down the continental slope and out onto the abyssal plains, which is the flat surface of a turbidite sediment accumulation.

The existence of turbidity currents was originally deduced from indirect evidence -- the breaking of submarine cables when strong ones pass, and the presence of layers of coarse-grained terrigenous sediments containing shells of shallow-water organisms sandwiched between layers of normal deep ocean sediment, indicating that the terrigenous material had been transported rapidly over long distances.

2. Benthic Storms [7]

Another sediment-moving phenomenon, discovered only recently, is the abyssal or benthic storm. Unlike a turbidity current, which is an underwater avalanche triggered by great forces such as an earthquake, the benthic storm behaves like a blizzard in the deep ocean. In a benthic storm, a rapidly-moving current sweeps over large areas of the deep seafloor, picking up sediment and causing dramatic changes in its path. In some areas, the storm scours the bottom; in other places, it deposits enormous loads of clay and silt. This "stormy" current may last for as long as two weeks at a time and lift sediment 300 feet off the bottom. In the western North Atlantic, scientists have recorded benthic storms moving as fast as 75 cm per second (30 in. per second).

---

[7]This section was written solely for use in this text by Victoria Kaharl, Woods Hole Oceanographic Institution, Woods Hole, Massachusetts

The recent discovery of benthic storms surprised most oceanographers because, with the exception of the infrequent turbidity current, it was thought that all deep sea currents moved very slowly. Scientists are not yet able to predict benthic storms; nor do they know what causes them. In addition to turbidity currents and the more frequent benthic storms, the deep seascape is affected by a steady, continuous circulation of frigid bottom water. In the polar regions, the surface waters sink because they are colder and denser than the water in the lower latitudes. As the cold waters at each pole sink, they begin to flow toward the equator, gradually mingle, and eventually rise to the surface, beginning the entire process anew. The paths these cold waters take is determined by various factors such as the topography of the ocean bottom. While this global current system moves slowly -- reaching speeds of only 5-15 cm per second (2 - 6 in per second) -- it constitutes a continuous, relentless movement of massive amounts of water that play a major role in sweeping bottom sediment into drifts in a peculiar pattern throughout the world ocean that has remained unchanged for millions of years.

Scientists theorize that the great quantities of sediment stirred up by benthic storms may be picked up and carried downstream by the less energeitc but persistent cold water currents. Some of the largest sediment accumulations, which are in the North and South Atlantic, are 1,000 km long, 200 km wide and 2 km thick (600 mi long, 120 mi wide, and 6,500 ft thick).

3. The Mapping of the Sea Floor

As we saw earlier in this unit, the shape of the ocean floor did not become known until fairly recent times. Even currently used maps are compilations of many thousands of echo sounding records along the tracks of ships. Some new techniques, however, promise to change this. The first is an adaptation of the usual echo sounder, called side-scanning sonar. Instead of pointing the sound pulse straight down below the ship, the sonar pulse is sent out to the side so it reaches the ground at a grazing angle. Those parts of the ocean floor that slope toward the sensor reflect sound more effectively than those that are horizontal or slope away. Shipboard electronics are able to convert the returned signal into a picture of the sea floor that may be viewed while the ship is still underway. Using this method, a swath several kilometers wide may be surveyed along the path of the ship that carries the sonar gear. Two different versions of this scheme are now operational: one, called GLORIA, is towed behind the research vessel and can obtain images of a swath 20 km (12 mi) or more in width along the path of the vessel, while the other, called SeaMARC I, is towed near the ocean floor and can obtain higher resolution images, though of a narrower swath. The images that these devices obtain of the sea floor look somewhat like landscapes lit by a low-angle sun.

Sea Beam is an elaboration of the usual echo sounder. Instead of a single acoustic beam aimed straight down below the ship, Sea Beam sends out 16 beams in a fan shape, each designed to measure the distance to a particular point on the sea floor along a line that is perpendicular to the ship's travel. With this system, a contour map showing the topography of the sea floor can be generated. With all of these methods, shipboard computers process the data received from the sonar devices and present them in forms that are easy to interpret. Scientists on board can examine the results and modify their cruise plans if necessary to maximize the time they spend in the most interesting areas. Cruise time is extremely expensive, and this ability to analyze results almost immediately is an important cost-saving measure.

Another technique uses measurements from an orbiting satellite and overcomes a significant shortcoming of all the other methods: the availability of data only along the tracks of ships. The satellite method promises to provide, for the first time, a broad view of ocean-floor topography featuring a uniform standard of resolution and accuracy throughout all the oceans.

The technique involves measuring, to extremely high precision, the elevation of the sea surface. This was accomplished by a satellite named Seasat, that used radar waves to make the measurements, accurate to centimeters. It may at first seem surprising that the ocean's surface mirrors (on a much reduced scale) the topography of the sea floor, but Figure V-6 shows how this works.

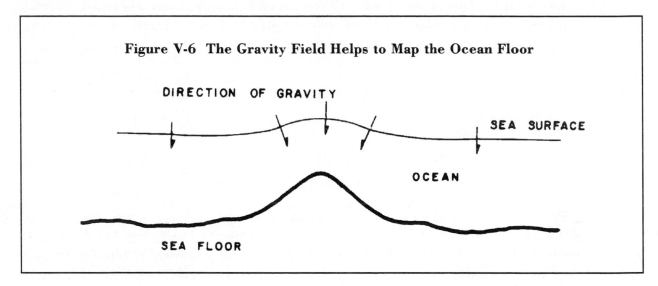

Figure V-6 The Gravity Field Helps to Map the Ocean Floor

Gravity is a function of all objects that have mass, acting to attract a mass to all others around it. A mountain or plateau on the sea floor is a massive object, and this excess of mass tends to attract the surrounding seawater to it. In effect, the mountain's mass distorts the local gravity field, as shown in the diagram, and this produces a mounding of seawater above the extra mass. The satellite sees this as a bulge on the ocean surface that is a much reduced version of the sea-floor topography.

There is, however, a significant difference in how these two methods (echo sounding and satellite ranging) view the ocean floor. The satellite measurements rely on the distortion of the gravity field by excess mass on or below the sea floor, and so it provides a view not only of the topography, but also of density differences that may exist in the oceanic crust. In combination, the two methods promise to give us a very detailed view of the shape and structure of the ocean floor, comparable to maps that have been available for the continents for the past hundred years. The exploration of the other two-thirds of the Earth's surface has begun in earnest.

STUDY QUESTIONS

V-9. Is sea level likely to be higher or lower when sea-floor spreading rates worldwide are low? Why?

V-10. What may be consumed at the oceanic trenches in addition to basaltic ocean floor?

V-11. Would you expect to find thicker sediments near the continents or in the middle of the Atlantic Ocean?

V-12. How long does it take on the average for one meter of sediment to accumulate in the deep ocean, far from sources of terrigenous sediment?

V-13. What is the principal advantage of side-scanning sonar over the usual electronic depth sounder?

V-14. Would the sea directly over a submarine canyon have a bump or a depression in its surface?

E. CHEMISTRY AND PHYSICS OF SEAWATER

1. The Composition of Seawater

Water is one of the most remarkable and unusual substances in the universe. Because it is so abundant on Earth and therefore so familiar to us, it has actually skewed our view of what constitutes normal behavior. For example, we all know that ice floats on water, and so we might guess that this is typical behavior for the solid and liquid phases of substances. In fact, the converse is usually true for most materials -- the solid is usually more dense than the liquid, and so, for instance, solid basalt would <u>sink</u> in a pool of its own lava rather than float on top.

Indeed, the fact that water is liquid at all on Earth is purely a function of its unusual properties. Most other substances made of similarly light atoms (methane, for example) are gaseous at room temperature and only liquefy at much colder temperatures. Another notable property of water is its ability to dissolve other substances. No other common liquid is able to dissolve so many materials.

All of these properties are a result of the atomic makeup of the water molecule. Composed of two hydrogen and one oxygen atoms (H_2O), the water molecule has a <u>polarized</u> structure because the hydrogen atoms have a positive electric charge while the oxygen atom has a negative charge (Figure V-7). This configuration, in which one "end" of the molecule is positively charged and the other "end" is negatively charged, accounts for many of water's unusual properties.

Unlike electrical charges attract, while like charges repel one another. For this reason, whenever a bunch of water molecules get together, they tend to line up in such a way as to mutually attract one another. This cohesion of the molecules explains why the melting and

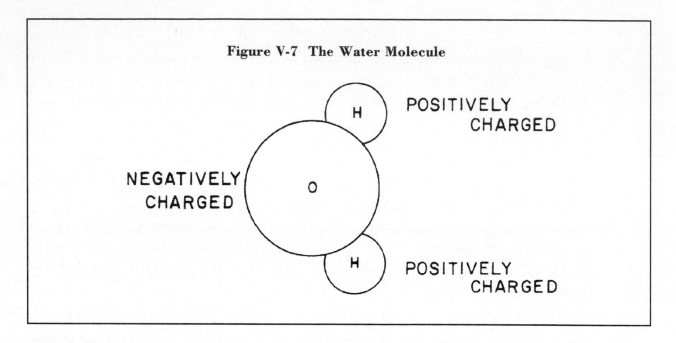

Figure V-7 The Water Molecule

boiling points of water are so low -- in a gas, the molecules fly about with little regard for one another, but in liquids, and especially in solids, they interact with one another in a more cohesive manner. When water freezes, however, the molecules arrange themselves in a very regular way that has an open hexagonal structure. Because the structure is so open, ice is less dense than water; because it is hexagonal, snowflakes assume a magnificently delicate six-fold symmetry.

But it is water's role as a solvent that most concerns us here. Seawater carries dissolved within it a large number of substances in the form of ions -- portions of molecules that have been torn apart to form charged atoms or groups of atoms. For instance, when common table salt (sodium chloride, symbolized as NaCl) is dissolved in water, the molecule is torn apart into a positively charged sodium ion (Na^+) and a negatively charged chlorine ion (Cl^-). These are each attracted to the negatively or positively charged parts of the water molecule. Because they are now separate from one another, the Na^+ and Cl^- ions no longer form a single salt molecule, but have been incorporated into the liquid water as individual ions. This is an important point because, as we shall see, it is not necessary for the sodium and chlorine to have come into the ocean in the form of salt -- indeed, each element may have come from a different source. But when seawater is evaporated, the ions are left behind and, without the polarized water molecules to keep them apart, they will combine to form salt.

Seawater contains nearly all of the elements, though many are found in extremely small concentrations. The four most important ions in terms of concentration are the chlorine (Cl^-), sodium (Na^+), sulfate (SO_4^{--}), and magnesium (Mg^{++}) ions, in decreasing order of importance. Next most important are the dissolved gases: carbon dioxide (CO_2), nitrogen (N_2), and oxygen (O_2). These are not torn apart, but are incorporated into the water as whole molecules. They are not held nearly as tightly by the water molecules as the ions, and so can escape fairly readily. A glass of water that has been allowed to stand in the open for a long time comes to taste "flat" -- what has happened is that the dissolved gases in it have escaped. Many faucets have an "aerator" attached that mixes air with the tap water to increase the concentration of dissolved gases, improving the water's taste.

You may have noticed that carbon dioxide was mentioned before nitrogen and oxygen, while in the atmosphere, carbon dioxide is present in much smaller amounts than the other two gases. In fact, carbon dioxide dissolves readily in water (as any soft drink or beer drinker knows) with the result that 62 times more carbon dioxide is dissolved in seawater than is present in the entire atmosphere. The oceans are thus an important regulator of carbon dioxide in the atmosphere -- an effect that we shall see in the unit **The Atmosphere** is of exteme importance to Earth's climate.

Next in order of concentration are a group of ions that are vital to biological processes: the nutrients. Chief among these are the nitrate (NO_3^-), nitrite (NO_2^-), and phosphate (PO_4^-) ions. These are present in concentrations typically of only a few parts per million, but they are essential to the presence of life in the oceans.

Finally, there are a large number of trace elements, present in concentrations of only parts per billion. Some of these, such as iodine, iron, lead, and mercury, have important effects on life in the ocean and on other organisms, such as ourselves, that consume seafood. Iodine is essential to human health, and before it was artifically added to table salt, seafood was a principal source. Mercury, on the other hand, is extremely toxic to humans, and increasing levels of this substance in some waters, due to industrial pollution, have posed serious health hazards.

An important result obtained from the voyage of the H.M.S. Challenger in the 1870's was the discovery that the relative proportions of the dissolved ions in seawater were the same regardless of where in the world the sample was taken. There may be more or less water in which the salts are dissolved, but that is the principal variation. It would seem from this, that ocean water has been pretty well mixed.

2. Why is the Ocean Salty?

You may recall from the unit **A Sense of Time** that Sir Edmund Halley in 1715 tried to determine the age of the Earth from the assumption that salts carried by the world's rivers were the source of the saltiness of the oceans. While his assumption that the oceans were continuing to become more salty was wrong, until recent years his other assumption was generally accepted. Even so, there has been a long-standing problem with this explanation in that there is a serious mismatch between the ions prevalent in the oceans and those found in river water.

The ions most commonly found in river water are the carbonate (CO_3^{--}), calcium (Ca^{++}), and sulfate (SO_4^{--}) ions, while those most prevalent in ocean water are the chloride and sodium ions. Some of these discrepancies are easily explained in terms of chemical reactions that take place in the oceans, such as the removal of the carbonate ion by biological processes. Others, such as an excess of chlorine and bromine in the ocean, cannot be explained in this manner.

In an attempt to find a solution to this problem, scientists turned to another source for the excess ions -- the mantle. It had been found that ultramafic rocks, formed deep within the Earth, contained gas with a composition that contained chlorine, bromine, and other elements that matched oceanic composition much more closely than did river water. An alternative explanation was proposed: the excess volatile ions, like chlorine, must have come from the mantle, while the rivers supplied some of the other ions like sodium and magnesium. But how did these ions find their way from the mantle into the oceans without using the rivers as a conduit? The only logical mechanism seemed to involve the spreading ridges, where material from the mantle was involved in the process of creating new ocean floor.

The most recent step in the development of this idea was taken in 1977 when the submersible Alvin visited the crest of a spreading ridge in the vicinity of the Galapagos Islands off the coast of Peru. Here torrents of hot water were found issuing from vents in the rock, showing that ocean water circulates freely through the fresh rock of the ridge, creating hot springs. The circulation of water through hot rock is known to geologists from such places as Yellowstone Park in Wyoming and is termed hydrothermal circulation. This was just the kind of mechanism that was needed to introduce the missing ions from mantle sources.

Soon, hydrothermal vents were found in other places along the ridge crests, and estimates based on the likely total activity of vents made it clear that the entire volume of water in the oceans could circulate through them in only five to ten million years. Because of the high temperatures involved, chemical reactions between mantle rock and ocean water can proceed rapidly and it is now felt that the oceanic vents may well be the dominant influence on the chemical composition of seawater.

The development of our ideas about the chemistry of seawater provides us with a fairly typical example of how scientific explanation evolves. As stated throughout this course, the best theory is that which best explains all the available observations, and Halley's use of river salts was quite valid for his time, since all that he knew was that both river water and seawater contained salts. Detailed observations of their compositions did not yet exist. Once these observations were made, however, it was realized that the river water theory was flawed and the search was on for a better explanation. Other possible sources for the missing ions were sought and a mantle source was hypothesized on circumstantial evidence -- mantle rocks contained the needed ions in approximately the right proportions. This new idea, however, remained in the relatively weak category of hypothesis until it could be confirmed by direct observation of the vent mechanism on the ocean floor.

Detectives and scientists have much in common in their approach, only for the scientist it is nature and not mankind that sets the mysteries.

3. Salinity, Temperature, and Density of Seawater

Because the relative proportions of dissolved ions are essentially constant, we may use the measure of salinity as one descriptor of seawater. Salinity is just the total amount of dissolved material that is present in seawater, expressed in parts per thousand. Salinity can vary from low values of about 10 parts per thousand found in the Baltic Sea to values of 40 parts per thousand in the Red Sea. The salinity of normal ocean water is in the range of 34 to 36 parts per thousand.

There are a number of factors that can affect salinity. For instance, a large input of fresh water near the mouth of a major river may produce a local zone of low salinity. The Baltic Sea in northern Europe is fed by many rivers and it has only very narrow connections to the open ocean of the North Sea. Many bays that are fed by large rivers, such as the Chesapeake in the United States, contain brackish water -- water of low salinity.

On the other hand, areas in which evaporation is very high, such as the Red Sea and the Mediterranean, often are characterized by water of high salinity. The water is evaporated but the salts are left behind. The freezing of sea ice also has the effect of increasing salinity of seawater. When ice is frozen from salt water, the ice selectively excludes salts. While seawater is too salty to drink, sea ice frozen from seawater is in fact potable, containing only a small amount of salt. In response to this withdrawal of salt, the seawater remaining below the ice becomes more salty than before.

The density of seawater is determined by the salinity, temperature, and pressure. Of these, the salinity and temperature are most important. Cold water is more dense than warm water, and saline water is more dense than fresh water.

The vertical structure of the ocean is determined by density differences, with denser waters occurring in the deep ocean and less dense waters being found near the surface. Vertical mixing is slow, with the result that we find a layering of ocean water based on the density at any given location, but the form of the layering is not consistent everywhere. Each layer contains water that reflects conditions at its point of origin.

As a rough generalization, we may divide the ocean into three layers: surface waters, the thermocline, and deep waters. Surface waters are well-mixed, of uniform salinity and extend only to depths of 50 - 100 meters. They vary substantially in temperature and salinity from one locality to another, according to the influences of evaporation, fresh water, and the formation of sea ice, discussed previously. The thermocline is a thin region of rapid change in temperature dividing the surface waters from the deep waters. This often begins a region of rapid salinity change as well, that may extend to greater depths.

The deep waters are characterized by relatively uniform salinity and temperature. They are quite cold, with the temperature generally within the range of 3 - 4°C (37 - 39°F). This is the temperature at which the density of water is the greatest. Deep bottom waters for the most part are formed in the Antarctic region, where surface waters are cooled by the frigid climate and made more saline by the production of sea ice. The combination of cold temperature and high salinity produces the densest seawater found in the world. Dense bottom water is formed in the Arctic Ocean as well, but because that ocean basin is ringed by continents, little of the bottom water produced there escapes.

The kind of complexity that can result is shown in Figure V-8. Cold water forming near the surface at high latitudes tends to sink and move toward the equator at depth, while water warmed in the tropics moves toward the poles. In this way, heat is transferred from the equator to the poles, and the oceans exert a very powerful moderating influence on world climate, distributing heat from the Sun in a far more equitable manner than if the oceans were not present.

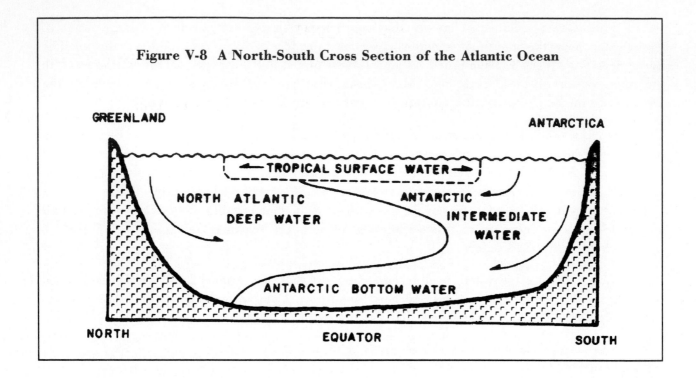

Figure V-8 A North-South Cross Section of the Atlantic Ocean

An important attribute of deep bottom water is that it is rich in oxygen and nutrients. In high latitudes, the cold water at the surface is able to dissolve more gases from the atmosphere, and everywhere nutrients such as phosphates are carried down from the surface layer in particles of organic matter. The organic debris decays on the ocean floor, releasing the nutrients into the bottom layer. We shall see the importance of this point in the next unit, when we discuss the El Niño phenomenon.

STUDY QUESTIONS

V-15. What are some of the unusual properties of water that are due to its polarized structure?

V-16. What are the three most common ions in seawater?

V-17. What is the most common dissolved gas in seawater?

V-18. Where do the sodium (Na+) and chloride (Cl-) ions that make up salt in seawater appear to come from?

V-19. What is the thermocline?

V-20. What is the origin of much of the world's bottom waters?

RECOMMENDED READING

David A. Ross,, <u>Introduction to Oceanography</u>, Third Edition, Prentice-Hall, Inc. (1982).

Keith S. Stowe, <u>Ocean Science</u>, Second Edition, John Wiley and Sons (1983).

These are both excellent and well illustrated texts.

UNIT VI THE BLUE PLANET:

DYNAMICS OF THE OCEANS

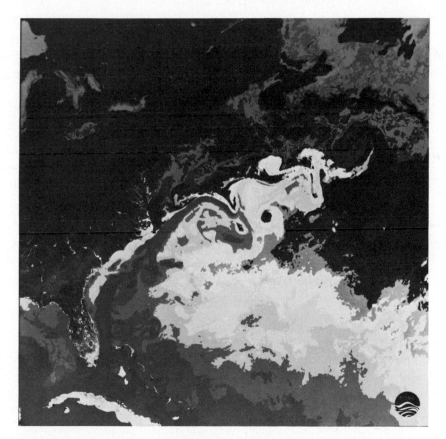

WESTERN NORTH ATLANTIC - GULF STREAM   Courtesy of NASA, Oceanic Processes Program.

A. INTRODUCTION

1. Overview

Ocean currents reflect the influences of surface winds and density differences and are measured by a variety of techniques. The oceans are turning in space along with the Earth as it spins on its axis, and this results in distinctive behavior that is referred to as the Coriolis Effect. You will have the opportunity to investigate it at first hand in an exercise that requires a phonograph turntable. Prevailing winds and Earth's rotation cause great gyres, or rotary motions, within each ocean basin. The Gulf Stream is an example of the tendency of ocean currents to intensify along the western edge of the gyre. This moving mass of water has a vital influence on the climate of much of Europe. The oceans in fact are an important influence on global climate, serving as a huge reservoir of thermal energy in close interaction with the atmosphere. Vertical mixing of ocean waters is limited in most places, but in localities where deep nutrient-rich water upwells, fish and fisherman prosper. The phenomenon of El Niño involves a shifting in the trade winds and the equatorial currents that suppress upwelling off the west coast of South America. When a strong El Niño occurs, the local fishing industry is hard hit. Worse, global climate reacts to the change in ways that can bring disaster to widespread areas of the planet. A new tool for the oceanographer is the satellite, from which a global view of the oceans may be obtained. Measurements from space tell us of the temperature and plankton content of seawater, of the flow of currents, and of the shape of the ocean floor. Circular eddies or ring currents have been observed to be a significant part of oceanic circulation, and their peculiar effects are just beginning to be understood. Marine life has been the subject of study for centuries, but until recently only that life in the uppermost few meters of the sea could be investigated in detail. Now marine biologists are extending their observations to greater depths, including the floor of the deep ocean. This has been the locale for a major new discovery in recent years -- of communities of creatures cut off totally from the Sun and dependent on the chemistry of hot springs in the spreading ridges of the sea floor.

2. Objectives

Upon completion of this unit you should be able to:

1. describe how ocean currents are measured, both by historical and modern techniques

2. relate ocean currents to its causes: wind, temperature, and density variations

3. relate the operation of the Coriolis Effect to the rotation of the Earth

4. relate global circulation patterns to atmospheric circulation and Earth's rotation

5. describe the origin, characteristics, and significance of the Gulf Stream

6. explain the occurrence of upwelling and downwelling currents

7. describe how the oceans and atmosphere interact

8. describe the El Niño phenomenon and its apparent consequences

9. describe how measurements made from Earth orbit can provide new information on oceanic behavior and on biological conditions that are of use to the fishing industry

10. relate the kinds of life found in the midwater and on the ocean floor to their environments

11. explain the operation of hot vents in terms of tectonic activity on and near oceanic spreading ridges

12. relate the presence of life around these deep-ocean vents to the physical and chemical conditions created by the vents.

3. Key Terms and Concepts

ocean currents
direct measurement of currents
indirect measurement of currents
prevailing winds
Coriolis Effect
gyre
Eckman layer
Eckman-layer flow
intensification of western currents
Gulf Stream
Sargasso Sea
upwelling and downwelling currents
residence time of seawater
heat capacity
El Niño
plankton

climatic effects of El Niño
Law of the Sea Treaty
satellite oceanography
phytoplankton
satellite radar ranging to sea surface
geoid
mesoscale eddies or ring currents
warm-core rings
cold-core rings
midwater biology
sea floor biology
ocean vent biological communities
chemosynthesis
black smokers
white smokers

4. Corresponding Video

Major new revelations about the sea will include an astronaut's view of the oceans from space, as well as recent dramatic footage shot during the flight of the space shuttle, Challenger. Startling findings about the perplexing El Ni ño current will be presented, and a variety of deep sea submersibles will take you ever deeper to preciously unexplored sites on the sea floor -- those that yield clues to early life on Earth. Animation and computer graphics greatly enhance our understanding of internal dynamics of the oceans and the geography of the ocean bottom. We will explore the midwater region, the largest and least understood habitat on Earth.

B. WAVES [8]

1. Wind Driven Waves

When you arrive at the seashore, one of the first things you note is the state of the sea; whether it is rough or whether it is calm. It is rather obvious that on a windy day the waves are higher than when there is a gentle breeze. You may also notice that the waves are rather smooth out to sea, forming swells, and that as they come on shore they take the form of turbulent breakers. The first thing we need to do is to make the term wave, which we have just used, a bit more precise. The more specialized term we would like to use is progressive water wave. This term is most easily explained using Figure VI-1 and is most applicable to the smooth swell. A technical description of breaking waves is more complex and does not need to be attempted here.

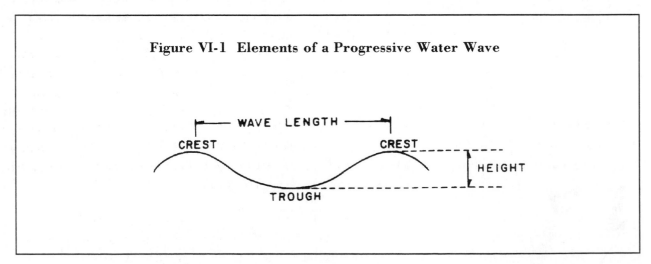

Figure VI-1 Elements of a Progressive Water Wave

The highest portion of the wave is called the crest and the lowest portion is called the trough. The horizontal distance between two crests or between two troughs is the wavelength

---

[8]This section was written solely for use in this text by Walter Pilant, Department of Geology and Planetary Science, University of Pittsburgh, Pittsburgh, PA

and the vertical distance between the trough and the crest is the <u>wave height</u> or <u>amplitude</u>. The pattern described by the figure travels over the surface of the sea at the <u>wave velocity</u>. The wave velocity depends in a rather complicated way upon a number of factors, the most important being wavelength and water depth. When the water is deep, the wave velocity is proportional to the square root of the wavelength rising to a maximum in excess of 200 meters per second (greater than 450 miles per hour). At such wave velocities, the wavelength becomes so long that, to the wave, the water no longer appears deep and in this case another relation applies. The rule for shallow water is that the wave velocity is proportional to the square root of the water depth. Since all oceans are of finite depth, velocities of the above magnitude are the highest observed. The reason that such rapidly progressing patterns are not commonly seen, is that the wave height is so small they are not noticed except by sensitive scientific measuring instruments. Only great earthquakes produce visually observable waves that move at such high speed.

Even though the wave pattern may move rapidly in such cases, the actual volumes of water do not travel very far. If you were to go beyond the line of breaking waves at the seashore and watch the swell as it goes by, the little bits of seaweed and other debris floating in the water at different depths would rise and fall as the wave passes by. On careful observation, you would see that they move shoreward as the crest passes and seaward as the trough passes. Figure VI-2 shows the <u>orbital motion</u> of small volumes of water (flagged by debris). Note that this motion dies away rather rapidly with depth, almost disappearing by the time the depth equals one wavelength. This decrease of water motion with depth has two practical applications. The first is that when a large swell comes in and is about to break, a swimmer may dive below the wave and experience much less commotion of the water. The second is that submarines can ride out the most violent of storms while submerged only 20 meters or so. Since water motion is confined largely to the surface, wave effects will have little influence on the sea bottom except near the shoreline. The principal water forces acting on the bottom of the deep ocean will be oceanic currents, to be described later in this unit.

EXERCISE

You can do a simple experiment to demonstrate the rapid decrease of wave energy with depth. All you need are a liter (quart) bottle, an empty tin can at least 15 cm (6 inches) in height, a kitchen sink or bathtub, and a towel (to wipe up afterwards). Cut both ends out of the tin can. Fill the sink two-thirds full of water and move the liter bottle rapidly up and down in the water to get a series of waves moving away from the point of application. Rapid motion is necessary so that the whole sinkful of water is not put into sympathetic oscillations. As the waves move away from the moving bottle they will decrease in amplitude. The next step is to insert the can with the other hand to a depth of 10 cm (4 inches) and hold it still. As the waves arrive in the vicinity of the can they will move around and beyond it, but the water in the interior of the can will hardly move at all. You might say to yourself, the rigid metal of the can is stopping the wave action. You would be right, but the wave energy should be able to enter the interior through the open bottom. According to the argument above, however, there will be little energy at depth. Now raise the can slowly so that less and less of it is beneath the surface of the water. As you do so, you will see the water on the interior of the metal surface become more and more disturbed until at the point where the can just grazes the surface (but never leaves it), the motion of the water inside the can will be nearly as large as outside the can.

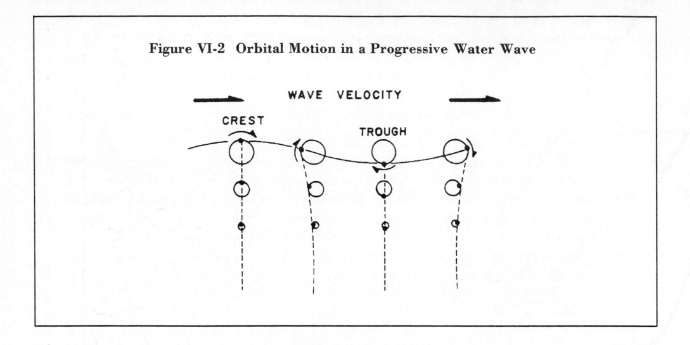

Figure VI-2 Orbital Motion in a Progressive Water Wave

Now that we have described wave patterns and motion, we should look into the causes of waves themselves. Most progressive water waves are driven by the wind, gradually picking up energy from the wind as it travels over the surface of the water. The longer the wind blows steadily, the more energy the waves acquire and the larger they become. Since the Pacific Ocean is larger than the Atlantic Ocean, this is the first reason why surfing is more successful in the Pacific; the waves are larger! A second reason, particularly applicable for North America, is that the prevailing wind direction is from the west. The highest sea waves generated during storms are estimated to be about 25 m (80 ft) in height.

2. Waves from Other Sources

Less frequent, but often spectacular, causes of ocean waves are great earthquakes that take place in the deep trenches of the ocean floor. Often, a dangerous seismic sea wave is triggered by the earthquake and moves out rapidly over the surface of the ocean. Since so much energy is released by the earthquake, even a small part of it can cause a large seismic sea wave; sometimes called a tsunami -- from the Japanese. The word tidal wave is a misnomer since the effect has nothing to do with tides. The rapid progress of the seismic sea waves associated with the 1960 Chilean earthquake is shown in Figure VI-2. This seismic sea wave traveled with wave velocities in excess of 200 meter per sec (450 miles per hour), crossing the entire Pacific Ocean in one day.

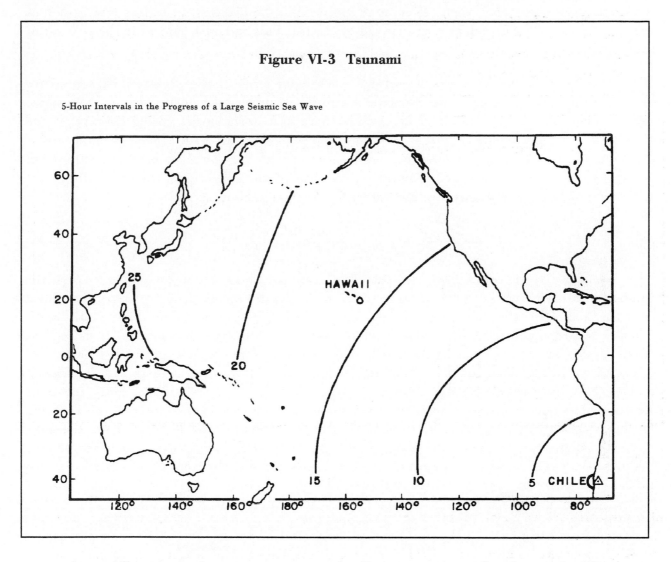

Figure VI-3   Tsunami

5-Hour Intervals in the Progress of a Large Seismic Sea Wave

As rapidly as seismic sea waves move, the P waves sent out by the earthquake move even faster. From a quantitative examination of seismograph records it is possible to estimate whether such a wave will be excited by a great earthquake taking place somewhere around the edge of the Pacific Plate. It is then possible to send out radio warnings of possible damaging waves so that harbor and coastal areas may be evacuated and/or made secure. Several nations lying around the Pacific Basin have cooperated to form into a warning service to minimize possible damage and loss of life. However, people sometimes respond strangely to warnings. Once, when warned of an approaching seismic sea wave, many persons did not evacuate but went down to the shore to watch it come in. Fortunately, the arriving wave was large, but not disastrous.

Earthquakes cause another type of water disturbance known as a _seiche_. This is found in smaller bodies of water such as lakes. The water is set into motion as a whole without a wave pattern being developed. As the water moves toward one end of the lake, the water level rises while falling at the other end. A half-cycle later the water has gone the other way, increasing the level at the other end of the lake. During the Alaskan earthquake of 1964, the ice in several small lakes near the source region was cracked and showed from which direction the earthquake energy had come. Even more spectacular, though, was the setting into motion of the entire body of water comprising the Gulf of Mexico. This can only happen in the greatest of earthquakes.

Lastly, wave patterns can be caused by changing atmospheric pressure. High pressure depresses the ocean surface, while low pressure raises it. In a major storm, we may get a storm surge. The wind can pile up water against the shore in a storm surge, raising the level of the sea. If we combine high winds, low pressures, and high tides (as with a passing hurricane or typhoon), we may get exceptionally high levels of water with very large waves. The combination is capable of doing considerable damage. As a great storm approaches, all these effects are taken into account in the warnings put out by the weather service.

3. Waves and the Shoreline

As the swell approaches the shoreline, two processes begin to act to change the fundamental nature of the wave pattern. As the water depth becomes less, the waves slow down in accordance with the rule governing the wave velocity in shallow water. That is, the wave velocity is proportional to the square root of the depth. Shallow, in this case, is defined as those areas where the water depth is less than one-half of the wavelength of the pattern of approaching waves. There is also a slowing force on the bottom comprised in part by friction and in part by returning water from the waves breaking on the shoreline. Since the same number of wave crests have to pass any given point during a certain time interval, we find that as they slow down the distance between crests (the wavelength) must decrease. Since the wave energy is more concentrated, the wave height must increase. For a while the wave pattern endures, the swell growing in amplitude and the crests and troughs more closely following each other. This is shown near the center of Figure VI-4.

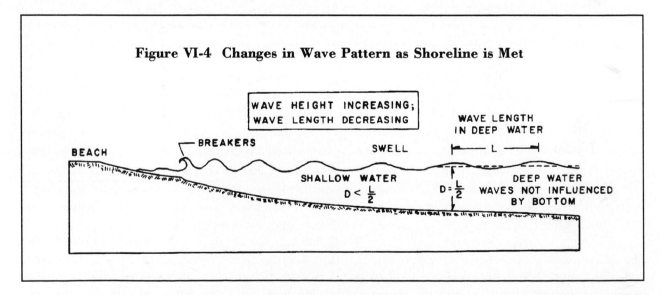

Figure VI-4 Changes in Wave Pattern as Shoreline is Met

As the swell comes even nearer the shore, the water cannot pile higher and higher. Water volumes in the crest tend to travel faster than water volumes in the trough and the crests then overtake the troughs, spilling over into them. If the sandy bottom is nearly level, the crests fill the troughs with a big rush of foam. If the bottom is somewhat steeper, we find breakers perfect for surfing. For very steep bottoms, the waves break right onto the sand itself. As waves break, the violently churning water can exert tremendous forces both upon the sand of the bottom and upon solid rock in the form of sea cliffs or isolated rocks lying above the sea surface. Some of the strongest erosional forces found in nature are generated at this energetic shoreline. However, it can be seen that as the shoreline is eroded away and the sand is distributed by the action of the breaking waves, the beach will become shallower and shallower. After a time, the incoming surf will no longer pound but merely break into foam. This foaming wave action is far less energetic and a wide shallow beach will be formed that is relatively stable. If all ocean waves approached straight in upon the shoreline, this would be all that happened. We might note at this point that the beaches of the eastern United States are of a more gently sloping nature compared to those of the western coastline where they are generally steep. Thus we have a third reason for better surfing on the west coast; breaking waves instead of foaming ones.

But the waves approach from the direction of the swell's source as modified by the local prevailing winds. This direction of approach may not be from directly out to sea. As the swell approaches the shoreline the early arriving crests begin to slow, allowing the following ones to catch up causing a bending of the wave crests. This process is known as <u>wave refraction</u> and is illustrated in Figure VI-5.

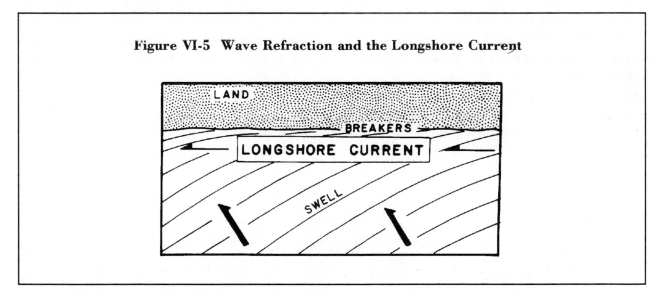

Figure VI-5 Wave Refraction and the Longshore Current

Even though the principal effect of refraction is to make the swell approach the shoreline in a more nearly straight-on direction, the figure shows that there will still be a component of wave velocity parallel to the shoreline. This gives rise to a longshore current that, together with the breaking waves, can move beach sand parallel to the shore. If to the right of the figure there was no source of sand (say that the shore was formed from a seacliff of extremely resistant rock) then the longshore current would tend, over a period of time, to remove much sand from the area of the figure and move it out the left-hand side. Then the beach would narrow and be less desirable from many viewpoints.

The major source of beach sand is not the erosion of sea cliffs and shorelines but the influx of sand from rivers and streams as they enter the sea (or lake). Two human actions (in densely populated countries) have tended to reduce this influx of sand in recent decades. People have extracted even more water from shorebound rivers for metropolitan water sources. This water is often impounded by dams far from the shore where the load of sand and silt that the river carries is deposited in the bottom of a reservoir. Dams are also constructed to control flood waters, but the flood control basins they create have a similar effect. This tremendous load of sand and silt never reaches the sea. Thus on many shorelines, lakes as well as oceans, we see a net loss of sandy beach. Efforts to stop this longshore transport by the erection of rock or steel barriers merely slows this flow, temporarily saving "your" shoreline while depriving your "neighbor" of his source of sand and depleting his shoreline even faster. It is in problem areas of this type that understanding and cooperation are necessary to create the most beneficial environment while considering the competing needs of water supply, flood control, recreation, and scenic beauty.

C. TIDES [9]

1. Tidal Forces

The earliest sailors knew that the Sun and the Moon were intimately linked to the tides but they did not have the theoretical basis for understanding their interactions in detail. Yet tides were so important to navigation that meticulous records of the tides were kept in every port so that ships would know when to set sail. The word "tide" appears in many current expressions in the sense of the best or safest time to do something. Today, with all our current scientific sophistication, there are still many questions to be answered about why tides have the heights and timing that they do. On the other hand, modern computers can predict (on the basis of previous observations) what the tides will do at any given port far into the future. Our first task will be to see how our nearest heavenly neighbors influence the tides.

Tides are caused by the gravitational attraction of the Moon and the Sun in combination with the movement of the Earth. As the Moon revolves about the Earth in its orbit, the Earth does not remain still, but revolves itself about the center of mass of the Earth-Moon system. Because the Earth is so much more massive than the Moon, this point is within the volume of the Earth itself. In either case, the pull of gravity is matched by the acceleration (change from a straight line path) needed to keep the Earth or Moon in nearly circular orbits. This acceleration is known as centripetal (center seeking) acceleration and is described by Newton's laws of motion. For a rigid body such as the Earth, this centripetal acceleration as it moves in its orbit will be the same everywhere within the body (accelerations from rotational motion are taken into account elsewhere). It may not seem that this should be so, but a simple experiment with a piece of cardboard pierced by three pieces of chalk will demonstrate the point. If the cardboard is moved, as in Figure VI-6, about a blackboard in any path (not just a circular one) but is not rotated, then the three pieces of chalk will all trace out the same path. Since the

[9]This section was written solely for use in this text by Walter Pilant, Department of Geology and Planetary Science, University of Pittsburgh, Pittsburgh, PA

centripetal acceleration is related to changes in direction of motion, it will be the same at all points of the moving cardboard. So it is with the Earth.

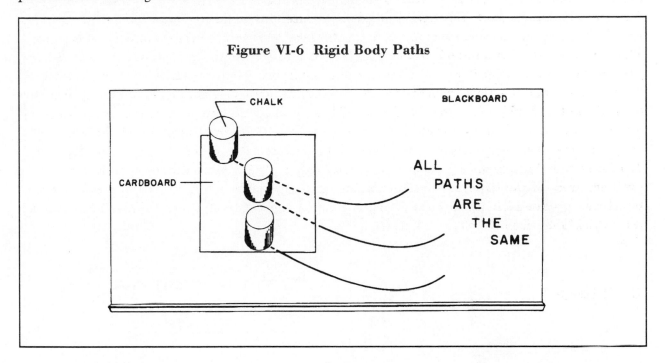

Figure VI-6 Rigid Body Paths

Even though the centripetal acceleration of a moving planetary body is the same everywhere within its rigid portion, this is not true of the gravitational attraction experienced by that body. Gravity lessens with distance, so that those portions of the Earth more distant from the Moon will experience slightly lessened gravitational attraction. For the Moon to maintain its (almost) circular orbit about the Earth, the centripetal acceleration it experiences will exactly match the Earth's gravitational attraction at the center of mass of the Moon. Likewise for the Earth. The centripetal acceleration necessary to follow its small orbit about the center of mass of the Earth-Moon system will equal the gravitational pull of the Moon only at the center of the Earth. On the side of the Earth nearer to the Moon the centripetal acceleration will be less than the pull of the Moon's gravity, while on the far side of the Earth the centripetal acceleration will be greater than the pull of the Moon. The effect is described in Figure VI-7.

2. Tidal Bulges

If gravitational attraction and centripetal acceleration match at the Earth's center, there will be a mismatch at the surfaces nearest and farthest from the Moon. On the side nearest to the Moon, gravity will be the stronger and it will tend to pull the water of the oceans toward that part of the Earth. On the farther side, the rigid Earth will tend to pull out from under the fluid water and leave it behind. The net effect of this is to have not one, but two tidal bulges on the surface of the Earth. This double tidal bulge will remain in fixed relation to the direction of the Moon, while the Earth rotates under it once each <u>25</u> hours. The extra hour arises because the Moon is moving in its orbit. During the course of this rotation any given point will see two high tides and two low tides. We call this a <u>diurnal (twice daily) tide</u>.

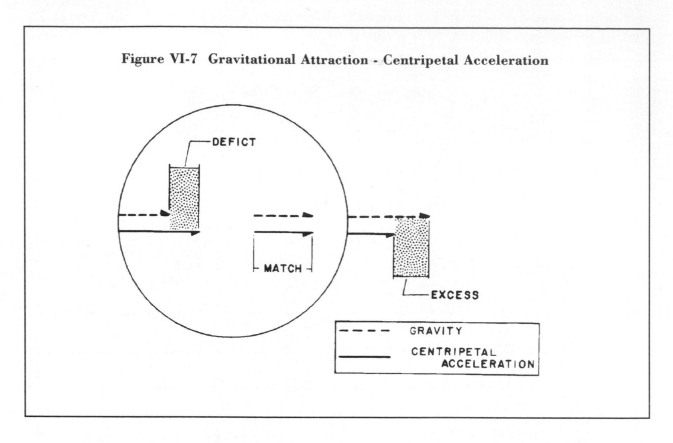

Figure VI-7 Gravitational Attraction - Centripetal Acceleration

The Sun also has a tidal influence, though it is only 46% as great as that of the Moon. Although the Sun is much more massive than the Moon, it lies at a much greater distance and its gravitational influence is much reduced. The net solar tide-producing force is thus less than the lunar tide-producing force. The motions of the Sun and Moon are such that the Sun appears overhead once every 24 hours while the Moon takes approximately 25 hours to return to an overhead position. Thus the Moon will appear to move in the sky relative to the Sun. So also will the tidal bulges produced by the Moon move in relation to the those produced by the Sun. The two twin bulges will be superimposed twice each month when the Moon is in the same portion of the sky as the Sun or when the Moon is directly opposite to the Sun. When the Sun and Moon are at right angles to each other, the twin tidal bulges will not coincide. When the tidal bulges do coincide, we have extra large tides known as spring tides while the smaller ones produced by lack of coincidence are known as neap tides. The two situations are shown in Figure VI-8.

The tidal pattern now becomes more complicated. The Earth rotates once daily beneath a tidal pattern that changes twice a month. The pattern that would be seen at the tidal equator would be as in Figure VI-9 and is known as the equilibrium tide. In it we see the diurnal tide slowly growing and lessening during a twice monthly cycle.

It would be nice if this were sufficient to fully describe oceanic tides. However we have left a lot of things out. First of all, the Earth's Equator is tilted by 23.5 degrees with respect to the orbits of the Moon and Sun. This makes every other high tide higher than the ones in between. In addition, the rise and fall of the level of the oceans requires that great masses of

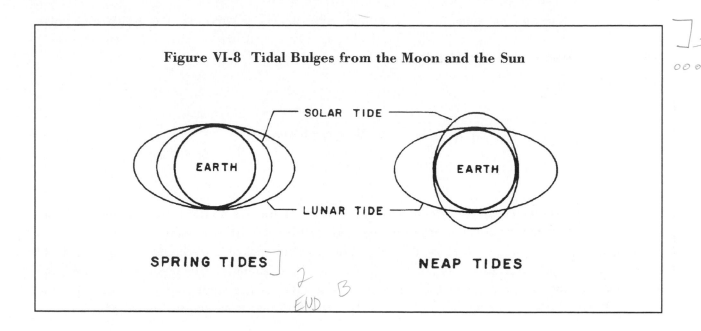

Figure VI-8 Tidal Bulges from the Moon and the Sun

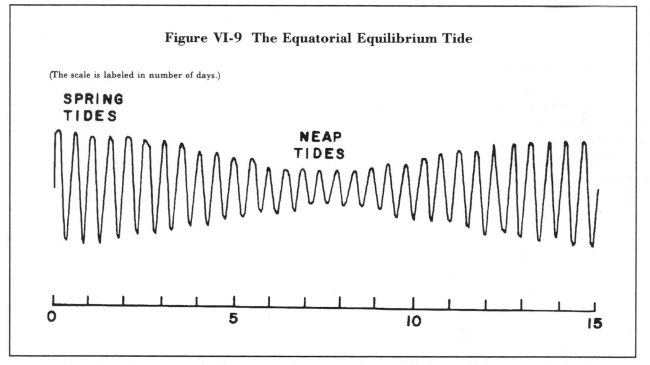

Figure VI-9 The Equatorial Equilibrium Tide

water must race across the Earth's surface. But, as we have seen in the previous section, there is a limiting velocity to the water's movement. Consequently the tides lag behind the tide-producing forces in various amounts according to the depth of the water. Land masses such as peninsulas and large islands also get in the way of the moving water and add further time delays. Not only is the physical mechanism of massive water movement incompletely understood, but the geometry is also very complicated. Since all these factors cannot be taken into account, we proceed from the knowledge that the Moon and Sun have a very regular (periodic) appearance overhead. If we can determine the local tides by accurate measurements, then they will always maintain the same amplitude and time delays relative to the lunar and solar positions and thus may be predicted. This makes accurate measurements of tides an important task in harbors and bays around the world.

3. Tidal Measurements

Measurements of the tides would be fairly easy if we did not have to contend with the ups and downs of the water level due to wave action. But, we can use what we have learned to make a simple tide gauge which will largely eliminate the wave effect. One merely needs to insert a long pipe into the water so that (as in the experiment of the preceding section) the water surface inside the pipe will remain calm. The level of the water surface inside the pipe will thus record quite accurately the level of the tide. A sketch is shown in Figure VI-10.

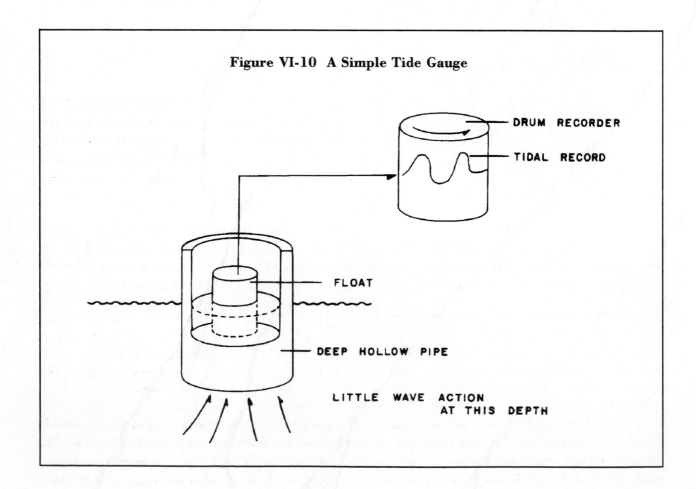

Figure VI-10 A Simple Tide Gauge

DRUM RECORDER

TIDAL RECORD

FLOAT

DEEP HOLLOW PIPE

LITTLE WAVE ACTION AT THIS DEPTH

4. Other Tidal Effects

The tidal forces described previously also act upon the fluid atmosphere and upon the solid Earth itself. The atmosphere experiences tides but they have no noticeable physical manifestations as do oceanic tides. However, the motions of the atmospheric tides can be detected in several ways and are useful in studying atmospheric properties.

In putting together our model of tide-producing forces, we have treated the Earth as a rigid body. This is largely correct, but there is a slight deformation of the solid Earth as well as of the oceans. This Earth tide, as it is called, is much less than the oceanic tide, but again it is definitely measureable using precise instruments. However, the existence of earth tides may have an important consequence. It has long been known that the rotational rate of the Earth has been slowing down due to frictional effects of the tides. For a long time it was supposed that this frictional effect was produced as ocean water poured through narrow channels on the Earth's surface. On re-investigation, it has been found difficult to make a case for ocean tides as the source of friction. This has led investigators to look elsewhere for a source of frictional loss. One possibility that has been suggested lies in the plastic zone, or asthenosphere, in the unit **Plate Tectonics**. Not only can such a partially melted zone dissipate energy, but the dissipation itself will provide part of the heat to keep it in a partially melted state. However, this remains a hypothesis and we need to know the properties of the asthenosphere much better before this idea can be accepted.

D. OCEAN CURRENTS

Early seafarers found that they had two different forces to contend with in their explorations: wind and ocean currents. Their large sailing ships did not tack well into the wind, and so they had to chart the prevailing wind directions and plot routes that would take best advantage of them. Ocean currents -- massive streams of moving water -- could speed them on their way if they took advantage of currents moving the way they wanted to go; or they could impede their progress if the current opposed them.

Fortunately, surface currents and winds tend to move in the same direction. There are many variations from this general scheme, however, as Benjamin Franklin discovered in his conversations with his Nantucket sea captain acquaintance. The boundaries of ocean currents can be quite sharp and the wise sea captain pays close attention to their whereabouts.

This of course led to an early interest in charting ocean currents that persists to the present time. Our interest now extends not only to the currents that exist near the ocean surface, but also to those that ply the depths of the sea. A number of methods have been devised to measure ocean currents, and these may be divided into two classes: direct and indirect.

In direct measurements, the motion of the water is monitored either by a device that is fixed with respect to the ocean bottom or by a device that floats with the current. In the former, a buoy may be anchored to the bottom and the flow of water past it is measured and recorded. Given a sufficient number of such buoys (or the relocation of the buoy to a number of different sites), the measured directions of water flow can be plotted and compiled into a general picture of the extent and direction of the current.

Floating buoys go where the currents take them, and if their motion is monitored over a period of time, their path provides a direct map of at least a portion of the current. The simplest such device is a note in a sealed bottle, cast from a ship. The note directs the finder to contact the researcher with information of where and when the bottle was found. This provides rough information on where the current ends up, but little detail on how it got there. Even so, this is a very inexpensive way of gaining a broad-brush view of oceanic circulation. More sophisticated floating devices use radio transmitters to allow them to be tracked during their travels and later retrieved for reuse. They may also contain devices that permit the measurement of temperature and salinity.

The currents near the surface tell us only part of the story, however. In many places, currents at depth travel in different directions and at different speeds than those above. These may be of little interest to the sea captain, but they are of great importance to the oceanographer. One way of measuring them directly is through the use of buoys that are designed to float at a specified depth. This is accomplished by carefully adjusting the density of the float.

An object whose density is greater than that of the surrounding water will sink, while one whose density is less than that of the surrounding water will rise. Because the density of water increases with depth, all that needs to be done is to adjust the density of the float to exactly equal that of the water at the desired depth. This can be accomplished by adding weights to the interior of the sealed, hollow float. The buoy will then sink until it reaches the depth at which its density matches that of the water and there it will stay, moving horizontally with the deep currents.

There are also indirect methods for measuring ocean currents. Because of its dissolved salts, seawater conducts electricity. Whenever a conductor of electricity moves within a magnetic field, electrical currents are generated. This, in fact, is the way in which all electrical generators operate to supply our homes and cars with electricity. In a similar manner, the motion of masses of salt water in the ocean within the Earth's magnetic field also generates electical currents within the water. These may be detected by metal plates towed behind a ship and the motion of the water deduced from the currents.

Still another indirect method for measuring ocean currents utilizes the measurement of density changes within the ocean. Once the dynmaics of ocean currents is understood, maps of density variations may be used to calculate the resulting currents at depth. As you saw in the previous unit, the density of seawater is determined mostly by the salinity and temperature. The measurement of salinity and temperature profiles can be used to construct maps of density variations, which in turn serve to indicate the directions and speed of currents within the body of the ocean.

For many years, this was the principal method of measuring deep ocean currents and was accomplished through the use of a simple but ingenious device, called a Nansen bottle, for bringing seawater samples up from the depths. The Nansen bottle is lowered on a cable to the desired depth in an inverted position at which point the bottle is flipped over, trapping the water sample inside. A thermal sensor on the bottle records the temperature of the water when it is captured. The sample bottle is then retrieved and the seawater is analyzed in a chemical laboratory to determine its salinity. Because of its simplicity, reliability, and low cost, the Nansen bottle is still in use today.

<u>STUDY QUESTION</u>

VI-1. What is the difference between direct and indirect measurement of ocean currents?

E. OCEANIC CIRCULATION

Waves on the ocean surface are caused by the wind. In addition, the waves produce a rough surface on the sea upon which the wind can get a purchase. This is sufficient to set in motion the surface layers of the ocean, and provides the essential driving force for the major surface currents that are diagramed in Figure VI-11.

The connection between the winds and the currents is not so simple as it may seem, however. For one thing, the major currents such as the Gulf Stream and the Kuroshio Current off Japan contain water flows that are far too concentrated and rapid to be explained by the direct driving force of the wind.

What the winds do accomplish is the establishment of large circular vortexes, called <u>gyres</u> in the ocean basins. In the tropics, the trade winds tend to blow toward the west, while at mid-latitudes, the westerlies on average blow toward the east. Nearer the poles, the prevailing winds tend to blow once again toward the west. Note that the gyres nearest the equator rotate in a clockwise sense north of the equator and in a counterclockwise sense south of the equator, but that small gyres nearer the poles (such as the one in the north Pacific) tend to rotate in the opposite sense.

To gain further understanding of the operation of currents such as the Gulf Stream, it is necessary to understand the influence that the Earth's rotation has on the circulation of the oceans. This is called the Coriolis Effect.

1. The Coriolis Effect

The Coriolis Effect describes the tendency of any moving mass on the surface of the Earth to be deflected from its path to the right in the northern hemisphere and to the left in the southern hemisphere.

The Coriolis Effect may seem at first like some kind of mysterious force, but in fact it is simply the result of the tendency of moving bodies to want to move in straight lines. Look at Figure VI-12a, which looks down on a rotating platform from directly above. You might visualize the device as similar to the rotating platforms or merry-go-rounds that are often found in children's playgrounds. The platform is rotating in a counterclockwise direction as shown by the circular arrows.

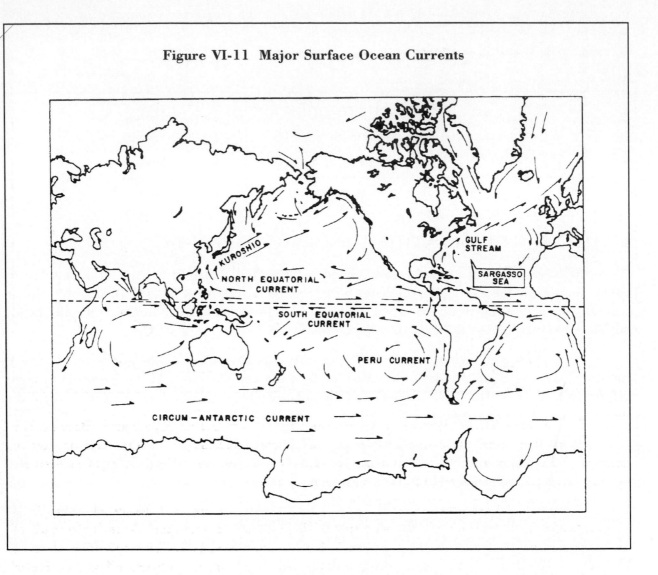

Figure VI-11 Major Surface Ocean Currents

Imagine that you are standing on point (Y) and you throw a ball directly toward a friend who is at point (F). The direction of the ball is shown by the heavy straight arrow. From the time that the ball leaves your hand it will move in a straight line. In the meantime, however, you will have moved with the platform to point Y^1 in Figure VI-12b, and your friend will have moved to point F^1. The ball will arrive where your friend used to be at point F, but now it is clearly going to miss its intended target. The ball will travel far to the right of your friend (as seen by you), in spite of the fact that you initially threw it directly toward him. From your point of view, the ball appeared to curve to the right.

From a vantage point above the rotating platform, it is clear that the ball did <u>not</u> curve to the right; both you and your friend traveled in curved paths to the left while the ball traveled straight ahead. But to someone on the platform (especially if they are not aware that they are in a rotating frame of reference), all moving objects will appear to veer off to the right of their intended paths.

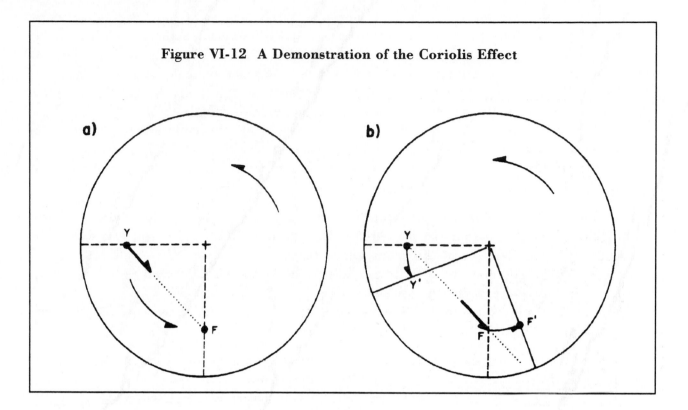

Figure VI-12 A Demonstration of the Coriolis Effect

The Coriolis Effect operates in a very similar way on the rotating Earth. Because the Earth rotates from west to east, the effect in the northern hemisphere is the same as that in Figure VI-12. A moving water or air mass appears to want to veer off to the right as seen by someone fixed on the surface of the rotating Earth.

Because the Earth rotates only once in 24 hours, we are not aware of being in a rotating frame of reference. Only the wheeling of the Sun, Moon, and stars through the vault of the heavens serves to remind us that we are living on a spinning ball. From our point of view, the Coriolis Effect acts like a curious kind of force that bends moving objects off to one side of their intended path. For this reason, it is often referred to as the "Coriolis Force".

If you understand the concept of the Coriolis Effect, you should be able to see from Figure VI-12 that it is less severe when the motion of the object is very fast compared to the speed with which the platform turns. If you throw the ball faster, it will miss your friend by a lesser amount. On the other hand, if the platform is turning very quickly and you throw the ball slowly, you will miss him by a greater margin. For this reason, the Coriolis Effect is most noticeable for slow-moving objects such as water and air masses, and less noticeable for fast-moving objects such as bullets (though it is still present to a slight extent).

To see that the Coriolis Effect works to veer moving objects to the <u>left</u> of their paths in the southern hemisphere, we would only need to reverse the direction of the rotating platform in Figure VI-12. Here is a simple exercise that you can try for yourself to see how it works. As seen from directly above the South Pole, the Earth now appears to be rotating in a clockwise direction. The Earth always rotates from west to east, but seen from the South Pole, east is to your right, while seen from the North Pole, east is to your left. Hence the appearance of an opposite sense of rotation.

Phonograph turntables rotate in a clockwise sense, and so we can use one to demonstrate the Coriolis Effect in the southern hemisphere. With a pencil, make a small hole in the center of a sheet of paper and place it on the turntable with the spindle through the hole. If your phonograph has a stationary spindle, you may have to make the hole large enough so that the paper will turn freely. Next trim the edge of the paper so it does not overlap the turntable edge -- you don't want the paper corners to strike the pickup stylus when the turntable is in motion. As a final step, you may need to tape the paper to the turntable in one or two places to keep it from slipping. Start the turntable, making certain that the phono arm does not contact the paper. If you have an automatic changer, you may have to intercept the arm as it drops toward the paper and gently return it to its rest.

Now take a marking pen, a crayon, or soft pencil, and while the turntable is moving, try to draw a straight line on the paper. That is, move the pen in a straight line across the paper. Do this in a variety of directions: move the pen toward the left, toward the right, toward the spindle and away from it. Try to note which is the beginning and end of each line. Stop the turntable and examine the marks on the paper. If you have trouble determining which is the beginning and which is the end of each mark, the beginning probably starts more abruptly, with the mark disappearing more gradually at its end.

Remove the paper from the turntable and put an arrow head on each line showing the direction that the pen moved, pointing from the beginning toward the end of the line. Note that in every case, the line bends toward the left, corresponding to the Coriolis Effect in the southern hemisphere.

2. The Gulf Stream and Other Major Ocean Currents

Now let us see how the Coriolis Effect influences the ocean currents. Let us consider the case where surface winds are blowing to the north in the northern hemisphere. Surface waters will be dragged along, but will be deflected to the right (east) by the Coriolis Force. Water at progressively greater depths is dragged along by the layer immediately above. But the Coriolis Force always acts to the right of the direction of motion, so water is progressively deflected farther and farther to the right as we go to greater depth. In fact, at some point, we will find that the water at depth is flowing in the opposite direction to that of the wind. Due to frictional losses, however, the return flow is much less than the forward flow near the surface. Even so, the net flow of water in the moving layer turns out to be to the right at roughly a right angle to the wind direction.

The layer of water that is set in motion is called the <u>Eckman layer</u>, after the Swedish oceanographer who described it around the turn of the century, and the net flow of water to the right of the surface winds is called the Ekman-layer flow. Its action greatly complicates the motion of ocean water in response to the driving forces of the persistent winds. We shall only outline the process here without going into much detail.

We may use the North Atlantic gyre as an example, which is shown schematically in Figure VI-13.

To proceed farther, you need to understand what is meant by the <u>horizontal component of the Earth's rotation</u>. This refers to the extent to which a horizontal plane somewhere on the

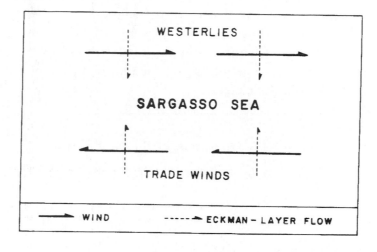

Figure VI-13 Converging Water in the Center of a Gyre

Because Eckman-layer flow always moves to the right of the wind direction, the prevailing winds cause a convergence and downweling in the Sargasso Sea in the North Atlantic gyre.

Earth's surface rotates about an axis that is perpendicular to the plane. You may think that a plane anywhere on Earth is rotating along with the Earth, and that is true, but at most places, only a portion of that rotation is contained within the horizontal plane. If you are standing on a horizontal plane at the North Pole, then it is easy to see that the plane is rotating horizontally, like a Frisbee in flight. If you are standing on a horizontal plane on the equator, you and the plane both are traveling with the rest of the Earth in a circle, but the plane of that circle is perpendicular to the plane upon which you are standing. For this reason, there is no horizontal component of rotation of the plane.

This, by the way, explains why the Coriolis Effect disappears at the equator. If the surface of the Earth is not rotating horizontally there, then there will be no tendency for a moving object to veer off to either the right or the left. As we travel from the equator toward either pole, the horizontal component of rotation increases and so does the Coriolis Effect. Take a moment and try to picture how this works in your mind's eye. Go back over the preceding two paragraphs if necessary.

As a result of this effect, any body of water that is traveling with the Earth has a particular rate of horizontal rotation that depends on its latitude, with the rate increasing from zero at the equator to a maximum at the poles. In the northern hemisphere, that rotation is counterclockwise, the same as that of the Earth itself.

A device on display in many science museums demonstrates the effect directly. A long pendulum (usually several stories high) is set in motion along a particular line in the morning. As the day goes on, the line along which the pendulum is observed to swing appears to rotate clockwise (in the northern hemisphere) with respect to the floor of the building. Actually, the line of the pendulum's swing is staying as fixed as it can, and the floor of the building is rotating counterclockwise beneath it. The device, called a <u>Foucault pendulum</u>, would be pointless to install in a museum on the equator, since it would show no rotation at all.

Now let us return to the water that is sinking beneath the center of the North Atlantic gyre. That water shares the horizontal component of the Earth's rotation that is appropriate to its latitude. As it sinks and spreads out, however, the rotation is slowed, just as a twirling ice skater slows her spinning by extending her arms outward. Because of this, the water mass is now rotating slower than the ocean floor below it. As a result, the body of water migrates south to where its slowed rotation rate fits that of the solid Earth. Recall that the horizontal rotation rate deceases as you move toward the equator.

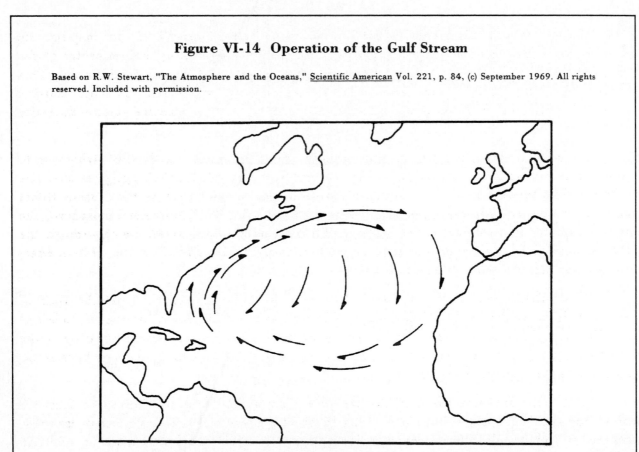

Figure VI-14 Operation of the Gulf Stream

This effect causes the entire center of the gyre to slowly move toward the equator (Figure VI-14). To avoid emptying the northern oceans, however, this southward flow must be balanced somewhere by an equal northward flow. Any water moving <u>away</u> from the equator, however, must increase its counterclockwise rotation rate in order to stay in sync with that of the solid Earth. This can be accomplished by water moving north along the east coast of North America, where friction with the coast to its left provides the necessary torque to give the moving water its necessary counterclockwise spin.

At last we have the reason for the narrow and powerful Gulf Stream. All the water that has moved south throughout the entire area of the North Atlantic gyre must now be funneled along the coast of North America in order to provide it with the additional rotation that it needs as it moves north.

The explanation is neither obvious nor easy to understand, so we shall summarize its most critical features: The prevailing winds create a vorticity or gyre that, combined with the Coriolis Effect, creates an Eckman-layer flow that causes downwelling in the center of the gyre. The descending water spreads out at depth and slows its rotation, causing the water to move toward the equator. The return path to the north is taken along the western margin of the ocean basin, where friction with the continent slows the western edge of the current. This imparts to the water the additional counterclockwise rotation that it needs in order to move north.

The last step in this process is not at all obvious from Figure VI-14, but may become clearer if you hold the page between thumb and forefinger at a point at the top center of the page (directly above Figure VI-14) and rotate the page counterclockwise about this point. This is roughly how the Earth rotates, and perhaps this will help you to see why water moving toward the north (that is, toward your fingers) will need to pick up additional counterclockwise rotation.

The same western intensification of ocean currents occurs in the Pacific, giving rise to the powerful Kuroshio Current off Japan. In the southern hemisphere, the large gyres circulate in the opposite sense, that is, in a counterclockwise direction. But because the Coriolis Effect and the resulting Eckman-layer flow are directed to the left in the southern hemisphere, the convergence and downwelling in the center of the gyre occurs once again. Going through the whole process again, we would find that the current would be intensified on the western edge, just as in the case for the northern hemisphere.

Of all these great currents, the Gulf Stream is the best known. From the Gulf of Mexico, it flows northeasterly off the coasts of Georgia and the Carolinas, casting out to sea at Cape Hatteras. Because the Gulf Stream carries warm tropical waters, beaches south of Cape Hatteras are noted for their warm water, but beaches just to the north of the cape have water temperatures that are distinctly cooler.

From Cape Hatteras the Gulf Stream sweeps across the Atlantic Ocean and fans out against Europe. Perhaps you have noted that much of Europe stands at rather high latitudes compared to comparable climatic zones in North America or Asia. London is farther north than the northern tip of Newfoundland in Canada; northern Scotland stands at the same latitude as Juneau, Alaska; and Scandinavia shares the same latitude range as Siberia. The habitability of the northern tier of western Europe may well be due to the Gulf Stream, bringing warm water from the tropics to its shores. A competing theory, however, holds that cold currents moving south through the Denmark Strait between Iceland and Greenland create a situation in which warm water from the Mediterranean is drawn to the north, along the coasts of Spain, France, and England. Whichever is the case, northern Europeans can thank the prevailing patterns of oceanic circulation for a hospitable climate.

Traveling at a speed of a few kilometers per hour (as noted by Franklin's sea captain friend), the Gulf Stream is 50 to 75 kilometers wide but less than two kilometers deep. Even so, the volume of its flow is more than 100 times the combined flow of all the rivers of the world.

The Sargasso Sea, in the center of the Gulf Stream gyre, results from the downwelling waters described earlier. It got its name from masses of floating seaweed that accumulate over the converging currents. Early seamen told stories of ships that became hopelessly entangled in the seaweed, but in fact the ships were merely becalmed in the dead center of the gyre, between the regions of east- and west-flowing prevailing winds.

3. Upwelling and Downwelling Currents

Compared to the horizontal motions of the surface and deep currents, vertical motion of seawater is surprisingly restrained. When we recall from the previous unit that a scale model of the oceans would resemble a sheet of paper, this seems even more curious. This behavior results from a stable arrangement in which the stratification or layering of the ocean is based on density, with the densest layers on the bottom. We have already seen earlier that the bottom waters are cold and saline, making them very dense. In most places, it would require an enormous amount of energy to raise these dense waters to the surface. In some places vertical mixing does occur, however, and these upwelling and downwelling currents have a number of important effects.

We have already discussed the downwelling that occurs in the center of the large oceanic gyres. Similarly, upwelling occurs in the center of the smaller gyres that go in the opposite direction, such as the Alaskan gyre in the northern Pacific Ocean (see Figure VI-11).

Along shorelines, we find that the Coriolis Effect and the Eckman-layer flow are important influences. Consider what can happen in the northern hemisphere when a persistent wind blows parallel to a shoreline, as shown in Figure VI-15. In the left diagram, the action of the wind is to force the surface water against the shore where it piles up and sinks, producing a downwelling current. In the right diagram, an upwelling current is formed.

It turns out to be possible to measure the residence time of water in the deeper layers of the ocean. The method used is radiocarbon dating, a radioisotope technique similar to those discussed in the unit **A Sense of Time**. The radioactive parent isotope is Carbon-14 (C^{14}), which decays with a half-life of 5730 years. Carbon-14 is produced in the upper atmosphere by the bombardment of cosmic rays from outer space. The carbon dioxide in Earth's atmosphere is in equilibrium with respect to Carbon-14, with just as many being formed as are decaying in a given unit of time. Hence the concentration of Carbon-14 in atmospheric carbon dioxide (CO_2) is essentially constant. The most common isotope of carbon is Carbon-12, which is not radioactive. It serves as a standard against which to measure the concentration of Carbon-14.

So long as the CO_2 gas is in the well-mixed atmosphere, its C^{14}/C^{12} ratio remains constant. As C^{14} atoms decay, they are replaced by new ones created in the upper atmosphere. Atmospheric CO_2 gas is dissolved in the ocean's surface waters, keeping the ratio in the upper layer nearly what it is in the atmosphere. However, once the water sinks into the deeper layers, it is cut off from the new supply of Carbon-14 in the atmosphere, and the C^{14}/C^{12} ratio

Figure VI-15 Downwelling and Upwelling Currents at a Shoreline

The situation shown is for the northern hemisphere. In the Southern Hemisphere, reverse the wind directions.

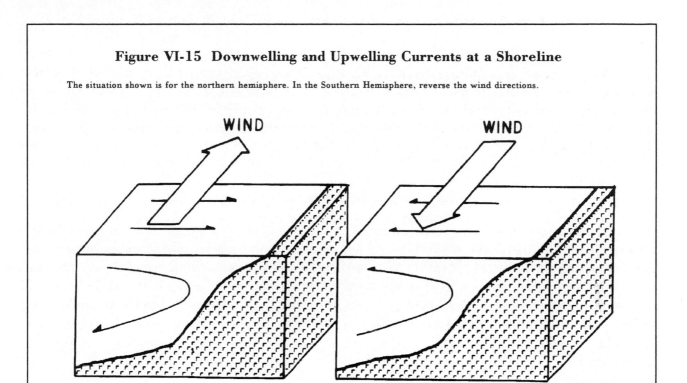

begins to decrease. In this way, it is possible to determine how long water has been away from the surface layer.

Radiocarbon dating of CO_2 in deep water yields ages on the order of 1000 years. This, then, gives us an indication of the time necessary for surface waters to circulate through the deep waters and to find their way back to the surface once again.

STUDY QUESTIONS

VI-2. In the southern hemisphere, a wind blows to the north along the west coast of a continent. Would this produce upwelling or downwelling?

VI-3. You are on a merry-go-round that is turning in a counterclockwise direction, and attempt to throw a ball to a friend that is riding on the horse just behind you. Which way will the ball appear to be deflected as seen by you?

VI-4. Can you see from Figure VI-11 on page 183 why the water temperature at the beaches of San Francisco is generally colder than at the beaches of New Jersey, though they are both at the same latitude?

F. CLIMATIC EFFECTS

1. Interactions Between the Oceans and the Atmosphere

In the previous unit we noted that because of its polarized nature, the water molecule has a number of unusual properties. One of these is that it has a very high heat capacity. <u>Heat capacity</u> is the amount of heat that must be put into one gram of a material in order to raise its temperature by one degree Celsius. Similarly, we might think of it as the amount of heat that escapes from one gram of a material when it cools by one degree Celsius. As its name implies, the heat capacity is a measure of the ability of a material to store heat when its temperature is raised and to release that same amount of heat when it cools.

Because of its high heat capacity, the water in the world oceans is an enormous reservoir for thermal energy. It serves to temper world climate, absorbing heat from the Sun in the tropics and redistributing it to more northerly regions via the great ocean currents. For this reason, the Gulf Stream is able to bring its warmth thousands of kilometers from the tropics to the European shores before cooling to the temperatures of Arctic waters.

Contrariwise, regions in warm latitudes that are visited by cold ocean currents enjoy a natural form of air conditioning. The delightful climate of the seacoast of California is due to this effect. The circum-Antarctic current (see Figure VI-11) on page 183 travels from west to east, completely encircling the continent of Antarctica with a cold current and forming a partial barrier to the transfer of heat from lower latitudes to the frigid Antarctic landmass.

A full discussion of the influence of the oceans on world climate must wait until we have learned something of the workings of the atmosphere as well: we will do this in the next unit. For now, we look at one particular case where the nature of the ocean currents is of prime importance in determining climatic impact.

2. El Niño

The El Niño phenomenon occurs in the Pacific Ocean and usually its effects are felt most noticeably at the coasts of Ecuador and Peru in South America. In Spanish, the name means "The Child", an allusion to the Christ child. It is usually around Christmastime that fisherman notice the warming of the ocean surface that is its first sign.

Historically, Ecuadorians and Chileans welcomed El Niño. Its altered patterns of ocean currents and winds brought them gifts from the tropics such as floating coconuts. It was only after the fishing industry became highly developed in the 1940's and 1950's that the occurrence of El Niño took on a more sinister connotation.

The coastal waters off Ecuador and Peru are normally the site of an upwelling of deep bottom waters, bringing cold water to the surface. When these are replaced by warm waters during an El Niño event, the local anchovy fishing industry is dealt a massive blow as catches decline to near-vanishing levels. More important than this local effect are the the widespread climatic changes that can accompany a particularly severe El Niño. Such was the case with the Child's visit in late 1982. In its wake torrential rains and flooding struck desert areas; parching droughts visited parts of every continent save Europe and Antarctica; tropical cyclones or hurricanes were spawned in record numbers in the Pacific; whole populations of sea birds were decimated; and even coral colonies on the sea floor were ravaged by the altered conditions. In a single event, a substantial number of the world's ecosystems were seriously disturbed.

To understand what happens during an El Niño event, it is necessary first to understand the normal situation just south of the equator in the Pacific Basin. Figure VI-16 presents a schematic view of the most important elements. The prevailing winds off the coast of South America are from the southeast. An upwelling of cold deep ocean water is produced by the mechanism diagramed in Figure VI-15, but note that the effect is reversed here. Off the coast of Peru we are in the southern hemisphere, and so a wind blowing from the south creates an offshore current and an upwelling, just as a wind blowing from the north would in the northern hemisphere. Don't forget that the Coriolis Effect works to deflect moving objects toward the <u>left</u> in the southern hemisphere.

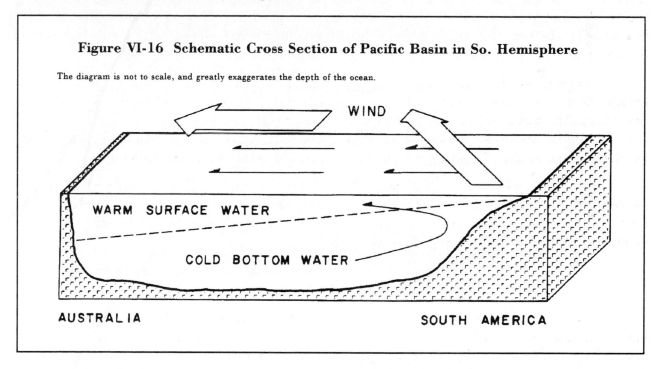

Figure VI-16 Schematic Cross Section of Pacific Basin in So. Hemisphere

The diagram is not to scale, and greatly exaggerates the depth of the ocean.

WIND

WARM SURFACE WATER

COLD BOTTOM WATER

AUSTRALIA   SOUTH AMERICA

In the previous unit, we mentioned that ocean bottom water is rich in dissolved oxygen and in nutrients such as phosphates, derived from decaying organic matter that has fallen from the surface layers above. Because of the relative scarcity of marine life at these depths, these life-sustaining resources are not consumed, and remain in the bottom waters. When upwelling occurs, however, the introduction of nutrient-rich water into the sunlit surface zone produces an explosion of life.

The process begins with <u>plankton</u>, marine organisms that simply float with the water and are not able to swim rapidly. Plankton range in size and variety from microscopic plants to jellyfish, and they form the base of the food web in the ocean. In the upwelling off Peru, anchovies are among the principal beneficiaries of planktonic abundance. The Peruvian fishing industries are based on the anchovy, with much of the catch going toward the production of high-protein fish meal. A portion of it, of course, goes to supply pizza-lovers all over the world.

At these latitudes, we are not far from the equator, and the Coriolis Effect is fairly weak. In fact, the Coriolis Effect goes to zero at the equator, because this is where the tendency to veer to one side switches from right to left. The westward movement of surface waters is aided substantially by the presence of the Trade Winds, which blow to the west across the entire width of the Pacific Ocean. The persistence of these winds causes warm surface water to pile up against the western edge of the basin, greatly thickening the surface layer there.

What we have drawn is a picture of the "normal" behavior of the ocean in this region, extraordinary though it may seem. The onset of El Niño is caused by a faltering of the persistent winds that drive the process. Once this happens, upwelling of cold water off the coast of South America ceases and the slanting dotted line in Figure VI-16 is no longer a stable configuration. Winds at the equator relax or blow eastward and surface currents actually reverse direction as the warm water piled up in the western part of the basin flows back toward South America. The dotted line separating warm surface waters from cold bottom waters becomes more horizontal, greatly thickening the surface water layer off the coast of Peru. In the process, ocean surface temperatures in the east may rise by as much as 8°C (14°F).

With a thick layer of warm water near South America, upwelling is contained entirely within the warm water. Nutrients from the bottom water no longer make it to the surface and the food web collapses. Each time a severe El Niño strikes, the fishing industry of Peru is crippled.

El Niño's effects can range far wider, however. During the 1982-83 occurrence, ocean surface temperatures were elevated in a band nearly a thousand kilometers wide stretching across much of the Pacific Ocean. The climatic repercussions were many and severe. The warm surface waters gave rise to heavy rains in Ecuador and the entire central Pacific region. Arid regions that normally see only centimeters of precipitation were deluged by more than three meters (nine feet) of rainfall, leading to widespread flooding and landslides. Thousands of homes were destroyed in Ecuador, and damages to crops and property amounted to more than 400 million dollars in a nation that could ill afford it.

Farther from the source, climatic effects were felt in many parts of the world. Flooding in the lower Mississippi valley was laid at El Niño's door. So too, were violent storms that brought a combination of flooding and severe wave damage to the west coast of the United States.

During El Niño, the tropical rain belt shifts from the eastern Pacific to the central Pacific. Thus, severe droughts afflicted portions of Australia, Indonesia, India, southern Africa, and Central America. Australia was particularly hard hit. The worst drought of the century turned much of eastern Australia to tinder and bushfires raged in many places, killing 75 people. Gigantic dust storms moved millions of tons of topsoil and the toll of livestock was staggering. Relief, when it finally came, was excessive. Heavy rains brought floods, severe soil erosion, and further decimation of the weakened herds.

In the Pacific, Hawaii's garden isle of Kauai was pummeled by a rare hurricane, and six major storms in five months raged through French Polynesia, which had not seen a single hurricane in the previous 75 years. Yet, surprisingly, it would appear that hurricane formation in the Caribbean region was actually suppressed. It is an ill wind that blows no good, and El Niño was no exception. Even as the anchovy population of the South American seas was being ravaged, other species moved in and flourished in the altered conditions. Shrimp, tuna, bonita, and dolphins multiplied and expanded their ranges, while on land deserts burst into unaccustomed bloom. But there were many trade-offs. Vast areas of coral reefs perished in the soaring ocean temperatures, leaving only bleached and lifeless remains. In many places sea birds virtually disappeared as populations of fish upon which they fed shifted.

Life on Earth is extremely adaptable. When the environment changes abruptly, some species suffer and others prosper, taking over altered and vacated ecological niches. Mankind, too, has adapted his lifestyle and survival techniques to individual environments, and sudden change is almost always a hardship. Disruption of the status quo, whether done at the hands of man or by the whims of nature, more often than not brings disaster in its wake. By the time El Niño had spent its fury, the toll stood at 1,100 dead and nearly nine billion dollars in damage.

What causes El Niño and when can it be expected? On average, they occur every four or five years, though not with any regular pattern. The previous one occurred in 1976-77 and, though not so severe, brought to the eastern United States a record cold winter and to California its worst drought. Clearly its effects are not always the same, since the 1982-83 visit brought the California coast storms and deluges.

The exact cause of El Niño is still not known. An immediate precursor to El Niño is a shift of a major persistent low-pressure cell normally found over northern Australia and Indonesia eastward into the central Pacific, a phenomenon called the Southern Oscillation. Its onset is apparently tied to the shifting of the monsoons of the Indian Ocean. There the winds and ocean currents change direction twice each year, alternately bringing drought and torrential rains to India.

The 1982-83 El Niño, perhaps the most severe of this century, prompted scientific investigations of international scope. The hope is to determine its causes sufficently to allow prediction of its occurrence and severity. A major ten-year program has been launched under the acronym TOGA (Tropical Ocean and Global Atmosphere). Should another El Niño occur during this period, it will be monitored by environmental satellites, research vessels, and a variety of instruments designed to sense oceanic and atmospheric conditions.

STUDY QUESTIONS

VI-5. Why do the oceans serve to temper world climate?

VI-6. Why would a strong El Niño affect global climate?

VI-7. What are some of the climatic effects of a strong El Niño?

G. THE DRYING OF THE MEDITERRANEAN

As we learn more about the present-day oceans and their workings, it seems only natural to inquire about their ancient history. Plate tectonic theory assigns a very mobile nature to the ocean floor; continental drift implies that the very shape of the ocean basins has changed throughout geologic time. The reconstructions in the unit **Plate Tectonics** allow us to see what shapes the oceans have had at various times; our knowledge of the principles of ocean dynamics should allow us to reconstruct the ancient current systems within the oceans. This should also help us to model ancient climates as well, and we will look at efforts to do so later on, in the unit **Climates of Earth**.

In a few cases we do not need to rely on theoretical models in order to determine the ancient state of the oceans. Direct evidence tells us of radically different oceanic conditions in the geological past. The Mediterranean Sea is a particularly interesting example.

Not every "sea" is an ocean. The North Sea, for instance, is a shallow body of water that covers continental rather than oceanic crust. Its waters communicate with the Arctic and Atlantic Oceans, but it seldom reaches depths greater than 100 meters (330 ft). The fact that oil has been discovered beneath its waters is a sure indication that we are dealing with continental and not oceanic crust. In essence, the North Sea is flooded continental shelf on Europe's northern border.

The Mediterranean Sea, on the other hand, is a genuine ocean basin. Its waters reach depths as great as 4,300 meters (14,000 ft), indicating that the crust beneath it is thin and oceanic in nature. Take a moment now and look at its situation and outline on the map in the **Geographical Review** unit on page 44. Disregarding the presence of the Suez Canal, note that its sole communication with the world oceans lies through the narrow Strait of Gibraltar, separating Spain from Morocco by only 15 kilometers (9 mi).

The Mediterranean is a very warm sea, with a hot and sunny climate. As a result, evaporation is very high. Even though a number of major rivers such as the Rhone in France and the Nile in Egypt pour fresh water into it, more water is evaporated than the rivers can supply and the Mediterranean is consequently very salty. At the Strait of Gibraltar there is a net inflow of Atlantic Ocean water to replace the evaporative losses in the Mediterranean Basin. Actually, surface water flows into the Mediterranean, but dense saline waters flow out at depth into the Atlantic.

When the Aswan High Dam was being built on the Nile River, boreholes were sunk to determine the nature of the foundation for the dam. To the astonishment of the engineers, it was found that there was a deep canyon cut into the bedrock that was now filled almost completely with sediment (Figure VI-17). In addition, the lower portion of the sediments contained fossils of marine (sea-dwelling) organisms. When this discovery was first made, no explanation was known, and the mystery joined that of a similar deep gorge cut in bedrock below the Rhone River in France.

The mystery was solved when oceanographic research discovered that the floor of the Mediterranean held thick layers of sediment containing salt and gypsum deposits up to one kilometer thick. The only way in which these deposits could form is by the evaporation of seawater -- hence their generic name evaporite. To form evaporite beds of such thickness, it

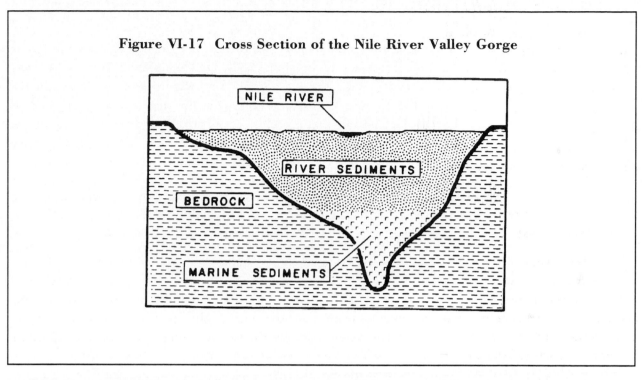

Figure VI-17 Cross Section of the Nile River Valley Gorge

would have required the evaporation of the entire Mediterranean Sea. Earlier research based on sudden changes in the fossil record and on geological indications of a sudden change in local climate had hinted at such a possibility. Now proof was at hand and the great canyons of the Nile and Rhone Rivers could be explained.

The African Plate is gradually moving to the north, encroaching on the European Plate. The Mediterranean Sea is a closing ocean, perhaps one of the last remnants of the ancient oceans that predated Pangea. Its very name means "sea in the middle of land", and with a lower-case "m" the term might well be applied to other oceanic fragments like the Black and Caspian Seas. Approximately eight million years ago, the jostling between the African and European Plates apparently folded up a ridge of land at the western end of the Mediterranean Basin, blocking the Strait of Gibraltar. Cut off from its supply of ocean water, the Mediterranean would have dried up in only a thousand years, leaving behind a bed of evaporites on its floor.

During its time, this great dried-up ocean basin would have presented a geological spectacle not to be seen anywhere on Earth today. With its floor standing five kilometers (3 mi) below sea level, it would have been three times the depth of the Grand Canyon of Arizona and vastly larger in area. From many places on its "rim" (the present shoreline), the opposite side could not be seen, and temperatures well in excess of those found in Death Valley would have presented a formidable barrier to travel by any but the hardiest of creatures.

Rivers such as the Nile and Rhone that emptied into the basin would do so in spectacular waterfalls, their waters spreading out and evaporating completely on the baking ocean floor. A huge waterfall of this type would have great powers of erosion, and it would soon migrate upriver, carving a deep canyon in the riverbed as it went, much as Niagara Falls is cutting a gorge as it retreats up the Niagara River. As far upstream as the Aswan Dam site, the bottom of the Nile gorge cut in solid granite stands at 210 meters (700 ft) below sea level.

About 5.5 million years ago, the great dam at Gibraltar was breached. This may have been accomplished by the normal processes of erosion in the newly uplifted land, or it may have been aided by earthquake activity along the plate boundary, which here cuts through the area of the strait. Whatever the cause, once a flow of water from the ocean was established, it quickly became a torrent. For a brief period this grandest of waterfalls fed the refilling of the Mediterranean, a process that may have lasted only a hundred years. Ten kilometers (6 mi) or more in width, with a flow a thousand times that of Niagara, it should have presented a truly impressive view. One wonders whether any ape-like ancestors to Homo sapiens that might have come across the scene had yet developed the capacity to appreciate it in an aesthetic sense.

H. OCEANOGRAPHY FROM SPACE

Consider for a moment the relative human population densities for the land and the ocean. Except within the most heavily traveled trade routes, a ship can travel for days and not see another vessel clear out to an unobstructed horizon. In the open oceans, the major shipping lanes are concentrated in a relatively narrow band of latitudes from N 30° to N 60°, and so it is only within these zones that frequent observations of phenomena such as wind and current speed can be carried out more or less continuously.

On the other hand, a single orbiting satellite may make several trips around the world each day, logging almost continuous records of a wide variety of observations. Seasat, the first satellite designed specifically for ocean measurements, failed prematurely after only 100 days. But during that time it produced a wealth of observations that have opened up whole new areas of study. For example, during its 100 days of operation, Seasat made approximately as many measurements of wind speed and direction as during the previous century of shipboard observations.

Another consideration has made satellite observation timely. In 1983 an international Law of the Sea Treaty was drawn up and signed by a number of nations (though not by the United States). The treaty restricts access to many coastal waters, a situation that ship-bound oceanographers find distressing. Satellite observations can help to alleviate this loss to some extent.

Oceanography from space developed somewhat indirectly at first, utilizing instruments on board satellites that were designed primarily for other purposes. Results from Landsat (land resources), Nimbus (weather), and the manned Skylab (space habitability and general science) satellites showed the potential for space-based ocean studies and led to the brief but highly successful Seasat mission.

Satellites are able to do far more than just take photographs of the ocean surface. Let us look briefly at some of the types of measurements that can be made from space. The temperature of the surface waters can be monitored by measuring radiation emitted from the ocean surface. All bodies emit electromagnetic radiation at wavelengths that depend on the temperature of the body. Measurements taken of radiation ranging from the infrared to the microwave region (see Figure VII-3 on page 218) make it possible to determine the ocean surface temperature to an accuracy of better than 1°C (2°F).

Photographs and measurements taken in visible light show that the color of the sea surface conveys quite a bit of useful information. Clear deep water appears as a dark blue, while murky waters appear lighter. The presence of phytoplankton lends a distinct greenish cast to the ocean surface. Phytoplankton are microscopic marine plants that are the starting point for much of the oceanic food web; they are so incredibly numerous that they alone account for half of total global photosynthesis. By mapping the worldwide abundances of phytoplankton, we can obtain a global perspective of biological productivity in the oceans.

One of the most useful and versatile measurements that can be made from space is that of the exact distance from the satellite to the ocean surface. This is accomplished by using radar or laser ranging, in which a pulse of radiation is sent down from the satellite and bounced off the ocean surface. The time taken for the roundtrip from satellite to ocean and back again is measured and this allows the distance to be determined to within 10 - 20 centimeters (4 - 8 inches). In the previous unit you learned that the sea surface mounds up over areas of high gravity, reflecting the presence of extra mass on the sea floor. In this way, massive structures on the sea floor such as volcanoes, plateaus, and the continental shelf can be seen clearly in maps that show the elevation of the sea surface.

Figure VI-18 shows the kind of detail that can be obtained by this method, where the edge of the continental shelf off the east coast of the United States and the undersea mountains supporting Bermuda show up as places where the contour lines appear very close together. This represents a relatively steep slope in the ocean surface.

When you stand on a beach and look out to the ocean on a calm day, you may find it hard to believe that the surface of the ocean is anything other than flat. Yet, from your first understanding of what a globe represents, you have known that the world is roughly spherical in shape, and that the ocean surface must share that shape. Now you can take your understanding a step further with the realization that the sea surface also contains undulations that reflect the mass distribution on the ocean floor. These undulations are not great, generally amounting to no more than a few centimeters of elevation change per kilometer of horizontal distance. Even so, in the roughly 2,400 kilometer (1,500 mi) horizontal distance shown in Figure VI-18, the total elevation change of the ocean surface amounts to more than 50 meters (160 ft).

The shape of the Earth as defined by the mean ocean surface is called the geoid. The shape of the geoid is determined by the distribution of mass within the Earth along with its rotation, and the gentle undulations that we have been discussing are determined mostly by the structure of the oceanic crust. There is, however, another but smaller effect on the elevation of the ocean surface and this is due to the currents. Currents produce bulges in the sea surface on the order of one meter (three ft), and the measurement of these small distortions allows oceanographers to deduce the presence of currents.

Because this effect is nearly 100 times smaller than that due to the geoid, the contribution of the geoid must be subtracted from the total measurement before the effect of the current can be seen. Fortunately, many ocean currents change with time, while the geoid does not, and this helps to separate the two effects.

Figure VI-18 Average Elevation of Sea Surface in NW Atlantic Ocean

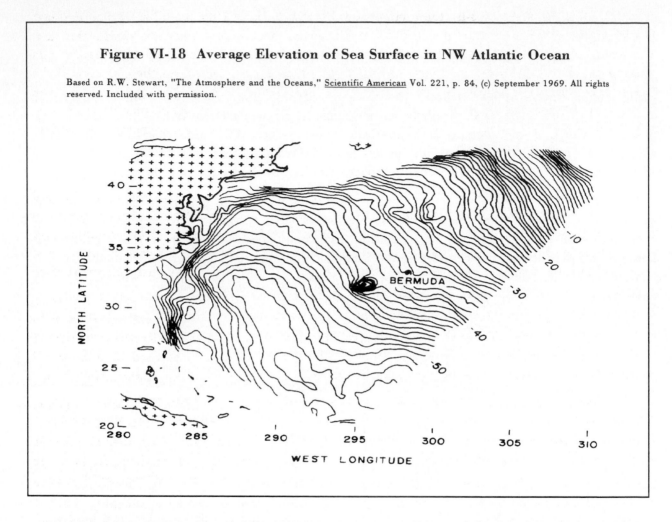

Wave height may also be measured by radar ranging, providing information of use to mariners. Equally useful are satellite measurements of wind speed, obtained indirectly by measuring the roughness of the sea. The tracking of cloud movement with time also allows measurement of wind speed and direction, though not directly at the sea surface.

In views from space, ice floating in the water reflects sunlight brilliantly and stands out starkly against the darker open ocean water. Measuring the extent of sea ice is useful not only for the tracking of icebergs that might be hazardous to shipping, but as a climatic influence and as a sensitive indicator of global warming. Having available up-to-date maps of sea ice coverage is an important new aid to any ships that venture into the polar regions.

The availability of satellite-based oceanographic observations has raised a formidable problem for the modern oceanographer: that of dealing with all the data. The voyage of H.M.S. Challenger cited at the beginning of the previous unit may put things into perspective. The results of that single voyage required 20 years to analyze and publish. Now we have satellites taking measurements continuously, filling the airwaves with data radioed back to laboratories on Earth. Stored on magnetic tape in digital form, the data are available for scientists of many nations to study and interpret. Without the aid of high-speed electronic computers, the task would be an impossible one.

Within the next decade, the quality and quantity of data from satellites should increase substantially. Four new satellites are on the drawing boards for launches within the next decade. The satellite NROSS will be designed to measure surface winds and temperature and the topography of the polar ice sheets. TOPEX will refine measurements of the elevation of the sea surface, hopefully providing more accurate measurements of ocean currents. OCI will look at ocean color and map distributions of chlorophyll and the all-important plankton. Finally, GRM will map the geoid in great detail, providing a foundation for the interpretation of data from the TOPEX mission.

Practical applications for all this data are easy to find. We have already mentioned several examples of aid to shipping and yachting, and the television series highlights critical aid given by the United States National Oceanic and Atmospheric Administration to participants in the Marion, Massachusetts to Bermuda yacht race as a colorful example. More mundane but also more important, the day-to-day operations of the merchant marine and armed forces naval vessels are influenced by the availability of up-to-date information on currents and sea surface conditions.

The view from space of plankton distribution provides new information of critical concern to the fishing industry. Waters that are rich in plankton are also likely to be rich in fish. Combined with data on seawater temperature, plankton maps can be used by commercial fishing fleets to increase the efficiency of their operations, with a significant reduction of costs.

Perhaps the greatest beneficiary of oceanography by satellite will be climatic studies. The ability to predict oceanic changes such as El Niño would pay handsome dividends in lives saved and property damage averted. The prediction of long- and short-term weather also stands to gain much from satellite observations of the oceans because the interactions between the oceans and the atmosphere are so important in understanding the causes of weather. We shall return to this subject in the next unit.

Not all the observations from space are being made by robot scanners. Paul Scully-Power became the first oceanographer to make observations from space in 1984 when he rode the space shuttle Challenger into orbit and observed the world's oceans through the cockpit windows. Among the features that he reported seeing were intricate circular eddies in the Mediterranean and elsewhere. During the past few years, large rotating bodies of water have captured the interest of oceanographers as major features of the world oceans. Because they are large and yet occur on a scale smaller than that of the great oceanic gyres that occupy most of the major ocean basins, these are called mesoscale eddies, or middle-scale eddies.

The Gulf Stream is a prolific producer of these eddies, and most studies so far have been concentrated in the northeastern Atlantic. Though their existence has been known for 40 years, detailed studies have received considerable impetus in recent years because of the large-scale view afforded by satellite observations in the infrared, which clearly show eddies. Figure VI-19 diagrams the creation of eddies on both sides of the Gulf Stream.

In (a), as the Gulf Stream pulls away from the continental shelf and moves to the northeast across the Atlantic, it sometimes develops a meander that intensifies. The loop can become cut off (b) and take off on its own (c) as a ring current. Note that in (a) and (b) the core of the ring consists of water that has been brought over from the other side of the Gulf Stream. In this part of the Atlantic the water on the continental slope is very cold, while the

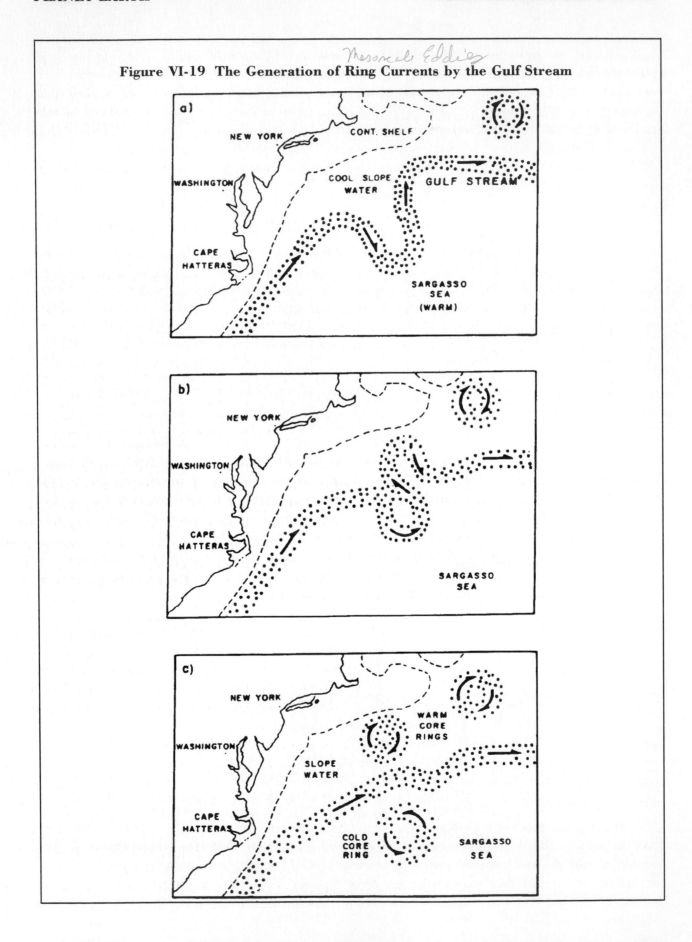

Figure VI-19 The Generation of Ring Currents by the Gulf Stream

water to the east of the Gulf Stream is part of the warm Sargasso Sea. As a result, the eddies consist of three distinctly different waters. In the case of the ring that appears south of the Gulf Stream in (c) the core of the ring consists of cold slope water, hence it is called a <u>cold-core ring</u>. The ring current itself is water that previously flowed in the Gulf Stream, while outside the ring current are the warm waters of the Sargasso Sea.

Now consider the ring that broke off to the north of the Gulf Stream. Its core waters came from the Sargasso Sea and are warm, hence it is called a <u>warm-core ring</u>. Once again, it is surrounded by a ring current of Gulf Stream water. Note that because of the way that they form, warm-core rings rotate in a clockwise direction, while cold-core rings rotate counterclockwise.

Cold-core rings range from 150 - 300 km (95 - 190 mi) across, while warm-core rings tend to be a bit smaller. One of these rings may rotate once every two to five days, and its total lifetime may range from months to a year or two. The temperature difference between the center of a cold-core ring and the surrounding Sargasso Sea can be substantial, amounting to 10°C (18°F). Interestingly, the core of a cold-core ring extends right down to the ocean floor, while that of a warm-core ring extends less than halfway to the ocean floor. Each year, between five and eight rings are produced on either side of the Gulf Stream.

Marine biologists have found the mesoscale eddies to be ideal natural laboratories in which to study the effects of environmental changes on complex ecosystems. The cold slope water contains a high oxygen level and many nutrients, supporting abundant phytoplankton and a large population of organisms, though the total number of species represented is not great. On the other hand, the Sargasso Sea supports a great diversity of tropical species, but because its waters are low in oxygen and nutrients, the total biomass is much lower.

When a ring current is formed, populations of organisms from the one ecosystem are transported into the other and cut off from their own. For a while, they feel no ill effects, since the core of the ring current serves as a closed aquarium preserving the original ecosystem on the other side of the Gulf Stream. But as the ring current weakens and begins to die, mixing occurs and the ecosystem begins to change. One change that was noted in the cold-core rings was that as warm water began to invade, cold-water species of a shrimp-like crustacean migrated to greater depths in an attempt to find colder water. Eventually, however, the trapped population of cold-water organisms began to starve in the nutrient-poor water of the Sargasso Sea.

Mesoscale eddies undoubtedly exist in conjunction with the other great currents of the world. It has been estimated that a sizeable fraction of the total energy of world oceanic circulation is tied up in them. The physics, chemistry, and biology of mesoscale eddies are just beginning to be worked out, and questions about them are far easier to come by than are answers. Nonetheless, they are a major feature of the oceans and oceanographers will be devoting considerable attention to them in coming years.

STUDY QUESTIONS

VI-8. How is the presence of phytoplankton determined from space?

VI-9. Why are cold-core rings found south of the Gulf Stream?

I. THE EXPLORATION OF INNER SPACE

Just as mesoscale eddies have become a fascinating new line of investigation for the marine scientist, so has the study of midwater organisms for the marine biologist. Until recently, marine biologists were restricted to the uppermost 30 meters of the open sea that is accessible to the scientist equipped with SCUBA gear. That, certainly, is the region with the richest and most diverse collection of life, and the one that is by far the most familiar to us. It is the region dominated by sunlight, chlorophyll, and phytoplankton. What was known of life in the greater depths was obtained by towing nets behind ships. Few creatures could survive such collection intact, let alone alive, especially the fragile life forms that tend to inhabit the midwaters.

Now marine biologists, using submersible research vessels, are venturing into the realm of the midwater to study at firsthand the life that inhabits the largest environment on Earth. Because of its depth as well as it breadth, the volume of the midwater is more than 200 times greater than the habitable space on all the landmasses put together. Richard Harbison, of the Woods Hole Oceanographic Institution, eloquently explains why he has chosen to study midwater organisms:

> ...the midwater environment is so different from the one we know that it is the best comparative system we biologists have until life is discovered on another planet. Imagine, if you can, what it must be like to live in a world without sun, where the only light is produced by other living things. A world without boundaries, where you can move in any direction, where there is virtually no turbulence, current, or motion, where there is no need to struggle against the force of gravity where there are no seasons. That's why I study midwater animals -- to try to understand what is possible for life in this most alien of all the environments on this planet. A better understanding of how animals live in midwater cannot help but teach us more about how animals live in general, and how they interact with their environment.

Because of the absence of sunlight, in the middle depths of the ocean there are no living phytoplankton -- chlorophyll is useless in the absence of light. There are, however, a wide variety of zooplankton -- animals that drift with the water -- ranging from microscopic creatures to jellyfish and including some fish and squids. The majority are transparent and gelatinous whose body tissues are mostly water. These fragile creatures are well adapted to feeding in the quiet, stable midwaters of the world oceans.

At present we know very little of these organisms. New species are being discovered with almost every dive, and knowledge of their behavior is almost totally lacking. We do not know how they capture prey and avoid predators, what their mating rituals are, or even how long they live. The use of research submersible vehicles such as <u>Alvin</u> now allows the biologist to visit them in their own habitat, but dive time is expensive and scarce. Marine biologists look forward to the possibility of constructing a deep ocean habitat from which observers can operate for extended periods of time. Without such a facility, a sizeable part of the marine population will remain shrouded in mystery for a long time to come.

For a number of years, efforts to investigate the life forms of the deep-sea environment have slowly eroded older notions that the deeper sea floor environment was a biological desert. The muds and oozes found on the sea bottom are in fact rich with a great diversity of life forms even though they are poor in resources. Rather than being a hostile environment, the ocean bottom is rather benign and stable, offering long-term security to its inhabitants, if not an abundance of food.

Ocean-bottom dwellers have adapted to their particular ecological niche, with small size, low metabolism and slow movement, long lifetimes, and slow sexual maturation among their characteristics. Like the midwater organisms, for the most part they are still dependent on the Sun as the primal source of energy. The base of their food web is organic material that rains down from the densely populated surface waters.

This view of ocean-floor life received a jolt in 1977 when the submersible <u>Alvin</u> settled to the floor of the Galapagos rift on the crest of an east-west spreading ridge off the coast of Ecuador. In the volcanic field of the rift it observed warm water emerging from the oceanic crust in a series of vents. Surrounding the vents was a rich biological community quite unlike anything observed to date elsewhere on the sea floor. Long tube worms tipped in red extended up to two meters (seven feet) into the water from their anchors in the basaltic rock. Large clams, beds of mussels, and even pale crabs as big as dinner plates were observed around the vents. Here was a community of life, totally cut off even from indirect dependence on the Sun, apparently living on the energy and nutrients of the vents themselves.

Virtually all life on Earth depends on the Sun as its basic source of energy. The process of photosynthesis is used by plants to create carbohydrates and other organic matter from inorganic raw materials: carbon dioxide, water, oxygen, and mineral nutrients such as nitrates and phosphates. Plant eating animals derive energy from the chemical process of oxidation of carbohydrates, and flesh eating animals merely continue the food chain extending from the Sun. When we speak of taking a walk after a heavy meal in order to "burn off" some food, we are in fact speaking with scientific exactness. The burning or oxidation of carbohydrates is what keeps us all going.

The remarkable thing about the vent community is that its prime source of energy comes from chemosynthesis, not photosynthesis. As you saw in the previous unit, seawater circulates through the hot oceanic crust at the spreading ridges, engaging in chemical reactions with the hot rock and dissolving minerals before emerging at the hot spring vents. The process that appears to be most important to the life forms in the vent community is the creation of hydrogen sulfide, the gas that gives hot spring waters and volcanic fumaroles their characteristic "rotten egg" odor. Hydrogen sulfide has the property that it may be oxidized readily to yield energy. Chemosynthetic bacteria that live in or on the vents utilize energy derived from the oxidation of hydrogen sulfide to produce carbohydrates from water and carbon dioxide. These bacteria form the base of the food web for the vent communities.

4
380A

The hydrogen sulfide is formed within the vent system from chemicals derived both from the seawater and from gases emerging from the young rock of the oceanic crust. Sulphate ions (SO_4^{--}) are abundant in seawater and are carried into the vent system where the high temperatures permit oxygen to be replaced by hydrogen to form hydrogen sulfide (H_2S). Temperatures within the vents can reach as high as 350°C (660°F). At the sea surface or on land, water at these temperatures would immediately flash into steam, but at the high pressures prevailing at the sea bottom, even water this hot remains in the liquid state. The pressure is too great to allow steam to form. In the cooler vents, the bacteria seem to live and multiply in chambers within the hot rock, sometimes surviving in temperatures well in excess of 100°C (212°F). They hold the record as the highest-temperature life forms found on this planet.

Many vents are cooler, some only at around 25°C (77°F). Their presence may be seen by the shimmering of their warm waters as they issue from the rock. Sometimes vents appear as "white smokers" -- water cloudy with chemosynthetic bacteria -- issuing from porous mineral structures encrusted with life. In hotter areas the water may be emitted with considerable force, laden with metal sulfides that turn the water black and smoky in appearance. In this case, the vent may consist of a chimney built of the sulfides and are called "black smokers". You will see examples of these on television.

The tube worms are perhaps the most exotic of the life forms found in the vent communities. They have neither mouths nor digestive tracts and appear to feed directly on the hydrogen sulfide dissolved in the water, though bacteria living symbiotically within their bodies may actually be responsible for the chemosynthesis that is taking place. They tend to live nearest the vents along with the mobile crabs that have been observed feeding on the worms. Farther out are the mussels, followed by the clams. These filter-feeders live on the bacteria, and are most abundant around the cooler vents.

Much interest has centered on these communities as models of how life may have originated on Earth. Life is known to have begun in the oceans, and environments similar in chemistry to that of the vents must have been present in the early life of our planet. This does not mean that the vent communities necessarily resemble the most primitive life forms, since the present forms depend on oxygen dissolved in the ocean bottom waters. As we shall see in latter units, oxygen was essentially absent from the early Earth and became abundant only at a much later time through the process of photosynthesis. Nonetheless, it is the chemosynthetic process that fascinates biologists searching for the origins of life. The process that gives birth to the ocean floor may also have spawned that most precious characteristic of Planet Earth -- life.

STUDY QUESTIONS

VI-10. What is a common characteristic of midwater organisms?

VI-11. What energy source takes the place of the Sun for the vent communities on the oceanic ridges?

VI-12. Why isn't the water in the very hot vents observed to boil?

RECOMMENDED READING

Samuel W. Matthews, New World of the Ocean, National Geographic Magazine, pp. 792-832 (Dec. 1981). Beautifully illustrated article on recent oceanographic research.

Thomas Y. Canby, El Niño's Ill Wind, National Geographic Magazine, pp. 144-183 (Feb. 1984).

Peter H. Wiebe, Rings of the Gulf Stream, Scientific American, pp. 60-7- (Mar 1982).

Charles D. Hollister, Arthur R. M. Nowell, Peter A. Jumars, The Dynamic Abyss, Scientific American, pp. 42-53 (Mar. 1984). Benthic storms and unique instrumentation designed to detect them.

John M. Edmond, Karen Von Damm, Hot Springs on the Ocean Floor, Scientific American, pp. 78-93 (Apr. 1983).

UNIT VII THE CLIMATE PUZZLE:
THE ATMOSPHERE

HURRICANE FROM SPACE

A. INTRODUCTION

1. Overview

The atmosphere of the Earth is a huge and very complex system that interacts with the land, the oceans, and the Sun to produce the weather and climate of our planet. In this unit we shall look at the structure and dynamics of the atmosphere and learn the essential causes of weather. The chemical composition of the atmosphere is an important determinant of Earth's overall climate through the action of what is known as the Greenhouse Effect. Consistent planet-wide circulation patterns determine the directions of prevailing winds, affect local climate, and determine a number of factors that have strongly influenced human history and habitat such as the development of trade routes and the occurrence of monsoons in Southern Asia.

2. Objectives

Upon completion of this unit you should be able to:

1. describe the chemical composition of the atmosphere

2. relate the properties of temperature, pressure, and humidity to the physical state of the atmosphere

3. relate the different divisions of the atmosphere to the elevation, temperature, and phenomena associated with each division

4. describe the relationship between the wavelength of radiant energy and the temperature of its source

5. describe the Greenhouse Effect and deduce the effect of changing concentrations of gases in the atmosphere such as carbon dioxide

6. relate the seasons in the northern and southern hemispheres to the tilt of the Earth's axis and its position in its orbit around the Sun

7. deduce how the input of solar radiation changes with latitude

8. distinguish between the different modes of heat transport on Earth: radiation, transport of sensible heat and latent heat, and ocean currents

9. understand the operation of the Coriolis Force on the motion of air masses in Earth's atmosphere

10. deduce the directions of winds produced by pressure gradients in combination with the Coriolis Force

11. relate the processes of convection and wind production to the global circulation pattern.

3. Key Terms and Concepts

temperature	convection
atmospheric pressure	sensible heat
relative humidity	latent heat
radiosonde balloons	hydrologic cycle
barometer	prevailing winds
troposphere	pressure gradient
tropopause	isobars
stratosphere	geostrophic wind
ozone layer	Coriolis Force
mesosphere	Hadley Cell
thermosphere/ionosphere	westerlies
radiant energy	trade winds
wavelength	intertropical convergence zone
transmission and absorption of radiant energy	doldrums
Greenhouse Effect	horse latitudes
the cause of seasons	monsoons
	cyclones

4. Corresponding Video

This program will help you answer the question, "What makes climate change and why?" You will see animations of the atmospheric and oceanic circulation systems, the twin engines that drive Earthly climate after the light and heat are received from the Sun. You will see satellite images of hurricanes and computer simulations of tornadoes forming. Computer animation will also illustrate the Milankovitch Theory, whereby slight variations in Earth's orbit change the amount of sunlight received on its surface. Finally, there will be a demonstration of climate modeling, a process by which scientists through a dry run in a computer can assess the impact of changing trends in the atmosphere without having to experience the devastating impact these changes might impose on Earth itself.

B. COMPOSITION AND STRUCTURE OF THE ATMOSPHERE

A summer thunderstorm brings high winds, bending tree limbs and swirling leaves, great splattering raindrops, lightning strokes, and booming thunderclaps. John Muir, one of America's early environmentalists, described how he once climbed a thrashing tree in the midst of a thunderstorm to better appreciate the spectacle of nature's show. Fortunately his tree was not struck by lightning, and those of us who share his love of natural phenomena, but do not share his daring, are content to watch from a porch or through a window. Others may prefer to draw the blinds closed and concentrate on something else.

Whether our reaction is one of awe or fear, all will agree that the weather exerts a powerful influence over our lives. The type of local agriculture is largely determined by the climate. Hurricanes, tornadoes, floods, and droughts can take lives and wreak havoc on local or even national economies. By 1985 a protracted drought had contributed to mass starvation in Ethiopia and other parts of Africa. On a longer time scale, it was only 10,000 years ago that the great ice sheets waned on the North American continent, uncovering much of Canada and the northeastern United States. What causes these changes in climate? Can they be predicted?

Before we can tackle such questions, we need to learn something about the weather and climate machinery of Planet Earth. The atmosphere forms the most important part of that machine. In this unit we will look at the composition, structure, and operation of the atmosphere.

Only two gases, nitrogen (N_2) and oxygen (O_2), make up 99% by volume of our atmosphere. Nitrogen accounts for 78%, oxygen for 21%, and one more gas, argon, accounts for most of the remaining one percent. Water vapor may locally account for as much as 3%, or it might be nearly absent. Water vapor is a clear, colorless gas -- what we see as steam or fog is really a collection of tiny droplets of liquid water suspended in the air. Other gases are present in minute amounts: neon, helium, methane (CH_4), carbon monoxide (CO), sulfur dioxide (SO_2), and ozone (O_3). Even carbon dioxide (CO_2), necessary to photosynthesis in plants, accounts for only 0.03% of the atmosphere.

The atmosphere also carries tiny particles of matter about on its winds. Larger particles settle out or are washed out by precipitation fairly quickly, but very small grains can stay aloft for months or years, kept suspended by the constant jostling of air molecules. Among the particulates found in the atmosphere are soil blown into the air as dust, sea salt resulting from the evaporation of ocean spray, smoke from forest fires and the combustion of fossil fuels, and fire ash or silicate dust and sulfate salts blasted into the atmosphere during volcanic eruptions.

Pollutants are gases or particulates in the atmosphere that we breathe that are undesirable from the standpoint of the health of plants and animals. Carbon monoxide, nitrogen oxides, and sulfur dioxide are all toxic even at fairly low concentrations. Carbon monoxide is created by incomplete combustion and is often a product of automobile exhausts, as are oxides of nitrogen. Sulfur dioxide is a common byproduct in the burning of coal and oil, and chemical reactions with water vapor can convert it into droplets of sulfuric acid, leading to the formation of acid rain.

Hydrocarbons such as gasoline fumes can react with nitrogen oxides to form complex organic molecules that remain suspended in the air, turning it a yellow-brown color and making breathing difficult and sometimes even hazardous. We call it smog, and catalytic converters have been installed in American cars to reduce the amount of unburned gasoline that exhausts from the tailpipe. In California you may see special fittings on gas pump nozzles that trap gasoline vapor from the gas tank before it can escape to the air.

Not all pollutants are man-made, however. Volcanoes are among the greatest pollutors, putting large quantities of sulfur dioxide, hydrochloric acid, and particulates into the atmosphere.

Water vapor varies strongly in concentration with elevation and temperature. Water exists as a solid (ice), liquid (water), or gas (water vapor) depending upon the pressure and temperature of its environment. Most water vapor is confined to the warm dense region of the atmosphere within ten kilometers of the surface.

Temperature, pressure, and humidity are the three most important properties that describe the physical state of the atmosphere. It is important to recognize the difference between temperature and heat. Heat and temperature both relate to the motion of molecules. Molecules and atoms are always in motion. In solids they vibrate about their fixed positions, while in gases they move about in a chaotic manner, interacting with their neighboring molecules and frequently changing direction. At high temperatures, molecules move more rapidly than at low temperatures. In fact, temperature may be regarded as a measure of the average kinetic energy of atoms in a region. When a hot region comes into contact with a cooler region, heat energy flows from one to the other, cooling the hot region and heating the cooler region until the temperatures of the two become the same. When this occurs, the flow of heat ceases. The tendency for heat to flow from warmer to cooler regions is the principal driving force for the weather machine.

The source of heat and the object being heated do not necessarily have to be in direct contact. The Sun is extremely hot and emits energy in the form of light. That light can fall on the much colder surface of the Earth and warm it, pumping heat into the ground and raising its temperature.

The temperature of the atmosphere can be measured by thermometers of one kind or another carried into the high atmosphere by rockets or into the lower atmosphere by instrumented balloons called radiosondes. Instrument packages attached to the balloons record the temperature, pressure, and humidity, and radio the data back to the ground as the balloon rises.

Atmospheric pressure refers to the force exerted on a unit of area due to the weight of all the air contained in a column directly above it. At sea level, average air pressure amounts to 1.01×10^5 nt/m^2 (14.7 lbs/in^2). That column of air, rising from sea level to the top of the atmosphere, weighs the same (and exerts the same pressure) as a column of water 10.3 meters (34 ft) high or a column of mercury 760 millimeters (29.9 in) high. In fact, a column of mercury forms the basis of the traditional barometer used in measuring atmospheric pressure.

A glass tube, closed at one end, is filled with mercury and inverted with its open end in a dish full of mercury (see Figure VII-1). If the tube is about a meter long, at sea level the mercury will be found to stand at a height of about 760 mm in the tube above the surface of the mercury pool in the dish. Above the mercury inside the glass tube is nothing -- a vacuum.

Above the mercury outside the tube is the atmosphere, exerting its pressure on the mercury surface and forcing it up into the tube until the column of mercury in the tube exerts the same pressure (force per unit area) on the mercury in the dish as the column of air. Changes in air pressure are reflected in changes in the height of the mercury column, which may be measured easily and accurately.

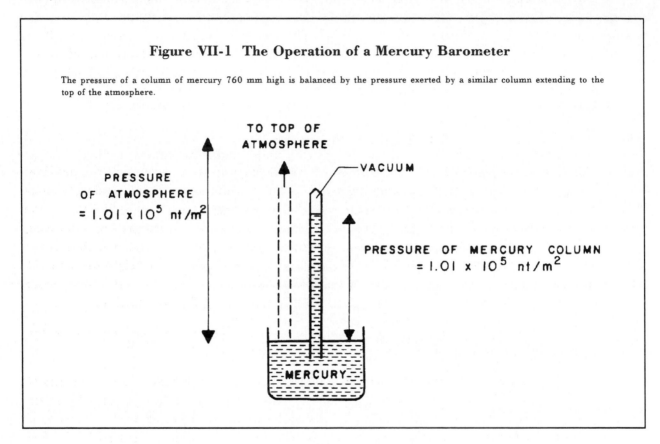

Figure VII-1 The Operation of a Mercury Barometer

The pressure of a column of mercury 760 mm high is balanced by the pressure exerted by a similar column extending to the top of the atmosphere.

The aneroid barometer uses a different method for measuring atmospheric pressure. A sealed bellows made of metal expands when air pressure falls, and contracts when it rises. Measuring the change in size of the bellows allows the recording of pressure changes.

Atmospheric pressure falls off rapidly with height above the Earth's surface. As a very rough guide, the pressure decreases by a factor of two for every five to six kilometers of elevation gained. A common unit of pressure is the millibar, where the pressure at sea level is approximately 1,000 millibars, or one bar. At 5.5 kilometers (18,000 ft), the pressure is down to about 500 millibars, at 11 km (36,000 ft) it is about 250 millibars, and so on. At 50 km (30 mi), the pressure is less than 0.1% of its sea-level value.

Relative humidity is a measure of the amount of moisture in the air (as water vapor, not droplets) compared to the carrying capacity of the air. If the relative humidity is 50%, then the air contains just half the water vapor that it can carry at that temperature. Cold air can carry much less water vapor than warm air, and so the relative humidity can vary considerably as the temperature changes. As the temperature drops toward the end of the day, the relative humidity can reach 100%, at which point liquid water drops begin to condense from the water vapor, causing dew to form. Morning fogs often form for the same reason. Winter air in our heated houses is often very dry because cold air from outside the house, already low in moisture content, is heated with a resultant lowering of the relative humidity, often to very low percentage values.

The interplay of temperature, pressure, and humidity causes many features of the weather system. A body of air that rises encounters lower pressures at height and expands, which cools the air. If the air is moist, its relative humidity will increase and can exceed 100%, at which point droplets of water will form, giving rise to clouds. Condensation of water droplets is aided by the presence of particulates in the air that act as seeds or nuclei about which water droplets or ice pellets can begin to form.

1. Layers of the Atmosphere

We have already mentioned that when air rises, it expands and cools, and this provides a partial explanation of why the air temperature is usually lower at higher elevations, such as in the mountains. At one time it was assumed that this trend would continue to the top of the atmosphere, but radiosonde balloons and sounding rockets showed that this was not the case. Figure VII-2 shows that the temperature indeed drops as we rise to 10 km (6 mi), but above that elevation it warms again, reaching a maximum of about 0°C (32°F) at a height of about 50 km (30 mi). Above this elevation it cools once again, then above 80 km (50 mi) it once again rises, this time reaching very high temperatures at a height of about 200 km (125 mi).

On the basis of these temperature swings, the atmosphere has been divided into layers, called the troposphere, stratosphere, mesosphere, and thermosphere.

The troposphere is the lowest layer, extending to an average height of about 10 km (6 mi). Its top ranges in height from 8 km (26,000 ft) at the poles to 18 km (59,000 ft) at the equator. It contains most of the moisture in the atmosphere, and so most of what we regard as weather -- clouds and storms -- is confined to it. The ground, as its base, is warmed by the Sun and serves as the source of heat for this layer. Much of the troposphere is in constant motion, giving rise to constantly shifting weather patterns because warm air lies below cooler air, which is an unstable situation. Warm air tends to rise and be replaced by denser, cooler air, which in its turn is heated by the warm surface of the Earth. This keeps the troposphere in constant motion and brings an endlessly changing pattern of weather.

Between 10 and 50 km, however, the temperature rises. This places warm air above cooler air, which is a stable arrangement. The result is a layer in which little vertical mixing occurs. That is, the layer is stratified, hence its name, the stratosphere. Normally, neither clouds nor storms reach into this region. The boundary between the troposphere and strato- sphere is called the tropopause. Its height ranges from 8 km (26,000 ft) at the poles to 18 km (59,000 ft) at the equator.

Why does the temperature increase in the stratosphere? In Figure VII-2, note that there is a layer of ozone extending throughout much of this region. Ozone is a molecule made up of three atoms of oxygen bound together (O_3) instead of the usual two (O_2). It happens to be a very strong absorber of ultraviolet rays from the Sun, and absorbs much of the energy of these rays. In the process the air is heated by the energy of the absorbed rays. In addition to

Figure VII-2 Temperature Structure of the Atmosphere

Adapted from R.S. Quiroz, Bulletin of the American Meteorological Society Vol. 53, pp. 122-133, 1972. Included with permission.

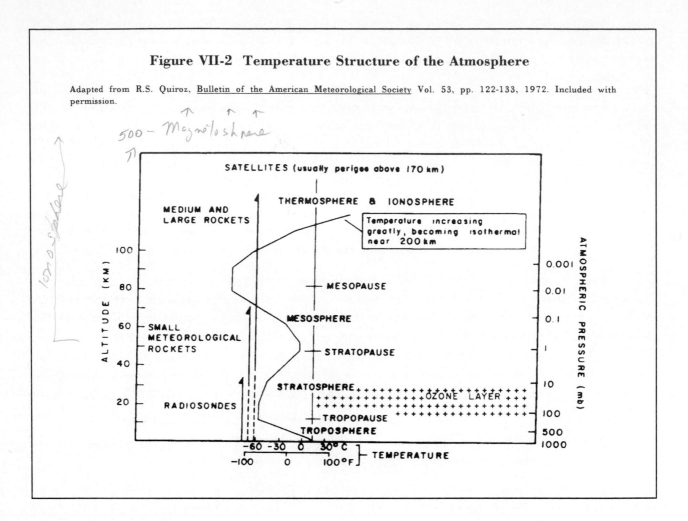

producing the temperature hump at about 50 km, it also blocks the shorter, more harmful untraviolet rays that reach Earth's surface. Ultraviolet rays are responsible for sunburn and can cause skin cancer. Were it not for the ozone layer, most present forms of life probably could not exist on the continents.

There has been much concern in recent years about the continued integrity of the ozone layer. Some concerns centered on widespread use of chlorofluorocarbons (CFC) such as Freon ™ as propellants in spray cans. Chlorofluorocarbons are very stable and so eventually diffuse into the stratosphere where sunlight can act on it, releasing chlorine which is capable of destroying ozone. CFC is also used as a refrigerant in air conditioners and refrigerators. Since 1977, use of CFC in spray cans has been placed under strict governmental regulation in the United States, but at present no satisfactory substitute for it has been found as a refrigerant.

Other concerns for the ozone layer have surfaced. High-flying jet aircraft such as the Concorde supersonic transport and some military aircraft operate within the ozone layer, releasing nitrogen oxides in their exhausts. Nitrogen oxide is capable of reacting with ozone and destroying it, and because the stratosphere is so stable, these exhaust gases would remain within the ozone layer for long periods of time, maximizing the chances for ill effects. Other human sources of nitrogen oxides are automobile exhausts and nitrogen fertilizers at ground level, and atmospheric blasts of nuclear bombs. The principal regulator of atmospheric O_3, however, is nitric oxide released from bacterial action in the soil. We will discuss this process in **The Balance of Nature** unit. Lightning also contributes nitrogen oxides to the atmosphere.

It is one of the many ironies of nature that, while ozone in the stratosphere is a highly beneficial shield blanketing the Earth, at ground level in our cities it is a toxic pollutant created by the action of sunlight on automotive and industrial emissions.

Above the stratosphere, temperature once again falls with increasing elevation, marking the <u>mesosphere</u>. The final rise in temperature occurs within the <u>thermsophere</u>, above 80 km (50 mi). This heating is caused by the action of high-energy radiation from the Sun on the extremely thin air. The effect is that a negatively-charged electron is knocked out of free oxygen (O), molecular oxygen (O_2), or nitrogen oxide (NO), leaving them as positively-charged ions. Their temperature is raised in the process, and because there are so few molecules per cubic centimeter at this height, it takes very little energy to produce large increases in temperature.

Because the lower portion of the thermosphere is characterized by the presence of ions, it is also referred to as the <u>ionosphere</u>. We will have more to say about this region and the next, nearly empty, one above it -- the magnetosphere -- in the **Impact of Man** unit.

STUDY QUESTIONS

VII-1. What are common sources of sulfur dioxide pollution in the atmosphere?

VII-2. What is the difference in the mechanical state between an object at high temperature and one at low temperature?

VII-3. A lowering of atmospheric pressure is often described by the statement: "The barometer is falling". Can you see where the expression comes from?

VII-4. Why does the presence of ozone cause heating of the air in the stratosphere?

C. HEAT FROM THE SUN AND THE GLOBAL ENERGY BALANCE

1. Radiant Energy

The Sun puts out a tremendous amount of energy, but because the Earth is small and far away, we intercept only a tiny fraction of the total. Nearly all of that energy travels to us as radiation. In this unit, we shall use the term "radiation" to mean radiant energy. In this context, it has nothing to do with radioactivity, but simply refers to the sending out (transmission) or receiving (absorption) of radiant energy.

Radiation travels as electromagnetic waves. Some of the kinds of electromagnetic waves are radio and television waves, microwaves, infrared, visible light, ultraviolet, X-rays, and gamma rays. Electromagnetic waves can traverse even the empty space that separates Earth from the Sun, traveling at the speed of light (299,000 km or 186,000 mi per second).

Different types of radiant energy are distinguished from one another by their <u>wavelengths</u>, or the distance between successive wave crests. Figure VII-3 shows the characteristic wavelengths of different kinds of radiant energy.

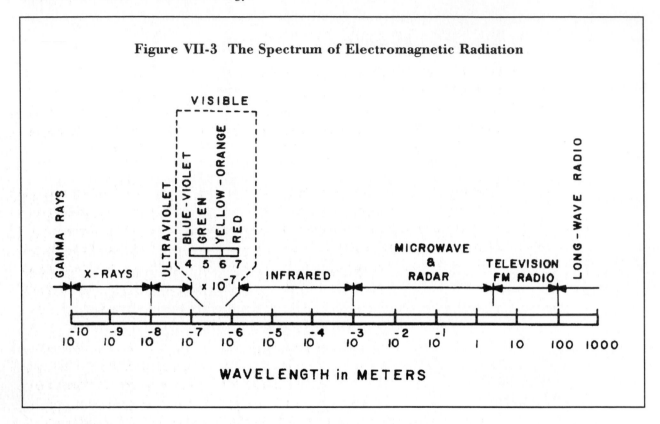

Figure VII-3 The Spectrum of Electromagnetic Radiation

All forms of radiation behave in a manner similar to that of light. Light rays travel through space in straight lines, and large objects that are opaque to light cast shadows. Some substances, like glass or water, are transparent and transmit light through them. Opaque substances do not transmit light, but either absorb or reflect it. Coal absorbs visible light and appears dark, while snow reflects it and appears bright. The transmission, absorption, and reflection properties of a material depend upon its structure and upon the wavelengths of light that strike it. For instance, ordinary glass is transparent to visible light but is nearly opaque to ultraviolet light: that is, it absorbs ultraviolet light. This explains why it is difficult to get a sunburn through a window, since ultraviolet rays are responsible for the effects of sunburn.

Radiation from the Sun comes to us in a broad range of wavelengths, but most of the energy is contained in a band ranging from the near infrared through the near ultraviolet, including the entire visible spectrum. For this reason, the Sun's light appears white, a mixture of all the colors.

In addition to what may be reflected from a body or transmitted through it, radiant energy is produced by a body and emitted by it into its surroundings. As a rule, the wavelengths of radiant energy emitted by a body depend on its temperature, with hot bodies emitting more short-wavelength (bluer) radiation, and cooler bodies emitting longer-wavelength (redder) radiation. While radiant energy is usually emitted in a broad band of wavelengths, a simple rule relates the wavelength for peak emission to the temperature of the body. This rule is known as Wien's Law and can be given in the form:

Rule 1

$$\text{Peak wavelength (m)} = \frac{2.88 \times 10^{-3}}{\text{Temperature } ^\circ\text{C} + 273}$$

This rule applies strictly only to what is called a "black body", which is an object with a surface that absorbs all radiation falling upon it and that emits all radiation appropriate to the temperature. Rough, dark surfaces satisfy this requirement better than light-colored or reflective surfaces. A lump of charcoal is a good approximation to a black body, but we shall apply Wien's Law to objects with other kinds of surfaces to get at least a rough idea of the peak wavelengths that they emit.

Using this rule, if we know the temperature of a black body, we can determine the dominant or peak wavelength of the radiant energy. Let us take some examples. The surface of the Sun has a temperature of about 5,700°C. Using Rule (1), we find that the dominant wavelength of its light should be about 4.8×10^{-7} meters. This falls within the visible spectrum. Note that this gives us only the peak wavelength -- the actual band of radiation emitted by the Sun's surface extends from infrared to X-rays.

Another example is yourself. Normal skin temperature is about 34°C (93°F) which, when put into Rule (1) and the arithmetic carried out, yields a peak wavelength of 9.4×10^{-6} m, or nearly 10^{-5}m. From Figure VII-3, you can see that this falls in the infrared part of the spectrum. Our eyes are not sensitive to infrared rays and for that reason you do not appear to "glow in the dark", but in fact you do. An instrument sensitive to the middle infrared part of the spectrum can literally take a picture of you in the absence of visible light. Your "glow" is entirely within the infrared.

Another familiar example of infrared emission is an asphalt street or other dark surface that has been heated in the sunlight all day. For a while after the surface becomes shaded, you can feel, but not see, the radiant heat being emitted from it in the infrared part of the spectrum.

We have spent so long on the concept of radiant energy because it is critical to understanding how the Earth interacts with sunlight, and this is the basis for all weather and climate. To be sure you understand it, let's take one last example. So far we have used the temperature of an object to find the wavelength of its radiation. We may turn that around and use the known wavelength emitted to obtain its temperature. We will need to rewrite Wien's Law as stated in Rule (1) into a new form:

Rule 2

$$\text{Temperature } (^\circ\text{C}) = \frac{2.88 \times 10^{-3}}{\text{Peak wavelength (m)}} - 273$$

Consider a brightly glowing coal in a furnace. What is its temperature? Because it glows red, we can choose a wavelength from Figure VII-3 of around 7×10^{-7} m. Placing this number in the denominator of Rule (2), we might assume that the temperature of the coal is approximately 3,800°C (6,900°F). In addition to the visible red light that it emits, you can feel the infrared radiant heat that accompanies the visible light. Remember that not just one wavelength, but a broad range, is emitted. Because of this effect, and the fact that our eyes cannot see the emitted infrared radiation, our estimate of the coal's temperature is on the high side, since the true peak of the radiation is in the infrared rather than in the red.

2. Light and Heat From the Sun and the Greenhouse Effect

When you leave a car parked in the Sun with its windows rolled up, the temperature can rise inside the car far above the outside temperature. A greenhouse works the same way, and even though a greenhouse may need a heater to keep the temperature above freezing in the winter, it takes less heat than if this natural heating effect were not in operation. This process in fact is called the Greenhouse Effect, and its operation depends upon two effects. The first is the fact that the enclosed structure traps heated air and prevents it from escaping to the outside air. The second relies upon the fact that glass is transparent to visible light but is relaively opaque to infrared radiation.

Let us examine this second effect more closely. Visible light from the Sun passes through glass readily and falls on objects inside the greenhouse, warming them and raising their temperature. These objects radiate part of the energy away from them at a wavelength that is determined by their temperature, according to Rule (1). Even if they become quite hot to the touch (around 70°C or 160°F), their peak emission is at around 8.4×10^{-6} m, well within the infrared range. Because glass does not transmit infrared radiation (that is, it is opaque to it), radiant energy cannot escape from the greenhouse and is trapped within it.

The Greenhouse Effect, then, is partly dependent on an enclosing substance being transparent to incoming visible radiation and opaque to outgoing infrared radiant heat. It is fortunate for us that the Earth's atmosphere does not form a very effective greenhouse. If it were as effective as glass, the average temperature would rise to the point that living on Earth would become intolerable. As it turns out, oxygen and nitrogen are largely transparent to infrared, allowing the warm ground to radiate a significant amount of infrared energy into outer space. In addition, atmospheric circulation carries heat away from the surface quite effectively.

On the other hand, water vapor, carbon dioxide, and methane are much more opaque in the infrared, and their presence tends to produce an additional Greenhouse Effect that is dependent on the concentration of these gases. Clouds are also extremely effective blocks to outgoing infrared radiation. This explains why the coldest nights are those clear, crisp times when there are no clouds and the humidity is low. As we mentioned earlier, carbon dioxide accounts for only about 0.03% of the atmosphere, but its concentration has been increasing for some time due to the activities of mankind. This has raised concerns about possible long-term effects on climate. We shall return to this subject in the next unit.

As it is, with all other factors held equal, the total Greenhouse Effect on Earth at the present time produces a warming of about 35°C (63°F) over what the temperature would be without any such effect at all. That much of a temperature difference, of course, is a critical one. Without it, Earth would be largely a frozen planet.

STUDY QUESTIONS

VII-5. The burner in a gas stove produces a blue-violet flame when it is properly adjusted. Roughly how hot is the flame?

VII-6. Why is it often possible to walk barefoot on light beach sand but not on an asphalt road in the bright sunlight?

VII-7. When temperature-sensitive citrus trees are in danger of freezing on a cold, clear spring night, orchard managers often use machines to generate banks of smoke or mist over the trees. Why?

D. CLIMATE AND THE SEASONS

Seasons are caused by the tilt of Earth's rotational axis. Like a navigational gyroscope, the axis points in the same direction in space from year to year. Over long periods of time, it can precess and has other motions similar to those of a spinning top. These motions need not concern us now. Figure VII-4 shows the situation as Earth rotates about the Sun during the course of a year. In the northern hemisphere, in June the Sun stands very high in the sky at noon and its rays beat directly down, efficiently warming the ground. In December, on the other hand, the Sun's rays hit the northern hemisphere obliquely, and the same energy is spread out over a much larger area. Note from the diagram that the exact opposite is the case in the southern hemisphere, resulting in a reversal of the seasons.

At the equator, the Sun is never far from overhead at noon, somewhat farther north in June and farther south in December, but making about the same angle with the ground in both months. As a result, the temperature is generally very warm and much the same all year. At the poles, on the other hand, the sun never rises very high above the horizon, resulting in a very cold climate.

There is one further effect worth noting. Earth's orbit around the Sun is not perfectly circular. It is somewhat elliptical, with the result that the Earth is 147 million km from the Sun in January but 152 million km from the sun in July. This might seem to contradict the seasons in the northern hemisphere (but not in the southern hemisphere), except that the difference in energy received from the Sun due to this effect is small compared to the effect of the tilt of the rotation axis. Nevertheless, it introduces some asymmetry into the climate, tending to exaggerate seasonality in the southern hemisphere and reduce it in the northern hemisphere. As we shall see in the next unit, changes in these orbital parameters have been implicated in long-term climate changes.

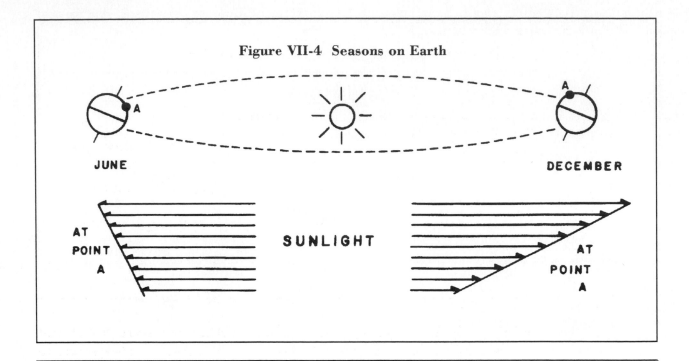

Figure VII-4 Seasons on Earth

STUDY QUESTION

VII-8. If you lived directly on the equator, when would you receive the most heat
from the Sun?

E. HEAT TRANSPORT MECHANISMS IN THE ATMOSPHERE

We have already discussed one method for transporting heat from one place to another:
radiation. This process is all-important in determining the net heat input from the Sun into the
Earth climate system. The remaining heat transport processes -- atmospheric convection, and
ocean currents -- serve to redistribute this energy throughout the world, determining local
weather and climate in the process.

Convection is the process by which heat is carried by a moving mass. In the unit
Continental Tectonics and Earth's Interior we discussed convection in the mantle as a means
of transporting heat from the Earth's interior to the surface. It might be well to mention at this
point that this quantity of heat coming up through the ground and into the atmosphere is
minuscule compared with the energy flux from the Sun, and so is very unlikely to have any
significant effect on climate.

There are many familiar examples of convection. A radiator heats cool air. The air
expands, becomes less dense and rises, displacing cooler air which sinks and enters the radiator
at its base to complete the convection cycle. A pot of water placed on a stove is heated from
below. Warm water tends to expand and rise over the hottest part of the pot bottom, displacing
cooler water that sinks near the edge of the pot.

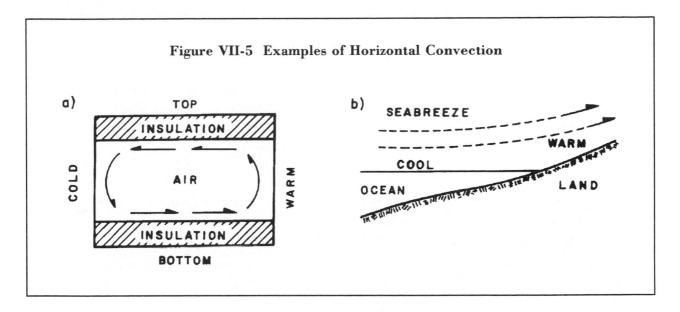

Figure VII-5 Examples of Horizontal Convection

Convection, then, is a cyclic motion of a freely-moving fluid in which a net flow of heat is carried from a warm to a cooler region by the fluid. Heat transport does not always have to be vertical in convection. Figure VII-5a shows a room with one cold wall and one warm wall. Its floor and ceiling are well insulated so that they will not affect the process. Air near the warm wall is heated and rises while air near the cold wall is cooled and sinks. Even though vertical motion always occurs in convection, the net result may be the horizontal transport of heat energy. In this case, heat is removed from the warm wall and carried to the cold wall. It is also important to realize that there is an actual transfer of heat. In the process the warm wall is cooled and the cool wall is warmed. Ultimately, thermal processes strive to reduce everything to the same temperature. They will continue to act only as long as there is a source of energy (such as solar radiation) to maintain a temperature difference.

On Earth, solar radiation maintains temperature differentials in many ways, but the two most important are the temperature differences between the equator and the polar regions, and that between continental and oceanic areas. During the summer the latter results in the phenomenon of the sea breeze along the ocean shore. The ocean does not change temperature so readily as the surface of the land and remains at a nearly constant temperature during day and night. The surface layer of the land, however, becomes quite hot during sunny days and cools noticeably at night. During the day, then, air over the land is heated and rises, being replaced by cool air from over the ocean. This produces an onshore breeze -- the sea breeze (see Figure VII-5). At night, the breeze dies as the land cools, or may even reverse direction if the land temperature falls below that of the sea, producing an offshore breeze.

On a larger scale, a similar situation is an important contributor to the monsoons that are so important to southern Asia and elsewhere. Seasonal temperature differences between continental landmasses and major oceans help to govern the annual alternation of rainy and dry seasons in these parts of the world. We shall return to this point in the next unit.

Moving air masses may transfer heat from one place to another in two different ways. One is by the movement of a warm or cold air mass, carrying its thermal energy with it. Meteorologists refer to this thermal energy as <u>sensible heat</u>. Heat transfer also may be accomplished using energy storage within the air mass due to the evaporation and condensation of water. In order to evaporate water, energy must be absorbed, and this energy input is called the <u>latent heat</u> of vaporization. That energy is effectively stored in the new physical state of the water (as gaseous water vapor). When you step out of a tub or swimming pool, your skin feels cool because of the water that is evaporating from it. Part of your body heat is being used to change the water from a liquid to a gaseous state. The latent heat is released when water vapor condenses back into the liquid state as rain or fog. The latent heat can then be transported if evaporation occurs in one place and the moist air mass then moves to another place before condensation occurs.

Heat transfer in the atmosphere is often a combination of both processes: the transfer of sensible heat and the transfer of latent heat. In the absence of external forces, the two taken together will act to transfer heat from a warm to a cooler region.

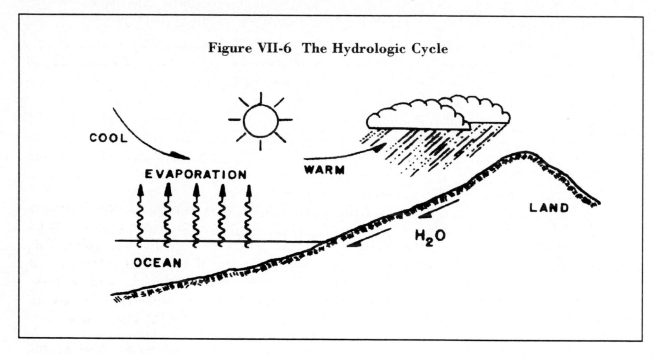

Figure VII-6 The Hydrologic Cycle

We can now see the operation of the familiar hydrologic cycle, diagramed in Figure VII-6. Both latent heat and sensible heat play important roles. The Sun supplies energy to evaporate water from the ocean (latent heat) and also supplies the temperature differentials needed to move the moist air mass onto the continents (transfer of sensible heat and latent heat). As the warm, moist air rises it expands and cools until its relative humidity reaches 100%, at which time condensation occurs and precipitation becomes possible. On the average, both latent and sensible heat processes tend to transfer energy from the warm surface of the Earth to higher, cooler regions of the troposphere, but more than 50% of solar energy reaching the surface is transported back into the atmosphere via latent heat transfer. Horizontal winds, updrafts (rising air), downdrafts (descending air), clouds, and precipitation are all byproducts of this energy transfer. Taken together, the whole scheme is a heat engine that produces motion and weather effects by attempting to reduce, on average, the temperature differences introduced by the Sun.

It takes a substantial amount of heat energy to raise the temperature of water. That is, water has a high heat capacity. Because of this, the world's oceans constitute an immense reservoir of heat that may be exchanged with the atmosphere at the sea surface. Warm and cold ocean currents can strongly influence climate locally, as you have seen in the units on oceanography. They are another example of the transfer of sensible heat, where this time the moving mass is water instead of air.

STUDY QUESTIONS

VII-9. At a seashore, the breeze is often off the land in the morning, but becomes a sea breeze in the afternoon. Why?

VII-10. A boiling pot of water is sitting on the burner of a stove. The burner is continually putting more heat into the water, yet the water remains at its boiling point: 100°C (212°F). Where is the heat from the burner going?

F. CIRCULATION OF THE ATMOSPHERE

1. Wind

What makes the wind blow? You already know part of the answer. Wind is simply motion of air masses in response to pressure differences, acting in such a way as to transport heat energy from warm regions such as the equator to cold regions such as the poles. This statement is fine as a general description, but it omits a number of complicating factors that must be taken into account if we are to be able to understand and predict winds. Many of these complications arise from the Coriolis Effect, which we found to be an important influence on oceanic circulation. We shall find it to be equally important in the case of the atmosphere.

In the previous unit you learned that oceanic circulation is largely governed by the prevailing winds. Now we need to determine what governs the directions of these persistent winds. Let us begin by considering the case where we ignore effects of the Earth's rotation and then introduce the Coriolis Effect to see what changes it introduces.

You are probably familiar with the way in which high and low pressure areas move across daily weather maps. Recall that the atmospheric pressure is just the weight of the column of air directly above a unit area. The greater the mass of air contained within the column, the higher the atmospheric pressure measured at its base. The total mass of air above a given area is a function of several variables: temperature, humidity, and the extent to which air masses are converging toward that region or diverging from it.

When you blow up a balloon, it is clear that the pressure inside the balloon is greater than in the atmosphere outside the balloon. If the mouth of the balloon is allowed to open, the air inside will rush out in response to the pressure difference. In general, air tends to accelerate from a region of high pressure toward one of lower pressure. In addition, it is not the value of the pressure that is important, but how rapidly the pressure changes over a given distance.

The pressure gradient is defined to be the difference in pressure over a unit distance, where that distance is measured in the direction of maximum pressure change. The direction of the gradient is just that direction of maximum change. Figure VII-7a shows how this works. At the bottom of the diagram the pressure is high (1,010 millibars) and at the top it is low (990 millibars). The horizontal lines are isobars: lines of equal pressure. The pressure gradient points from bottom to top, perpendicular to the isobars.

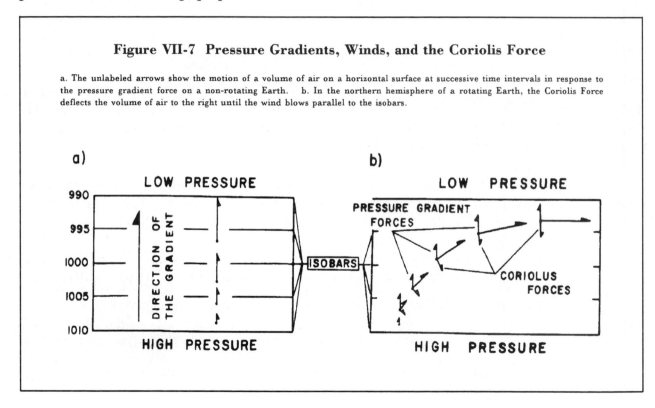

Figure VII-7 Pressure Gradients, Winds, and the Coriolis Force

a. The unlabeled arrows show the motion of a volume of air on a horizontal surface at successive time intervals in response to the pressure gradient force on a non-rotating Earth.   b. In the northern hemisphere of a rotating Earth, the Coriolis Force deflects the volume of air to the right until the wind blows parallel to the isobars.

A volume of air in this diagram will experience a force due to the pressure gradient in the same direction as the gradient, that is, toward the top of the diagram. Newton's Second Law of Motion states that a mass acted on by a force will accelerate in the direction of the force, moving ever more rapidly with time. This is shown in the same diagram for a small volume of air at successive time intervals, starting at rest and accelerating in response to the pressure gradient force.

Once we introduce a rotating Earth, however, the situation changes dramatically, even though the pressure gradient may remain unchanged. In Figure VII-7b, the volume of air begins from rest and initially is accelerated toward the top of the diagram by the pressure gradient force. As soon as it begins to move, however, it is deflected to the right by the Coriolis Force.[10] So long as the air moves toward lower pressure, it is accelerated to higher velocity. The Coriolis Force increases with velocity, however, and bends the motion ever more to the right.

[10] In the northern hemisphere. In the southern hemisphere the deflection is to the left.

Eventually the air will be moving parallel to the isobars. At this point, the Coriolis Force will exactly balance the pressure gradient force and the air will continue to move parallel to the isobars at a constant velocity. Such a wind is referred to as a geostrophic wind. A geostrophic wind is attained when an air mass has reached a sufficient velocity that the pressure gradient force is exactly balanced by the Coriolis Force. Winds associated with high and low pressure zones come close to being geostrophic, especially at altitudes high above the ground where there is little to impede the air's motion. Near the ground, however, friction with the land surface slows the air and prevents it from becoming truly geostrophic. The result may more closely resemble not the rightmost set of arrows in Figure VII-7, but the one or two just preceding it to its left, in which the motion of the air still has a component moving it from high toward low pressure in addition to the motion to the right. Because of this frictional effect, there is a net flow of air away from high pressure regions and toward low pressure regions.

From the figure, we may state a general rule: in the northern hemisphere, if the wind is blowing into your left ear, then your nose points toward the low pressure region. In the southern hemisphere, the same is true if the wind blows into your right ear, since the Coriolis Force works in the opposite direction south of the equator.

If the isobars are curved, the same general rules hold. In Figure VII-8 we see how winds circulate around high and low pressure regions in each hemisphere. Note that the "ear and nose" rules continue to work regardless of where you happen to be standing.

A large circulating air mass associated with a low pressure region is referred to as a cyclone, and the direction of rotation of any cyclone is determined by the Coriolos Effect. The term cyclone also has a colloquial use, however, in referring specifically to tropical cyclones such as hurricanes.

Figure VII-8 Wind Directions around High and Low Pressure Regions

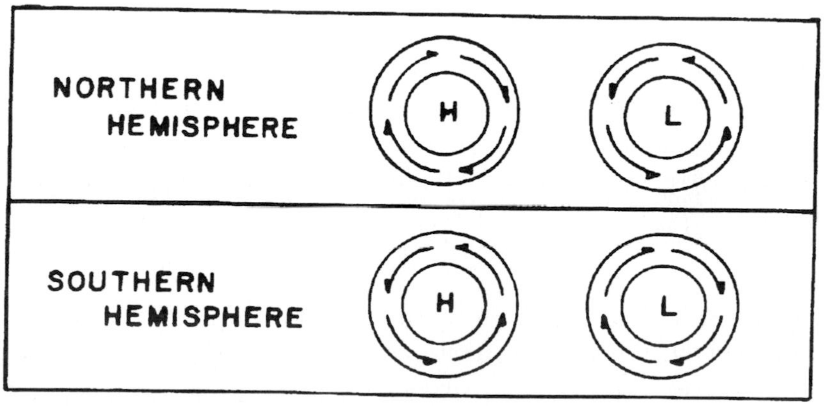

Major storms are often associated with low pressure regions, and hurricanes (called typhoons or cyclones in some parts of the world) invariably have winds that circulate in a counterclockwise direction in the northern hemisphere and in a clockwise direction in the southern hemisphere. One reason that storms can have lifetimes of days or weeks before they finally dissipate is that the Coriolis Effect prevents the winds from moving air directly into the low pressure zone and wiping it out by the mass of air converging upon it. Storm systems are really long-persisting eddies in the atmosphere.

2. Global Circulation Patterns

In our description of atmospheric circulation as a heat engine, we mentioned that heat is transported from the equator to the poles. You might expect that this would be accomplished by a single convection cell in which warm air at the equator rises, travels at height to the polar region, cools and sinks, and returns at low elevations to the equator. This simple kind of convection circuit is called a Hadley Cell. For a number of reasons, this picture does not work. The depth of the troposphere is very small compared to the distance from equator to pole, and this would make the convection very inefficient. The warm air would have to spend so much time in the high troposphere that it would cool and begin to sink. This in fact occurs at a latitude of about 30°N (approximately the latitude of New Orleans in the United States or Cairo in Egypt). There are other, more complex reasons involving the Coriolis Effect and the conservation of angular momentum of the Earth. The result is a breaking up of the circulation pattern into at least two zones between equator and pole (see Figure VII-9).

Subtropical latitudes are dominated by the Hadley Cell, with the southerly-flowing surface winds in the northern hemisphere deflected somewhat to the right (west) by the Coriolis Effect, which is weak at these low latitudes. Note that the northerly-flowing tropical surface winds in the southern hemisphere are deflected to the left, which is also to the west. The descending cooled air at the northern edge of the Hadley Cell generates persistent high-pressure regions at about 30°N latitude. North of these, the Coriolis Effect becomes stronger and more dominant in its effects, largely replacing the vertical circulation of the Hadley Cell with horizontal circulation about low-pressure regions. This serves to continue the process of moving heat from the warmer temperate zones to the colder polar regions. A weak vertical circulation pattern, suggested by the dashed loops in Figure VII-9, also appears to exist at mid- and high-latitudes, but horizontal convection is the dominant process. A similar situation exists in the southern hemisphere.

The high-pressure regions at about 30°N indicate that the principal circulations in the zone just to the north of them should be from west to east, and in fact they are. The reason for this westerly flow is more complex, and once again has to do with conservation of angular momentum of the Earth. These persistent winds in the mid-latitudes are known as westerlies while those that blow from the northeast or southeast in the tropics are called trade winds. In the days of sailing ships, trade routes between Europe and North America would take advantage of the pattern, using a more southerly route for travel from Europe to America and a more northerly route for the return.

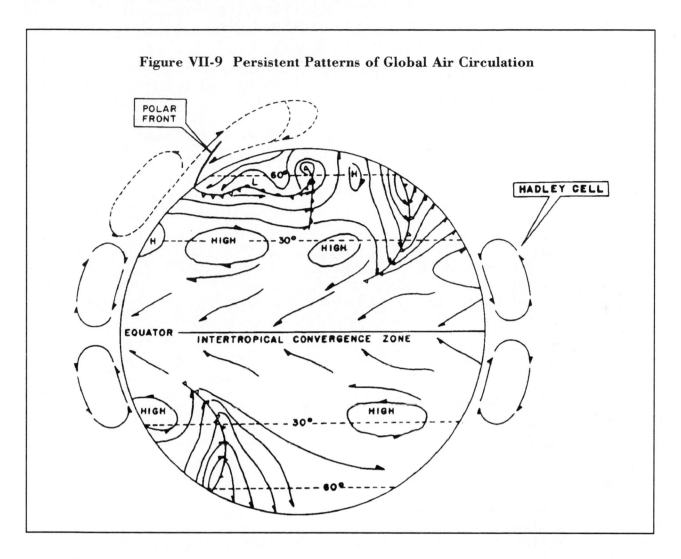

Figure VII-9 Persistent Patterns of Global Air Circulation

From the direction of the trade winds in Figure VII-9, perhaps you can see why Columbus, setting sail from Spain, made landfall not on the continental United States, which is at the same latitude as Spain, but farther south, in the West Indies.

Figure VII-9 shows that the trade winds in the northern and southern hemispheres converge near the equator in a low pressure region called the <u>intertropical convergence zone</u>. Here warm, moist air rises and cools, producing heavy rainfall characteristic of equatorial regions. Because the pressure tends to be uniformly low in the convergence zone, the pressure gradient is small and the result is a general lack of strong winds. Early sailors labeled this zone the <u>doldrums</u>. Other zones of generally light winds occur at the <u>horse latitudes</u>, in the high pressure regions separating the trade winds from the westerlies.

The general pattern described above is an average, over all seasons, and for the entire Earth. A more detailed description requires that we modify it to account for the seasons and for the effect of Earth's landmass distribution. Because the temperature of the ocean does not change very much from winter to summer, the bands of high and low pressure shown in Figure VII-9 tend to be persistent over the oceans, giving rise to features such as the Pacific High off the west coast of North America and the Bermuda High in the Atlantic Ocean.

Persistent downwelling masses of dry air associated with these high pressure zones are responsible for the existence of the Sonora Desert in Mexico and the Sahara Desert in northern Africa. In the southern hemisphere, the zone of persistent highs, governs the existence of the Peru and Atacama Deserts seaward of the Andes Mountains, the Namib Desert of southwestern Africa, and the deserts of western Australia.

The situation is very different on the continents, however. They tend to be hot in the summer and cold in the winter. In summer, they warm the air above them, which rises and produces a local low pressure region. In winter, the air is cooled and sinks, resulting in a high pressure zone. The huge mass of the Asian continent displays this effect very strongly, giving rise to an annual phenomenon that affects the lives of over two billion people: the monsoons.

During the northern hemisphere winter, a large high pressure zone forms over the central Asian continent, pushing the intertropical convergence zone south of the equator (see the heavy line in Figure VII-10.) Cool, dry winds blow out of central Asia from north to south bringing rainless winter monsoon winds to China and much of southeast Asia. In summer, a low pressure zone forms over southern Asia, and the intertropical convergence zone shifts far north, looping up over Nepal and China. The result is a flow of warm, moist air from the persistent high pressure zone over the Indian Ocean. The summer monsoons bring deluges of rain that make farming, and especially rice culture, possible in these regions. A failure or even delay of the summer monsoon can bring drought and famine to vast numbers of people.

The monsoon phenomenon is most active in the eastern hemisphere because it is caused by the asymmetrical arrangement of landmasses -- Asia north of the equator and the Indian Ocean south of the equator. The western hemisphere, however, has North and South America more symmetrically placed on either side of the equator, with the result that wet and dry seasons are less prominent.

STUDY QUESTIONS

VII-11. If the Hadley Cell ends at about 30°N. latitude, how is the poleward flow of heat continued?

VII-12. The north coast of Australia is also affected by a monsoon-like phenomenon, but to a lesser extent than India because Australia is a much smaller continent than Asia. When would you expect the rainy season to be in northern Australia?

G. FRONTS AND STORMS

We tend to take for granted the rapidly shifting patterns of weather in most parts of the world. Around the Rocky Mountain area, notorious for its rapid changes, there is an expression, "If you don't like the weather, just wait ten minutes." Most natural processes, as you have seen, operate on vastly longer time scales, with change coming slowly. Why is weather so changeable?

Figure VII-10 Origin of the Monsoons

Light arrows show prevailing winds; heavy lines show isobars in millibars in excess of 1000. The heavy line shows the location of the intertropical convergence zone. For additional information see L.J. Battan, Fundamentals of Meteorology, Prentice-Hall, 1979. Source: Herbert Riehl, Introduction to the Atmosphere, McGraw-Hill(c) 1972. Included with permission.

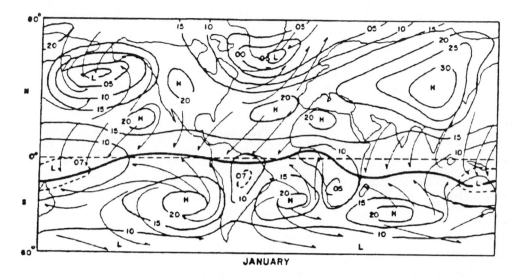

JANUARY

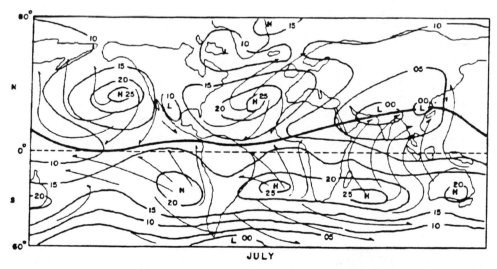

JULY

Weather as we know it, with its clouds and storms, is confined to the lowermost layer of the atmosphere, the troposphere. Look again at Figure VII-3. We have already noted that warm air underlies cold air in the troposphere, and that this potentially unstable situation accounts for the constantly shifting patterns of weather. Warm air tends to rise and be displaced by descending colder air. Global circulation tends to keep this process going at all times, and local instabilities can develop into storm systems, fed energy by the heat engine of the atmosphere and guided by the patterns of convection and the Coriolis Effect. Many storm systems are generated where large masses of warm and cold air interact at fronts. We first will examine the structure of different kinds of fronts and then go on to relate them to the development of storms.

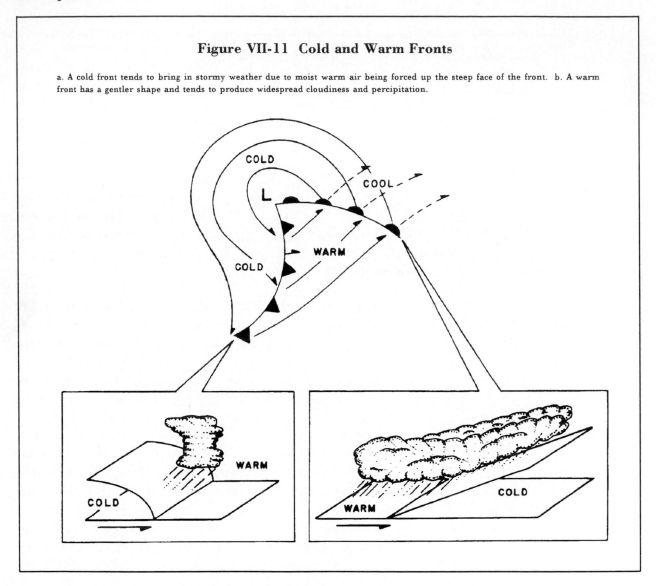

Figure VII-11 Cold and Warm Fronts

a. A cold front tends to bring in stormy weather due to moist warm air being forced up the steep face of the front. b. A warm front has a gentler shape and tends to produce widespread cloudiness and percipitation.

Figure VII-11 illustrates the differences between cold and warm fronts. The views in the lower part of the diagram show the structures of the fronts while the symbolic representation in the upper part shows how they are depicted on weather maps. In both cases the fronts are moving from left to right.

When a cold air mass displaces a warm air mass, the cold air forces itself under the warm air in a blunt wedge that rapidly forces the warm air to greater height. The warm air cools as it rises and its moisture begins to condense into clouds. This process of condensation releases the latent heat of the water vapor, which warms the air and encourages it to continue rising. Powerful updrafts can result in the production of thunderstorms that are often associated with cold fronts impinging on warm, moist air.

The effect of the latent heat is a very important one in this case. It accounts for the fact that in temperate climates, thunderstorms tend to occur in the spring and summer and not in the winter. Moist summer air provides the extra energy of latent heat to fuel the violent updrafts of a thunderstorm. This effect is usually much reduced in the dryer winter air.

A warm front behaves differently from a cold front because its structure is different. The warm air usually slides up and over the resident cold air mass along a gently sloping plane (see Figure VII-11b). Cloudiness and precipitation often occur, but in a far more extended version than in the cold front model. As a warm front approaches, high clouds are often seen. The clouds thicken and lower, and precipitation in the form of rain or snow may follow, perhaps lasting for some time.

A third type of front may form when a fast-moving cold front approaches and overtakes a warm front. The result is an occluded front, in which the warm air mass is lifted up between the colliding cold masses and no longer contacts the ground. At the ground surface the temperature may not drop much as an occluded front passes, but cloudiness and precipitation in the high warm air mass may result nonetheless.

Cyclones develop around low pressure regions, and are especially abundant in the polar frontal system separating the cold polar air masses from the warmer mid-latitude westerlies (see Figure VII-9 on page 229). These extra-tropical cyclones, as they are called, are not usually similar to hurricanes and typhoons, which are tropical cyclonic storms. In general they are massive circulating low-pressure weather systems that tend to move from west to east in the mid-latitudes. They develop as the result of wave-like instabilities in the polar frontal system.

More spectacular are tropical cyclones that develop over the open oceans in low latitudes. Called hurricanes or typhoons (or called cyclones in the South Pacific and Indian Oceans), their structure is similar to but more compact than extra-tropical cyclones and they derive their often devastating power from the warmth and moisture of the oceans. They are characterized by strong updrafts associated with a low-pressure region. Condensation and the release of latent heat intensifies the updraft, causing a convergence of air around its base. The Coriolis Force twists the air currents into cyclonic motion, while the converging air streams speed up as they spiral in toward the central updraft, much as a twirling ice skater spins faster as she draws her arms in closer to her body.

Tropical cyclones can grow in strength while they are over warm ocean and become of hurricane strength with wind speeds in excess of 33 m/sec (73 mi/hr) around a central region called the eye of the storm. In this central low pressure zone, winds are more gentle and cloud cover is light and sometimes absent. The reason is that the principal updrafts occur in a band surrounding the center of the storm. Warm dry air from elevations over 10 km is sucked down into the eye of the storm, producing the deceptively clear air of the eye. When a hurricane passes directly overhead, people who have taken shelter will sometimes emerge, thinking the storm has passed, only to find the fury of the winds renewed once the eye has passed. Their ordeal is only half over but now, of course, the winds will come from the opposite direction. Hurricanes begin to weaken and dissipate once they are over land because the land surface exerts a greater frictional drag upon them and also because they are deprived of moisture input from the ocean.

Another feared type of storm occurs on a much smaller scale than hurricanes. Tornadoes occur almost always in conjunction with thunderstorms and can develop violent twisting winds in their funnel-shaped vortices with speeds of 100 m/sec (220 mi/hr). The funnels range in size from a hundred meters or so up to a kilometer across. So much energy packed into such a small package makes them capable of immense damage, albeit restricted to the small areas actually touched by the funnel.

Tornado funnels apparently form aloft and propagate downward, and seem to be associated with the strong downdrafts accompanying thunderstorms. Even so, the worldwide distribution of tornadoes shows that they form mostly in land areas where conditions are particularly favorable. They are found in the mid-latitudes in Europe, Japan, northeastern India, South Africa, coastal Australia, New Zealand, and a small area of central South America. But by far the largest number are found in the United States, mostly east of the Rocky Mountains, where they most commonly result from the interaction of warm, moist air sweeping north from the Gulf of Mexico beneath cool, dry air moving east at a greater height. Tornadoes are essentially unknown in the huge landmass of Asia north of the Himalayas.

The common summer thunderstorm is a much less imposing affair than a hurricane or tornado, but it has its own majesty and can be capable of wreaking damage as well. Thunderstorms develop when warm, humid air near the ground wells upward in a strong updraft, causing condensation and the familiar latent heat intensification of upwelling. Because there is so much water condensing and falling as rain (and sometimes as hail), strong downdrafts develop in parts of the storm, producing high winds in advance of the storm cell where the downdrafts encounter the ground and spread out.

The presence of both up- and downdrafts in thunderstorms can toss around even large airplanes, and because a thunderhead can extend to heights of over 12 km (39,000 ft), jetliners are usually routed around them. Infrequently it is necessary for a commercial airliner to land through the fringes of a thunderstorm, providing a memorable experience for its passengers. When conditions are too turbulent, flights are forced to circle aloft until the storm clears or are rerouted to other airports.

Small, local thunderstorms may last only an hour or less before they dissipate, but a series of storms may become organized ahead of an advancing cold front in what is called a squall line. These tend to be more dangerous and can spawn hail and tornadoes.

Large "thunderstorm" systems sometimes form over North America, typically 160 to 400 km (100 to 250 mi) across, and account for more than 50% of the summer precipitation in the farm belt. These super storms are called mesoscale convective clusters.

Lightning and thunder add considerable drama and some danger to the performance put on by a thunderstorm. The exact mechanism by which lightning is generated is still not known for certain, but the process appears to be related to the precipitation of ice and water droplets within the thunderclouds. In this process, negatively-charged electrons are stripped from atoms and transported downwards to the lower reaches of the cloud, leaving the upper portions positively charged. Lightning bolts occur as a means of reducing the separation of electric charges that build up, either within the cloud or between the ground and the cloud.

Lightning is simply a flow of a large electric current through the air, as Benjamin Franklin demonstrated with his famous (and dangerous) kite-flying experiment. These currents can reach 50,000 amperes or more in a single stroke, and can electrocute people. Prompt treatment with cardio-pulmonary resuscitation (CPR) techniques can sometimes revive a victim of lightning stroke.

The electric current spreads out once it hits ground, taking whatever routes good electric conductors may afford it. For this reason, safe shelter from lightning may be found in automobiles or houses, where metal walls or pipes will provide a route for the electric current that generally will detour it away from you. For this reason it is best to stay away from plumbing or television antenna leads during thunderstorms.

All in all, more people in the United States are killed by lightning than by the effects of hurricanes and tornadoes combined. If you are indoors and are not part of the electrical circuit, you have little to fear from lightning. On the other hand, persons outdoors are in exposed positions, and should attempt to find shelter in a nearby building or auto. Trees, however, are poor choices for shelter, since trees are frequent targets for lightning strokes. Golfers, boaters, and mountain climbers often find themselves in exposed situations during thunderstorms, and these enthusiasts are advised to consult weather forecasts before setting out.

STUDY QUESTIONS

VII-13. Check the weather map each day in a newspaper. Look for features that illustrate some of the principles we have discussed. The best weather maps will show pressure isobars and wind directions as well as fronts. If your newspaper's map does not, you might check major or capitol city newspapers in your library. This will enable you to look at a week or more of development in the weather system at one time. Some of the things to look for: circular wind patterns around low and high pressure cells; the relation between precipitation patterns and cold and warm fronts; and the development of extra-tropical cyclones.

VII-14. Why is widespread cloudiness more likely to be associated with a warm front than with a cold front?

VII-15. Can you think of a reason why thunderstorms are more prevalent during the summer than during the winter?

H. WEATHER FORECASTING

1. Observations and Methods of Forecasting

Few scientific activities are as visible to the layman as weather forecasting. Evening news programs feature "meteorologists" who stand before colorful weather maps and radar pictures and deliver detailed forecasts for the next 24 hours and more general forecasts for the next few days. Some of the television personalities on your screen are trained scientists who participate actively in preparing the forecast while others are simply delivering forecasts prepared by private meteorological firms or the National Weather Service. All, however, rely on data collected worldwide under the guidance of the World Meteorological Organization in a carefully organized and standardized program.

At hundreds of weather stations around the world, instrumented weather balloons called radiosondes are released twice each day at synchronized times -- at whatever local time corresponds to midnight and noon in Greenwich, England. These balloons carry standardized instruments that measure atmospheric pressure, temperature, and humidity as they rise through the troposphere and into the stratosphere. As the balloon rises to regions of lower pressure, the gas in the balloon expands. The balloon eventually bursts and the instrument package is lowered by parachute to the ground. In many cases, the motion of the balloon is tracked during its travels so that information on wind speed and direction is also obtained.

Today, weather maps are constructed from information compiled from the radiosondes, along with observations made from the ground and from satellites. Published maps showing whole continents usually display atmospheric pressure at sea level, temperature, winds, and frontal systems (Figure VII-12). These large-scale maps are called synoptic maps, because they give a synopsis or overall view of the weather in a large region at a particular time. Weather forecasters also make extensive use of similar maps showing conditions aloft, allowing them a more three-dimensional view of the atmosphere.

Two more recent techniques have added to the forecaster's tools -- satellites can scan the atmosphere from above it and determine profiles of temperature, humidity, and ozone concentration as well as images of cloud cover. Weather radar is now mostly used to determine local patterns of precipitation.

Even with this background of worldwide data, however, the weather forecasting task remains profoundly complex. Several prominent meteorologists have maintained that the weather forecasting problem is the most difficult that exists in nature except for that of understanding the human condition. The atmosphere is big, has many external and internal interactions, and is constantly changing on scales that range from a few meters to the span of the entire Earth. The physical laws governing atmospheric behavior are reasonably well understood, but the problems associated with applying them to all parts of the atmosphere simultaneously and throughout an extended period of time are among the most formidable found in all of science. It is not surprising that weather and climate researchers are among the best customers for manufacturers of the world's largest and fastest supercomputers.

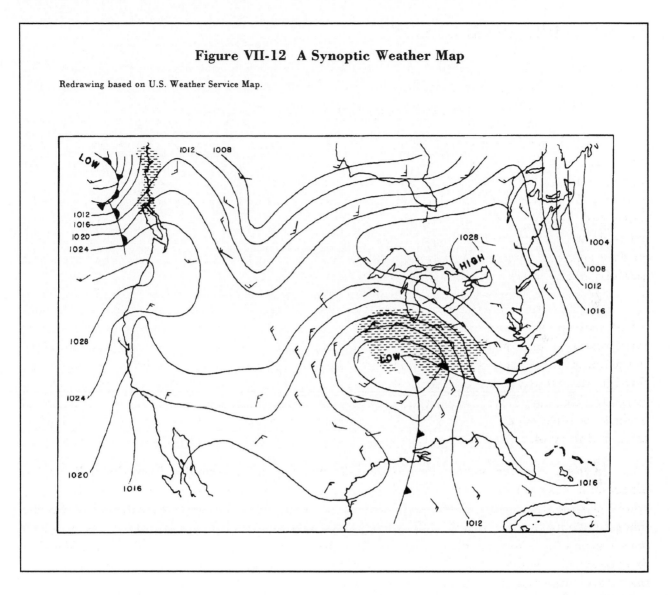

Figure VII-12 A Synoptic Weather Map

Redrawing based on U.S. Weather Service Map.

As it is practiced today, weather forecasting utilizes a number of different approaches. The traditional method relies on the synoptic maps and so is referred to as the <u>synoptic approach</u>. From the existing pattern of highs, lows, fronts, temperature and humidity data, and the relation between surface conditions and those at height, the meteorologist applies his experience and knowledge of prevailing and historical weather behavior to predict the motion and development of fronts and storms. Initially, this practice was as much art as science, as it often relied upon the forecaster's experience and intuition, and the best practitioners often found it difficult to teach their skill to others.

Another approach uses what climatologist Stephen H. Schneider calls a kind of meteorological uniformitarianism -- weather records are searched to find close matches to the present synoptic maps. The weather that resulted from the past situation can then be used as a guide to forecasting the weather at the present time. This method is called the <u>analogue approach</u>. It simply assumes that similar conditions will give rise to similar weather. While this is often the case for a broad-brush view of the weather, differences in details of past and present conditions can sometimes lead to substantial differences in the resulting weather.

A somewhat different method is the statistical approach that uses statistical analysis to relate different kinds of data to valid predictions on the basis of past records. Present data are then used to make predictions using the statistical guidelines.

More recently, numerical weather models have been used to supplement the other three approaches. Here, a synoptic description of the state of the atmosphere in terms of pressure, temperature, humidity, wind velocity and other variables measured at several different elevations are fed into a computer which then calculates the evolution of the weather system as time goes on. In order to do this, the computer uses the physical laws that govern interactions between all the different parts of the atmosphere and between them and external influences such as the Sun and the oceans. We shall discuss numerical models and how they are constructed in some detail in the next unit.

At present, fairly reliable weather forecasts can be made covering a span of about two days, but only large-scale events such as temperature changes and the arrival of fronts can be predicted with accuracy. Smaller scale phenomena such as local thunderstorms and showers, hail, and downpours can only be predicted in a statistical sense. A forecast that predicts a 20% chance of precipitation for your area means that there is one chance in five that you will encounter measurable precipitation during the time covered by the forecast wherever you happen to be. Very transient and localized phenomena such as tornadoes can hardly be predicted at all, but once they have formed and are sighted, they are tracked and tornado warnings are sent out to people in their vicinity. Similarly, conventional weather radar can track thunderstorms and help to predict where the most severe activity is likely to occur during the next hour or so.

Noise in the data actually makes the first 24-hour forecast less reliable than that for the second 24-hour period. Useful but less accurate forecasts may be made out to five to seven days.

RECOMMENDED READING

Louis J. Battan, Fundamentals of Meteorology, Prentice-Hall, Inc., 1979.

Kendrick Frazier, Storms on a Deadly Scale, Mosaic, Nov.-Dec. 1982.

UNIT VIII THE CLIMATE PUZZLE:

CLIMATES OF EARTH

TAYLOR GLACIER

A. INTRODUCTION

1. Overview

In the last unit you learned the essential mechanisms that govern the operation of Earth's atmosphere and of the weather that results. This unit will focus on the more general subject of climate, in which our concern is not the day-to-day weather variations, but the longer-term picture of temperature and precipitation. Climatologists reconstruct ancient climates by use of climate proxies -- records of any sort that serve to indicate what the prevailing climate was in times long past. With the aid of climate proxies plus some kinds of direct geological evidence, a rough picture of climate throughout geological time is beginning to emerge. Through the use of numerical models of climate systems, it is now becoming possible to test various hypotheses that purport to explain the onset of the Ice Ages and that suggest that they are far from over. These models also allow us to predict the consequences of mankind's emissions into the atmosphere, and to give us warnings of changes that we are unintentionally bringing about in Earth's great climate systems.

2. Objectives

Upon completion of this unit you should be able to:

1. distinguish between the use of the terms "weather" and "climate"
2. relate some examples of local climate to large-scale features of atmospheric circulation
3. describe different methods used to reconstruct past climates
4. describe major episodes of climatic history throughout geologic time
5. relate climatic change to influences on the development of human cultures
6. describe climatic conditions during the Ice Ages
7. discuss possible causes of the Ice Ages
8. understand the nature of numerical models and relate their operation to actual climate behavior
9. relate projected increases in atmospheric carbon dioxide to possible future climate changes
10. discuss the possible effects of a number of man's activities on global climate.

3. Key Terms and Concepts

climate

rain shadow

climate proxies:

    historical records

    glaciers and moraines

    archaeological evidence

    tree rings and dendrochronology

    pollen

    oxygen isotope ratios

    sea-level changes

Little Ice Age

Maunder Minimum

Ice Ages

interglacial

Holocene epoch

Pleistocene epoch

Permo-Carboniferous glaciation

Volcanic Dust Veil Effect

Spectrum analysis

The Year Without a Summer (1816)

applications of numerical models:

    weather forecasting

    south Asian monsoons

    Cretaceous climate

    effects of El Niño

    diversion of Russian rivers

    Nuclear Winter

Milankovitch hypothesis

orbital ellipticity

precession of the equinoxes

numerical models

heat transfer

algorithm

initial conditions of a model

boundary conditions of a model

resolution of a model

climatic effects of increasing concentrations
    of carbon dioxide and trace gases

4. Corresponding Video

In this program, you will see how critical the monsoons are to the survival of pre-agricultural civilizations. You will see how scientists have constructed records of past climates through such diverse sources as stalactites in limestone caves, cores from the 4,000-meter thick East Antarctic ice sheet, and ocean bed cores chock full of the skeletal remains of dead microbial creatures. You will examine the role that volcanism has played in Earthly climates, and contrast that to the role played by carbon dioxide and other trace gases, the so-called Greenhouse Effect. You will explore how the warming trend from the Greenhouse Effect is first evidenced in the extreme sensitivity of alpine glaciers. And finally, you will examine future climate trends on Earth, and how man may someday affect these trends through conscious planning, or terraforming.

B. WEATHER AND CLIMATE

There is an expression used by inhabitants of Denver, Colorado: "The climate is great, but the weather is awful!" The expression indicates that while the general or average weather in that area is delightful, with abundant sunshine and frequent clear, crisp days, the weather is very changeable, with occasional high winds, violent thunderstorms, snow squalls, and sometimes even tornadoes bursting upon the scene.

Weather, of course, is the moment-by-moment and place-by-place description of the state of the atmosphere, with events taking place on a small scale in both space and time. A fall day may begin warm and rainy, clear by midday, and finish with snow flurries. A thunderstorm may drench Minneapolis but leave St. Paul dry and calm. Such are the vagaries of weather.

Climate, on the other hand, encompasses the totality of weather effects, accumulated over an extended period of time. In order to describe the climate of a region, it is necessary to know the extremes as well as the averages. It is one thing to know the mean annual temperature and rainfall of a place, but it is much more meaningful to know how hot it gets in the summer and how cold it gets in the winter; whether the rains tend to fall in thunderstorms or in extended drizzles; whether there are wet and dry seasons; and whether the place is subject to occasional high winds, tornadoes, or hurricanes.

In the last unit we described the effects of the relatively stationary high pressure cells positioned over parts of the oceans. For the most part, these highs produce an outflow of dry air that travels with the prevailing winds. Downwind of these highs, rainfall is scant, with many of the world's bleakest deserts attributable to their effects. Other dry climates may result from a combination of atmospheric and geographic variables. A mountain range forces air masses traveling over it to greater heights, where they cool and condense water vapor in the form of clouds yielding rain or snow. When the wrung-out air descends on the other side of the mountain, it warms and its relative humidity drops. Mountains often produce a "rain shadow" downwind of them, and dry climates are found east of the Rocky Mountains and north of the Himalayas.

The interiors of large continents are often arid when hemmed in by mountains, as in the case of central Asia, and the effect persists throughout much of the western United States. West of longitude 100°W (cutting through the middle of Nebraska), irrigation is generally necessary to large-scale farming of crops other than wheat in the United States. East of this line, moisture flows north from the Gulf of Mexico and results in a wetter climate that supports agriculture without the need for irrigation.

Climate remains the ultimate limiting factor in the production of food, and even fairly small shifts in climate can affect the nourishment of tens of millions of people. As this is being written, large parts of northern Africa remain in the grip of a devastating drought that in many areas has been in progress since at least 1970. In Ethiopia and other African countries starvation and malnutrition are rampant, and a whole generation of children will probably bear the lifelong burdens of deformity and retardation as a result. Countless others are perishing for want of the life-giving rains. The problem is especially acute in those regions where the population has expanded to the limits of the best-case climate. When natural climatic fluctuations decrease rainfall and diminish food production from optimum levels, disaster is inevitable.

VIII-1. Cite examples of more-or-less permanent natural factors that can produce arid regions.

C. RECONSTRUCTING PAST CLIMATES

Were the climates of Earth always as they are today? There is much evidence in historic geological records that says "no"! In recent times, we may simply consult meteorological records from stations around the world. Figure VIII-1 shows that in the northern hemisphere, the mean annual temperature rose somewhat from the late nineteenth century until 1938, then fell until around 1970, and has been rising again since then. Your great-grandfather may well have been correct in maintaining that winters today aren't what they used to be. Or was his memory influenced by a lack of central heating and snowplows in his youth? This is in fact a serious problem for climatologists who would like to reconstruct climate records from historical descriptions when precise measurements are lacking. Often it is difficult to separate the cultural from the climatic. Direct measurements of temperature are preferable, but they extend back only to about 1700. In their absence, we must gain what information we may from geological climate recorders such as ocean sediments and ice or from a variety of climate indicators, which are called climate proxies. Historical examples of the latter might include records of first and last days of frost and of crop yields.

In spite of difficulties in using them, historic (and even prehistoric) cultural evidence may be used to help reconstruct major shifts in climate. Such a shift apparently occurred during the interval 1500-1850 AD which is now referred to as the Little Ice Age. Entries in diaries and tax records mention the deteriorating climate as shepherds were forced out of their high mountain pastures in the Alps by advancing glaciers. In some cases the mountain glaciers pushed into lower alpine valleys, crushing houses and causing great hardship. Mountain glaciers are sensitive indicators of climate because their length is dependent on the amount of precipitation and on average summer temperatures.

Glaciers form wherever large amounts of snow from the previous winter fail to melt during the summer. The next winter's snowfall is deposited on top of the old snow, steadily building up layer upon layer. The weight of the pile, together with some melting and refreezing, finally compresses the snow into ice. It is a property of ice that it is plastic in nature and can flow downhill under its own weight almost like a thick liquid. We previously encountered this kind of solid-state creep in the unit **Continental Tectonics and the Earth's Interior** when we discussed convection in Earth's solid mantle.

A glacier moves downhill at speeds that are on the order of a few centimeters to one meter per day. Many may extend far into lower valleys, where snow falls much less and summer temperatures are warmer. Here, summer melt exceeds winter snowfall, and the glacier

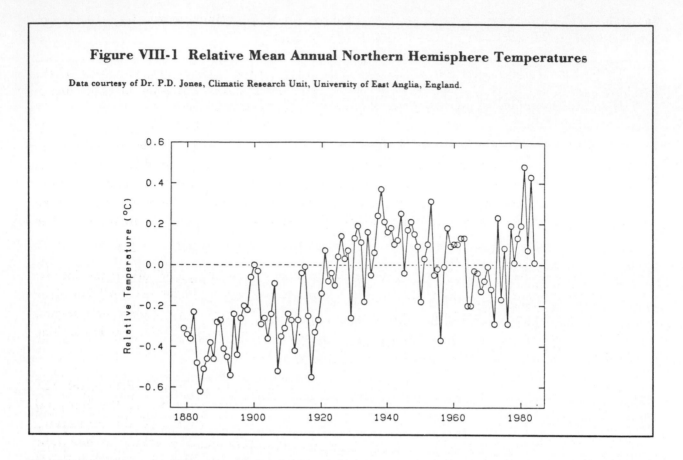

Figure VIII-1 Relative Mean Annual Northern Hemisphere Temperatures

Data courtesy of Dr. P.D. Jones, Climatic Research Unit, University of East Anglia, England.

begins to melt, eventually reaching a point where the advancing ice melts completely and the glacier ends in what is called the toe of the glacier.

If winter snowfall in the glacier's source area increases or if summer temperatures drop, the toe will advance farther downhill. Contrariwise, if snowfall decreases or summer temperatures rise, the toe will retreat up the valley as more ice melts than is coming down to replenish it.

Glaciers are massive and destructive. They can pluck huge boulders from surrounding rock and carry them along in their bottom layer, gouging and scouring out more rocky material. Rocks and gravel are transported down to the toe of the glacier by the moving ice where they are deposited in great heaps fringing the toe, called moraines. If a glacier has advanced and then retreated, we can determine from the location of the moraine just how far the toe of the glacier had reached. If the moraine can be dated, perhaps by Carbon-14 dating of wood in the moraine, it can serve as a useful climate proxy, indicating the coldest and wettest point in time.

In the case of the Little Ice Age, there are numerous direct references to colder-than-average temperatures in the historical records. One set consists of records of the numbers of days that the canals in the Netherlands were frozen. Because the canals were used for transportation, reliable records have been kept since 1633. Today, they hardly ever are frozen, but during the seventeenth century, freezes were much more common, sometimes lasting three months. Contemporary paintings show the citizenry enjoying outings on the ice, skating, or simply crossing the canals without benefit of bridge or boat.

The French, meanwhile, were concerned for the safety of their vines, and some of the harshest winters did great damage. Records of grape harvests serve as another kind of climate proxy, with late harvests indicating cooler summer temperatures.

It is not known what caused the Little Ice Age. One hypothesis capitalizes on astronomical records that seem to indicate that sunspots virtually disappeared from the face of the Sun during the period 1645 to 1715. In recent years, sunspots, which are actually cooler regions of the Sun's surface that are the sites of vast magnetic storms, wax and wane in an eleven-year cycle. They are clearly indicators of solar activity, as we shall see in the unit **The Sun**.

Could this period of low sunspot activity, named the Maunder Minimum for its discoverer, have been responsible for the low temperatures of the Little Ice Age? No clear connection between sunspot activity and climate has been established, but there is no question that even a slight decrease in solar energy output could produce a drastic cooling on Earth. The possibility remains intriguing.

Somewhat earlier, during the Middle Ages in the period from about 900 to 1200 AD, another major climatic shift occurred -- an abnormally warm period known as the Medieval Optimum. During this period, wine was produced in English vineyards and the far-ranging Vikings established a colony on the southwest coast of Greenland. The colony at one time boasted a population of 3,000 and lasted for four hundred years, but by 1450 AD these hardy pioneers became isolated by increasing sea ice and perished at the close of the Medieval Optimum.

Climate is obviously a potent influence on civilization. The entire social fabric can be severely stressed by prolonged drought, cold summers, and consequent crop failures. In fact, the collapse of some early civilizations has been blamed on climatic stress. Near the junction of Arizona, New Mexico, Utah, and Colorado in the American southwest many ruins have been excavated of the Anasazi Pueblo Indians. Thousands of people lived in elaborate pueblo structures, some built into shallow caves beneath overhanging cliffs. These spectacular cliff dwellings are a major feature of Mesa Verde National Park in Colorado.

Decreasing rainfall in the region at first forced this agricultural people to develop sophisticated irrigation techniques in order to grow the corn, beans, and squash upon which their diet was based. But by 1300 AD, the cliff dwellings were deserted, and the Anasazi dispersed for reasons unknown. A substantial drought lasting from 1276 to 1299 may have been at least partly responsible, though it has been argued that these people had successfully weathered earlier, more severe droughts. There is some archeological evidence that the populations of the pueblos had increased to the limits of the ability of the agricultural system to supply food during a relatively benign climatic period. Perhaps when the long drought set in, a desperate famine resulted, not unlike the situation developing today in parts of Africa, India, and Bangladesh. For the Anasazi, however, there were no prospects of humanitarian aid from other, more fortunate parts of the world, and their civilization collapsed.

How do we know that a drought occurred three thousand years ago in the American southwest? Here we make use of still another kind of climate proxy: tree rings. As a tree grows, it adds a layer of cells to the outside of its trunk, building a distinctive new layer, or ring, each growing season. The age of a tree can be determined by counting its rings. In semiarid regions like the American southwest, trees respond to the climate, adding a thick ring in favorable years and a thin ring when stressed by harsh conditions. At lower elevations ring thickness is most influenced by rainfall, while trees growing at high elevations increasingly feel the effects of temperature, which regulates the length of the growing season.

The study of tree rings, or dendrochronology, has been employed to date the Anasazi dwellings and to monitor the climate during their heyday. Because the patterns of thick and thin growth rings is common to most trees in a given region, the patterns contained in timbers used in the Anasazi buildings can be matched to those found in living trees, establishing the crucial link to the present.

Anthropologists at times have evoked climate change to explain the sudden decline of ancient civilizations. At the moment there is a growing suspicion that this explanation may have been overworked in some cases, but there is no denying either the importance of climate as an influence on society or that drastic climate changes have taken place at critical times for some early cultures.

The ancient Harappan culture in the Indus River Valley of Pakistan and India flourished around 4,500 years ago and was based on a varied agriculture that may have been the first to cultivate cotton. They apparently did not practice irrigation but relied on natural rainfall. Today the ruins of the Harappan civilization are being uncovered on the fringes of the Thar (or Rajasthan) Desert of Pakistan. It is a dry place, except for annual flooding of the Indus river, and demonstrates dramatically how climate can change in one place over a short period of time. In fact, there is evidence that the drying of this region took only a few years.

As climatologists have pushed further back in time to reconstruct ancient climates, they have turned increasingly to diverse climatic indicators. One of these uses pollen as an indicator of the kinds of plants that grew in a region. Pollen is often preserved in sediments, and different species of plants produce recognizably different pollen. As a result, a cooling climate may be indicated by a decrease in pollen from oak and hickory trees and replacement by pollen from pine trees.

Global climate changes also affect the amount of glacial ice held in the polar regions. Piled atop Antarctica and Greenland, this mass of freshwater ice is effectively being withheld from the oceans. The more polar ice there is, the lower the level of the oceans. These changes in worldwide sea level can often be detected by wave-cut beaches and terraces, or benches, cut into seashore areas. If these terraces can be dated by some independent method, they can serve as useful climate proxies.

One of the most useful of all climate indicators is the oxygen-18/oxygen-16 ratio measured in organic sediments and in ice. Nearly all oxygen in our atmosphere has an atomic weight of 16 (that is, there are eight protons and eight neutrons in its nucleus). A small percentage (0.2%), however, has two extra neutrons, giving it an atomic weight of 18. Both kinds of oxygen behave the same chemically, combining with hydrogen to form water, for example. The only difference is that oxygen-18 is 12.5% heavier than oxygen-16.

This mass difference influences the physical behavior of water. For instance, water molecules containing the heavier oxygen-18 do not evaporate as readily as those containing the lighter oxygen-16, and this effect is more pronounced at lower temperatures. This fact may be used in two distinct ways.

The first involves measuring how the oxygen-18/oxygen-16 ratio varies in ice cores drilled from the Antarctic or Greenland ice sheets. Like sediments, these ice sheets are also layered or stratified, with the youngest layers nearer the top of the ice sheet. In many cases annual layers can be counted like tree rings. Because the ice is formed from snow which in turn is derived from evaporation from the oceans, a higher oxygen-18/oxygen-16 ratio in the ice indicates warmer conditions, allowing a greater fraction of oxygen-18 to be evaporated. Thus, ice cores may provide important climatic records extending back 120,000 years or more.

At the same time, growth of continental glaciers and ice sheets preferentially removes oxygen-16 from the oceans, leaving behind water that is slightly enriched in oxygen-18. This in turn finds its way into the skeletal and shell structures of sea creatures that contribute their hard parts to oceanic sediments. Calcium carbonate ($CaCO_3$) is the principal constituent of these fossil remains, and the oxygen in these molecules is drawn from the seawater in which they lived, recording the oxygen-18/oxygen-16 ratio that prevailed during their lifetimes. In effect, the creatures of the sea were unwitting scientists recording climatic conditions around them for our later perusal.

The use of oxygen-18/oxygen-16 ratios in calcium carbonate (limey) sediments is complicated by another effect: When creatures incorporate oxygen from seawater into their skeletal parts, the resulting oxygen-18/oxygen-16 ratio depends not only on that of the water but also on the water temperature, with a greater proportion of oxygen-18 being deposited at lower temperatures. Thus, the oxygen-18/oxygen-16 ratio in limey fossils reflects two different climatic variables: the temperature of the ocean waters and the amount of freshwater ice tied up in continental glaciers. Separating these two effects can be difficult, especially since they are related: colder temperatures tend to result in more polar ice. Nevertheless, because of the great span of geological ages represented by limestones from which oxygen-18/oxygen-16 ratios may be derived, this method has proved to be a key element in deciphering the climate puzzle for the last 500 million years. With it and the use of the various proxies mentioned throughout this section, we may push our climate inquiries back into the vastness of geological time.

STUDY QUESTIONS

VIII-2. How can the toe of a glacier retreat back up a valley? Does the ice move uphill?

VIII-3. How can logs used in ancient Indian buildings be used to reconstruct climatic conditions at that time?

VIII-4. How can those logs, cut down hundreds or thousands of years ago, be used to date the events recorded in their rings?

VIII-5. Would you expect the oxygen-18/oxygen-16 ratio to be higher or lower in rainwater than in the oceans?

VIII-6. The effects of temperature and ice volume are difficult to separate in the oxygen-18/oxygen-16 ratio studies. How then can we use these results with confidence in determining ancient climate trends from limey shells?

D. CLIMATE IN THE GEOLOGICAL RECORD

In 1966 a United States team drilled through the northern Greenland icecap at Camp Century, finally hitting bedrock 1,370 meters (4,500 ft.) down. Even more remarkably, they were able to recover an essentially continuous core of the entire thickness of the ice sheet. A Danish team analyzed the core using the oxygen-18/oxygen-16 method, obtaining the results shown in Figure VIII-2. Time runs from the top down in thousands of years, and oxygen-18/oxygen-16 ratios indicating colder climate are to the left, while warmer climates are recorded where the curve moves to the right.

Notable in the diagram are the tremendous variations extending throughout the record compared to the slight changes found during the most recent 10,000 years. This last epoch, called the Holocene, represents the relatively warm interval in which we now live. Clearly shown in the Camp Century ice core is the most recent of the great Ice Ages. It is shown in the figure as extending from 10,000 to about 70,000 years ago, preceded by a mostly warm interval.

The latest glaciation was only one of a number of advances and retreats of great ice sheets that covered most of Scandanavia and Canada as well as northern Europe and Russia and the northern tier of the United States. Two kilometers or more in thickness, these vast expanses of ice withdrew sufficient water from the world oceans to drop sea level by 100 meters (330 ft) below present levels. Under such conditions the seashore would have withdrawn in many places, uncovering nearly half of the present continental shelves. A person could walk across the English Channel or across the Bering Strait separating Russia from Alaska with dry feet -- and some may have, introducing Asian stock into the Americas and establishing the forebears of the present Native Americans.

During the Ice Ages, the world was a different place. Even where there was no ice (about 70% of the total continental area), climate was often very different. Many of the continental interiors may have been quite arid, but at least for a while the Sahara Desert was in bloom, enjoying a wetter climate than now. Recent Satellite investigations of the Sahara have disclosed river valleys and other water-eroded features buried beneath the sand dunes.

In addition to the oxygen-isotope data, there is a wealth of more direct evidence of the existence of Ice Age glaciation. A two-kilometer ice sheet is a mighty landscape architect, gouging out lakebeds, carrying huge boulders hundreds or even thousands of kilometers, carving out broad U-shaped valleys in the mountains, and depositing extensive moraines at its

Figure VIII-2 Last 120,000 Years Recorded in a Greenland Ice Core

O^{18}/O^{16} ratios indicating colder climates are to the left, warmer to the right. Time from top down in thousands of years. Reprinted from W. Dansgaard, S.J. Johnson, H.B. Clausen and C.C. Langway, Jr., K.K. Turekian, editor, <u>The Late Cenozoic Glacial Ages</u>, 37-56. New Haven: Yale University Press, 1981, and in L.J. Lockwood <u>Causes of Climate</u>, Halstead Press, Fig. 5.6, p. 149, 1979. Included with permission.

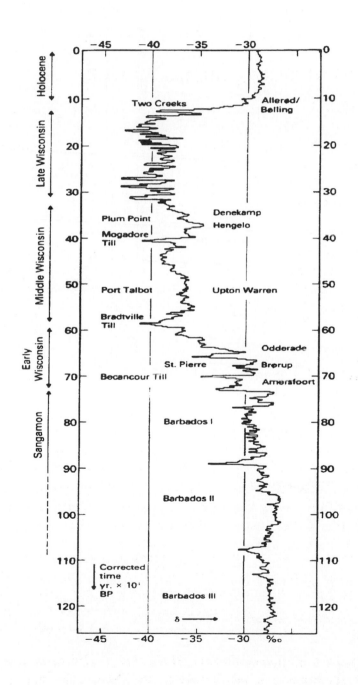

Figure VIII-3 Oxygen Isotope Record from Fossil Foraminifera Shells

Oxygen isotope record from fossil foraminifera shells extending to before the last magnetic polarity flip 730,000 years ago. Cold climates appear to the left; warmer climates to the right. Reprinted from N.J. Shackelton and N.D. Opdyke, <u>Quarternary Research Journal</u> Vol. 3, pp. 39-55, 1973. Included with permission.

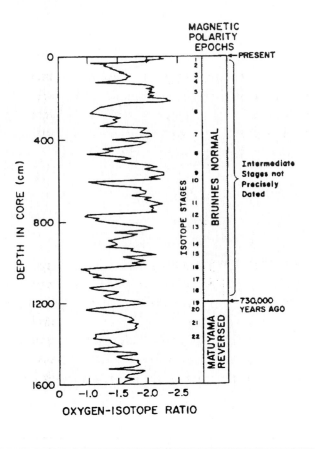

terminus. Great scratches or grooves can be found in bedrock that once felt the abrasion of house-sized rocks being dragged along by the moving ice sheet.

These kinds of this direct evidence of glaciation were observed and interpreted in 1836 by a Swiss professor of natural history, Louis Agassiz. Prior to his explanations, except for a few Swiss naturalists who had glaciers to study in their own backyards, the scientific world had interpreted boulders displaced far from their sources, glacial grooves, and moraines, as more work of the Biblical flood. Pursuing his studies and spreading his ideas in both Europe and America, Agassiz proposed the existence of the Ice Ages.

For the better part of two million years the ice sheets have come and gone defining what has been called the <u>Pleistocene epoch</u>. Originally lumped into four major Ice Ages on the basis of the direct evidence of their effects, the ice sheets probably advanced and retreated a number of times, as may be seen in Figure VIII-3. This shows a record of oxygen-18/oxygen-16

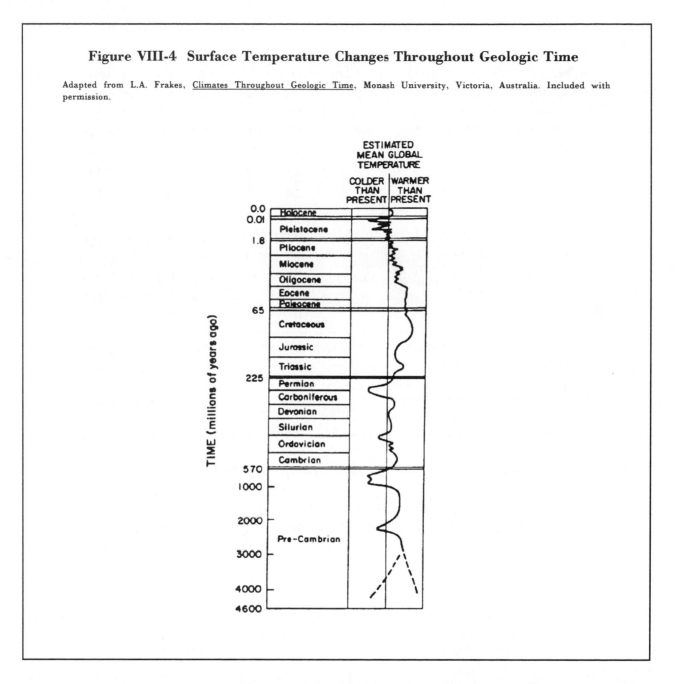

Figure VIII-4 Surface Temperature Changes Throughout Geologic Time

Adapted from L.A. Frakes, Climates Throughout Geologic Time, Monash University, Victoria, Australia. Included with permission.

changes recorded in the limey shells of fossil foraminifera, tiny shelled creatures that lived on the ocean floor.

    This core from the Pacific Ocean floor was subjected to magnetic analysis and was found to contain a record of the most recent flip of the Earth's magnetic field that occurred some 730,000 years ago (see Figure III-20 on page 75). This established one firm date in the core (in addition to the top of the core, which records the present time). As a result, the core provides us with a detailed record of approximately the last half of the Pleistocene epoch. It would appear from this record that climate fluctuations occurred rapidly and relatively frequently during the Pleistocene, with major peaks (or dips) occurring with a periodicity of approximately 100,000 years. We shall return to the significance of this periodicity in a little while.

Taking a longer view, evidence of many kinds may be compiled to produce a rough picture of average world temperature changes (Figure VIII-4). From it we can see that Pleistocene climates have not been at all typical. In fact, throughout the entire Mesozoic (225 - 65 million years ago) and Cenozoic (last 65 million years), except for the Pleistocene, temperatures have been distinctly warmer than they are today.

Only when we look to the Paleozoic and earlier (more than 225 million years ago) do we see evidence for global cooling sufficient to trigger widespread glaciation. One such period occurred near the Permian-Carboniferous boundary while a smaller one occurred in the Ordovician. At least two Ice Ages appear to have occurred in the Precambrian.

What, then, is "normal" Earth climate? This question is more complex than it might seem. By now you already know that the Earth is a very dynamic planet, and that its geography has been changing constantly throughout time. Should we expect it to have one "normal" climate?

About Precambrian climate we can say little. For much of it life existed in the oceans, and so we can say that the oceans did not freeze and they did not boil. In all likelihood the range of temperatures in at least the latter half of the Precambrian was not very different from what it has been since.

Perhaps most remarkable was the climate during the Mesozoic and early Cenozoic. Though some cryptic evidence exists for rapid sea-level fluctuations in the Mesozoic, it is generally believed that there were no ice sheets present during this era, not even in the polar regions. In addition, climate was probably less extreme in general, with less temperature difference between the poles and the equator than today. This was the Age of the Dinosaurs, in which reptiles dominated the web of life. Because reptiles are mostly cold-blooded, a warm and even climate would have provided a favorable ecological situation for them.

STUDY QUESTION

VIII-7. How can a slow-moving glacier carve a valley into a U-shaped profile?

E. CAUSES OF THE ICE AGES

Our foray into climate history allows us to recognize the present time as one of the anomalous periods of Earth history. Figure VIII-3 on page 251 gives us every reason to believe that our present warm period is simply another interglacial in the long alternation of glaciations and interglacials. Barring climatic changes introduced by humans, it appears highly likely that the ice will advance once again in a few thousands or tens of thousands of years. We can at least take comfort in the slow march of geologic time.

Why does Figure VIII-4 appear to show only five major periods containing Ice Ages? There is, of course, the very real possibility that there were many more that have escaped notice, but even so, the question remains: Why do some periods, like the Pleistocene, show frequent Ice Ages and others, like the Jurassic, apparently none at all?

The patterns seem to show both long-term and short-term behavior. On the long term, some periods, lasting for tens of millions of years, favor glaciation while others do not. There have been many explanations advanced to explain the onset of an Ice Age, but only one so far has gone relatively unchallenged, and that one applies only to the Ordovician and Permo-Carboniferous glaciations. Fragmentary evidence for those glaciations has been found in widely separated places: India, equatorial Africa, southeastern South America, and Australia. When these continents are reassembled into Gondwana, which was in existence at those times, the glaciated areas form a polar cap at about the locations of the Ordovician and Permian South Poles.

Even so, it is necessary to explain why there was any polar ice at all. One theory holds that an excess of land area near the poles helps to bring on a glacial episode by providing a base for the accumulation of snow, which then reflects the Sun's energy back into space, producing further cooling, more snow yet, and so on in a self-reinforcing cycle.

Of equal importance may be the geography of the ocean floor. Ocean currents, as we have seen in the unit **Circulation of the Oceans**, are efficient transporters of heat energy from the equator to higher latitudes. When the poles are accessible to wide-ranging ocean currents, the result may be a lesser temperature differential between equator and poles and a failure to build up significant ice in the polar regions. On the other hand, when the poles are isolated from the major ocean currents, then they tend to cool and perhaps spawn an Ice Age. This was certainly true for the Ordovician and Permo-Carboniferous glaciations, and may be true today, where access to the Arctic Ocean has been nearly cut off by the encircling continents. In addition, the presence of Antarctica astride the South Pole prevents currents from reaching that pole as well.

Many other suggestions have been put forward, including decreases in carbon dioxide content of the atmosphere causing a reduction of, or changes in the dust content of the stratosphere due to volcanic eruptions. Fine dust and sulfate particles can remain suspended for a long time, decreasing the solar energy that can penetrate and warm the lower atmosphere. As an example of this effect the "Year Without a Summer" is often cited.

In 1815 the volcanic island of Tambora in Indonesia exploded violently, blasting the island nearly to oblivion and killing tens of thousands of people. It injected huge quantities of volcanic dust into the atmosphere, changing the surrounding region into a darkness that lasted several days. Sunsets around the world were brilliant during the next year -- an effect of the dust cloud that encircled the world. Whether because of the eruption, or by coincidence, or both, the summer of 1816 was in many places the coldest on record. In New England, midsummer frosts occurred, destroying crops and prompting the name, "The Year Without a Summer."

Studies of other massive volcanic eruptions and succeeding weather patterns indicate that there may be a Volcanic Dust Veil Effect on climate. During the last few years, we have seen both El Niño and the eruption of El Chichon volcano in Mexico, resulted in a long-lasting dust veil. Even so, it is difficult to relate these events to local climatic effects such as droughts. Climatic changes on the scale of a few years have been far easier to describe than to explain.

Returning to our discussion of the Ice Ages, there are also the shorter-term variations to explain, such as those shown in Figure VIII-3 on page 251. In particular, there appears to be a prominent cycle lasting about 100,000 years in the records of Pleistocene glaciation. Oxygen isotope records of this type have been subjected to mathematical analysis to determine the mix of cycles present in them.

We may use as an analogy a chord played on a piano, with four notes played simultaneously. If the waveform of the sound were displayed on an oscilloscope, we would tend to see a complex trace that might resemble the curve in Figure VIII-3. When we hear the sound, however, our ears have no difficulty in separating the complex sound waveform into its four component notes. A mathematical procedure known as spectrum analysis allows us to do the same thing with any complex waveform as our ears do naturally with sound.

When climate records of the past one million years are subjected to spectrum analysis, three prominent cycles emerge from them: a strong one with a period of 100,000 years and three lesser ones with periods of 43,000, 24,000, and 14,000 years. These cycles are so clearly defined in the record that they suggest we should search for natural causes that operate with the same set of periodicities. In fact, as long ago as 1920 the Yugoslav astronomer Milutin Milankovitch produced calculations showing that energy input from the Sun may have been affected by regular changes in Earth's orbit that have much the same periodicities.

At the time Milankovitch published his work, it was not taken very seriously as an explanation for the Ice Ages because the supporting evidence was not yet at hand. By now it should no longer surprise you that a scientist may come up with an essentially correct explanation and be effectively ignored by colleagues until supporting evidence builds up (often at a much later time). The scientific method demands that we test hypotheses against all the observational evidence that is available. Often when a hypothetical explanation is first advanced, data are sparse and inadequate to make a confident decision. The existence of a hypothesis generates an important feedback mechanism within science, however. The hypothesis may suggest experiments or procedures that will generate new evidence that may have a bearing on the hypothesis and serve as a test of it. Thus hypotheses and theories not only provide explanations for available evidence, but also help to identify the kinds of new observations that need to be made in order to enlarge the relevant body of available evidence. In this manner a hypothesis is subjected to ever more stringent tests: Weaknesses are found and refinements made to the explanations until either the hypothesis is discarded and replaced by a new one, or it emerges as a powerful new theory that can take its place as a generally accepted explanation of natural phenomena.

This process is good for science, but it can be rough on a scientist who has the foresight and imagination to be too far ahead of the available evidence. If he or she has managed to publish the hypothesis then eventually fame will come, though all too often posthumously. Of course it should be noted that a very large number of published hypotheses do not stand up to the test of time and eventually are forgotten by all but a small number of specialists, who maintain knowledge of the failure only in order to avoid repeating it.

The delayed fame of Milankovitch is due to his use of small, regular changes in Earth's orbit around the Sun in order to calculate corresponding changes in the amount of energy reaching the polar regions. These changes are in the ellipticity of the orbit, the tilt of the Earth's axis, and the precession of the equinoxes. The first describes the extent to which the orbit departs from a perfect circle. The second describes the angle by which the axis tilts away from the perpendicular to Earth's orbital plane and determines the severity of seasons. The third describes changes in the direction of the Earth's axis (but with the same tilt), and determines where the Earth is located in its orbit during winter or summer in a particular hemisphere. At the moment, for instance, we are closest to the Sun during northern hemisphere winters, leading to slightly milder winter and summers in the northern hemisphere but harsher winters and summers in the southern hemisphere. These three changes are diagramed in Figure VIII-5.

The period of ellipticity changes is 100,000 years, that of axis tilt is 40,000 years, while the precession changes in a more complex manner with two combined periodicities of 23,000 and 19,000 years. All four periodicities match those in the oxygen-isotope data within the uncertainties that apply to each. Thus the circumstantial evidence in favor of the Milankovitch hypothesis is very strong, but not all climatologists are convinced that a cause-and-effect relation has been established. For one thing, when the actual energy changes in the polar regions are calculated for each orbital effect, it is found that the ellipticity changes should be the least prominent and yet it is the 100,000 year ellipticity period that in fact dominates the data.

One possible explanation for this seeming discrepancy is that the Earth climate systems are more sensitive to changes with a period of 100,000 years than to changes with shorter periods. In effect, the climate system has a built-in resonance. The effect is similar to a very loud rattle in your car that might be excited by much smaller vibrations produced by the engine. Another automobile example is the vibration of an out-of-balance wheel that is excited only at certain speeds (that is, when the period of revolution of the wheel matches the natural vibration period of the wheel and suspension). In a similar manner, a natural sensitivity on the part of the Earth's climate system to changes with a period of 100,000 years may exaggerate the influence of the orbital eccentricity.

Climatologists will only be fully convinced of the validity of the Milankovitch mechanism when the cause-and-effect link has been fully established between orbital changes and the climatic response. Progress has been made along this line by the use of numerical models of climate that attempt to reproduce the essential workings of the climate system in complex computer simulations. The advent of faster and more powerful computers in recent years has led to the emergence of the numerical model as a major new research and prediction tool in studies of weather and climate. In the next section we will explore the nature of numerical models, and you will even get a chance to try your hand at one.

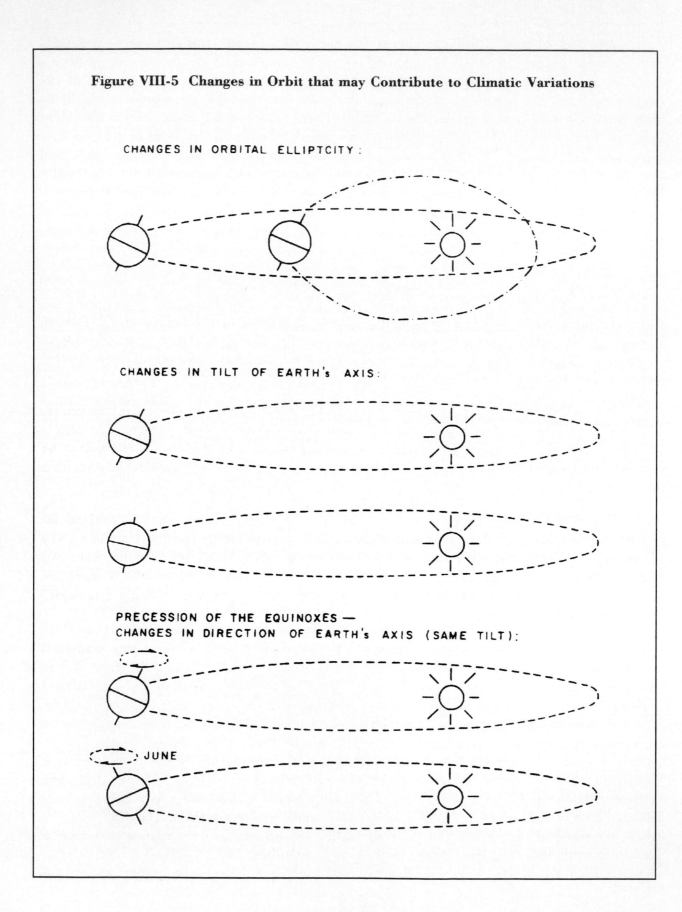

Figure VIII-5 Changes in Orbit that may Contribute to Climatic Variations

CHANGES IN ORBITAL ELLIPTCITY:

CHANGES IN TILT OF EARTH's AXIS:

PRECESSION OF THE EQUINOXES —
CHANGES IN DIRECTION OF EARTH's AXIS (SAME TILT):

JUNE

F. CLIMATE MODELING

1. The Concept of a Numerical Model

A climate model is not really different from all the other models that we have discussed. It is an artificial (in this case mathematical) representation of a natural system that is subject to the laws of physics and chemistry and, if constructed properly, behaves like the natural system it represents.

In the oceanography unit, **Dynamics of the Oceans**, you used a phonograph turntable as a model for the rotating Earth in order to demonstrate the Coriolis Effect. This was a tangible scale model that relied on natural laws acting similarly on it and on the real thing: the Coriolis Effect works for both the large, slowly rotating Earth and for a small, rapidly rotating turntable.

Scientists frequently have used scale models of this type to help them to understand physical phenomena. During the last century, for instance, Lord Kelvin constructed a miniature Alpine valley from wood and placed in it a representation of a glacier made of pitch (a very viscous pine resin). His model very nicely demonstrated the qualitative behavior of glaciers, but the scaling-down can introduce difficulties, and in many cases it is hard to obtain meaningful quantitative predictions from such models.

This is the case with climate. We might try to model climate using the air mass in a fishbowl or terrarium, for instance, and some general features might be reproducible, such as convection cells and latent heat transfer through evaporation and condensation. Other features -- thunderstorms and Earth's global circulation patterns of trade winds and westerlies -- simply cannot be modeled convincingly on such a scale.

Perhaps the most remarkable fact in all of science is that nature appears to obey laws that can be expressed mathematically. On one level, mathematics may be regarded as an arcane manipulation of symbols and numbers, a pure product of the human imagination. Yet it remains the scientist's most powerful tool precisely because it is the only means by which we may precisely describe the workings of nature. The great discoveries of Kepler, Galileo, and Newton were the mathematical statements of the laws that governed the motions of the planets. Those same laws (along with others) also govern the motions of air masses in Earth's atmosphere.

During the nineteenth century and the first half of the twentieth century, great emphasis was placed on manipulating the mathematical forms of natural laws so that measurable phenomena could be described by exact mathematical expressions. For instance, Newton's laws of gravitation and motion can be manipulated to derive an exact expression for the distance D that an object will fall in a given time t under the influence of Earth's gravitational acceleration g:

Equation 1

$$D = 1/2 \ gt^2$$

This algebraic expression is a mathematical model of how an object falls after it is released (neglecting the effects of air resistance). The value of g at the Earth's surface is approximately 9.8 m/sec² and can be measured very precisely at any point on Earth. Using simple algebra, you can calculate the distance D for any value of t that you choose. Hence, the mathematical model represented by the equation has the ability to make a prediction. For example, it predicts that two seconds after an object is released, it will have fallen 19.6 meters, using the average value for g. The experiment can be tried, and it is found that the equation models reality quite accurately, so long as the object falls in a vacuum (no air resistance). If we want our model to truly represent a dropped object in our atmosphere, then we will have to introduce additional mathematical laws that govern air resistance, manipulate the model further, and come up with a new (and more complex) equation that describes how far a dropped object falls in a given time.

So long as we are dealing with only a single object or with interactions between two objects, it usually is not too difficult to manipulate the mathematical laws into an exact expression that describes reality. But difficulties begin to set in when we have to deal with large numbers of interacting objects. In such a case statistics is often used to give an average or probabilistic expression, but if we want to know in detail what is happening to a particular element of our model, it often happens that it is not possible to obtain an exact mathematical expression to describe it.

The Earth's atmosphere consists of a large number of air masses that interact in many different ways with one another and with external influences such as the oceans, the land, energy from the Sun, and so on. It is far too complex a system to describe exactly in a single mathematical expression.

And so we come to still another kind of model: the numerical model. It is identical to the mathematical model discussed above except that once we realize that an exact general expression for the entire model is not attainable, we simply omit the mathematical manipulation and stop at the point where we have described the laws governing the interactions between the elements of our model. Then, instead of manipulating the mathematics to obtain a general algebraic expression, we use actual numeric values in the laws that govern the interactions and see how these values change with time.

First, the individual elements that make up the model must be specified along with measurable quantities that define the state of each element. In the case of a weather or climate model, an element may be a block of air above some point on Earth, so many kilometers by so many kilometers in area and so many kilometers thick. Other blocks of identical size and shape are defined on all sides of the original block and above or below it so that the totality of blocks (elements) in our model encompasses the entire atmosphere that participates in climatic processes (usually encompassing the troposphere and perhaps part of the stratosphere), every-where on Earth. In essence, we take the entire atmosphere and slice it up into a series of identical blocks.

The state of each element, or block, in our model is specified for a given instant of time by a series of numbers that define its temperature, pressure, density, humidity, wind direction and speed, and so on. We begin the operation of our model by specifying all these numbers for every block in the model. This is the <u>initial condition</u> of the model and defines the state of the model at the starting time. From here on, the model runs itself. The mathematical laws governing the interactions between elements allow us to calculate how the temperature, pressure, etc. of any block will change due to the influence of its neighbors. We do this for each block, updating the conditions in the block to reflect the effects of the interactions. At this point we have a slightly changed model from the initial condition, and we can repeat the process, calculating a new set of changes based on the new state of the model. These can then be used to produce a third version of the model, and so on, as long as we wish to continue the process. What we end with is a numerical model that evolves with time, hopefully mirroring changes that take place in the actual atmosphere.

G. AN EXERCISE USING A NUMERICAL MODEL

The concept of a numerical model is best explained with the aid of a simple example. At first glance, the material of this section may appear formidable to you, especially if you are among the mathematically shy. Don't despair! Nothing more complex than simple arithmetic is required, and you need only follow the instructions, step by step, in order to complete it. As you work on it, you will learn more if you try to understand what is happening in each step. The wonderful thing about it is that if you understand the individual rules that govern a scientific model, then the mathematics will keep track of the details for you. In this way, a set of simple understandings (that is, of the rules that govern a scientific model), may enable you to arrive at quite complex conclusions. That is the power of mathematics as used within scientific models.

Our model is designed to demonstrate heat transfer between three blocks or elements (Figure VIII-6). Each block can be anything that is capable of storing heat: a jar of water, a brick, a hunk of metal, or almost any other object. We will concern ourselves only with the temperature of each block and how these three temperatures change with time. Having defined the elements of the model, we now need to state the physical laws or rules governing the interactions between the blocks in a mathematical form. Fortunately, these are fairly simple.

In Figure VIII-6, the temperatures of each block are specified by T_1, T_2, and T_3. In addition, two arrows labeled dQ_1, and dQ_2 link the blocks and represent heat flowing between them, transferring thermal energy from one block to the next. The physical law governing heat flow between two blocks can be stated as follows: the heat energy (dQ) transferred per unit time is proportional to the temperature difference between the two blocks. The law is stated mathematically by:

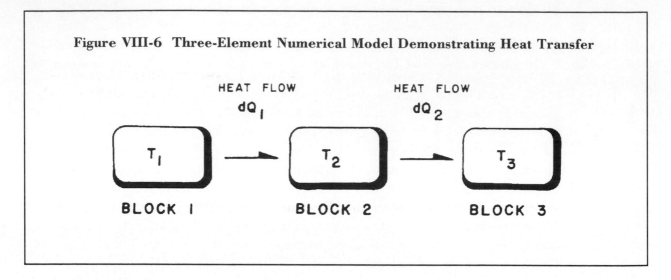

Figure VIII-6 Three-Element Numerical Model Demonstrating Heat Transfer

Equation 2

$$dQ_1 = B \times (T_1 - T_2)$$

and

$$dQ_2 = B \times (T_2 - T_3)$$

The symbol B represents a number that describes the efficiency of heat transfer: the larger the number, the more efficient the heat transfer process. In our model we rather arbitrarily set B to have the value 0.3.

Now we need to know how heat energy flowing into or out of a block changes its temperature. The appropriate physical law states that the change in temperature of a block is proportional to the heat energy that flows into it. If energy flows into a block, its temperature increases; if energy flows out of it, its temperature decreases. This temperature change (TC) can be represented mathematically by:

Equation 3

$$TC = C \times dQ$$

To keep things as simple as possible, we will set the symbol C to always have the value 1. Now look again at Figure VIII-6 and recall that dQ_1 and dQ_2 represent the flow of heat between the three blocks. Block 1 can only communicate with Block 3 through Block 2. How does the temperature of each block change?

The temperature of Block 1 will drop if heat dQ_1 is removed from it:

Equation 4

$$TC_1 = - C \times dQ_1$$

The temperature of Block 3 will rise if heat dQ_2 is added to it:

Equation 5

$$TC_3 = C \times dQ_2$$

Finally, the temperature of Block 2 will rise as heat dQ_1 is added to it and drop as heat dQ_2 is taken from it, so its temperature change is determined by the difference between dQ_1 and dQ_2:

Equation 6

$$TC_2 = C \times (dQ_1 - dQ_2)$$

As the final step in the process, we update the temperature of each block:

Equation 7

$$T_1(\text{New}) = T_1(\text{Old}) + TC_1$$

$$T_2(\text{New}) = T_2(\text{Old}) + TC_2$$

$$T_3(\text{New}) = T_3(\text{Old}) + TC_3$$

and repeat the cycle starting with Equation 2.

If you are having trouble understanding what's going on here, try thinking of our model as three bricks of different temperatures that have been brought into contact with one another. The hottest brick will lose heat to the cooler bricks, raising their temperatures while its own temperature drops. The dQs show how heat energy moves between the bricks and the TCs give the resulting temperature changes. You don't really need a full understanding of the model in order to work the following exercise, but you will learn more from it if you have at least a feel for what is going on in each step.

We can now lay out a plan for our model. Such a plan is a list of procedures needed to make the model work and the order in which they are to be executed. It is called an <u>algorithm</u>, and one for our model is diagramed in Figure VIII-7.

Now to make the model work. On a following page is a worksheet that will help to guide you in each step, and the first few entries have been filled in as a demonstration. Only addition and very simple multiplication are required.

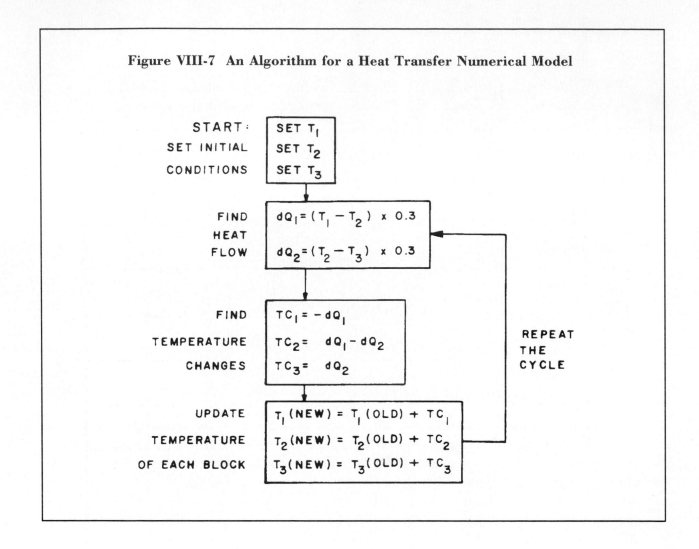

Figure VIII-7 An Algorithm for a Heat Transfer Numerical Model

Figure VIII-8 Numerical Model Worksheet

(Round off all entries to one place after the decimal point.)

STEP	T_1	T_2	T_3	$dQ_1 =$ $(T_1\text{-}T_2) \times 0.3$	$dQ_2 =$ $(T_2\text{-}T_3) \times 0.3$	$TC_1 =$ $\text{-}dQ_1$	$TC_2 =$ $(dQ_1\text{-}dQ_2)$	$TC_3 =$ dQ_2
0	100.0	0	0	30.0	0	-30.0	30.0	0
1	70.0	30.0	0	12.0	9.0	-12	3	9
2	58	33	9	7.5	7.2	-7.5	.3	7.2
3	50.5	33.3	16.2	5.16	5.1	-5.2	.1	5.1
4	45.3	33.4	21.3	3.57	3.6	-3.6	0	3.6
5	41.7	33.4	23.9	2.49	2.85	-2.5	-.3	2.8
6	39.2	33.1	26.7	1.83	1.92	-1.8	-.1	1.9
7	37.4	33	28.6	1.3	1.3	-1.3	0	1.3
8	36.1	33	29.9	.9	.9	-.9	0	.9
9	35.2	33	30.8	.6	.6	-.6	0	.6
10	34.6	33	31.4	.4	.4	-.5	0	.5

The three boxed temperatures represent the <u>initial condition</u> of the model. In this case we start with Block 1 at 100°C, Block 2 at 0°C, and Block 3 at 0°C. This is purely a matter of choice, and we could have chosen any three other temperatures as our initial condition. We choose the initial condition and then the model takes over and tells us how the system will change with time in reaction to the initial condition.

Now go from left to right on each line starting from the boxed temperatures. First we calculate dQ_1 using the values of T_1 and T_2 on the same line to the left. Do the same for dQ_2. Then calculate TC_1, TC_2, and TC_3 using the values of dQ_1 and dQ_2 that you just found. Then start the next line by updating temperatures T_1, T_2, and T_3 by adding TC_1, TC_2, and TC_3 to the old values of T_1, T_2, and T_3 that are directly above.

Repeat the procedure for each line until you have completed all 10 lines. Work carefully, rounding off your answers to one place after the decimal point and be careful to observe signs. Note that you added -30.0 to 100.0 to get 70.0 for the value of T_1 in line No. 1. From time to time you will get small negative numbers for TC_2 as well, which need to be treated in the same way.

Once you have completed all ten lines in the worksheet, look at your results and note how the temperatures of the three blocks have changed with time.

Your intuition might already have told you that the three blocks would exchange heat until they all were at the same temperature. Note that that is just what is happening. Temperatures change quickly while the temperature differences are great but more slowly once they are small. In our numerical model each step (completing a new line on the worksheet) represents a certain period of elapsed time. After the ten steps that you have done, there is still a three-degree spread in temperature between Blocks 1 and 3. If you were to continue for another thirteen steps, you would have reduced the spread to less than 0.1 degree, with all three temperatures essentially the same at 33.3°C.

By the way, congratulations! You are now officially a computer, where the term is used in its pre-electronic age connotation as "one who computes or calculates." Before the advent of electronic computers, many people were employed to carry out such arithmetic tasks using slide rules, logarithm tables, and eventually, slow mechanical calculating machines. You can appreciate the tedious nature of the work of these human computers, but they were the unsuing heroes of much of pre-1950 science.

In the exercise, you used only straightforward addition, subtraction, and multiplication to solve the problem. Modern electronic computers are very good at doing arithmetic fast and accurately, and because of this, numerical models have become a potent tool for all areas of quantitative science. The algorithm in Figure VIII-7 may be used as a blueprint for a computer program, instructing the machine exactly what step to do and when to do it.

The model you have worked on has three blocks and two mathematical relations (Rules 2 and 3) that govern its operations. Only two variables (temperature and heat flow) were involved. In order to model world climate, the atmosphere must be divided into thousands of blocks with something like a dozen or more relations and variables. A meaningful prediction using such a model may involve millions of steps. Perhaps now you can appreciate why we did not give you the real thing to attempt by hand.

1. Applications of Weather and Climate Models

There are two kinds of conditions that we may impose on a model: <u>boundary conditions</u> and <u>initial conditions</u>. Boundary conditions specify the environment in which the model is to operate. Examples of boundary conditions might include: the amount of energy entering the atmosphere from the Sun; the amount of carbon dioxide in the atmosphere; the amount of ozone in the ozone layer; the effects of continental landmasses in generating high and low pressure regions. There are of course many more that may be specified. In the model you just worked with, the values of the constants B and C in Equations 2 and 3 are the boundary conditions placed on the model, specifying the efficiency of heat transport between the blocks in heat capacities.

Initial conditions are the starting conditions that are specified when the model begins to run. The starting temperatures of the three blocks in your exercise were examples of initial conditions. Other examples of initial conditions in a model might include: the atmospheric pressure in each block or element of the model; the humidity in each block; the percent of cloud cover in each block. Note that the initial conditions deal with variables that are inherently transient, and whose values will likely change as the model runs. The boundary conditions, on the other hand, are likely to be treated as fixed and imposed on the model by the experimenter, and their values generally do not change while the model is running.

With these points in mind, we can now make a distinction between two types of models: the <u>numerical weather prediction model</u> and the <u>general circulation model</u> of the atmosphere. The weather prediction model sets boundary conditions appropriate to the real world at the present time, and then sets a particular set of initial conditions that, for instance, might describe this morning's weather. The model is then set running to simulate the passage of 24 hours, and the result is the model's prediction of what the weather will be for tomorrow morning. The model of the exercise was of this type: you specified the initial conditions (the temperatures of the three blocks) and the model then predicted how those temperatures would change with time. In a weather model, you specify the initial conditions by in effect supplying the model with the current weather map. It then predicts how the weather map will change with time.

In terms of forecasting weather, numerical models are already in use and are increasing in importance as predictive tools. There are, however, two types of limits on the development of these models. The first is available computer power. To be of any use whatever, a computer program must be able to predict tomorrow's weather and take less than 24 hours to do it. As a result, the memory and speed of the computer places a practical limit on the numbers of blocks or elements that can be handled in a numerical model. In today's models, a single block may cover an area the size of the entire state of Colorado. The size of the element defines the resolution of the model. Phenomena on a smaller scale than that of one block cannot be treated.

Because of this, large-scale features such as frontal systems are better served than small transient features such as thunderstorms, tornadoes, and local rainshowers. This is one reason why weather forecasts are generally more accurate in dealing with temperature than with precipitation in a particular area -- temperature does not vary so much throughout a large air mass, but precipitation can be very local.

The second limit on numerical weather models is on our ability to specify the exact starting conditions. This limit is closely related to the first, since how do you specify the weather conditions in a block, half of which is clear and sunny and the other half of which contains clouds and rain? Each variable is allowed only one value throughout the entire block. As computers advance and the size of each block shrinks, then it will be necessary to be able to measure the conditions in each of thos blocks. In developed countries there are many manned and unmanned weather stations, reporting conditions at least twice a day, but this is not true over the oceans and in lesser developed countries. This will become more of a problem as the resolution of weather models improves.

A general circulation model works in the same manner as a weather prediction model and may be run as one, except that if it is allowed to run for a long enough time, it arrives at a steady-state condition. In effect, the initial conditions are no longer of any importance, because the final, steady-state average condition usually depends mostly on the boundary conditions and very little or not at all on the particular initial conditions chosen. The general circulation model may be used as a climate model as opposed to a weather model. Another difference is that the climate model generally encompasses the entire Earth, while the weather model may be limited to one continent.

The value of a general circulation model is that it allows us to determine the consequences of changes in the boundary conditions (that is, in the overall atmospheric environment). By experimenting with a climate model of this sort, a scientist can learn about the operation of climate systems and determine what changes are most likely to have an influence upon Earth's climate. Figure VIII-9 shows the interactions that are treated in a climate model in which the atmosphere, oceans, ice, and landmasses are all considered as one coupled system. In such a case, the conditions are set and the model is allowed to run until it reaches a stable condition. If everything of consequence has been accounted for, then the result should mirror the state of the real Earth climate.

As an example, general climate models have successfully reproduced the seasonal shifts in pressure and rainfall that accompany the monsoons of the Indian subcontinent. Indeed, many of the global climatic characteristics discussed in the previous unit have been reproduced by numerical climate models. Another use for the models has been to attempt to reproduce some of the ancient climates discussed earlier in this unit. For instance, the reconstruction of the continental positions during the warm Cretaceous period has been incorporated into a climate model. The model does indeed predict a warmer climate for the Cretaceous, but not as warm as the geological record would indicate. This suggests that some unaccounted-for factor must be taken into consideration. Perhaps the carbon dioxide content of the atmosphere was higher then, leading to a stronger Greenhouse Effect.

Climate models are being applied to a wide variety of problems: the Volcanic Dust Veil Effect, the El Ni ño ocean current anomaly, the climate at the close of the Ice Ages 9,000 years ago, the effects of vegetation on regional climate, and many others.

With these encouraging results, it is natural to turn to numerical models for their predictive abilities. Can the models be used to predict the consequences of environmental changes brought about by mankind?

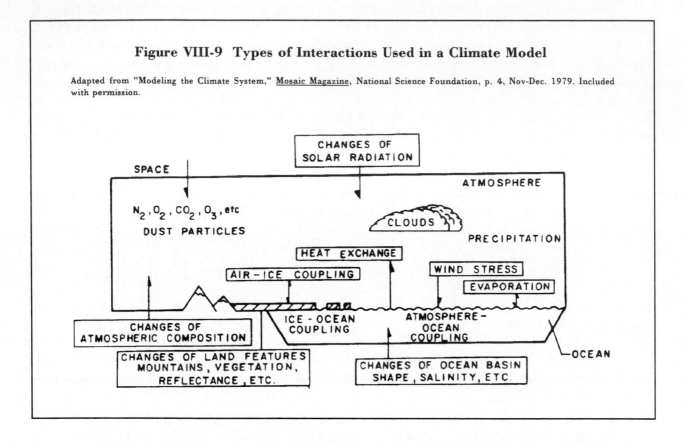

Figure VIII-9 Types of Interactions Used in a Climate Model

Adapted from "Modeling the Climate System," Mosaic Magazine, National Science Foundation, p. 4, Nov-Dec. 1979. Included with permission.

A recent major proposal by the U.S.S.R. for large-scale environmental modification is a case in point. In order to increase the agricultural productivity of its arid regions to the south, the U.S.S.R. is planning to divert major quantities of water from four rivers that flow into the Arctic Ocean: the Yenesey (Jenesei), Ob, Dvina, and Pechora rivers. What effect will this have on the Arctic Ocean? The amount of sea ice in the Arctic Ocean is an important influence on northern hemisphere climate because of its high reflectivity. Sea ice can participate in a rather unstable process in which the white ice reflects solar energy back into space producing local cooling, which creates more sea ice, and so on. Changes in the sea ice coverage of the Arctic Ocean have been proposed as one of the possible triggers for the onset of the Ice Ages.

Concern has been expressed that diversion of fresh water from the four rivers might upset the salinity of the Arctic Ocean and hence the amount of sea ice. A numerical model was constructed of the Arctic Ocean and its major tributaries. The tentative conclusions from use of the model were that there appeared to be little cause for concern and that effects on the Arctic Ocean and its sea ice coverage would be extremely slight. Nevertheless, further studies were recommended because of the great cost of the project and its implications for worldwide climate, and because of the relative simplicity of the model used.

The concept of a Nuclear Winter has received much attention recently, and its development has relied extensively on the use of climate models. Briefly stated, in the event of a major thermonuclear exchange between the U.S.A. and the U.S.S.R., so much dust and soot from burning cities and secondary forest fires would be injected into the atmosphere that major climatic effects might be expected. Recent modeling indicates that the effects might be sufficient to produce drastically lower temperatures for a period of weeks to months. The result could be widespread and prolonged freezing (even in summer) with resulting crop failure and famine on unprecedented scales. Equally severe might be the effect of the dust and soot cloud on sunlight, decreasing it by 95% or more and effectively bringing photosynthesis to a halt. Virtually every aspect of Earth's climate could be affected, including global circulation patterns and the vertical structure of the atmosphere.

With introduction of the concept of a Nuclear Winter, our perception of the horror of nuclear war has been drastically increased. We shall return to this vitally important subject for a fuller discussion in the final unit of this telecourse.

STUDY QUESTIONS

VIII-10. The laws that govern the interactions between the elements of a numerical weather model are always the same for a given model. How is it possible for such a model to give different results? Why aren't the predictions of a numerical weather model always the same?

VIII-11. Look at your worksheet from the numerical model exercise. What would have been the effects on subsequent results if you had chosen your starting temperatures to be those that you obtained in step 5?

H. CLIMATIC FEEDBACK MECHANISMS AND MODEL VERIFICATION [11]

Clouds, being very bright, reflect a large fraction of sunlight back to space, thereby helping to control Earth's temperature. Thus, predicting the changing amount of cloudiness over time is essential to reliable climate simulation. But indivudual clouds are smaller than even the smallest area represented by the grid box element of a climate or weather prediction model. A cloud is typically a few kilometers in size, not a few hundred -- the size of many high-resolution model grids. Therefore, no global climate model available now (or likely to be available in the next few decades) can explicitly resolve individual clouds. These important climatic elements are therefore called sub-grid-scale phenomena. Yet, even though we cannot explicitly treat individual clouds, we can deal with them and their effects on the grid-scale climate. The method for doing so is known as parameterization, a contraction for "parametric representation". Instead of solving for sub-grid-scale details, which is impractical, we search for a relationship between climatic variables we do resolve (for example, those whose variations occur over larger areas than the grid size) and those we do not resolve. For instance, climatic modelers have examined years of data on the humidity of the atmosphere averaged over large areas and have related these values to cloudiness averaged over that area. It is typical to choose an area the size of a numerical model's grid -- a few hundred kilometers. While it is not possible to find a perfect correspondence between these averaged variables, reasonably accurate

---

[11]This section written by Stephen H. Schneider, National Center for Atmospheric Research, Boulder CO. For additional information see Stephen H. Schneider, Coevolution of Climate and Life, Sierra Club Books, 1984.

relationships require a few factors, or <u>parameters</u>, which are derived empirically from observed data, not computed from first principles. For example, to represent the average amount of cloudiness in each grid box we only need to give box-averaged values of temperature, winds, and humidity. The latter values are explicitly resolved in the models. Individual clouds, of course, are not resolved, but the effects of such parameterized clouds can be included objectively in the model through an empirical relationship between the amount of averaged clouds and other averaged variables. The big question is, are these empirical parameterizations accurate enough for the particular predictions being made with our model?

The most important parameterizations deal with processes called <u>feedback mechanisms</u>. This concept is well known outside of computer-modeling circles. The word feedback is common vernacular. For instance, you may ask a friend for feedback on a piece of work you are doing. Then, you can if you want, modify your project in response. As the term implies, information can be "fed back" to you that will possibly alter your behavior.

So it is in the climate system. Processes interact to modify the overall climatic state. One important feedback mechanism is easily visualized. Suppose a warm wind blows over a valley covered lightly with snow. The temperature will rise, of course, melting the snow cover and replacing a bright, highly reflective snow surface with a much darker, more sunlight-absorbing meadow. Thus, the temperature rise caused initially by the warm wind will be further enhanced by the <u>positive-feedback effect</u> on temperature of disappearing snow cover. Similarly, a cold snap that brings on a snow cover tends to reduce the amount of solar heat absorbed, subsequently thereby intensifying the cold. This interactive process is known to climatologists as the snow-and-ice/albedo/temperature feedback mechanism.[12] Its destabilizing, positive-feedback effect is becoming well understood and has been incorporated into the parameterizations of most climate models.

Unfortunately, other potentially important feedback mechanisms are not usually as well understood. The most difficult one is the so-called <u>cloud feedback</u>. Earlier we mentioned that clouds are highly reflective of solar radiation and that more cloudiness usually means less solar energy absorbed by the Earth. Let's construct a feedback process using this fact. A warm wind blows over a large lake. We know that warm, wet surfaces usually evaporate more water than cooler or drier surfaces. Thus, the warming lake puts more moisture into the air. Clouds form, blocking some sunlight from reaching the lake's surface, thereby cooling it back down a bit. This is a stabilizing, or <u>negative-feedback</u> example: the cloud formation, in response to initial surface warming, "feeds back" on that warming in a manner that inhibits large temperature changes.

But this is not the only scenario. Clouds can get taller, enhancing the Greenhouse Effect, and further warming the surface. This positive feedback effect could be competing with other positive or negative feedback processes, the sum of which determines the overall sensitivity of the climate to forcing factors such as CO_2 increase or a change in the energy output of the Sun. Unfortunately, at present the climatic sensitivity to such forcings is not established to much better than a factor of two.

How, then, can we have confidence in model predictions? At least three verification methods are used, and none by itself is sufficient. First, we must check overall model-

[12]North, G.R., 1975: Theory of energy-balance climate models, <u>Journal of Atmospheric Sciences</u> 32:2033-2043.

simulation skill against the real climate for today's conditions to see if the control experiment is reliable. The seasonal cycle is one good test. But even if the model does a good job in this test, there is no guarantee that the model will correctly predict long-term future climates unless the net effect of all the important causal factors of long-term climatic change are properly included. The seasonal-cycle test doesn't tell us how well the model simulates slow changes in ice cover or deep ocean temperatures, since these changes do not affect the seasonal cycle, though they do influence long-term trends.

A second method of verification is to test in isolation individual physical subcomponents of the model (such as its parameterizations) directly against real data. This still is no guarantee that the net effect of all interacting physical subcomponents has been properly treated. Third, some researchers express more confidence a priori in a model whose internal makeup includes more spatial resolution or physical detail, believing that "more is better." In some cases and for some problems this is true, but by no means for all.[13] The optimal level of complexity depends upon the problem we are trying to solve and the resources available to the task.

All three methods must constantly be used and reused as models evolve if we are to improve the credibility of their predictions. Perhaps the most perplexing question is whether we should ever consider our confidence in their forecasts sufficient reason to alter our present social policies -- on CO_2-producing activities, for example. Viewed from this angle, the seemingly academic field of computer climate modeling becomes a fundamental tool for public policy.[14] If the public is totally ignorant of the nature, use, or validity of climatic (or many other kinds of) models, then public policymaking based on model results will be haphazard at best. In this case, the decision-making process tends to be dominated by a technically trained elite.

In The Genesis Strategy I put the issue this way:

> The processes at work in determining climatic change are simply not yet fully understood. Despite the uncertainty in measurements and in theory, estimates must be given and difficult decisions may have to be made on the basis of the available knowledge.... In any case, efforts to develop better estimates and models of potential effects will be absolutely necessary to help us reduce the uncertainties in decision making to a tolerable minimum. Improvement of the quality of these estimates is the responsibility that atmospheric scientists and their funding agencies owe to long-range planners, for the climatic effects of human activities are self-evidently the outer limits to growth. The real problem is: If we choose to wait for more certainty before actions are initiated, then can our models be improved in time to prevent an irreversible drift toward a future calamity? And how can we

---

[13]Schneider, S.H., 1979: Verification of parameterizations in climate modeling. In Report of the JOC Study Conference on Climate Models: Performance, Intercomparison and Sensitivity Studies, World Meteorological Organization, Geneva, October 1979, pp. 728-752

[14]For example, see the July 30, 1979 testimony before the U.S. Senate, Committee on Governmental Affairs (Senator Abraham Ribicoff, Chairman), Carbon Dioxide Accumulation in the Atmosphere, Synthetic Fuels and Energy Policy, a Symposium (Washington, D.C.: U.S. Government Printing Office). In a subsequent hearing in 1981 Tennessee congressman Albert Gore, Jr. commented, "Many of our witnesses today are from the academic environment. But the Greenhouse Effect is not merely an academic question. The paths we take today I hope will help us determine whether we have disasters or merely manageable problems in our future." This appeared on page 5 of Carbon Dioxide and Climate: The Greenhouse Effect, Hearing before the Subcommittee on Natural Resources, Agriculture Research and Environment and the Subcommittee on Investigations and Oversight of the Committee on Science and Technology, U.S. House of Representatives, 97th Congress (Washington, D.C.: U.S. Government Printing Office, 31 July, 1981)

decide how much uncertainty is enough to prevent a policy action based on a climate model? This dilemma rests, metaphorically, in our need to gaze into a very dirty crystal ball; but the tough judgment to be made here is precisely how long we should clean the glass before acting on what we believe we see inside.[15]

Nothing has happened since I first wrote these lines that would have me substantially modify that statement, except perhaps to mention that the real climate is now ten years closer to a direct verification of model predictions of human impacts on climate. And since the present northern hemisphere climate on land is about 0.5°C warmer than a century ago, this is at least suggestive that observations are roughly (to some factor of 2) consistent with them.[16] Problematically, we are performing the experiment on the laboratory of Planet Earth, with us and other living inhabitants along for the ride.

1. Pre-Pleistocene Paleoclimatic Methods [17]

We know a lot about the paleoclimatology of the Ice Ages, the most recent part of geologic history, because the geologic record is so good for that time. Indeed, Pleistocene (Ice Age) paleoclimatology is almost a separate subdiscipline because, compared to earlier parts of Earth history, paleoclimatologic changes over much shorter time intervals and for much smaller geographic areas can be discerned for the Pleistocene. For the rest of Earth history during the last 600 million years, climatic changes that occurred over longer time intervals and in broad regions are all that can be revealed with current methods. Prior to 600 million years ago, the geologic record is so poor that climatic change of only the coarsest scales can be described.

You have already learned of one method that is beginning to be used for the study of pre-Pleistocene climates, that is, numerical modeling. This development is largely very new, however, and paleoclimatology still is mostly studied using other methods. These include qualitative circulation models and studies of the changes in distribution of proxy climatic indicators.

Qualitative circulation modeling is much easier than numerical modeling because it does not require complex programs or vast amounts of computer time. The objective in qualitative modeling is to predict the general patterns of oceanic and atmospheric circulation and rainfall for the past. Such models cannot provide details of temperature, wind speed, and precipitation, as can the numerical models. However, qualitative models can take into account changes in global paleogeography brought about by continental drift and sea-level fluctuations, which are considered to be the primary causes of global climatic change. Like numerical models, qualitative circulation models are based on our current understanding of the factors controlling the major features of atmospheric and oceanic circulation, such as land-sea temperature contrasts at mid-latitudes. Qualitative models have helped explain some patterns in the proxy climatic indicators that were not well understood previously.

An example of the usefulness of qualitative paleoclimatic models is the explanation of the distributional patterns of clams, brachiopods, and other marine animals during the Permian Period, from 285 to 240 million years ago. At that time, almost all continental crust was

---

[15]Schneider, S.H., with L.E. Mesirow, 1976: The Genesis Strategy: Climate and Global Survival Plenum, New York.
[16]Gilliland, R. L. and Schneider, S.H., 1984: Volcanic, CO_2 and solar forcing of Northern and Southern Hemisphere surface air temperatures, Nature 310:38-41
[17]This section was written solely for use in this text by Judith Totman Parrish, U.S. Geological Survey, Denver, CO

aggregated into a large continent stretching from pole to pole, called Pangea. The west coast of this continent was nearly straight north to south but the east coast curved eastward at mid-latitudes in both the northern and southern hemispheres, surrounding a large, equatorial ocean called Tethys. During the early 1970's, when continental drift was still hotly debated, some geologists observed that the distribution of tropical marine animals from the Permian was better explained by present geography than by the postulated Permian geography. As expected, the tropical faunas are found near the Permian equator on the west coast of Pangea. In Tethys, however, the tropical faunas extend nearly halfway to the poles in both hemispheres. The regions that surrounded Tethys are much closer to the equator now than they were in the Permian, which led geologists to question the validity of the Permian continental reconstructions. However, this peculiar geography would have had some very strong effects on the oceanic circulation patterns. At the equator, the globe was ocean for all but 60 of the 360 degrees of longitude.

This means that there was a very long fetch (room to travel unhindered by landmasses) for the equatorial current and, potentially, the water in that current would have been very warm by the time it arrived in Tethys. Upon reaching the east coast of Pangea, the water in the equatorial current would have been deflected north and south in currents resembling the Gulf Stream. Just as the Gulf Stream contains enough heat to warm northern Europe today at high latitudes, the Tethyan currents would have warmed the continental shelves bordering Tethys to high latitudes. Thus, tropical animals could have lived at the latitudes required by the Permian continental reconstructions.

Apart from qualitative climatic models, pre-Pleistocene paleoclimates are studied using a variety of proxy climatic indictors, including sediments of different type and biogeographic patterns. Sediment type that are good indicators of climatic conditions include the following:

evaporite	salts of various kinds, including halite (common table salt), formed by evaporation of seawater over long time periods and indicative of dry and usually warm conditions
coal	accumulated and compressed remains of abundant plant life, indicative of wet surface conditions, either cool or warm
laterite	red tropical soil rich in iron and aluminum, indicative of wet and hot conditions
tillite	glacial deposits including grooved (striated) stones, indicating the presence of glaciers and ice sheets

eolian sand	sandstones deposited in sand dunes, indicative of dry conditions
reef carbonate	coral reefs today occur only in ocean waters warmer than about 20°C (68°F), and many lines of evidence show that this was true of ancient reefs as well, even when they were not built by corals, but by other organisms

There are many other proxy climatic indicators in sedimentary rocks, including different types of clay minerals; phosphate-rich rocks, chalks, and so on.

The distribution of the stable isotopes of oxygen, which you learned about in the section **Reconstructing Past Climates**, and of carbon have proved very helpful in understanding climatic change, particularly that of temperature. Stable isotopes are to paleoclimatology what the paleomagnetic signatures of rocks are to plate tectonics -- both provide strictly quantitative data, the one for reconstruction of past climates and the other for determining past continental positions.

The patterns of distribution of ancient plants and animals also can provide some clues to paleoclimatic patterns, as seen in the above example from the Permian. For example, tropical and subtropical plants of types that would have required abundant rainfall are very widespread in the Eocene Epoch (55-38 million years ago), indicating that rainfall worldwide was much more abundant during that time than at most other times in Earth history. We still do not have an explanation for this pattern; numerical climate models that can test the effects of variations of CO_2 in the atmosphere, for example, perhaps will provide explanations.

I. FUTURE CLIMATES

In the previous unit we discussed the Greenhouse Effect, in which certain gases in the atmosphere such as water vapor and carbon dioxide allow visible light from the Sun to enter the atmosphere but prevent infrared radiation from the warm surface from escaping. The result is an elevation of temperature. At present levels, this is clearly a beneficial effect. Without it, Earth would be a frigid planet. Unfortunately, whenever we burn wood, coal, or hydrocarbons, more carbon dioxide is released into the atmosphere.

Some of the carbon dioxide (CO_2) is removed from the atmosphere by the oceans and converted by chemical and biological processes into carbonates such as limestones that eventually precipitate out onto the ocean floor. This, however, is a slow process that cannot keep pace with our production of CO_2. It has been estimated that approximately half of all the CO_2 produced by man's activities since the beginning of the Industrial Revolution is still in the atmosphere.

Figure VIII-10 shows the measured trend of CO_2 concentrations that have been measured at the Mauna Loa Observatory in Hawaii in the 25 years since 1958. In addition to a steadily rising trend, there is a seasonal component as well, producing a sinusoidal "ripple" in the curve. In spring, plants take up more CO_2 than is produced as they become more active, but in fall and winter respiration of plants and animals, decay of organic matter, and combustion, more than compensates by producing more CO_2.

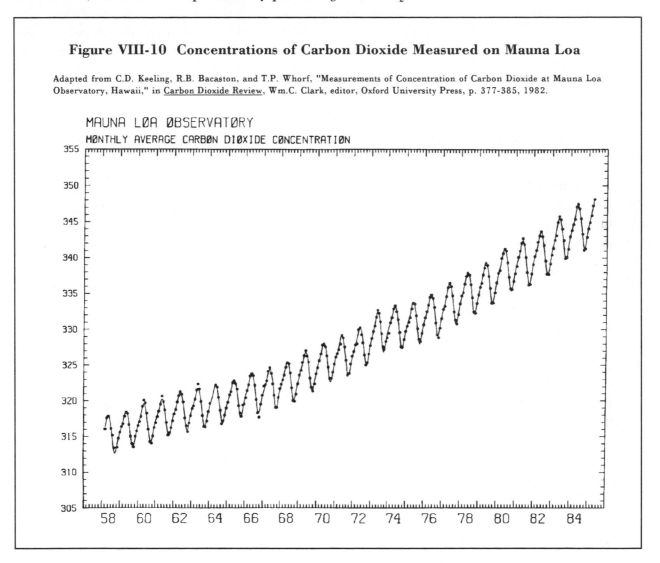

Figure VIII-10 Concentrations of Carbon Dioxide Measured on Mauna Loa

Adapted from C.D. Keeling, R.B. Bacaston, and T.P. Whorf, "Measurements of Concentration of Carbon Dioxide at Mauna Loa Observatory, Hawaii," in Carbon Dioxide Review, Wm.C. Clark, editor, Oxford University Press, p. 377-385, 1982.

Present estimates of the climatic effect of this steady increase in CO_2 based on model studies project an increase in average global temperatures of about 0.5-1°C (1-2°F) within the next 20 years and by 1-3°C (2-5°F) by the middle of the next century. This may not seem like very much, but in fact if it occurs it will constitute a more important climatic effect than any others known since the recent Ice Ages.

Carbon dioxide is not the only gas that contributes to the Greenhouse Effect. Other gases, present in trace amounts, can have very similar effects. Chlorofluorocarbons, which are used in spray cans and air conditioners; nitrogen oxides, a byproduct of the use of nitrogen fertilizers and combustion; and methane (CH_4) either have their own Greenhouse Effect or can enter into complex interactions with ozone in the atmosphere, which is another greenhouse gas. The combined effects of trace gases may be climatically as important as the effects of CO_2.

A team of NASA scientists studying the greenhouse problem concluded that during the decade of the 1970's, climate warming on the order of 0.2°C (0.4°F) has occurred. About half of this increase was ascribed to carbon dioxide and the other half to trace gases. On the other hand, they warned that due to the accelerating pace of mankind's influence on the production of greenhouse gases, we can expect to see an additional increase during the decade of the 1980's -- 0.2 to 0.3°C (0.4 to 0.5°F) -- which is large enough that we may begin to see some significant effects.

What effects might we observe from a substantial warming? At first these would be slight, but with time more severe changes might be expected. Some of the polar ice in existence today appears to be in a rather delicate state of balance. We have already mentioned the way in which sea ice can exert a kind of feedback effect on climate by altering the reflectivity of the ocean surface. In particular, the ice sheet that covers West Antarctica may be susceptible to rapid movement off the supporting land and into the ocean. The melting of sea ice would not affect the level of the world oceans, but melting of continental ice or calving of continental glaciers into the oceans would raise sea levels everywhere. Should the West Antarctic ice sheet melt, sea levels would rise by about five meters (16 feet), inundating vast tracts of valuable lowlands, including a number of heavily populated coastal cities.

Perhaps even more critical would be the effects of climatic changes on agricultural patterns. Any major shifts in climate would produce winners and losers. In the event of a significant warming, one major loser may be the great breadbasket of the American Midwest. Climate models have tended to indicate that a very real possibility is an extension eastward of the relatively dry prairie lands west of the 100th meridian into what is now highly productive cornbelt that does not require irrigation. Other areas of the world may benefit from warmer and perhaps wetter climate as global temperatures increase, but extreme disruption of the status quo would be inevitable.

What can we do about it? Any answer will take us well beyond the boundaries of our study of climate, and for that reason we will defer further discussion until the final two units of the telecourse, in which we will attempt to bring together many of the threads that run through this course.

STUDY QUESTION

VIII-12. What activities of mankind release large quantities of carbon dioxide to the atmosphere?

276

1. Melting of the West Antarctic Ice Sheet [18]

A question of crucial practical importance is: How fast could the West Antarctic ice sheet disappear? A partial answer is that it could not happen in a time measured in decades -- at the very least, several centuries would be required. The reason for this is related simply to the vast volume of ice that must be removed -- about 2 million cubic kilometers. To melt such a vast amount of ice in place would take thousands of years, even without allowing for replenishment by snowfall. Therefore, if the ice is to melt rapidly, it first must move off as icebergs into the ocean, which provides an essentially limitless reservoir of heat for melting. To accomplish this in less than several centuries would require at least a 20-fold increase in the speed of glacier flow over the present speed. Even if this were to occur, too rapid a breakoff of icebergs would clog the sea in front of the glaciers. A rate of sea level rise of a meter or two per century, which is as fast as it could occur, would be extremely serious in its consequences, but would involve no sudden catastrophe. And we must emphasize that we do not <u>predict</u> the actual occurrence of such an event. Too little is known yet about the effect of climatic warming on the Antarctic ice sheet to make any prediction -- it is even possible that the ice sheet could grow, due to increased snowfall!

2. Climate Sensitivity to Carbon Dioxide Increases [19]

Although the reality of the Greenhouse Effect has been proven beyond doubt from study of several planetary atmospheres, there remains a very large uncertainty about how sensitive the climate is to a given change in atmospheric composition. Thus, even if we had foreknowledge of how atmospheric carbon dioxide, methane and other greenhouse gases will change in the next century, we cannot yet predict the magnitude of the climate warming with a high degree of confidence.

Several study committees of the National Academy of Sciences have estimated that doubling of atmospheric carbon dioxide, expected to occur sometime in the next 50 to 100 years, is likely to cause an eventual global mean warming of 3° (plus/minus 1.5° that is, someplace between 1.5°C and 4.5°C). Even if this factor of three were the full range of our uncertainty, it covers a great range -- from a global temperature approximately equal to the warmest ever experienced by mankind to a level hotter than has occurred for many millions of years, perhaps since the Mesozoic, the age of dinosaurs.

The estimate of 1.5°C to 4.5°C as the climate sensitivity to doubled CO_2 was based initially on climate model simulations of the type discussed earlier in this unit. The climate models developed by different groups studying the problem are in close agreement on the magnitude of the radiative forcing of the climate system caused by an increase of CO_2. They differ, however, in their calculations of feedback effects, such as changes in snow and ice cover and changes of clouds that may accompany a climate change. Thus, one way to reduce the uncertainty in climate sensitivity will be to develop improved understanding and more precise modeling for the feedback processes in climate models.

---

[18]This section was written solely for use in this text by Charles R. Bentley, Geophysical and Polar Research Center, University of Wisconsin, Madison, WI
[19]This section was written solely for use in this text by James Hansen, Goddard Institute for Space Studies, New York, NY

Recently there have been attempts to estimate climate sensitivity empirically, on the basis of past measured climate changes. One of these empirical tests is to compare the warming that has been observed to occur between 1850 and 1980 with the warming computed by climate models for the increase in CO_2 which is estimated to have occurred in that period, from about 270 ppm to 338 ppm. This test of course depends upon the assumption that the CO_2 change has been the principal cause of climate change in the past century. The principal complication in applying this empirical test is that the full warming due to the added CO_2 is not expected to have been realized yet, because the oceans have such a large heat capacity that many years may be required for them to come to equilibirum with changed radiative forcing.

The effect of the ocean heat capacity in delaying the warming can be computed in very much the same way as the heat transfer between blocks, in the exercise earlier in this unit. In reality, the ocean is more complex than a simple block of heat capacity because of ocean dynamics which mixes heat down into the ocean in ways which are poorly understood. With present simple models for representing this heat storage it can only be concluded that the warming of about 0.5°C in global temperature between 1850 and 1980 is consistent with a climate sensitivity of 1.5°C to 4.5°C for doubled CO_2. As the CO_2 amount continues to increase in coming decades and as our understanding of ocean dynamics improves, this empirical test should narrow the range of uncertainty about climate sensitivity.

Other empirical tests of climate sensitivity can be obtained from study of the climate changes which have occurred on long time scales, such as from glacial to interglacial periods. There are uncertainties in the climate forcings, such as changes in atmospheric composition, which have occurred on paleoclimate time scales, but knowledge of these is improving as paleoclimate data, including that contained in long polar ice cores, is analyzed. As of today, the paleoclimate studies are able to confirm that the climate sensitivities in the models are of the right order. It is hoped that in the near future, study of paleoclimate history will provide us with the opportunity to quantitatively measure climate sensitivity without waiting for the final results of the current 'global experiment' which man is conducting by increasing atmospheric CO_2 and trace gases.

3. Carbon Dioxide and Future Climate [20]

After projecting the amount of CO_2 in the atmosphere, we must then turn to either empirical methods or computer models to estimate potential climatic effects. We have no direct empirical case in which, for example, CO_2 doubled and <u>at the same time</u> we have accurate measurements of the climatic response. Thus, we must instead model the Greenhouse Effect, since there is no definitive empirical test. What the Greenhouse Effect essentially represents is the preferential ability of the atmosphere to allow solar radiation to penetrate to the surface, in comparison to the more difficult escape of infrared radiation from the atmosphere and surface to space. The CO_2 Greenhouse Effect, however, really works in Earth's atmosphere not so much by trapping radiation emitted from the surface, but rather by reradiating infrared radiation in significant amounts downward to the surface, causing a warming.

---

[20]This section was written solely for use in this text by Stephen H. Schneider, Center for Atmospheric Research, Boulder, CO

Figure VIII-11 Temperature Difference Shown in Both Nature and GCM

The geographical distribution of the difference in surface air temperature between August and February in degrees Celcius. (This is a measure of the strength of the annual climatic cycle and a surrogate for climatic change from other causes.) Top: computed distribution from a global general circulation computer model. Bottom: the observed distribution. Source: Manabe and Stouffer, 1980.

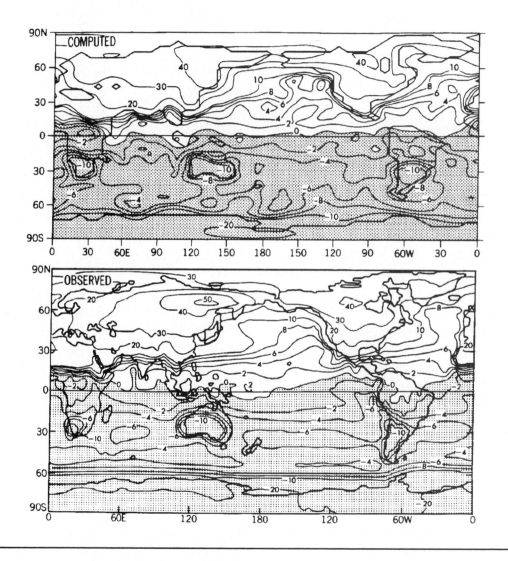

Heating of the surface leads to more evaporation of water, and water vapor is an even more important greenhouse gas than CO_2. Thus, there is a strong positive feedback effect between increasing temperature and increasing water vapor in the atmosphere. The water vapor/greenhouse feedback effect will significantly amplify the surface temperature response initially caused by a CO_2 increase. Many other climatic feedback processes (e.g., the change in surface solar energy absorptivity as a function of snow cover, and changes in cloud height and amount) must be considered when estimating the surface temperature response to a CO_2

increase.[21] It takes complex physically-based mathematical models to be able to produce meaningful estimates of the climatic effects of CO_2 increase, simply because we have no unambiguous historical evidence for simultaneous increases in CO_2 and temperature.

The temperature of Earth's surface over the past hundred years has been estimated from thermometer and other proxy or instrumental data. The data suggest a global warming of some 0.5°C from 1880 to about 1950, a cooling of about half that magnitude in the high latitudes of the northern hemisphere during the 1950s and 1960s, and a substantial rewarming during the 1970s and early 1980s. (Even these estimates of global temperature trends are uncertain to a few tenths of a degree C, largely due to vast areas of missing data in remote land and oceanic areas.) Thus on average the period from 1955 to 1985 is about 0.5°C warmer than the period from 1870 to 1900. This is indeed consistent with model predictions for the consequences of a CO_2 increase of some 20 percent to 30 percent which is believed to have occurred over this time span. However, Earth's surface temperature warmed more rapidly through the 1930s and 1940s than can be accounted for by CO_2, and the subsequent cooling in the northern hemisphere clearly cannot be related to a CO_2 Greenhouse Effect warming.

Investigators have invoked the existence of dust veils from either human activities or explosive volcanoes, or variations in the energy output of the Sun, as well as variations in CO_2 or other trace gases, to explain the temperature records of the past hundred years. Unfortunately, we only have good quantitative knowledge of the variation of volcanic dust veils and solar energy output -- the so-called solar constant -- for the past decade or so. Thus, accounting for the effect of these alternative forcing mechanisms on climate for the bulk of the past century is largely a speculative exercise. For these reasons no one can argue that the CO_2 signal has been "reliably detected" in the climatic record without also implicitly invoking certain assumptions about what has happened to other plausible causes of climatic change. It is certainly true that one can claim that the observed climate of the past century is "consistent with" CO_2 greenhouse calculations, but there is room for at least a factor of two difference in the estimated long term, global temperature effect of a CO_2 doubling (from 1.5°C to 4.5°C).

Even though most global climatic models are within a factor of two agreement among themselves and with empirical data for the global temperature effects of CO_2, is this conclusive evidence that the CO_2 signal is in the climate records? Well, certainly it is circumstantial evidence, but whether it is "conclusive" depends upon one's criterion for "acceptable evidence". Fortunately, there are other forms of evidence available. For example, Mars and Venus both have CO_2 atmospheres, and the same kinds of physical models that we use to predict Earth CO_2 effects have been successfully used to predict the radically different surface temperatures on these two planets.

Furthermore, terrestrial climate models are not completely unvalidated. For example, our most complex climatic models, the General Circulation Models (GCMs), are able to reproduce with reasonable fidelity the largest climatic signal that has been observed since the Ice Ages: the annual cycle of the seasons. Figure VIII-11 , from Manabe and Stouffer[22] shows the temperature difference between February and August both in nature and in their GCM.

---

[21](See, e.g., Chapter 8 of Schneider and Londer, The Coevolution of Climate and Life, 1984 for further details.)

[22]Manabe, S., and R.J. Stouffer, Sensitivity of a global climate model to an increase in CO_2 concentration in the atmosphere, J. Geophsy. Res. 85, pp. 5529-5554, (1980).

The annual cycle, as simulated and in reality, is a more than an order of magnitude larger surface temperature change than is implied by CO_2 buildup. The fact that climate models can successfully simulate such a "climate change" (i.e., seasonality) certainly strengthens the arguments of those who believe a global CO_2-induced warming is both already present in the system and will be even more obvious over the next two decades.

Each of the above bits of evidence are still circumstantial; none by itself is highly convincing. But taken together they lead many investigators, myself included, to be fairly confident that scientific consensus on the presence of a global CO_2 warming will be reached by the end of the century.

J. SUMMARY: CLIMATES AND CLIMATE MODELS [23]

If there is one overriding principle in weather and climate, it is that of change. One need only look out the window from day to day, sometimes even hour to hour, to see changes in the weather that can occur on those timescales. As a year passes, seasonal changes are apparent in virtually every location on the earth. The most dramatic seasonal changes in the weather occur at mid- and high latitudes where the biggest variations in solar input occur through the course of the annual cycle. Over the years, it is clear that a great deal of variability in the weather exists from one year to the next. Famous "terrible" winters and "disastrously dry" summers are remembered for how unusual they were compared to some perceived "normal" seasonal range. However, even such a "normal" state is not constant and instead is slowly fluctuating over decades and centuries. Ice ages are the most widely cited evidence for extreme climate fluctuations over very long periods of time. Less well known but just as important were periods when the planet was even warmer than it is now.

Clearly the climate of the Earth is capable of variability on many different timescales. However, it is difficult to analyze conditions of past climates thoroughly because only scattered evidence is available to reconstruct conditions in those times. Similarly, clues as to the future climate of the Earth are even less available. Some type of tool is necessary to study past, present, and future climates. That tool is the climate model, and various manifestations are available and in use today.

Climate models are collections of mathematical equations that describe processes that take place in the physical world. By combining these equations in various ways and programming them on a computer, they can be solved to obtain simulations of climate states. Using the present distribution of the continents, solar input and atmospheric composition, climate models can produce a reasonable simulation of today's climate. Results from such a climate model simulation can then be analyzed in detail to see what processes, described in its various equations, are important for producing certain climatic features.

Past climates also can be reconstructed using climate models. For example, it is known that in the distant past the continents were in a much different configuration than at present. In a climate model the positions of the continents can be changed easily, and the model can

[23]This Section was written solely for use in this text by Gerald A. Meehl, National Center for Atmospheric Research, Boulder, CO

produce its own climate with the continents rearranged. This model climate then can be analyzed for clues as to conditions that existed all over the globe during those times.

It is also possible to simulate future climate states with climate models. In **The Atmosphere** unit, it was explained how carbon dioxide (CO_2) is dramatically increasing from man's burning of fossil fuels and Earth's climate may warm since CO_2 is a so-called "greenhouse gas." This property of CO_2 is well-known, but the climatic consequences of a large increase of CO_2 due to man's activities could only be guessed if it were not for climate models. In a climate model, the amount of CO_2 can be increased in the particular equation that pertains to it, and the entire model can then be run on the computer to produce a new climate simulation with that increase of CO_2. Then, just as when studying past and present climates simulated with climate models, this future climate with increased CO_2 can be analyzed to give an indication of how the weather and climate in certain regions will be different. For example, food production zones could shift, plant growth and vegetation could become altered, and certain regions could become much warmer.

Climate models present the opportunity to more quantitatively assess conditions of past climates, understand in more detail the workings of the present-day climate system, and estimate climate fluctuations in the future, some of which may be the consequence of mankind's activities.

RECOMMENDED READING

Stephen H. Schneider, Randi Londer, The Coevolution of Climate and Life, Sierra Club Books, 1984. A readable and comprehensive discussion of past, present, and possible future climate and its interaction with life. Highly recommended.

Roger Revelle, Carbon Dioxide and World Climate, Scientific American, August 1982.

Anon., Modeling the Climate System, Mosaic, Nov/Dec 1979.

UNIT IX TALES FROM OTHER WORLDS:

THE SOLAR FAMILY

JUPITER AND ITS MOONS

A. INTRODUCTION

1. Overview

The advent of the Space Age has brought about a great advance in knowledge of our neighbors in the solar system. Our studies of the Moon, and the results of the Apollo program in particular, have enabled us to construct a history of the early solar system that is wholly missing on Earth. We will take a tour of the solar system, noting the processes that have shaped each world. In doing so, we will gather the observational evidence upon which rest theories for the origin of the solar system and its evolution that will be developed in the next unit.

---

2. Objectives

Upon completion of this unit you should be able to:

1. describe individual planets and classes of satellites and small bodies of the solar system as we know them today

2. describe how the physical and chemical characteristics of planets change with distance from the Sun

3. describe the different processes that have acted on and that have affected the surfaces and interiors of the planets

4. synthesize and explain relationships that exist between the processes and characteristics of each planet.

---

3. Key Terms and Concepts

solar system
planets
satellites
ecliptic
geocentric model
heliocentric model
elliptical orbits
Martian "canals"
observations from spacecraft
Apollo missions to the Moon
comparative planetology
terrestrial planets
Jovian planets
asteroids
lunar maria
lunar highlands
mare basalts
ages of lunar rocks
impact craters
multi-ringed impact basins 316

rilles
lunar seismicity
early decay of Aluminum-26
asymmetry of the Moon
age dating using crater densities
radar mapping
permafrost
counter-flowing wind bands 304
stable eddy
planetary ring systems
tidal heating
shepherd moons 307
meteors
meteorites
chondrites
achondrites
iron meteorites
comets
Oort cloud of comets

4. Corresponding Video

Voyages to other planets have shown us that Earth is unique. This program will reconstruct the origins of our solar system and will explore Earth's neighbors through the latest technology -- Apollo lunar orbiter photographs of our moon; Voyager images of Saturn and its rings, and of Jupiter and its moons Io, Callisto, Ganymede, and Europa; animations of Uranus, Neptune; as well as simulations of ancient events on the moon. Computer-enhanced radar images will help reveal the secrets of Venus.

B. EXPLORATION OF THE SOLAR SYSTEM

The most stunning sight that is available to every human on this planet must surely be the midnight sky, seen on a clear night far from sources of artificial light. It is not surprising that astronomy was one of the earliest and best developed of sciences. Against the background of the fixed stars the ancients noted the motion of a number of celestial bodies: the Sun, Moon, and a group of five "stars" that moved about, some so rapidly that they were in a different place each night, and others so slowly that their motion could only be discerned over a period of months. These five objects were called planets, after the Greek word for "wanderer."

To the eye, the planets appear no different from the stars, but because of their motion and their brightness, the ancients quickly guessed that they are much closer to us. Two of them, Mercury and Venus, appear to move very quickly, but never leave the immediate vicinity of the Sun. As a result, they are seen as evening or morning "stars," but are never observable at midnight. The others -- Mars, Jupiter, and Saturn -- move about the sky in a band on either side of the ecliptic, which is a line that marks the path taken by the Sun. While these three range far from the Sun and may often be seen in the middle of the night, their motions are complex. They seem to travel at varying speeds and sometimes even appear to turn around and travel in the opposite direction for a while.

The early Greeks tried to explain planetary motions using a model in which the Earth stood fixed and unmoving in the center of the universe. Hipparchus, who lived in the second century before Christ, proposed such a geocentric (Earth-centered) model. His work was extended by Claudius Ptolemy in the second century AD Ptolemy was sufficiently clever about it that his model could make predictions about planetary motions that were accurate enough for the purposes of his time.

It is not at all unusual for an incorrect model to serve a long period of usefulness in science. In fact, the presently-accepted heliocentric (Sun-centered) solar system model was not taken seriously until 1543 AD, when the Polish astronomer Nicolas Copernicus published his revolutionary model. Copernicus was not the first to suggest that the Earth moved around the Sun. That distinction belongs to Aristarchus of Samos (310-230 BC), who wrote some 1,800 years earlier, but his model was considered too unnatural by astronomers of that time.

Copernicus' model really didn't do a much better job of explaining the observations because he based his model on circular orbits with the planets traveling in small circular paths called epicycles in addition to the larger circular orbital motion. This was necessary to explain observations of planetary positions. It was not until Johannes Kepler published his three laws of planetary motion in the early 1600's that highly accurate astronomical predictions became possible. Kepler replaced Copernicus' circular orbits and epicycles with slightly elliptical ones and specified the mathematical rules by which planets or comets move more slowly when they are far from the Sun and more quickly when they are close to the Sun.

Strict biblical interpretation required a moving Sun and a fixed Earth and Kepler's contemporary, Galileo, suffered denunciation and imprisonment for espousing the heliocentric view. It was a futile gesture on the part of the churchmen, however, for Galileo had already taken the next great step in the exploration of the planets: he had built a telescope and turned it heavenward.

1. Telescopic Observations

Galileo's telescope enlarged his view by about thirty times -- only a few times the power of today's typical binoculars. He turned it on the Moon and observed craters and mountains. He looked at Mars and noted that it did not appear to be perfectly round -- an illusion caused by the bright and dark markings on the Martian surface. He examined Jupiter and discovered its four largest moons, and noted appendages on Saturn that would later be recognized as rings. The era of telescopic astronomy had begun.

From this point on, observations and discoveries increased as telescopes became larger and more powerful. Uranus was accidently discovered by the German-English astronomer William Herschel in 1781. Neptune was discovered after its existence and location were predicted by two mathematicians who had noted irregularities in the orbit of Uranus. Around 1900, the American astronomer Percival Lowell predicted the existence of a ninth planet on the basis of slight irregularities that remained in the orbit of Uranus even after the effects of Neptune had been taken into account. That planet, named Pluto, was finally discovered in 1930. The discovery of Pluto was actually an accident, and we now know that Pluto is too small to affect the motion of Uranus to the extent originally predicted. Astronomers today continue to search for a tenth planet.

Many features of the planets were described and photographed, but a profound difficulty soon set a sharp limit on the resolution of our planetary vision. The Earth's turbulent atmosphere introduces unsteadiness in the magnified views of the planets, the same phenomenon that causes the apparent "twinkle" of stars. Telescopic photographs of the planets had to be made with short exposures to minimize the unsteadiness of the image, both the atmosphere still introduced distortions similar to those encountered when photographing an object in the bed of a rapidly flowing stream from above the water surface. For this reason, detailed planetary maps were constructed primarily from visual observations. The human eye combines high resolution and great sensitivity with an effectiveness that is still unmatched by any instrument, but its combination with the human brain can raise questions about the objectivity of its observations.

The story of the Martian "canals" illustrates the difficulties with visual observations. In 1877 Giovanni Schiaparelli produced a map of Mars showing a system of long straight lines that crisscrossed the surface. Percival Lowell, who was instrumental in the discovery of Pluto, charted some 500 canals, and insisted that these features must be the work of intelligent beings! Other reliable astronomers also claimed to have seen the canals, and William Pickering of Harvard Observatory described dark spots at the intersections of canals and called them oases. The matter of canals remained controversial, for many contemporary astronomers failed to see them, and challenged their reality.

Today, of course, we have high resolution photographs of the Martian surface taken by spacecraft. Surprisingly, these photographs show no trace whatever of the canals and oases. There is not even any correspondence between the major surface features of Mars and the prominent canals shown on Lowell's map. Martian canals and beauty both are in the eye of the beholder, it seems. Visual observations, especially transient ones made difficult by atmospheric twinkle, can fall prey to suggestion and a host of other interpretations imposed by the human mind. For this reason, visual observations unsubstantiated by photographs or other instrumental observation have always been regarded as the weakest kind of scientific evidence. This is also the principal reason for the reluctance of the scientific community to accept the reliability of numerous visual sightings of UFO's.

Partly because of the resolution limits on planetary details imposed by the atmosphere and partly because new and larger telescopes were opening up the vast reaches of the universe to study, planetary astronomy went into something of an eclipse after the Second World War. Valuable observations continued to be made, but the emphasis shifted toward spectrophotometry, in which the spectrum of light from the planets is analyzed in order to determine the composition of the surface or atmosphere from which the light is reflected.

The next major advance in our knowledge of the planets would require us to leave the veiling envelope of the Earth's atmosphere.

2. We Step into Space

The radio beeps of Russia's basketball-size Sputnik in 1957 marked the beginning of the space age as the first man-made device to be placed in orbit around the Earth. Shocked by this early lead that the Russians had taken, the United States responded with its own Earth orbiters, and in 1961 President Kennedy made his historic commitment to placing a man on the Moon within that decade. That commitment was fulfilled on July 20, 1969, and during the span of only three years, six successful Apollo missions landed on the Moon, set out instrument packages, and returned with a total of 382 kg of rocks and soil for later study. Russia sent three robots to the Moon that retrieved 310 g of material -- a small but important addition to the inventory. The last lunar landing was one of these robots, which returned in 1976. These missions revealed a vast store of information about the Moon that has helped us to increase our understanding not only of the Moon but also of the origin and evolution of all the planets.

In 1960, the United States began a series of missions aimed at a comprehensive study of the solar system. Spacecraft flew past Venus and Mars and returned the first detailed photographs and other information. By 1967 Russia joined the effort by landing a robot on Venus, and in 1971 they landed another on Mars. The United States spacecraft Mariner 10 flew past Mercury in 1974. Pioneer 10 visited Jupiter in 1973 and Saturn was encountered in 1979. With this mission a milestone was reached -- we had finally examined all of the planets known to the ancients.

The Voyager mission not only returned new information on the planets, but gave us our first glimpses of strange and fascinating new worlds -- the satellites of Jupiter and Saturn. Because of their great distance, telescopic observations had done little more than hint at the richness and diversity of geologic settings that would be found on these moons. The Jupiter and Saturn missions have been likened to the great voyages of discovery of Magellan and Vasco de Gama.

3. Prospects for the Future

The two decades of the 1960's and 1970's marked a Golden Age of space exploration. In contrast, the 1980's have been marked by a dramatic withdrawal from planetary exploration on the part of the United States. Present activity is mostly confined to the near-Earth environment of the Space Shuttle. The final encounters from launches of the last decade will be those of Voyager 2. It will visit Uranus in January 1986 and then go on to view Neptune in 1989. No planetary probes have been launched by the USA in this decade, though at this writing, the Galileo mission is planned for a tentative 1985 launch. This spacecraft will go into orbit around Jupiter and send a probe into the Jovian atmosphere. Space exploration has continued by the Soviet Union, however. Two Venera spacecraft made landings on Venus in 1982 and two others went into orbit around Venus carrying radar devices for mapping the cloud-shrouded surface of that planet. Vega probes are scheduled for launch toward encounters with Venus and Halley's Comet in 1985-86 and to one of the moons of Mars in 1988.

In 1980 a blue-ribbon panel, the Solar System Exploration Committee, was appointed by NASA to study the state of planetary exploration and suggest missions that combined high scientific priority and modest budgets. The panel recommended four missions with dates in the period 1988-1992. The first would orbit Venus and produce detailed maps of its surface using radar to penetrate its dense atmosphere; another would orbit Mars and analyze its surface and atmosphere in a manner similar to the Earth-orbiting Landsat and weather satellites; a third would fly past an asteroid and rendezvous with a comet; and the final mission would visit Saturn and probe Titan, its largest moon.

Other nations have joined the effort, with the European Space Agency, Russia, and Japan planning to send probes to Halley's comet in 1986.

Finally, NASA plans to place the 2.4 m diameter Space Telescope in orbit in 1986, finally lifting the telescopic eye above the atmospheric veil. Though designed primarily for deep-space astronomy, the Space Telescope will also make significant planetary observations.

Planetary scientists are frustrated by the current hiatus, and yearn for samples from Mars and Venus. The lunar sample program has shown the tremendous value of laboratory study of actual rocks from other members of the solar system.

It is not conceivable that planetary exploration should cease. It would be as if the exploration of our own planet had ceased after Magellan's voyage in 1522. The greatest discoveries -- and benefits therefrom -- surely lie ahead of us.

right of the diagram, it would be far smaller than the width of the line. The dominant characteristic of the solar system is emptiness.

The planets of the solar system may be placed into two groups: the <u>terrestrial planets</u> (Mercury, Venus, Earth, Mars), and the <u>Jovian planets</u> (Jupiter, Saturn, Uranus, Neptune). The terrestrial planets typically are close to the Sun, rocky, small, and have no or few satellites or ring systems. The Jovian planets, on the other hand, are far from the Sun, gaseous, large, and typically have many satellites and rings. Pluto does not fit into this scheme nicely, and may resemble an icy Jovian satellite more than a Jovian planet. It is so far away that we really know very little about it.

1. The Moon

As Earth's closest companion, the Moon is a logical place for us to begin our tour. The Moon rotates in such a way as to always turn the same face toward Earth. As a result, we knew nothing of the far side of the Moon before a Russian spacecraft photographed it in 1962.

The most prominent features of the Moon as seen from Earth are dark splotches called <u>maria</u> that form the crude facial features of the "Man in the Moon." In Latin the word <u>mare</u> means <u>sea</u> and early observers thought they saw in these dark areas counterparts to Earth's oceans. They gave them fanciful names like Sea of Rains (Mare Imbrium) and Sea of Tranquillity (Mare Tranquillitatis). In fact they are broad, relatively smooth plains cloaked by a thick lunar soil made up of a mixture of fragments and small grains of rock.

The surrounding bright areas of the Moon are the <u>lunar highlands</u> which stand considerably above the elevation of the dark mare plains. Other features of the Moon's surface include the craters and mountains Galileo saw through his telescope.

The following exercise will help you to become familiar with these features. You will need binoculars, opera glasses, or a small telescope.

EXERCISE

The Moon is best observed during the period from a few days before first quarter until shortly after full moon. Viewing is also good around last quarter, but you may have to get up before sunrise to take advantage of it. The Moon can also be seen during the day, but the bright sky tends to wash out its features, even as seen through binoculars, and there is also the danger of accidentally catching the Sun within your binocular's field of view. (<u>WARNING</u>): NEVER look at the Sun using binoculars or a telescope, not even during a total eclipse. Even a momentary glimpse of the magnified Sun can result in blindness or eye damage. Please confine your examinations of the heavens to the nighttime sky.

Most calendars identify the phases of the Moon. Mark the next first quarter and then wait for a clear night during the ensuing week. Locate a space outdoors where the Moon will be visible and where there is not too much interference from nearby lights. You will need something on which to prop your elbows such as the arms of a chair or a balcony railing in order to steady the binoculars.

Draw a large circle on a pad of paper and pencil-sketch the prominent features visible to you. If you are observing at other than full moon, draw in the terminator (the line between day and night on the Moon).

You should be able to see mountain ranges and craters highlighted by shadows near the terminator. Sketch the outlines of the dark lunar maria. Sketch in large craters, including any that have bright "rays" extending out from them. Note which have the larger number of craters per unit area: the lunar maria or the bright surrounding highlands. If you are observing at full moon, individual craters may be difficult to see.

A small flashlight may be useful to illuminate your sketch pad. When you are done, compare your sketch with the map in Figure IX-2.

The lunar highlands are formed of very rugged terrain, pockmarked with overlapping craters and long arcuate mountain ranges. It is not at all obvious why the highlands should appear so much brighter than the smooth maria. The reason turns out to be a matter of rock type -- the maria are composed of dark basalt while the highlands are composed of lighter-colored rocks that tend to be richer in the mineral feldspar. The highland rocks are less dense than the mare basalts.

A striking feature of the lunar surface is a complete absence of sedimentary rocks that require water for their formation, such as limestones, shales, and sandstones. There are sediments on the Moon, in the sense that much of the debris on its surface was formed by erosion (mostly breaking of the rock during impacts), transport (by being tossed from the point of impact), and deposition (landing on the surface). Even lithification exists in some rocks that consist of fragments that have been welded together into solid rock by the force of impact.

Nonetheless, the original igneous origin of lunar rock materials is apparent in all the returned samples, largeley due to the absence of chemical weathering, which in turn reflects the total absence of water on the Moon. Basalts and the lighter rocks of the lunar highlands are common on Earth, too, but the lack of water produces distinct differences between Earth rocks and Moon rocks. Most Earth rocks contain more or less water in minerals that are said to be hydrated -- water molecules are chemically bound into the mineral's atomic structure. No water is found in lunar rocks.

One thing is found in lunar rocks that is not generally found in Earth rocks: metallic iron. On Earth, exposure to oxygen and moisture produces oxidation of iron to common rust and other oxides. (Have you looked at the underside of your car lately?) As a result, iron in rocks usually appears on Earth as an oxide or combined with other elements. Moon rocks, on the other hand, do contain pure iron as small metallic grains. The absence of both oxygen and water, or any atmosphere for that matter, means that the kinds of weathering and erosion that

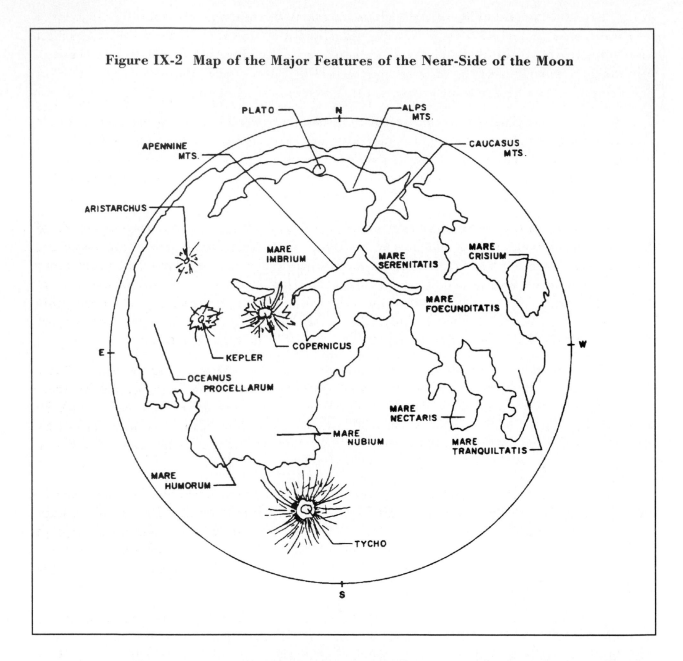

Figure IX-2 Map of the Major Features of the Near-Side of the Moon

are such important processes on Earth play no role whatever on the Moon. When viewed under a microscope, lunar rocks appear as fresh as if they had been formed only yesterday; while in fact, they are extremely ancient.

Radioisotope age dating of moon rocks reveals this antiquity -- the youngest rocks returned from the Moon are more than 3 billion years old. In contrast, the vast majority of Earth rocks are younger.

The mare basalts date in the range from 3.1 billion to 3.8 billion years, while the highlands rocks are significantly older: 4 to 4.3 billion years, for the most part. A very few samples have yielded a remarkable date: 4.6 billion years -- the presumed age of the Earth.

The amazing antiquity of the Moon's surface means not only that it has been a geologically dead planet for most of its life, but also that it preserves a record of the earliest periods of geological history that have all but vanished from the Earth's surface. Let us look more closely at some of these very ancient features of the Moon.

After the maria, the lunar craters are the most conspicuous features visible in lunar photographs. At one time they were thought to be due to volcanic activity, since they resemble volcanic craters or calderas on Earth. The lunar visits confirmed that they are in fact due to meteorite impacts. A rocky object falling onto the surface of the Moon will arrive at a speed of at least 2.4 km per sec (5,400 mi per hr). If it is the size of a house, it will do a great deal of damage when it hits. The result is a circular crater with a raised rim.

Millions of impact craters are found on the Moon, ranging in size from hundreds of kilometers in diameter, through dimples in the lunar dust obliterated by astronaut footprints, down to microscopic pock-marks on individual mineral grains. An important hint about lunar history is contained in the observation that the mare plains have far fewer and smaller craters than the highlands. The meteorite bombardment must have been tapering off 3.9 billion years ago, before the creation of the mare plains. This would explain why we see so few large impact craters on Earth -- no rocks of that age have survived our erosive environment.

The mountains of the Moon tend to appear as arcuate ranges ringing the maria. These are now recognized as impact features as well, but they must have been formed by truly colossal impacts. When you did the exercise earlier in this unit, you may have been struck by the rough circularity of the maria. If the basins now occupied by the maria were formed by impacts, then the bodies that produced them would fall into the size range that we associate with asteroids rather than meteorites.

One of the best examples of a huge impact basin is Mare Orientale. Not fully visible from the Earth, its striking nature can only be seen from spacecraft photos. It looks like a huge target on the moon with the bulls-eye a mare plain surrounded by four concentric rings of mountains, the outermost having a diameter of 900 km (560 mi). These immense structures speak to us of an early lunar history punctuated by planet-wracking collisions of unimaginable violence.

At this point you might suspect that the mare basalts formed at the same time as the collisions, but this is probably not the case. The mare basalt formed hundreds of millions of years later, as may be seen from their younger ages and from the few craters pocking their surfaces.

We know more about the maria than about the highlands because the smooth maria were favored over the rugged highlands as safe landing sites for spacecraft. On the later Apollo missions, a lunar "rover" was used to transport the astronauts some kilometers from their landing craft. On the Apollo 15 mission, the astronauts visited Hadley Rille, a long sinuous valley carved into the mare basalt. Rilles, rifts, and sinuous ridges are prominent features of the mare plains. The lunar rilles are likely to have formed as lava channels that have been incised into the surface.

Just as on Earth, the rifts signify a region under tension. The ridges appear to be areas where the basalt plains have been compressed and faulted, with one basaltic sheet thrust over the other. Unlike the Earth's spreading ridges and subduction zones, these tectonic features show very limited displacement. Nothing has moved very far, and both rifts and ridges probably originated due to the bending or shrinking of the dense mare basalts as they deformed the underlying layers. Plate tectonics apparently has not been active on the Moon at any time, further contributing to the static nature of our cosmic companion.

It was in Hadley Rille that we had our first look at what appeared to be layered bedrock -- lava flows exposed in the walls of the rille. Everywhere else that the astronauts walked, the lunar surface was mantled in a deep soil made up of ground-up rocks and tiny fragments shattered by meteorite impacts. This layer of loose soil extends from about 1 to 20 meters in depth. To learn something about the Moon's interior, astronauts on several missions set out sensitive seismographs that listened for and recorded moonquakes.

The seismicity of the Moon is far less than that of the Earth, in keeping with its reputation as a nearly dead planet. Moonquakes proved to have three sources: slippage along faults, similar to the sources of earthquakes; impacts of meteorites on the Moon's surface; and man-made events, such as the deliberate crashing of abandoned rocket stages onto the Moon. Because it was possible to determine just when and where events of the last kind would take place, they could serve in the construction of travel-time curves that could help to reveal the interior structure of the Moon, just as we used them for the Earth in the unit **Continental Tectonics**.

Natural moonquakes proved to originate mostly at depths ranging from 600 to 800 km. This is quite different from the case on Earth, where the majority of earthquakes originate at depths of less than 100 km. Because some of these moonquakes occur at the same time each month, tidal forces exerted on the Moon by the Earth may be at least partly responsible for them.

Analysis of the travel-time curves reveals a crust of about 60-km thickness overlying a somewhat denser mantle. The question of whether or not the Moon has a core has not yet been resolved. If it does have a dense metallic core, it must be fairly small in order to satisfy both the seismic data and the fact that the average density of the Moon is only 61% that of the Earth.

Putting all of the evidence together allows the construction of a fairly detailed picture of the Moon's early history. We will consider the formation of the Moon and the other members of the solar system later, but for now we are interested in the first 1,000 million years.

The obvious igneous nature of the old rocks -- or at least of the fragments of which many are composed -- implies that the outer layers of the Moon were molten at one time. Since the Moon is not volcanically active now, we have to ask what was the source of heat to accomplish this. One possibility is the heat energy released during meteorite impacts, but many lunar rocks show evidence of slow cooling at some depth. This could be produced by the largest impacts, but another answer might lie in evidence that suggests that certain radioisotopes such as Aluminum-26 may have been abundant in the early solar system. Aluminum-26 has a half-life of only 720,000 years, so even if it was abundant 4,600 million years ago, it would be virtually all gone by now. Its presence is suspected because its daughter product (the isotope to which it decays) is unusually abundant in the most ancient meteorites. Radioactive heating may well have been an important process in the first few millions of years of the solar system.

Widespread melting formed the silica-rich lunar highlands crust, which at that time may have covered the entire surface of the Moon. After the outer layers had solidified, the Moon was subjected to intense bombardment by large and small objects. The result is the rocks were smashed, fragmented, and in many cases the pieces were welded back together to form rocks composed of angular fragments of yet older rocks. The great impact basins that would later form the maria were created by stupendous collisions with asteroid-sized bodies that obliterated the original crust in those areas.

The celestial onslaught slackened after about 600 million years, and the shaping of the Moon's surface became dominated by a wholly new process -- the flooding of the great impact basins with molten rock. By the time the mare basalt plains had formed, the rain of meteorites had slowed to a trickle, punctuated by an occasional blockbuster. This occupied something like the next 1,000 million years. By 3,000 million years ago, the face of the Moon would already have been familiar to a present-day astronomer. In contrast, not a single continent on Earth at that time would have had a familiar shape, or been in the same position that it occupies today. At about the same time that the familiar part of Earth's history began to unfold, the Moon's geological evolution ground to a halt.

One of the most striking features of the Moon is its asymmetry. Most of the maria are confined to the near side, while the far side is dominated by highland terrain. It has been suggested that this is due to a difference in crustal thickness, which averages only 60 km on the near side but increases to 100 km on the far side. If the mare basalts formed because of the buildup at depth of radioactive heat (this time due to longer-lived radioisotopes like uranium, thorium, and potassium), the flooding of the impact basins would have occurred preferentially where the crust was thinnest. Another explanation requires a concentration of radioisotopes in the near-side hemisphere. Both hypotheses beg the question: Why is the Moon asymmetric in the first place? The fact that the Moon always turns the same face to Earth is likely involved, but a clear explanation of the asymmetry is not yet at hand.

As our nearest neighbor, the Moon has received the most detailed study in our first forays into the solar system. Its record of early solar system events provides an important key to understanding the development of all the other members of the solar family, including Earth.

STUDY QUESTIONS

IX-4. Why are Mercury and Venus never visible at midnight?

IX-5. What are some of the differences between Earth rocks and Moon rocks?

IX-6. Why are there fewer craters on the lunar maria than in the highlands?

IX-7. Why do we see so few impact craters on Earth?

IX-8. In what way is the Moon asymmetric?

2. Mercury

Mercury is the closest planet to the Sun, and the temperature on its surface ranges from a high of 430°C (800°F) at midday to a low of -180°C (-290°F) at night. Lead would melt in the Mercurean noonday sun.

The airless surface of this small planet closely resembles that of the Moon. Although we have no samples, only spacecraft photos, the story that Mercury has to tell us sounds familiar: a densely cratered terrain with at least one huge impact basin and extensive plains. But there are significant differences as well. Some of the plains are heavily cratered, and while smooth plains appear to be younger than the rest of the surface; observations seem to indicate a composition more like the lunar highlands than the lunar maria. Mercury does not seem to have anything quite like the lunar maria, and perhaps basaltic volcanism has not been an important process for it. In addition, the crater forms found on Mercury are not quite the same as those found on the Moon due to the higher density of the planet and hence a higher gravity field. Material ejected from the craters during impact, for instant, did not travel as far as on the Moon.

Crater densities (number of craters per unit area) may be used to date the Mercurian surface. If the rate of crater-producing impacts has decreased throughout the solar system with time, as indicated in the discussion of the Moon's history, then areas with the same density of craters on the Moon and Mercury should be roughly the same age. Because the lunar rocks from representative areas of the Moon have been radioisotopically dated, this provides us with a potent method for dating planetary surfaces. Some care needs to be exercised in applying this method to dating surfaces on widely separated planets or satellites, however, since we know that the rate of cratering depended on the location in the solar system. Nonetheless, it gives us a first-order system for dating cratered surfaces that have not yet been visited by sample-collecting astronauts. On this basis, most of the Mercurean surface appears to be more than 4,000 million years old. Mercury may well have lapsed into geological inactivity even earlier than the Moon.

There is evidence of tectonic activity in the presence of a unique planet-wide system of curved scarps, or cliffs. These range in length up to 500 km (310 mi) and in height up to 3 km (9,800 ft). The cliffs appear to be faults caused by compression due to early shrinkage of the planet's interior as it cooled during the first 1,000 million years of the solar system.

To date we have only photographs taken by Mariner 10 in 1974 and 1975 as it encountered the planet three times. Those photos cover only about 35% of the surface and are of lower resolution than the lunar photographs. The Soviets have now mapped about 30% of the surface at about a one- to two-kilometer resolution. Mercury's history probably differs in important respects from that of the Moon, but clarifying the picture may have to await future missions.

STUDY QUESTIONS

IX-9. How is it possible to date areas of Mercury's surface?

IX-10. Aside from crater and impact-basin formation, what seems to have been the dominant tectonic process active on Mercury?

---

3. Venus

Venus, next out from the Sun, is the first other world in our journey that has an atmosphere. It is a very thick and inhospitable atmosphere -- pressure at the surface is 90 times Earth's, and the principal gas is carbon dioxide, with some nitrogen and only trace amounts of oxygen and water. Venus' upper atmosphere contains clouds of concentrated sulfuric acid droplets. Considering this along with a surface temperature of almost 460°C (860°F), perhaps Venus is not such a fair lady after all!

The cloud cover is so dense and continuous that the planet's surface is hidden from visual observation. It can, however, be penetrated by radar. Early maps of the Venusian surface were obtained by radar measurements from Earth-based instruments, but maps with 100 km resolution are now available for about 93% of the planet from radar based on orbiting spacecraft.

The surface features on Venus are very intriguing. They should be interpreted, however, in the light of realization that 100 km is about what can be resolved on the surface of Mars using an Earth-based telescope.

Figure IX-3 is a map of the major features of Venus. The light areas are the lowlands while the dark shaded ones represent the areas that stand 2 km (6,600 ft) or more above the average radius of the planet. The latter comprise only about 5% of the total surface area. The overall impression is that most of Venus is really quite flat, but the Maxwell Mountains in Ishtar Terra (upper left of the map) rise to a height of 11 km (36,000 ft) above the average elevation of the planetary surface. This may be compared with the 8.8 km (29,000 ft) height of Mt. Everest above sea level on Earth.

The Maxwell Mountains sit on a large, high plateau called Ishtar Terra. There is another large plateau near Venus' equator called Aphrodite Terra. On the map, Ishtar Terra looks larger, but in fact Aphrodite Terra is the larger of the two -- recall that rectangular maps grossly expand the appearance of areas near the poles. These two highland plateaus look suspiciously like continents and are comparable in size to Australia. They are bordered by relatively steep scarps dropping down to the lowland plains. Are we seeing here an equivalent to Earth's continents and ocean basins?

If the water were removed from Earth's oceans, our planet would present a similar view, though our continents are more widespread and greater in number. Liquid water cannot exist on Venus, since it would quickly boil away in the intense heat. All of Venus' water must reside in its atmosphere as a gas. The question remains: Could the lowland basins be great basaltic sheets that formed at spreading ridges by plate tectonic mechanisms? If the answer is

Figure IX-3 Major Surface Features of Venus

Shaded areas represent the highlands of Venus. Adapted from Stuart Ross Taylor, <u>Planetary Science: A Lunar Perspective</u>, Lunar and Planetary Science Institute, Fig. 2.25, p. 56, 1982. Included with permission.

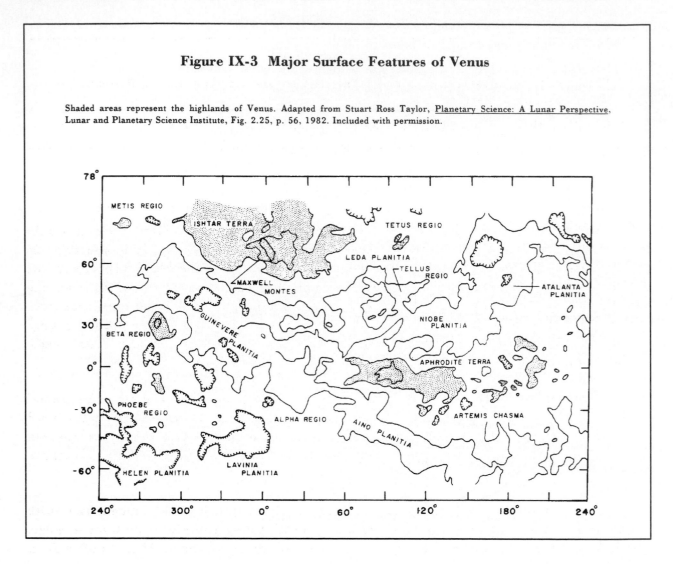

yes, then Venus, like Earth, truly has continents and ocean basins. Unfortunately, the question cannot be answered at this time due to the poor resolution of the radar images.

Two large isolated mountains exist in Beta Regio and Phoebe Regio (left of map) which may be large volcanoes similar to those found on Mars. Arguing against plate tectonic mechanisms on Venus is the absence of lines of volcanoes similar to the Hawaiian Islands or those associated with subduction zones. Rift valleys in Aphrodite Terra tell of tensional tectonics while the Maxwell Mountains in Ishtar Terra may speak of compressive forces at work.

Higher resolution views are needed, and two Russian spacecraft did provide that by landing on Venus and photographing a few square meters of rock-littered ground around their landing pads. More recently, two Russian spacecraft went into orbit around Venus and began a medium-resolution radar mapping of a portion of its surface. Radar pictures released to date show both impact craters and what appear to be volcanic features. When the gap in resolution between one meter and 100 km is filled, many of the questions we have raised will undoubtedly be answered.

STUDY QUESTIONS

IX-11. Why is our knowledge of Venus so much more limited than that of Mercury, the Moon, or Mars?

IX-12. Might there be oceans of water on Venus today? Explain your answer.

---

4. Mars

Photos sent back by the Mariner 4 spacecraft in 1965 showed a bleak cratered terrain resembling the surface of the Moon -- a sharp disappointment to the watching scientists. In 1971, Mariner 9 went into orbit around Mars and proceeded to map and photograph the planet in a detailed and methodical fashion. Recording images at a resolution of less than one kilometer, Mariner 9 sent back pictures that surprised and delighted the Earth-based scientists. The earlier views of a dead cratered surface were supplemented with ones showing huge volcanoes, canyons dwarfing anything on Earth, stratified deposits near the polar caps, channels resembling dry riverbeds, and intricately sculpted sand dunes.

About half of Mars is covered by an ancient cratered terrain that appears to be around 4,000 million years old according to the density of craters. At least three large multi-ringed basins have been found on Mars, reflecting the early period of heavy bombardment recorded on Mercury and the Moon. Curiously, this old terrain is largely confined to Mars' southnern hemisphere.

The northern hemisphere shows a greater variety. Volcanic plains predominate, with clearly visible lava flows and occasional volcano-like structures. The plains are light in color, not dark like the lunar maria, and crater densities indicate ages ranging from 3,600 million years down to 300 million years.

Several very large volcanoes are found on Mars (Figure IX-4). The largest is Olympus Mons which is 550 km (340 mi) across and 21 km (69,000 ft) high! We may compare this with Earth's largest volcano, Mauna Loa in Hawaii, which is about 160 km (100 mi) across and rises about 8 km (26,000 ft) above its base on the Pacific Ocean floor. A very curious feature of Olympus Mons is a 4-km (13,000 ft) high cliff that truncates the lower slopes of the mountain and completely surrounds it. Recent basaltic flows appear to have cascaded over this cliff in some places.

There is strong evidence that water, or some other fluid, once flowed on Mars. Channels and valleys abound that show the kinds of erosional features associated with water-carved terrain on Earth. These range in size from small branching tributary channels to the immense Valles Marineris -- three times the depth of the Grand Canyon and so large that it would stretch from New York to San Francisco. The channels clearly show the influence of running water, although there is none to be seen in them today. Where has all the water gone? Mars' atmosphere is thin and clear, and no large amount of water vapor is held within it.

Figure IX-4 Major Features of Mars

Key:CU=ancient cratered terrain -- PC,PM=cratered plains -- PV=volcanic plains -- V=volcanoes -- C=channel deposits --
P=plains. Adapted from Stuart Ross Taylor, Planetary Science: A Lunar Perspective, Lunar and Planetary Science Institute,
Fig. 2.21, pp. 48-49, 1982. Included with permission.

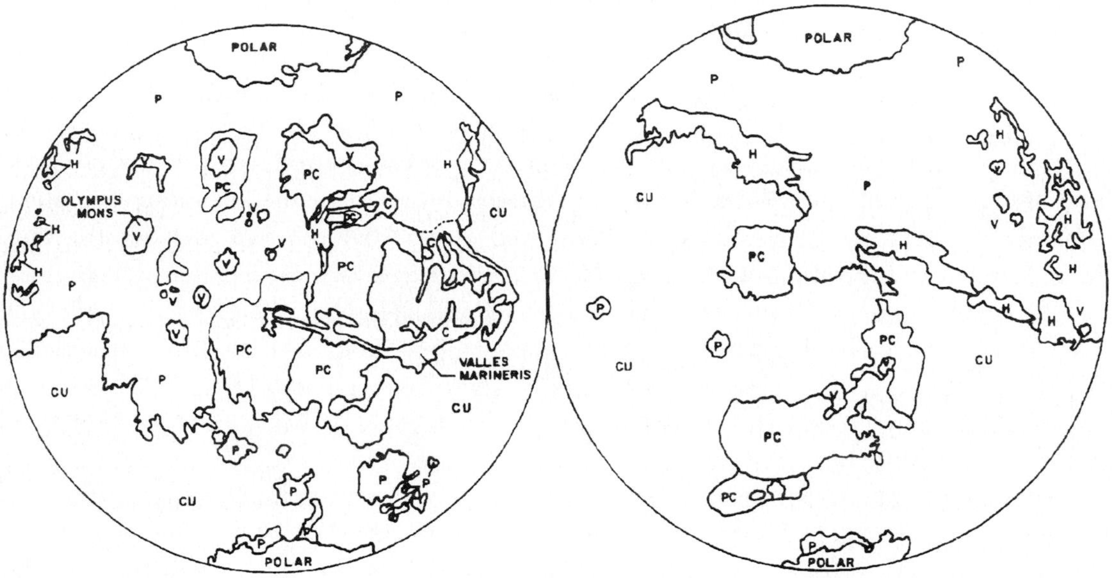

Part of the water is surely stored in the Martian polar ice caps, easily visible from Earth-based telescopes. The rest may reside as ice in the soil -- a kind of planet-wide permafrost. Martian temperatures range from -120°C (-180°F) to a relatively balmy high of -20°C (-4°F). As a result of the low atmospheric pressure, liquid water cannot exist the present Martian surface. Ice is the natural state of water on Mars.

Indeed, many of the channels begin abruptly, with the liquid that formed them apparently emerging from springs at the heads of valleys. This kind of behavior is typical of the melting of permafrost, but what caused the melting? Early speculations ran to models in which a sudden burst of solar activity provided the warmth to melt the ice, but recent measurements of cratering in the floors of channels indicate that the channels are of many different ages. As a result, water must have flowed on Mars many times, over a span of ages. Perhaps Mars, like Earth, had its Ice Ages, and water flowed on its surface only during warm interglacials. The orbital eccentricity of Mars is greater than Earth's, and the slow orbital changes that are thought to affect the cyclic climate of Earth should be even stronger on Mars.

If water has flowed on Mars, could life have begun there? This was an important question for the two Viking landers that set down on the Martian surface, peered at the landscape, and sampled and analyzed the soil beneath them. Three biology experiments looked for life in the soil samples and came up with enigmatic results.

One looked for the ability of an organism to use carbon dioxide (CO_2) or carbon monoxide (CO) as a source of carbon, as Earth plants do. This appeared to give a weakly positive result, but the effect could not be verified on repetition. The second looked for the possible use of carbon in a nutrient solution and its exhalation as CO_2 or CO, as in the case of Earth animals. This, too, gave ambiguous results. The final experiment involved moistening the soil, then feeding it a rich nutrient solution ("chicken soup", the scientists called it). Gases given off were monitored, and a puff of oxygen was noted when the soil was exposed to moisture. In addition, cameras looked for macroscopic life and a mass spectrograph looked (in vain) for long-chain organic molecules.

The results of these experiments suggested either that Martian life forms are totally unlike terrestrial forms in their chemistry, or, more likely, that Martian soil has an unusual inorganic chemistry that produced the observed effects. Most scientists involved with the project feel that the presence of life on Mars is unlikely at this stage in the investigation, but not totally ruled out. A final decision will have to await the return of actual samples to Earth.

STUDY QUESTIONS

IX-13. What is the evidence that Mars may once have enjoyed a warmer climate than at present?

IX-14. Why did carbon dioxide (CO_2) figure so prominantly in the Viking experiments that searched for life on Mars?

---

5. Jupiter and its Satellites

Jupiter is the giant in the solar family and is the best-known of the four Jovian planets: Jupiter, Saturn, Uranus, and Neptune. These planets differ from the inner terrestrial planets in fundamental ways. They are primarily gaseous in composition, with extremely low densities. Saturn, in fact, has an average density less than that of water -- a small object with the same density would float buoyantly in a tub of water.

The compositions of Jupiter and Saturn are very similar to that of the Sun. Their primary constituents are the gases hydrogen and helium, with methane, ammonia, and water present in small amounts. In all likelihood, these planets have no well-defined surface at all. The gaseous atmospheres become more and more dense with depth and gradually transform into a liquid hydrogen and helium envelope. The changing atmosphere is really all that we can see of these planets, and only the cloud-tops at that.

On Earth, clouds form where gaseous water vapor cools and condenses into liquid water droplets or ice crystals that are small enough to remain suspended in the air. On Jupiter and Saturn, the top-most cloud layers are formed from ammonia. Below these probably are clouds formed from a compound of ammonia and hydrogen sulfide, and lowest are water clouds. From here a clear atmosphere probably extends downward, increasing in pressure to such an extent that the hydrogen and helium are eventually squeezed into the liquid state. There is no sharp change from gas to liquid because the pressure is too high. Distinct differences in density and appearance between a gas and its liquid can occur only at low pressures. If the atmospheric pressure were high enough on Earth, the oceans would have an indistinct, fuzzy surface, grading upward from liquid to a dense atmosphere with 100% humidity.

On such a planet as Jupiter, geology is supplanted by meteorology. The surrealistic cloud patterns that you will see in the film are ever-changing, yet retain large-scale features whose mechanisms are familiar from our study of the Earth's atmosphere. On Earth, the trade winds blow from east to west near the equator, while the westerlies flow from west to east at the higher latitudes. Jupiter also has bands of counter-flowing winds, but it has half-a-dozen or more in each hemisphere. This results in the colored bands so prominent in the pictures. These bands are actually pastel-colored, but appear brightly colored in some computer-enhanced pictures.

In the shear zone separating two counter-flowing wind bands, stable eddies can form and persist for long periods of time. These eddies, with their continually shifting patterns, account for the surreal aspect of Jupiter close-ups.

The largest and most persistent eddy is the Great Red Spot, which has been observed telescopically for 300 years. A huge hurricane-like vortex, the Great Red Spot has drifted several times around the planet during this interval, staying roughly at the same latitude. It is so large that the entire planet Earth would fit easily within it.

It is not surprising that Earth-based meteorologists are intensely interested in the Jupiter observations. Here is a laboratory for atmospheric dynamics on a truly grand scale, and it is no exaggeration to say that studies of the winds of Jupiter are helping to improve our ability to track hurricanes on Earth.

The colors of the clouds of Jupiter are related to altitude and temperature. Red colors are associated with high, cool clouds; white, brown, and finally blue tints occur in progressively lower and warmer layers. The actual colors are probably due to small amounts of sulfur, phosphorus, or carbon present in the layers.

For many years it had been assumed that in the solar system only Saturn possessed a ring system. In 1977, however, Earth-based telescopes detected the rings of Uranus, and the suspicion grew that ring systems might be common among the Jovian planets. As a result, the Voyager 1 spacecraft was programmed to search for a ring around Jupiter, and it found one.

Jupiter's ring is faint and relatively transparent, being more easily seen when it is between the Sun and the observer. This accounts for the fact that it is hard to see from Earth, because our planet is always closer to the Sun than Jupiter. The ring is made up of millions of tiny grains of material, all orbiting around Jupiter. As a result, it is mostly empty space.

At present count, Jupiter is known to have 16 moons. All but the inner four discovered by Galileo are small, however. The Jupiter photos and other remotely-sensed observations returned by the four spacecraft would have been justification enough, but the views of the four Galilean satellites were a dividend that has paid handsomely in new knowledge about the solar system. Here are four worlds that differ from one another and from anything else so far observed in the Sun's family.

Io is closest to Jupiter and, like all the Galilean satellites, turns at such a rate that it always presents the same face to Jupiter, just as our Moon does to Earth. The effect is due to tides. Tides in Earth's oceans (as well as in the solid Earth) are raised by the gravitational attraction of the Moon. Because the Earth does <u>not</u> always present the same face to the Moon, each tide travels around the world as the Earth turns. This dissipated energy which is drawn from the Earth's rotation, is slowing it down and gradually lengthening the day. Eventually, the Earth will slow down until it presents the same face to the Moon, creating with it a cosmic mutual admiration society. In this arrangement, the tides still exist, but they don't move, and so don't dissipate energy. This happened to our Moon and to all four Galilean satellites a long time ago.

Nonetheless, the tides in the body of Io are not quite stationary. Because Io's orbit around Jupiter is not perfectly circular, Io alternately approaches and recedes from Jupiter each time it traverses its orbit. This produces a small tidal change, but because Jupiter's gravity field is so strong a large amount of energy is released in Io's interior.

The result is the most geologically active body yet encountered. Even before the Voyager spacecrafts visited it and photographed its six or seven volcanoes in active eruption, it had been theorized that the tidal energy released in Io's interior might be sufficient to keep at least part of the satellite in a molten state. A relatively thin solid crust provides the stage for a volcanic pageant of epic proportions.

The surface of Io is mottled greenish-yellow, though many published views have had the color enhanced and enriched in reds. Indeed, the whole planet looks rather like a pale cheese pizza. The colors are apparently due to sulfur and its compounds, and there is debate about whether the volcanoes are erupting nearly pure sulfur compounds or sulfur-rich rock. No impact craters are visible on Io. Its surface appears to be very young and may be only a few millions, or even thousands, of years old. The satellite is so active that in the age of the solar system, an amount of lava equivalent to the entire mass of Io may have been channeled through its volcanic vents.

Because Io has almost no atmosphere, the volcanic eruptions jet up from the vents without hindrance, forming graceful umbrella-like plumes that can reach heights in excess of 300 km (190 mi). That is an impressive height for an eruption, reaching more than six times the elevation of the top of Earth's stratosphere. Lava flows extend outward from the volcanic vents, and there is speculation that lakes of molten sulfur may exist on its surface, together with geysers of sulfur dioxide (SO_2).

There are substantial mountains on Io as well, some extending to 10 km (33,000 ft) in height and raising the question of how such edifices can be supported isostatically on a thin crust. There seems little question that Io will provide fascination for generations of scientists, and perhaps some day, for explorers as well.

Next out from Jupiter is Europa, a buff-colored moon whose surface is crisscrossed with a tangle of long, narrow streaks that reminded the Voyager spacecraft scientists of Lowell's Martian canals. Europa has neither active volcanoes, mountains, nor large impact-basins. A few small craters that may be due to impacts are visible leading to suggestions that the surface may be as old as 3,000 million years or as young as 100 million years.

Like Io, Europa has a relatively high density, indicating that most of its interior is rocky. Its surface, however, seems to be dominated by water ice, and the streaks may be frozen cracks or pressure ridges due to motion of blocks of ice in this outermost layer. The ice thickness has been estimated to be about 75-100 km (50-60 mi), but these figures are poorly constrained.

Jupiter's tidal forces are much reduced at the distance of Europa, but perhaps they, combined with radioactive decay in its rocky interior, are sufficient to allow liquid water to exist on icy Europa, perhaps in a layer beneath the solid ice surface. Might Europa harbor some forms of life in its depths? At this time, such questions are purely speculative, and firm answers must await far more data than we have so far.

Ganymede is Jupiter's largest moon. It and the fourth Galilean satellite, Callisto, differ from Io and Europa in having distinctly lower densities. These moons are ice balls, with rocky materials accounting for only about half of their bulk.

Craters in significant numbers appear on the surface of Ganymede, with the youngest appearing brilliant white against the browns of the rest of the surface. These impacts have apparently exposed fresh ice to view. The rest of Ganymede's surface is divided between older, heavily cratered terrain and a peculiar "grooved" terrain that looks like the work of a berserk plowman. In fact, the "grooves" are alternating ridges and valleys, tens of kilometers wide and a few hundred meters deep. Because of the numerous craters found on both types of terrain, the surface of Ganymede is very old, perhaps 3,000 million years or more.

The grooves seem to hint of tectonic activity of some sort in Ganymede's early history -- perhaps an icy version of plate tectonics that flourished while Ganymede was younger and warmer. In any case, the age of its surface indicates that Ganymede is now frozen to rock-like hardness, and present-day activity seems unlikely.

Callisto carries this progression to its extreme. Its surface shows almost no evidence of tectonic activity, but impact craters of all sizes abound, together with a few large multi-ring structures. It differs from the Moon and Mercury in that it has no maria plains -- just craters superimposed upon craters over its entire area. The youngest craters are white, as with Ganymede, exposing fresh ice from the depths.

The four large satellites of Jupiter echo the characteristics of the solar system itself. The moons closest to Jupiter are dense and rocky, while more volatile materials such as water become important in the outer, lighter moons. In the same way, the inner terrestrial planets are dense and rocky, while the outer Jovian planets are composed largely of volatile substances such as hydrogen, helium, methane, ammonia, and water.

STUDY QUESTIONS

IX-15. Why is there no well-defined "surface" on Jupiter?

IX-16. What would be the closest terrestrial analog to Jupiter's Great Red Spot?

IX-17. Why does Io have a molten interior and violent volcanism indicative of high heat flow from its interior, while Callisto is a cold world of ice, showing no signs of tectonic activity?

---

6. Saturn, Its Rings, and Its Satellites

Saturn differs from Jupiter in that its color bands are far more subdued, and it has fewer and less-pronounced long-lived eddies. There is nothing on Saturn equivalent to the Great Red Spot. Its clouds are more deeply buried in its atmosphere and are obscured by overlying haze. Nonetheless, it is not a dull planet -- the winds in its atmospheric bands can attain steady velocities of two-thirds the speed of sound on Saturn!

Saturn is smaller and even less dense than Jupiter and appears to have a similar composition. But it is hard to concentrate on the planet itself when the spectacle of its rings is in view.

The rings are amazingly flat. Two hundred thousand kilometers broad and only two kilometers thick, they are made up of dozens of small ringlets, giving the appearance of a phonograph record. This fine structure was unknown prior to the visits of spacecraft. From Earth-based telescopic views, they appeared as a few broad, nearly featureless bands, named the A, B, and C rings going from outside in toward Saturn. The A and B rings were seen to be separated by a gap, called Cassini's division, while the C ring was faint and diaphanous. Recent spacecraft photos have shown Cassini's "gap" to be nearly filled with particles and ringlets.

Like Jupiter's, the rings of Saturn are made up of large numbers of bodies in orbit around the parent planet. Collisions between these bodies tend to confine them to one very flat plane, and also to spread them out within that plane. The individual ringlets are thought to be formed by the gravitational influence of Saturn's larger moons plus the effect of smaller bodies that travel in or near the rings themselves. Acting like dogs herding sheep, these "shepherd" moons circle on either side of a ring, using their gravity fields to keep the ring material in line.

The rings seem to be made up of particles in the centimeter-to-meter range, and the A and B ring materials seem to be made mostly of ice. The C ring may have a different composition, but just what it might be is not known yet.

It is very hard to distinguish just what to call a "moon" in the case of Saturn. There may be a large number of shepherd moons in and around the rings, only a few of which actually have been seen. To date there are 17 satellites whose orbits have been well-defined. The actual number will undoubtedly increase dramatically when future missions take a closer look.

Only one of Saturn's moons is of a similar size to the Galilean satellites of Jupiter. Appropriately named Titan, it is larger than Mercury, Pluto, and Callisto, but slightly smaller than Ganymede. Its density tells us that it is another ice body, but what makes Titan so interesting is that it and Neptune's satellite, Triton, are the only moons known to possess an atmosphere.

Remarkably, that atmosphere is rich in nitrogen, like Earth's. But unlike our atmosphere, Titan's contains methane and no oxygen. The surface of Titan is so cold (-180°C or -290°F) that there may be lakes or oceans of liquid methane and the nitrogen atmosphere itself is very near the point of liquefying. Hydrocarbons and nitrogen compounds produced by the action of sunlight on methane and nitrogen produce a kind of smog that so far has prevented us from seeing any surface features on Titan.

STUDY QUESTIONS

IX-18. What are the characteristics that distinguish the Jovian planets from terrestrial planets?

IX-19. In views of Saturn's rings, the surface of the planet sometimes may be seen through them. Why?

IX-20. What characteristic does Earth share with Saturn's moon Titan?

---

7. The Atmosphere Of Titan [24]

How can a satellite have an atmosphere? The answer is that it should be both massive and cold. Massive bodies have gravitational fields that are strong enough to keep gas molecules from flying away into space. If the body is also cold, the gas molecules move slowly, making their escape even more difficult. Titan satisfies both of these criteria, being more massive than Mercury or Pluto and so far from the sun that its surface temperature is nearly 300 degrees below zero on the Fahrenheit scale. The high abundance of nitrogen on Titan invites comparison with the Earth. Our atmosphere is mainly nitrogen because living organisms and weathering by liquid water have removed the huge amount of carbon dioxide that would otherwise be present. (Venus offers us an example of an inner planet on which life and water are absent; its atmosphere is 97% CO_2 and 90 times as massive as ours.) On Titan, the dominant carbon-containing compound is methane (CH_4) not CO_2. Both the methane and the nitrogen that are present on Titan were originally trapped in the ices that formed this satellite. The nitrogen may have been captured in the form of ammonia (NH_3) which was subsequently converted to N_2 by solar ultraviolet light. Methane is also broken apart by ultraviolet light, but the resulting chemical products are easily condensed or form aerosols, and thus they leave the atmosphere. The hydrogen produced in these reactions escapes from Titan and forms a flattened cloud surrounding Saturn. The result is the preponderance of N_2 that we observe.

A further difference between these two nitrogen-containing atmospheres is the original source of the nitrogen. On Earth, the nitrogen and other gas-forming elements were probably brought to the planet in the rocky material that formed it. We find these elements in various compounds incorporated in meteorites, and these meteorites are considered to be representative of the materials that accreted to form the Earth and the other inner planets. During the

---

[24]This section was written solely for use in this text by Tobias Owen, Earth and Space Sciences, State University of New York, Stony Brook, NY

accretion process, the resultant heating broke down and evaporated these compounds, producing a gaseous atmosphere around the forming planet. A similar process must have occurred on Titan, but here the dominant carrier of volatiles was ice rather than rock. Since the materials that formed Titan were held at much lower temperatures than those that formed the Earth, there is even the possibility that the nitrogen we find in Titan's atmosphere today was simply trapped by the ice from the gases in the solar nebula. In other words, the nitrogen in Titan's atmosphere may be original gas from the cloud of materials that collapsed to form the solar system, whereas the nitrogen in our atmosphere has passed through several different chemical compounds formed under various conditions.

How do we know that Titan's atmosphere wasn't simply captured directly from the solar nebula? That would save the steps of condensing or trapping the nitrogen and then releasing it again. It is not a trapped atmosphere because it contains very little neon. Hydrogen, helium, oxygen, carbon, nitrogen, and neon (in that order) are the six most abundant elements in the universe. Helium and neon are chemically inert, so our main interest is usually in the other four. But if Titan's atmosphere were simply a captured relic from the solar nebula, it would contain almost as many atoms of neon as nitrogen. In fact, the abundance of neon is less than 1% on Titan. This is just what we would expect for an atmosphere whose constituents were originally trapped in ice, since neon will not undergo this process except at temperatures much lower than those prevailing in this part of the solar nebula.

This astronomical perspective about the elements from which atmospheric gases are formed also helps us understand why Titan's atmosphere is so different from those of the inner planets. In a cosmic mixture of the elements -- such as existed in the solar nebula -- hydrogen is so much more abundant than anything else that the other elements simply form compounds with it -- e.g., H_2O, CH_4, and NH_3. This should be the normal chemical situation everywhere in the universe under conditions of low (less than 500 degrees) temperature, and this is just what we find today in the atmospheres of the four giant planets. Jupiter, Saturn, Uranus, and Neptune are all sufficiently massive that even hydrogen cannot escape from their atmospheres. But on a small planet or satellite, the hydrogen is steadily leaving. The remaining compounds are then able to interact with each other, forming new substances. Here is where the big difference between Titan and the inner planets occurs.

On Venus, Earth, and even Mars, it is warm enough that water will evaporate from the liquid or solid state and water vapor can be present in the atmospheres of these bodies. Ultraviolet light from the sun has enough energy to break molecules of water apart. The hydrogen atoms that are produced can escape, but oxygen remains behind. This chemically active element readily combines with carbon compounds and converts them to carbon monoxide (CO) and carbon dioxide (CO_2). The result is that any small inner planet will inevitably end up with a CO_2-rich atmosphere, unless life develops as it did on Earth.

On Titan, it is so cold that water vapor cannot evaporate from the icy surface in significant amounts. (The low temperatures at Saturn's distance from the sun also explains the presence of icy moons and the thousands of millions of icy particles making up the magnificent system of rings. Even exposed to the vacuum of outer space, water ice simply does not evaporate (sublimate is the technical term) when it is kept at 310 degrees below zero Fahrenheit.) This means that there is no significant source of oxygen available in the atmosphere of Titan. The result is that this satellite has maintained a chemical balance that reflects conditions in the earliest phases of the solar nebula. It is as if Titan is a lost world that is frozen in time, a place where we can examine conditions and processes that existed at the time the solar system formed.

This opportunity is particularly interesting for studies of the origin of life on Earth. Biological evolution must have been preceded by chemical evolution. It is generally assumed that this primitive chemistry on our planet took place in an environment where no free oxygen was present. Otherwise the simple organic molecules would have been quickly converted to carbon dioxide instead of evolving toward the ultimate goal of a self-replicating molecule. On Earth, these early oxygen-poor conditions gave way to the present situation as hydrogen escaped, life began, and primitive organisms developed oxygen-producing photosynthesis. On Titan, these primitive conditions still exist, and we can hope to study what chemical pathways nature prefers to follow in proceeding from the cosmically abundant simple gases, methane and nitrogen, to more complex substances.

A tiny amount of carbon monoxide (CO) recently discovered on Titan offers the possibility of some oxygen compounds in addition to the dominant carbon-nitrogen-hydrogen based substances. The CO may be either primordial -- captured from the nebula like the N_2 and CH_4, or secondary -- the result of chemical reactions between infalling ice crystals and the atmospheric methane. The oxygen supplied in this way is insufficient to change the chemical balance of the atmosphere, but it is adequate to produce an additional family of interesting compounds.

Studies are currently underway for a possible joint United States - European mission to explore this lost world in detail. If all goes well, a probe will enter Titan's atmosphere and send back messages about its composition sometime during the year 2000.

8. The Outer Solar System

In speaking of Uranus, you may have the author of these notes at a distinct disadvantage. In January of 1986, the Voyager 2 spacecraft will visit Uranus and, if all goes well, will undoubtedly multiply manyfold our knowledge with the first closeup views of that planet. You already may well know more about Uranus just from reading newspapers and news magazines than all the world's planetologists know as of this writing. Such is the extent to which the space probes have added to knowledge of our cosmic neighborhood.

A remarkable feature of Uranus is the curious orientation of its axis of rotation. It lies almost in the plane of its orbit instead of nearly perpendicular to it, as in the case of most of the other planets. As a result, once every 42 years, the Sun appears directly over one of its poles, an occurrence that can never happen on Earth.

Like Jupiter and Saturn, Uranus also has rings. There are nine narrow rings, orbiting in the equatorial plane of Uranus. Ring systems, it would appear, are common features of the outer solar system.

Uranus and Neptune are nearly the same size, distinctly smaller than Jupiter and Saturn, but still massive compared to the terrestrial planets (see Figure IX-1). They, too, are gaseous planets of low density. They are so far from Earth that telescopic photographs give only the barest hint of what they look like.

Uranus has a clear outer atmosphere, with no clouds. Methane in the hydrogen and helium-rich atmosphere gives it a greenish tint. Neptune, on the other hand, frequently displays a haze or cloud cover that forms and dissipates over a period of weeks in its bluish-green atmosphere.

Neptune has one large moon, Triton, and at least one tiny one. It may also have a discontinuous system of rings. The orbit of Neptune is so far from the Sun that the intensely cold surface of Triton may well contain oceans of liquid nitrogen.

After its encounter with Uranus in 1986, Voyager 2 will continue on to Neptune and Triton, reaching them in August of 1989. It will have traveled twelve years since its launch from Earth and will have visited all four Jovian planets enroute.

The ninth planet, Pluto, does not seem to fit into the scheme of things. A small, icy world, Pluto is no bigger than our Moon and travels in an eccentric orbit that takes it far above and below the orbital planes of the other planets and within the orbit of Neptune at times. Like Uranus, Pluto has its rotational axis in the plane of its orbit. Some astronomers prefer to regard it more as an errant asteroid than as a true planet, but the discovery in 1978 of a Plutonian satellite, Charon, gave it a better claim to its traditional status. Charon is about a third the size of Pluto, making this something of a double planet system.

STUDY QUESTIONS

IX-21. In what characteristics does Pluto differ from Uranus and Neptune?

IX-22. What three planets are known to have rings?

---

9. Asteroids, Comets, and Meteorites

The largest asteroid, Ceres, is only 1,022 km (635 mi) in diameter. All the many thousands of asteroids taken together would make up a mass less than that of our Moon. In the past, however, they may have been much more numerous. Small and rocky, these cold, airless microworlds generally inhabit the gap between Mars and Jupiter in what is called the asteroid belt. Not all asteroids are found within this belt, however. At least one is known to travel outside the orbit of Saturn, while there is a class, called Apollo asteroids, whose orbits cross that of Earth.

Meteorites -- bits of extraterrestrial rock that fall blazing to Earth as "shooting stars" -- have been linked to asteroids for many years. It has proved possible to photograph the tracks of three incoming meteorites and to calculate their orbits. It was found that they came from the vicinity of the asteroid belt.

Studies of the reflectance and spectra of asteroids show that they may be grouped into eight distinct types, some of which correspond to known meteorite types. Meteorites also may be grouped into a number of different types. Some meteorites, called <u>chondrites</u>, are agglomerations of rocky fragments and melted or partially melted spherical particles callen <u>chondrules</u> and have compositions similar to that of the Sun (minus the gases, like hydrogen and helium, which would rapidly leak away into space from so small a body). Others, called <u>achondrites</u> are igneous in origin and are more similar to Earth rocks and are rich in the same minerals that are probably abundant in Earth's mantle. There are also <u>iron meteorites</u>, which are actually mixtures of iron and nickel. Some rare meteorites consist of approximately equal portions of iron and rock. Recently, three meteorites recovered from the Antarctic ice cap have been shown to be bits of the Moon, perhaps blasted into space during an impact on the lunar surface and later swept up by the Earth's gravity field. Similarly, a small number of achondritic meteorites are believed by some scientists to have originated as lavas on Mars.

There is a high probability that the iron meteorites represent the remaining fragments of the iron cores of disrupted asteroids, while the achondrites may represent their mantles or magmas produced in their silicate mantles. The chondrites may be derived from asteroids that were too small to separate into core and mantle, and their compositions reflect this, being a mixture of mostly silicate minerals with small iron grains.

If this picture is correct, then asteroids have probably suffered repeated collisions with one another, some gradually being reduced to rubble. A few bits and pieces fall on Earth, continuing on a vastly reduced scale the bombardment that began in the early history of the solar system.

In astronomical parlance, if an extraterrestrial rock falls to Earth's surface, it is called a <u>meteorite</u>; if it burns up completely in Earth's atmosphere, it is called a <u>meteor</u>. The difference is mostly a matter of size -- only the larger survive the fiery passage.

Meteorites have been termed "messengers from space", bringing us extraterrestrial samples without our having to escape Earth's gravity field in order to get them. In addition to the information that meteorites have brought us about the compositions of asteroids, they have also brought us some of the oldest materials in the solar system.

In 1969 a sizeable chondritic meteorite that may have totaled four tons entered the atmosphere above the town of Allende, Mexico. It flashed across the sky and broke into pieces in a series of spectacular explosions that startled people living nearby. The sky literally rained rocks.

When fragments of the Allende meteorite were dated using radiosotope methods, it was found that they ranked among the oldest bits of matter ever found -- 4.57 billion years old, the probable age of the solar system itself. Along with other meteorites of similar age that have been found, the Allende is not only a messenger from space, but a messenger from time as well, telling us of conditions in the early years of the solar system. We will return to the story they have to tell us in the next unit.

Meteor showers are usually associated with the orbits of comets, suggesting that these particular tiny grains searing the upper atmosphere were supplied by the comets. They are so small that they do not reach the ground before burning up. Most comet travel in highly elongated elliptical orbit that alternately carry them close to the Sun and beyond the orbit of Jupiter.

Our present model of a comet is of a small nucleus, perhaps hundreds of meters to a few kilometers across, consisting of a mixture of ice and fine grains of rocky material. As the comet approaches the Sun, it heats up and some of its ice begins to vaporize. The gas is expelled through its porous surface, carrying dust grains with it. These encounter the solar wind and are blown by it back away from the Sun, forming the comet's tail. Comets do not develop tails until they approach the Sun, where the heat causes a portion of its mass to escape into space. As a result, comets lose mass on every approach to the Sun, and tend to be short-lived. Sometimes comets break up and disappear, but continue to be seen as showers of meteors that occur when Earth crosses the disrupted comet's orbit.

Where do they come from? It is now thought that their source is in a spherical cloud of cometary nuclei that exists far beyond the orbit of Pluto, perhaps one-third the distance to the nearest star. Named for the theorist who proposed it, the Oort cloud may be disturbed by a passing star, sending comets in toward the Sun in fleeting but spectacular performances.

The best-known comet of all -- Halley's -- travels in an orbit that returns it to view once every 76 years. It has been observed since 240 BC, and is due to return again in 1985-86. Unfortunately, it will not be situated well for northern hemisphere observers, but watchers in the southern hemisphere should be treated to a grand show of its tail sweeping across the heavens.

STUDY QUESTIONS

IX-23. What appear to be the sources of meteors that enter the Earth's atmosphere?

IX-24. What group of meteorites is most like the Earth's core in composition?

IX-25. Why do the tails of comets stream away from the Sun?

D. GEOLOGICAL PROCESSES IN THE SOLAR SYSTEM

1. Meteorite Impacts, Craters, and Multi-Ring Basins

Not only Mercury, the Moon, Ganymede, and Callisto show the effects of near-saturation bombardment by meteorites. Most of the small bodies visited and photographed so far display a pocked visage as well. The two tiny moons of Mars, Phobos and Deimos, look like cratered potatoes. Saturn's small moons, Dione, Rhea, Thethys, and Mimas show distinct cratering.

Mimas sustained one huge impact that left a disk-shaped crater fully one-third the diameter of Mimas itself. Voyager scientists were amused to note the strong resemblance between the dimpled Mimas and Darth Vadar's Death Star in the Star Wars movies. It is hard to understand how Mimas could have withstood such an impact and remained in one piece.

Saturn's moon Hyperion may not have been so lucky. Measuring 205 km (127 mi) across, but only 100 km (62 mi) thick, it looks like a hamburger patty and may have reached that state by having sizeable chunks blasted away in a collision. Additional evidence for a collision with Hyperion is furnished by the satellite's apparently chaotic rotation. It is not synchronized with its orbital revolution, which would be expected for an undisturbed satellite this close to Saturn.

The history of the Moon that we outlined earlier shows that impacts, large and small, were ubiquitous during the first 1,000 million years. The solar system must have been an exciting place during those days.

During this stage of development, impacts were the dominant geological process acting on the rocky and icy planets. The way in which they produce the distinctive crater landform can be studied today in a variety of ways. One way is to directly model the cratering process using projectiles impacting a variety of materials. Another way is to mathematically model the process and use a computer to carry out the calculations. Both methods yield the result that high-velocity impacts can produce craters similar in shape to those found on the Moon.

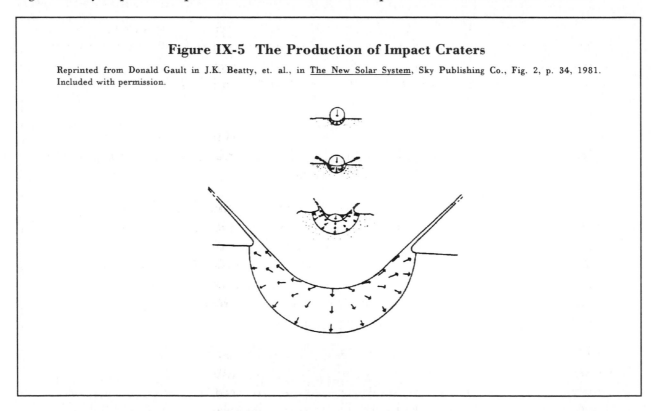

Figure IX-5 The Production of Impact Craters

Reprinted from Donald Gault in J.K. Beatty, et. al., in <u>The New Solar System</u>, Sky Publishing Co., Fig. 2, p. 34, 1981. Included with permission.

The essential process is shown in Figure IX-5. A meteorite falls onto the surface and impacts, giving up its energy of motion as heat and the production of a shock wave that travels through the ground, fracturing the rock layers as it goes. The heat at the point of impact may be so great that the impacting body is largely vaporized.

The shock wave spreads out through the surrounding rock, compressing and fracturing the rock. In the process, a conical jet of material is ejected from the crater, spreading a blanket of severely shocked rock in all directions. For some of the larger lunar craters, these ejecta blankets can extend considerable distances from the crater, forming bright splotches on the

Moon. In some cases, "rays" of ejected material may extend a quarter of the way around the Moon. You may have been able to see some of these bright rays in your binocular exploration of the Moon earlier in this unit.

The relative sizes of craters produced on an airless body depends mostly on the sizes of the meteorites that produce them. Not surprisingly, larger craters are produced by larger meteorites. The shape of the resulting crater changes with size, however. Small craters tend to be fairly clean, bowl-shaped depressions with a rim raised somewhat above the surrounding terrain, as seen in Figure IX-5. Larger craters (20 km or 12 mi and larger on the Moon) tend to have a mountain peak located in the center of the crater.

The central peak is due to rebound of the bedrock directly below the impact. The initial shock wave penetrates deep into the ground and severely compresses the rock -- sometimes to only a third of its original volume. After the shock wave has passed, the compressed rock rebounds upward, especially in the very center of the crater. It then collapses to form a low irregular peak. The process is shown in Figure IX-6.

Figure IX-6 Crater Shape as a Function of Size

Reprinted from Stuart Ross Taylor, <u>Planetary Science: A Lunar Perspective</u>, Lunar and Planetary Science Institute, Fig. 3.20, pp. 82-83, 1982. Included with permission.

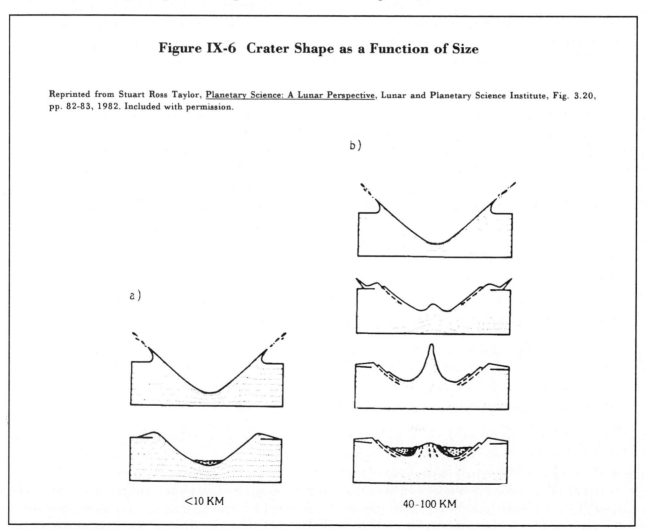

You can demonstrate this process for yourself very easily. Take a glass nearly full of water or milk and hold it under a faucet that has been adjusted to produce a single drop of water every so often. Watch what happens as the drop impacts the surface of the liquid. Immediately after the drop hits, the water rebounds and produces a small peak or spout that rises up some distance. If you lower the glass so that the drop falls farther, you will see that the peak elongates to such an extent that it breaks into one or more little droplets that are hurled back up into the air by the force of the rebound. This droplet then falls back down and impacts a second time.

The same thing apparently can happen when asteroid-sized bodies impact a surface. The result can be an oscillating or repeated impact that produces a series of crater rings of decreasing diameter. This is the appearance of the large multi-ringed basins on Mercury, the Moon, and Callisto that extend up to a thousand kilometers across.

STUDY QUESTION

IX-26. What kind of impact craters tend to show central peaks?

---

2. Planetary Volcanism

Volcanic processes are widespread in the solar system. Their effects are observable on Io, Earth, Mars, the Moon, Mercury, and perhaps Venus. It occurs wherever the temperature rises high enough to melt the local rock, usually at some depth. The magma then rises due to its buoyancy and erupts onto the surface.

If the point of eruption is very localized, then a volcano builds up from the solidified lava and ash produced by the eruption. On the other hand, large volumes of lava may be released through fissures extending over a wide area, producing floods of magma that cool to form nearly flat sheets of rock. Flood basalts of this kind make up the lunar maria and are also found in places on Earth, such as India's Deccan Traps and the Columbia River plateau of the northwestern United States.

Io and Earth continue to be volcanically active today, while the Moon's volcanic fires died out over 3,000 million years ago. The source of heat for both Earth and Moon is probably natural radioactivity, while most of the heat powering Io's furnaces is derived from the giant tidal forces of Jupiter.

The huge volcanoes of Mars such as Olympus Mons raise an interesting question: If Earth is the more volcanically active planet, why are our volcanoes smaller?

The observation tells us something important about the tectonic regimes of each planet. Earth is dominated by plate tectonic processes and as a result, volcanic activity tends to be spread out in linear or arcuate chains rather than being concentrated in one spot. Oceanic spreading ridges and subduction zones both produce long chains of volcanism. Hot spots (see Figure III-7 on page 51) are concentrated spots of volcanic activity, but the volcanoes produced by the hot spot are carried away by plate motion before they can grow to the scale of Martian volcanoes.

Perhaps if all the bulk of the Hawaiian Island chain volcanoes extending from Hawaii through Midway Island and on through the Emperor seamounts were lumped together into one mountain, we would see an Earthly rival to Olympus Mons. In other words, Mars may have hot spots, but its entire surface appears to be composed of only one plate which does not move. As a result, the products of its hot spot volcanism continue to pile up in one place, producing the solar system's greatest solitary mountains.

The volcanoes of Io are in a class by themselves. Nine of them have been observed to be in eruption during the visits of the two Voyager spacecraft, and if this is a typical level of activity, calculations indicate that the outpourings of these vents is sufficient on average to coat the entire surface of Io to a depth of one millimeter of new material in each Earth year. The nature of the magma is still not certain, but its color indicates that sulfur is far more abundant than is usually the case in the volcanism observed on other planets.

STUDY QUESTIONS

IX-27. Where in the solar system is plate tectonics known to be active?

IX-28. Why do the ice worlds such as Europa and Genymede have smoother surfaces than rocky worlds such as Mercury?

---

3. Other Kinds of Planetary Tectonics

Are plate tectonic processes such as sea-floor spreading and subduction unique to the Earth? It may be too early to attempt an answer to this question. Certainly there is no evidence for plate tectonic activity in the terrestrial sense on the Moon, Mercury, or any of the other heavily-cratered worlds. A consequence of plate tectonics should be the distinction between high continents and low-lying ocean basins. Neither the Moon nor Mercury display features that could be termed continental or oceanic.

The same is probably true of Mars. Venus, however, may be a different story. At the time of this writing our view of the Venusian surface is not clear enough to be able to resolve such features as spreading ridges or oceanic trenches. If the high plateaus such as Ishtar Terra and Aphrodite Terra are truly continental, their small area (only 5% of the surface of Venus) might indicate that plate tectonic processes on Venus are more subdued than on Earth.

In view of what we have learned, it is probably a mistake to try to force the other planets into Earth's tectonic mold. It is one of the fascinations in comparative planetology that each planet and major satellite may have its own unique mix of tectonic processes. For instance, the tectonics of Mars is dominated by a bulge on the surface that is 5,000 km (3,100 mi) across, 10 km (6 mi) high, and centered on the volcanic region of Tharsis. Around the bulge is a vast array of radial fractures. These affect almost a third of the planet's surface and probably formed as a result of stresses in the lithosphere induced by the bulge. One of the major features of Mars is the great canyon of Valles Marineris. Though certainly parts have been modified by the running water, other parts have no outlet. The canyon appears to have formed mainly by subsidence along faults radial to Tharsis.

Io presents the other end of the tectonic scale. It renews its surface so quickly that the slow and orderly processes of plate tectonics may not suffice. Io bears the appearance of a world whose entire surface has been buried under floods of lava.

Some worlds may be dominated by compressive forces associated with the shrinkage of a planet as it cools. Mercury and the lunar maria may have been affected by this process. Ironically, prior to the introduction of plate tectonics, planetary shrinkage was the favored mechanism for explaining the origin of Earth's mountains. Now discredited, the idea may yet have application elsewhere in the solar system.

The great fracture patterns on Europa and the grooved terrain on Ganymede probably represent processes unique to worlds made almost entirely of ice. Expansion of the crust through internal heating may have triggered the formation of these features. Here, planetary volcanism may be replaced by the flooding of water and its subsequent freezing. Because ice flows more readily than rock, the impact craters still visible on Ganymede and Callisto are much flatter than those on the Moon. Over the geological ages, these worlds have smoothed down to a near billiard-ball smoothness.

One thing is clear: the geological processes active on Earth do not constitute a complete catalog of all the possibilities. They are merely a subset of the larger geology of the solar system.

E. ATMOSPHERES OF THE TERRESTRIAL PLANETS

Of the four terrestrial planets, only three -- Venus, Earth, and Mars -- have appreciable atmospheres. Mercury and our Moon are essentially airless. Compared to Earth, the atmosphere of Venus is very thick and that of Mars is very thin. The pressure differences are impressive. Venus' atmospheric pressure at its surface far exceeds the maximums in normal pressure-cooker or boiler systems (and is also far hotter) while the pressure at the surface of Mars is equivalent to what we would find on Earth at an elevation of 30 km (98,000 ft).

Why do Mercury and the Moon have no atmospheres? This is because their sizes are small and they are in close proximity to the Sun. Small bodies have relatively low escape velocities, which is the speed that must be imparted to an object to cause it to leave the planet's gravity field with no further assistance. Heating by the Sun imparts kinetic energy to molecules near the top of atmospheres and may get them moving fast enough to escape into space if the temperature is high enough and the gravity is low enough.

In spite of the vast difference in pressure between Venus and Mars, the compositions of their atmospheres are very similar. Figure IX-7 compares the relative proportions of the major gases found in each atmosphere. Carbon dioxide makes up the bulk, with nitrogen a very secondary constituent. This contrasts sharply with Earth, in which nitrogen accounts for 77% and oxygen for 21% of the whole. Oxygen is a very minor constituent of Mars' atmosphere and is almost totally missing from Venus'.

Earth's atmosphere is unique in the solar system, and the presence of so much oxygen can be due only to one thing: photosynthesis. Life has transformed our planet in many ways,

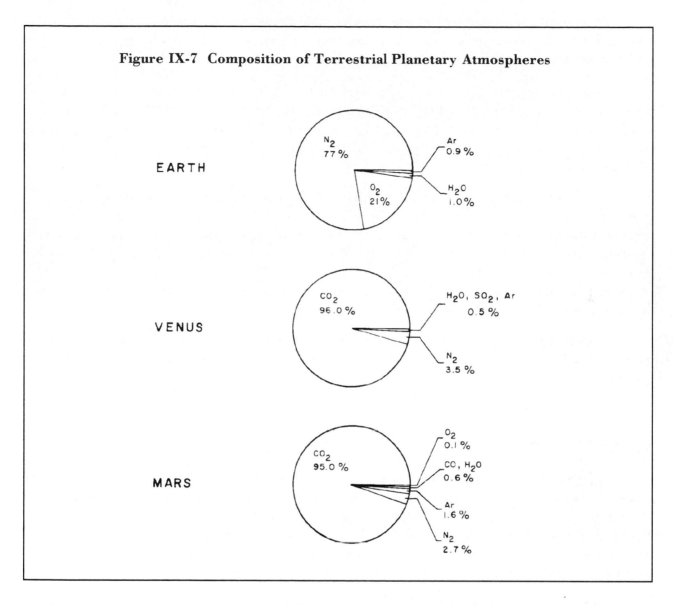

Figure IX-7 Composition of Terrestrial Planetary Atmospheres

but it is the atmosphere that has been most changed. The photosynthetic and oceanic processes discussed in the units **Climates of Earth** and **The Blue Planet** have both worked to remove carbon from the atmosphere and lock it into sedimentary rocks, as coal or as carbonate rocks such as limestone. Oxygen is a by-product of these processes. Nitrogen does not participate in these geological processes, and so has remained in the atmosphere.

It seems highly probable that Earth's early atmosphere was similar in composition to those of Venus and Mars -- heavy in carbon dioxide and deficient in oxygen. The result was a very different world from what we know today. Obviously, there could be no oxygen-breathing animals. Wood could not burn in an open fire (burning is simply the process of oxidizing), and because ozone (O_3) is formed from oxygen molecules (O_2) in the upper atmosphere, there would be no ozone layer to screen out dangerous ultraviolet rays from the Sun. Hence the continents, exposed to these rays, would be barren of most life.

For more than half the age of the Earth, oxygen was a minor constituent in our atmosphere and life flourished only within the oceans whose waters could screen out ultraviolet light. These same atmospheric conditions, however, were conducive to the production of life in the primitive oceans. Once free oxygen became abundant, chemical evolution could not occur except in environments that are protected from the tendency to oxidize. Living organisms themselves provide such environments, but in an oxygen-rich atmosphere such as today's, the building blocks of life would probably never get the chance to assemble into successful replicating forms in order to create life where none had existed before.

Where do planetary atmospheres come from, and when do they form? There are several possibilities here, but the most likely ones place the formation of the atmosphere soon after that of the planet itself. Heating associated with the assembly of the planet is in direct relation to the size of the planet, and so would be the production of heat from short-lived radioisotopes such as Aluminum-26. Volatile materials such as gases and water initially were combined with the rocky material that formed the primitive planets. As they heated up, these volatiles were "outgassed" or boiled off to form atmospheres.

As we have seen in the case of the Earth, vigorous tectonic activity can incorporate both the oceans and the atmosphere into its processes, recycling them through the lithosphere. On Earth, carbon and oxygen are removed from the atmosphere and locked into rocks. When these rocks are subducted, however, these volatiles become available once again through volcanic activity. Volcanoes emit large quantities of carbon dioxide and water vapor, much of which has probably been recycled from the atmosphere through tectonic processes.

Planets with a lower scale of tectonism may cause more of their atmosphere to be locked into rocks and less to be recycled back into the atmosphere. The low atmospheric pressure on Mars may be due in part to its low level of geological activity. It may also be due to its smaller size and resulting less-complete early outgassing.

Does this mean that Venus must have a very high level of geological activity? Not necessarily, because there is another effect that can account for at least part of its dense atmosphere. Unlike Earth and Mars, Venus has no water trapped on its surface as liquid or as ice. All of its water would have to exist in a gaseous form. Venus is closer to the Sun, and its CO_2-rich atmosphere causes a strong Greenhouse Effect. Light from the Sun enters the relatively transparent atmosphere and is converted into heat which becomes trapped because carbon dioxide and water vapor in the atmosphere prevent the heat from being radiated back into space. As a result the surface temperature is much higher than we would expect if there were no Greenhouse Effect.

Because of the high temperatures on Venus, water cannot exist there as a liquid. If Venus ever had oceans, they long since would have boiled away. There is, in fact, some evidence that Venus may have once possessed quantities of water comparable to the oceans of Earth. Deuterium, a heavy form of hydrogen, is naturally present in ocean water (forming what is called "heavy water"). It has been detected in Venus' atmosphere in quantities that suggest that ordinary water once may have been abundant. Because of the high temperatures, the hydrogen and oxygen in water molecules (H_2O) could have been split apart by the action of sunlight in the upper atmosphere. Because the hydrogen atom is so light, it was easily lost to space, while the oxygen was used to form carbon dioxide or incorporated into the rocky surface. This extra source of carbon dioxide may account for part of Venus' dense atmosphere.

In this way Venus may have lost its water. Because deuterium is heavier than hydrogen, it would not be lost as quickly as hydrogen, and so some of it has remained to provide us with a hint of what may have been oceans on an earlier, cooler Venus.

Mars tells quite a different story. Being farther from the Sun, the surface of Mars is always below the freezing point of water. Hence the water may still be there, but locked into the solid state as ice. The bright polar caps of Mars contain large quantities of ice and the Martian soil probably also contains ice in a kind of permafrost.

It seems likely that water once was abundant on all three planets. Due to the differing temperatures, it would have been present on Venus as a gas, on Earth as a liquid, and on Mars as a solid. On Venus, the water was broken up and its hydrogen lost forever to space. As far as we can tell, Earth has had its oceans and rain clouds throughout geological time, nurturing the origin and evolution of life, and providing the mechanisms of erosion by water that have shaped our landscapes. Mars, with its ancient cratered terrain, has seen liquid water on a much more limited scale, but the history of water on Mars is still very unclear.

The nature of the Martian river channels seems to indicate that water flowed only briefly at various times and never attained the kind of hydrologic cycle found on Earth in which the oceans supply water to clouds which feed water back to the rivers as rain. And yet those channels attest to floods of fairly large proportions. What caused them, and when? The question is a larger one, and probably involves all three planets.

When you studied climate changes on Earth in the unit **Climates of Earth**, you found that these changes have been substantial on the scale of geologic time. Mars clearly shows similar evidence of past climates different from today's. Could these changes have had similar causes?

Since Mars has neither oceans nor continents, the effects of continental positions on climate do not apply there, but the orbit of Mars is substantially more elliptical than Earth's, and the tilt of its axis is very similar. Orbital climatic effects on Mars, then, should be stronger than on Earth, where it seems fairly certain that orbital effects (the Milankovitch cycles) are responsible for at least the recent timings of the Ice Ages. At times, these may have been sufficient to trigger widespread but short-lived flooding on Mars.

A final planetary climatic effect needs to be taken into account. Well-established models for the operation of the Sun indicate that it should have gradually increased its output by several tens of percent since the formation of the solar system. A weaker early Sun may have allowed a more pleasant climate on Venus, with the possibility of liquid water oceans. Eventually, increasing temperatures would have triggered a runaway Greenhouse Effect, producing the pressure cooker environment prevailing there today.

On the other hand, Earth and Mars show no evidence of a cooler earlier climate. If anything, the opposite seems to be the case. For Earth, its early abundance of carbon dioxide with a resulting enhanced Greenhouse Effect may provide the answer, while a thicker early atmosphere containing additional carbon dioxide may suffice for Mars. As time passed, Earth's atmosphere exchanged carbon dioxide for oxygen, which does not contribute so strongly to the

Greenhouse Effect. The combination of the Sun's increasing strength and the lessened Greenhouse Effect may have resulted in a more stable climate than would otherwise have been the case.

STUDY QUESTIONS

IX-29. What gas is the most abundant in the atmospheres of Venus and Mars?

IX-30. What happened to the carbon dioxide that originally was in Earth's atmosphere?

IX-31. Why is the temperature at Venus' surface so high?

F. EVOLUTION OF INNER PLANET ATMOSPHERES [25]

Looking at Figure IX-7, one gets the impression that the atmospheres of Venus and Mars are totally different from ours. This was certainly the initial conclusion of some scientists who were studying these planets in the 1960's. But there is an underlying unity that becomes apparent once we study the geological record on Earth. This unity makes us appreciate our planet's unique qualities even more.

The rocks tell us that almost as much carbon dioxide has passed through the Earth's atmosphere as we find today in the atmosphere of Venus. If we vaporized all of the carbon dioxide found on our planet today in the form of deposits of carbonate rocks, the Earth would have an atmosphere with 70 times its current surface pressure. This atmosphere would be 98% CO_2, with the present amount of nitrogen accounting for most of the remaining 2%.

The reason we don't have such an atmosphere is that we have an abundance of liquid water and life instead. Water dissolves carbon dioxide and forms limestone by reactions with silicate rocks. Ancient limestones (carbonate rocks) thus represent CO_2 that was once in the atmosphere. At the present time, most limestones are formed from the shells of organisms, which are also produced from CO_2 dissolved in water. The Great Chalk Cliffs of Dover are a famous example of a calcium carbonate deposit formed from the fossil remains of minute sea creatures. Without life, the atmosphere could contain more carbon dioxide than it does today, and our planet would be warmer. Without water and life, Earth would have an atmosphere that closely resembled that of present-day Venus, a planet that manifestly lacks these two agents for removing CO_2. Apparently both Venus and Earth started with approximately the same reservoir of volatiles. So why did they turn out so differently?

The answer seems to center on the greater proximity of Venus to the Sun. Let's try a conceptual experiment. Suppose we moved the Earth to the orbit of Venus. What would happen?

---

[25]This section was written solely for use in this text by Tobias Owen, Earth and Space Sciences, State University of New York, Stony Brook, NY

Being only 0.72 times its former distance from the Sun, the Earth would receive sunlight that was $1/(0.72)^2$ or approximately 2 times as intense as it does in its present position. The oceans, which comprise nearly three-fourths of our planet's surface, would begin to warm up. More water vapor would cause a more efficient Greenhouse Effect, trapping heat radiation in the lower atmosphere and causing the surface of the Earth to warm up further. This would lead to more evaporation, and an additional increase in temperature. In other words, a positive feedback loop becomes established which continues to raise the surface temperature until the oceans boil away and the atmosphere is filled with water vapor. This is called a Runaway Greenhouse Effect.

At this point, the atmosphere itself is so hot that water vapor can easily rise to high levels where it can be broken apart by solar ultraviolet light as described in the text. On the Earth we inhabit, at its true distance from the sun, water vapor is kept in the sheltered lower atmosphere by the low temperature of the stratosphere (see the units on climate). Hydrogen escapes, deuterium also but less easily, and oxygen is left behind to form CO_2 and combine with rocks. Without water, there is no life. We have converted Earth to Venus!

What about Mars? Here the situation is less clear. If we move the Earth to the Orbit of Mars, it will become colder, the polar caps will increase in size and there will be less water vapor in the atmosphere. These two effects -- greater reflectivity caused by more extensive polar caps and less trapping of heat radiation by the smaller amount of water vapor -- will again produce a positive feedback loop and we can envision a runaway refrigerator! But in this case, there is a way out. We simply need to increase the amount of carbon dioxide in the Earth's atmosphere. This gas produces an efficient Greenhouse Effect all by itself (consider Venus!). The large amount of CO_2 would warm the planet sufficiently to mobilize the water vapor, further adding to the Greenhouse Effect. It turns out that a clement climate could be maintained on Earth at the distance of Mars if our planet's atmosphere were 95-100% CO_2.

There seems to be no way to keep Venus cool; liquid water simply can't persist on a planet that close to the sun. But if Mars had an atmosphere of nearly pure CO_2 with a surface pressure similar to ours today, it would have a climate that would allow liquid water to be stable on its surface.

This may provide the answer to the formation of some of the smaller Martian channels. The big ones can be made by episodic events -- a large meteorite impact melting the subsurface permafrost with resulting massive floods. Or a body of magma similarly warming subsurface layers, the Martian equivalent of Yellowstone Park. But some of the narrow, branching channels seem to have been cut by slow moving water that would require both a higher atmospheric pressure and a warmer surface than we find on Mars now. In fact, there is other evidence that Mars once had such a climate.

Careful studies of isotope ratios and noble gas abundances in the Martian atmosphere both point toward the same conclusion: like the Earth, Mars outgassed a larger volume of gas than its present atmosphere contains. It was probably less than 5% of the amount we find evidence for on Earth or Venus, but it was enough to have given Mars a CO_2-rich atmosphere with a surface pressure similar to our own. Under this early atmosphere, Mars could indeed have maintained the kind of water activity whose ancient signs we still see on the surface.

It is important to stress here that no one knows if this is really what happened. It is simply a model that fits the observations we have at hand thus far.

If Mars did start off this way, how did it change? Here we must be even more speculative. One possibility is that the existence of the water essentially led to its own demise. Liquid water dissolves CO_2 and allows the formation of carbonate rocks. Unlike Earth, Mars is too small to have the tectonic activity required for large-scale continued outgassing and/or recycling of volatiles that have left the atmosphere and formed rocks. As the atmospheric CO_2 abundance diminished, both the surface pressure and the Greenhouse Effect would be reduced. Ultimately we end up with what we find today, a planet so cold that CO_2 forms seasonal polar caps of dry ice.

So the basic problem with Mars is not that it is too far from the Sun, although that certainly doesn't help! The basic problem is that it is too small. A larger planet with more vigorous outgassing might have been able to sustain a thicker atmosphere. It is then a question of maintaining a sufficient partial pressure of CO_2 to provide an adequate Greenhouse Effect.

But is this really what happened? Is there such a simple correlation between the size (mass) of a planet and the thickness of its atmosphere? It would be helpful to examine some planets in other solar systems, associating their characteristics with their sizes and their distances from the stars they orbit. We are an extremely long way from having this capability unless we would make radio contact with a civilization living in such a system. Meanwhile, we need to go back to Mars, to see if we can find the carbonate rocks that should be present if this model is correct. We could even search for evidence of ancient fossils produced by micro-organisms that might have evolved during the first thousand million years of Martian history. If there really was a warm, thick atmosphere during this early epoch, we might expect to find something like the 3,500 million year-old stromatolites that have recently been discovered on Earth. While not as exciting as finding living organisms on Mars, such a discovery would constitute the first evidence that ours is not the only world on which life began. This would have a major impact on attempts to estimate the prevalence of inhabited planets in the galaxy, since at the present time, life on Earth is the only life we know.

RECOMMENDED READING

J. Kelly Beatty, B. O'Leary, A. Chaikin, eds., The New Solar System, Cambridge University Press, 1981.

Rick Gore, What Voyager Saw: Jupiter's Dazzling Realm, National Geographic Magazine, January 1980.

Rick Gore, Voyager 1 at Saturn: Riddles of the Rings, National Geographic Magazine, July 1981.

UNIT X TALES FROM OTHER WORLDS:

ORIGINS

BARRINGER METEORITE CRATER, ARIZONA

A. INTRODUCTION

1. Overview

Data obtained from the space program and from other studies tell us that conditions in the early solar system were very different from today. The dominance of massive impacts as a geological process gives a strong hint of the manner in which the planets assembled from a primordial cloud of dust and gas. Abundance of certain isotopes provides clues to early energy sources that might have helped to partially melt the interiors of large bodies in the solar system, while smaller ones went relatively unaffected. As a result of these processes, the larger rocky bodies underwent a chemical and physical fractionation process that separated them into core, mantle, and crust. Finally, we may have to accept the possibility that the gradual uniformitarian processes of geological and biological change have been occasionally punctuated by the collision between Earth and an asteroid or comet, with catastrophic consequences.

2. Objectives

Upon completion of this unit you should be able to:

1. identify the kinds of evidence that can tell us what happened in the early solar system

2. identify the principal kinds of models for the development of the early solar system

3. relate these models to the observational evidence

4. explain why there are alternative models in some areas or why some models are considered to be speculative in nature

5. identify conditions in the early solar system that led to the differences between the terrestrial and Jovian planets and that set the stage for the development of the characteristics of Earth.

---

3. Key Terms and Concepts

hydrogen	accretion
helium	planetesimals
nova	planetary fractionation
Aluminum-26	collision with asteroid
nebula	globergina *311*

---

4. Corresponding Video

In this program, you will focus on the relationship between Earth and the other planets. Animation, computer modeling, and enhanced NASA photographs of the Martian surface will underscore the strange terrain of the Red Planet. And aerial photography of Earth's Washington Scablands will help to unlock the mystery of water on Mars. The uniqueness of Earth in our solar system will be highlighted by satellite photography. The evidence for regular, cyclical mass extinctions on the Earth will illuminate how dinosaurs may have been wiped out and will prepare you for respected scientists' theories of a Death Star.

B. THE ORIGIN AND EARLY HISTORY OF THE SOLAR SYSTEM

In the unit **The Solar Family** we were concerned with observations -- many of them recent -- about the solar system and its members. In this unit you will focus on hypotheses intended to explain those observations by constructing a history of the solar system. Some of the theories enjoy broad support among planetologists, while others are regarded as speculative in nature. As stressed in all the units you have studied, that theory is best which best explains all available observations.

In this case, the testing of theories dealing with the origin of the solar system raises difficulties in that they deal with events that occurred 4,600 million years ago, while our observations are limited to historical times. Our own planet is so dynamic that few traces of early conditions have survived. When data are few, there are few constraints placed on theorizing, and any explanations are likely to be regarded as speculative.

The advent of the Space Age brought us immense quantities of new data. These in turn have served to clarify and discipline our explanations, because much of the data sheds new light on conditions that prevailed in the early years of the solar system. We are fortunate that the Moon is a repository for much of this early history. As our nearest neighbor, it has of course received the most study, including extensive laboratory work done on lunar rock samples. Photography and other data from spacecraft visits to the other planets have given additional information.

Putting it all together, along with our knowledge of physics and chemistry, allows the construction of a model for the early solar system that does a reasonably good job of explaining present-day observations. Even so, in a number of areas the model is ambiguous and different workers have different versions. We are still not to the point where there is a consensus among all planetologists, and that of course means only one thing -- more data are needed to resolve the controversies.

Why bother? What is the point of this vast detective story that has already consumed billions of dollars in the space program and taken up the lifetime efforts of so many people?

It is tempting to give a lofty answer similar to explorer George Mallory's when asked why he was willing to go through so much agony and effort to climb Mt. Everest, a quest that eventually took his life. He simply shrugged and replied, "Because it is there." In a similar manner, many people feel that the exploration of space is a natural consequence of our humanity -- that the urge to explore is a part of us that should not be denied.

But the space program (including those of all nations) is a venture of an entirely different magnitude, and the answer to our question has to acknowledge more (literally) down-to-Earth concerns. How can the exploration of the solar system benefit us Earthlings?

In the long run, the answers may be the same as those given to questioners of Columbus and Magellan: new resources and opportunities for colonization together with the development of new technical skills necessary to attain them. More immediately, however, we have new knowledge of processes that affected the early Earth and that set in motion the dynamics of the planet to which most of us are currently confined and upon which we are totally dependent. Can we afford not to understand them?

The oldest radioisotope dates obtained so far are for meteorites and lunar highlands rocks, with the oldest of each indicating 4,600 million years. Indirect age determinations for Earth yield the same number. Astronomers, on the other hand, regard the universe as being much older -- on the order of 15,000 million years. That date was obtained by noting that all the galaxies are receding from one another in an expanding universe. Roughly 15,000 million years are needed to move the galaxies to their present positions and velocities from the initial "big bang" that started everything. In that scenario the universe began with a colossal explosion in which all the mass that currently exists was created at one point in space and began an expansion that continues today.

The recognition that our Sun, and its solar system along with it, is something of a Johnny-come-lately in the universe and is important in understanding the chemical makeup of the planets. The original "big bang" is expected to have produced only very light elements, and indeed hydrogen and helium, the lightest and next-lightest elements, are by far the most abundant in the universe. The conversion of hydrogen into helium in the process of nuclear fusion is the principal energy source for the universe.

If this is the case, then where did the heavy elements like silicon and iron come from that are so abundant in the solar system? It turns out that the heavy elements are synthesized in the "red giant" stage of a star and in nova explosions, which occur at the end of a star's lifetime. Our own world, it would appear, is built of the ashes of a star's violent death.

There is more direct evidence that a nova was involved in the birth of the solar system. The anomalous presence in meteorites of certain stable isotopes such as Magnesium-26 hints at high energy events accompanying the solar system's birth. Magnesium-26, is the daughter product in the decay of Aluminum-26, a radioactive isotope with a half-life of only 720,000 years. Created in a nova, essentially all the Aluminum-26 would decay radioactively during the first few tens of millions of years of the solar system. Its incorporation into the matter of the solar system gives us direct evidence for the involvement of a nova.

Our story begins with an interstellar cloud of dust and gas located in one of the spiral arms of our galaxy. With all the new evidence collected in recent years, you might expect that our model for solar system development is a wholly new one. Yet, the model that follows is really an adaptation and extension of a broad outline first suggested by Rene Descartes in 1644 refined in 1755 by the Philosopher Immanuel Kant, and in 1796 by Pierre Simon de Laplace. Science is not always a continual overthrow of old ideas.

The original gas and dust nebula, or cloud, was massive and very large. Its total mass may have been thousands of times that of the Sun. As it collapsed under the influence of gravity, the cloud probably broke up into a large number of individual clouds, each of which would give birth to a star or system of stars. Most stars are not lone individuals like our Sun, but are members of stellar systems in which two or more stars orbit about a common center of mass. According to the usual view, our cloud coalesced in such a way that only one star, the Sun, was formed.

In some cases a cloud may collapse more or less directly into its center, provided that the nebula is not spinning rapidly. If you stand still with your arms extended and let them fall freely, they will fall to your side rapidly. If, however, you are spinning like an ice skater, your arms will tend to remain flung outward by centrifugal force. A slowly-spinning cloud would collapse quickly to form a massive star. The act of compressing a gas raises the temperature to the point where nuclear ignition takes place, and at that moment it may truly be said that a star is born.

The rate at which a star uses up its hydrogen fuel is strongly related to its mass, with the result that large stars actually have shorter lives than smaller ones. A star whose mass is twenty times that of the Sun might burn 40,000 times more brightly and last for less than one million years. Passing through the red giant phase, in which the star expands enormously, it then explodes as a nova, having lived fast and furiously. What is important to us is that these nova explosions are common and fairly frequent in young stellar systems. Perhaps one may have occurred every 100,000 years in the vicinity of the forming Sun.

These novae produced two effects. They contributed heavy elements (and short-lived Aluminum-26) to the solar nebula, and they also may have compressed it with the force of their explosions, helping the nebula to condense under the influence of its lesser gravity. The momentum of its rotation flattened it into a spinning disk, just as relaxed arms tend to be flung outward when you spin around. The next stage of the process was the condensing (or accreting) of the dust and gas in the nebular disk into the planets and satellites, but just how this was accomplished is still a matter of some controversy.

One scheme has the disk breaking into a series of rings that coalesce into early forms of the present planets, while another forms a multitude of small micro-planets of asteroid size, called planetesimals. Whichever scheme is used, the nebular disk stage of the solar system was quite short, lasting only about 100,000 years. During this time the Sun ignited, sending out an intense flood of particles called the solar wind that swept the remaining gas from the solar system.

Recent careful telescopic work has succeeded in photographing in the infrared part of the spectrum just such a nebular disk surrounding a star in the southern skies. It is possible that we are witnessing the very beginning of another planetary system that someday may be similar to our own.

Another source of controversy among the theoretical model-builders is the manner in which the differences in composition between the planets came to be. The solar nebula had three kinds of constituents: rocky materials (mostly silicates and iron), ices, and gases. In one view, the nebula was hot near its center and cooler toward its edges. In the central portion of the nebular disk the temperature was so high that nearly all solids were vaporized. As the nebula began to cool, only the iron and rocky materials could condense into solid form in its inner portions, while the ices and gases tended to condense farther out where the temperature was lower. Thus the separation into rocky terrestrial planets near the Sun and gaseous Jovian planets and icy satellites farther out might be explained.

An alternate view has a cooler nebula in which planetesimals (small, asteroid-sized bodies) form from relatively homogeneous nebular dust and gas. The accretion process releases heat, as does the early decay of Aluminum-26, with the result that the larger planetesimals become quite hot inside. The more volatile gases and liquids are boiled off to the surface, while iron, which has a lower melting temperature than silicate rocks, liquefies and sinks to the center of the planetesimal to form a core. This fractionation process, in which different chemical fractions are separated into their own regions of the planetesimal according to their density, would occur mostly in the larger planetesimals. The smaller ones would not be able to build up sufficient heat in their interiors for the process to go all the way to core formation. Collisions between these bodies would produce fragments of widely different compositions, such as we observe among the asteroids and meteorites today. Eventually most of these fragments would be swept up into the larger and still growing planets where fractionation would occur a second time, building core, mantle, crust, and atmosphere.

The idea of large numbers of planetesimals in the early solar system derives support from observations of massive impact features on the Moon and the other airless bodies. An early makeup of this sort would naturally lead to immense early collisions grading into smaller and fewer impacts as the growing planets swept the solar system clear of wandering debris.

Whatever mechanism acted to produce chemical differentiation between the terrestrial and Jovian planets also acted similarly in the case of the moons of Jupiter, where the closest moon, Io, is rocky in nature while the outer moons are low density and icy in composition. These moons are too small to have retained a significant atmosphere of any kind.

Some of the curiosities of the solar system may also be due to this history of intense bombardment. The extreme tilt in the axis of Uranus may have been caused by a late collision with a large object, and perhaps the orbits of the planets are slightly tilted with respect to one another for the same reason.

Indeed, the process may not yet be complete. Asteroids are not confined to the asteroid belt, but have been found in orbits that cross Earth (the Apollo asteroids), that follow Jupiter around in its orbit (the Trojan asteroids), and one, Chiron, that orbits beyond Saturn. The solar system has been swept clean but not spotless, and that could have important implications for Earth and its load of life. The impact of an asteroid-sized body on Earth today would have catastrophic consequences. One of the more fascinating suggestions of recent years has been a tying of a possible asteroidal collision to the great extinctions at the end of the Cretaceous period. At that time, 75% of the species living on this planet became extinct. Could an impact by a large extra-terrestrial body have caused such biological chaos? If the impact occurred on a continent, the resulting emplacement in the atmosphere of vast quantities of dust from the impact and soot from forest fires could result in a scenario similar to that of the Nuclear Winter discussed in the unit **Climates of Earth.** If, on the other hand, the impact point was oceanic, the enormous amount of water vaporized could have started a temporary severe warming because, like carbon dioxide, water vapor in the atmosphere contributes to the Greenhouse Effect.

While public attention has focused on the catastrophic impact hypothesis as an explanation for the final killing-off of the dinosaurs, the fact of the matter is that they had already been in decline for some time. More remarkable is the wholesale slaughter of tiny organisms that lived in the sea, such as the single-celled globergina species abundant in the Cretaceous. It is not so hard to imagine events that might kill off large not-very-adaptable beasts like dinosaurs, but much harder to imagine processes of less than catastrophic proportions that could simultaneously destroy the myriads of globigerina inhabiting the world's oceans.

What does this do to Hutton's uniformitarianism, with which we began this course? One approach would be to recognize that uniformitarian principles are adequate to explain most, but not all, geological events, and that catastrophist processes need to be invoked to explain the rest.

Another, more intriguing approach, would be to regard catastrophic events such as extra-terrestrial collisions as essentially uniformitarian processes. They, after all, have occurred ever since the formation of the solar system and continue today, though now they are few and far between due to a steadily decreasing supply of errant bodies. Collisions of this type may appear to be catastrophist in nature only because the extremely limited span of human history makes them seem more than commonplace.

STUDY QUESTIONS

X-1. What evidence points to the age of the solar system as being 4,600 million years?

X-2. What was the likely origin of the heavy elements that make up the silicate rocks and iron core that make up much of the Earth?

X-3. Why is it assumed that the novae explosions occurred shortly before the creation of the solar system?

X-4. The Jovian planets are composed mostly of hydrogen and helium, but these gases are rare on Earth. Why?

X-5. Compare and contrast the effects of an asteroidal collision with the model of a Nuclear Winter discussed in the unit **Climates of Earth**. What are the similarities, and what might the differences be?

C. EVOLUTION OF THE EARTH

You may have noticed that existing models for the early solar system allow for considerable variation from planet to planet, and that different versions of the model might produce very different scenarios for the early development of Earth. Nonetheless, we have enough information to construct a tentative outline, using information gained from the Moon and from our other companions in the solar system for the very early millenia for which the record on Earth is missing. We may combine this with the earliest geologic record to provide a comprehensive view of our planet.

First, we must ask a very fundamental question: At what stage did the Earth become layered into core, mantle, and crust? One suggestion envisions the Earth as accreting (combining together) as a series of layers of different compositions, depending on the available material and the temperature of the solar nebula at that time. Another proposes Earth accreting with a nearly uniform composition -- approximately that of chondritic meteorites (refer to Chapter VII Section C) -- with separation into layers occurring later on. P. 312

An initially chondritic Earth would heat up from two possible sources. Large colliding bodies would release vast quantities of energy as heat when they impacted, likely melting the outer portions of the Earth, at least near the impact site. Radioactive Aluminum-26, left over from the primordial novae, might release sufficient nuclear energy to at least partially melt the deep interior.

As the temperature rose, metallic iron would have been the first to melt, and, because of its high density, would have begun to work its way downward through the hot and plastic, but still solid rocky silicates. This act of "falling" toward the center of the Earth on the part of the dense iron actually would have released additional gravitational energy as heat, further raising the temperature. Once begun, the process would tend to reinforce itself, with more heat melting more iron that then falls toward the growing iron core and releases more heat in turn. It is, then, very likely that the separation into core and mantle occurred early in Earth history, probably before many half-lives of Aluminum-26 had elapsed. The later we have core formation happening in our model, the less radioactive heat is available to make it happen. The decrease in the numbers of parent Aluminum-26 atoms would proceed rapidly during the first few millions of years after the nova that produced them.

Our model for the early Earth is not one of slow, gradual evolution, but one of surprisingly rapid development. Accretion and outgassing probably proceeded together, building up a dense primitive atmosphere. If significant accretion preceded ignition of the Sun, some of the early atmosphere may have been blown away in the early, energetic phases of the Sun's new life. If the core did not exist from the very earliest stages of accretion, then it began forming as soon as the Earth reached a sufficient size that the heat trapped in its interior could raise the temperature to the melting point of iron. All of this could have occurred in the first twenty million years.

Now the Earth continued to grow in size by sweeping up the remaining planetesimals that wandered near its orbit. The surface of the Earth would have been in continual turmoil, with each new impact reworking the existing material, excavating to a considerable depth and bringing up denser material from below. On the Moon, we know this process continued until it began to slacken at around 3,900 million years ago.

It probably is no coincidence that the most ancient dated rocks on Earth are around 3,800 million years old. For many years, geologists have searched for remaining bits of the Earth's "primitive crust". If the planetesimal accretion model holds, then there probably was no such thing. Any early crust would have been temporary, being blasted and reworked with each new major impact. Old material would be mixed with new and final separation of the newly accreted material into components destined for the core, mantle, and crust would not be completed until accretion had nearly ceased.

Outgassing would have continued throughout, building up a substantial atmosphere of carbon dioxide, carbon monoxide, water vapor, and perhaps some methane and ammonia. The Earth may well have been too hot for the water vapor to condense to form oceans, and the dense, cloudy atmosphere would have resembled Venus' at this stage, with a Greenhouse Effect produced by the water vapor adding to the other sources of heat. Eventually, as the atmosphere cooled, the great rains began. The bulk of the water shifted from the atmosphere to the oceans and Earth's surface would have been visible from space for the first time.

When the rains came, were there already ocean basins waiting to receive their waters? The answer depends on many factors. We do not know just when plate tectonics in the present sense began to operate on Earth. Because the interior of the Earth was hotter than at present, with the likelihood that not only the core but substantial portions of the upper mantle were molten, volcanism must have been widespread. Perhaps the tectonic regime on Earth at that time may have resembled that on Io, in which the surface is being renewed from a great many sources at once, and the lifetime of any portion of the crust is quite short.

In contrast, plate tectonics is a slower, more orderly process, in which oceanic plate is created only along a few well-defined spreading ridges. Plates so created may live for 100 million years or more before being subducted and destroyed. Continental material is created as a byproduct of subduction and can enjoy an extremely long life due to its buoyancy and resulting resistance to subduction.

Perhaps plate tectonics is an intermediate stage in tectonic activity, between the hyperactivity of Io and the static state of the Moon and Mars. After all, if radioactive heating is a significant contributor to the planetary heat engine, then we would expect tectonic processes to be active at first while radioisotopes with short half-lives are still abundant, and then to gradually become less active as all the radioisotopes begin to diminish in number.

In any case, the continents must have begun to accumulate by 3,800 million years ago, because the oldest rocks are continental in nature. By this time the bombardment from space had slackened considerably, though occasional blockbusters would continue to arrive from time to time.

In the first three units of this course, you learned that continents consist largely of low density felsic rocks which presently are being formed as the byproducts of subduction zone volcanism. The origin of the ancient cores of the continents is still not well understood. One view has the area of continents growing gradually with time as new material is added to them; another has a rapid early growth in the total area of the continents, followed by a long period extending to the present in which continental area is growing very slowly or not at all. If the latter is correct, perhaps plate tectonics was a more rapid process early in Earth history or possibly the initial process of separation of continental material from the mantle was significantly different from the plate tectonics of today.

Questions of the very early history of the continents stand at the shadowy edge of present research. We cannot turn to sources outside the Earth for answers, and the rocks that hold the necessary evidence are very ancient and have often been heavily metamorphosed -- changed substantially from their original state by the actions of heat and pressure during the intervening millenia. Definitive answers will not be easy to obtain.

STUDY QUESTIONS

X-6. One model for the early Earth envisions it as beginning with a uniform composition similar to that of the chondritic meteorites. Why do you think the chondrites were chosen as a starting composition?

X-7. At what stage in Earth's development did the iron core probably form?

D. CATASTROPHISM AGAIN

1. Impact Structures on Earth

We have seen that meteorite or asteroidal impacts were a dominant process in the early history of the solar system. Earth, too, must have seen its share of these violent events. In the ancient Precambrian crust of the Canadian shield we can still find scars of impacts similar to those that excavated the craters of the Moon. But not all are very ancient.

If you have an atlas, or a map of Canada, get it out and look in eastern Quebec province, at latitude N 51°, longitude W 69°. Depending on the age of the map, you will see two arcuate lakes just above a prominent fork of the Manicougan River. These are Lake Manicougan and Lake Mouchalagane. Recent maps may show these as one ring-shaped lake, since the Manicougan River has been dammed and both lakes are now joined into one very peculiarly-shaped reservoir. This circular structure, 70 km (44 mi) in diameter, marks an impact that occurred some 210 million years ago, in the Triassic. All but the fractured roots of the Manicougan Crater have been erased from the ancient granite in the intervening years.

For a fresher and more recent example, we can travel to the Barringer Crater, popularly known as the Meteor Crater of Arizona. Located some 150 km (90 mi) southeast of the Grand Canyon, this crater is easily accessible by car and is viewed by many tourists who can drive to its rim and peer into its vast bowl, 1,200 meters (3,900 ft) across and 200 meters (660 ft) deep. Twenty-five thousand years ago, an iron meteorite 50 meters (160 ft) or more across blazed through Earth's atmosphere. It partly fragmented in flight, showering the region with small meteorites that survived the landing. The main mass or cluster of masses impacted the ground with an energy release of around ten megatons of TNT, leaving the state of Arizona with yet another tourist attraction. The main mass of the meteorite has never been found in spite of excavations in the floor of the crater, but microscopic spheres of nickel-iron mixed in the shattered rock and soil indicate that virtually all of the original mass vaporized on impact.

In the dry northern plains of Argentina is Campo del Cielo -- the Field of the Sky. A series of shallow craters on the order of 100 meters (330 ft) across give a hint that the site was aptly named. Excavations revealed an iron meteorite estimated to weigh 24 tons that fell from the sky some 4,000 years ago. That it survived intact is due to its relatively small size (compared to the one that formed the Barringer Crater, for instance) and its resultant lower velocity of impact due to atmospheric drag. It is one of the two or three largest meteorites found to date.

Iron meteorites, because of their tendency to rust in Earth's oxygen-rich atmosphere, cannot survive the passage of long stretches of geological time. Most of those on display in museums fell to Earth during the past 100,000 years. Medieval swordsmiths may have incorporated meteoritic iron in some of their weapons, perhaps giving rise to legends of enchanted swords presented to warriors by the gods. It is quite possible that some of the meteorites so used were actually observed to fall.

If impacts of relatively small bodies have been frequent in recent geological time, what about larger events? And if they did happen, what effects might they have had on life on

Earth? These questions have aroused intense interest in the scientific community during the past few years.

Substantial extraterrestrial collisions have been rare events during historic times, though small meteoritic chunks have been observed to fall in hundreds of cases, sometimes punching holes in houses, landing in swimming pools or just falling near astonished observers. Small pieces will be slowed down considerably by friction with the air and so will arrive at the ground with far less energy (and less effect) than asteroid-sized bodies.

A hint of what might occur in a larger encounter was given near the Tunguska River in southern Siberia on the early morning of June 30, 1908. A contemporary account describes the event:

> At 7:20 a.m. a mighty noise was heard resolving into thundercracks, though the sky was cloudless at the time. The noise caused houses to shake. Many inhabitants saw "a fiery body like a beam" shoot from the northwest above the ground before they heard the thunder. Immediately afterward the bang was heard, and in the place where the fiery body had disappeared, "a tongue of fire" appeared, followed by "smoke".

The blast has been estimated to have been equivalent to about 12 megatons of TNT, comparable to that of the Barringer Crater event, but its effects were quite different. No significant crater was formed, though thousands of trees were felled and stripped by the force of the blast. Observers in a town 60 km (40 mi) away were knocked down, seismographs throughout Asia registered the event, and the incoming fireball was seen throughout much of southern Siberia.

Expeditions to the site beginning in 1927 found no evidence of a meteoritic body, though more recent searches have turned up tiny spheres of metal and glass with a high nickel content, the likely remnants of the vaporized intruder. Detailed analysis of these spherules has shown an abnormal concentration of the element iridium, which is abundant in meteorites but rare in the crust of the Earth.

Why was no crater formed? It appears that the body remained intact down to a height of 8.5 km (28,000 ft) when it suddenly disintegrated, exploding with great force before it reached Earth.

What was it? Probably not an iron meteorite, because chunks of one that size would probably have reached the ground. The composition of the Tunguska meteorite has been a subject for debate for many years, with speculation including even such exotic possibilities as a body made of antimatter or a small black hole. Two hypotheses have received the most attention, however. One sees the Tunguska event as a collision with the head of a small comet. The icy parts of the comet would have evaporated while the dust-like rocky material would have vaporized to form the spherules found near the site.

An alternative explanation is that the meteorite was a stony chondrite. Unlike an iron meteorite, a stone is less likely to survive its passage through the atmosphere, but if it is big enough it will be slowed down only slightly in its fall. The result is a tremendous compression of the air ahead of the meteorite until finally the forces are great enough to disintegrate it almost completely. Adherents of this model point out that the icy head of a comet is much weaker than a chondrite and would be unlikely to survive in its flight to so low an altitude. If the Tunguska object was a chondrite, it may have been 160 meters (525 ft) in diameter and weighed seven million tons. Whatever its composition, it was fortunate that this object chose a very remote and unpopulated site as its target and that it disintegrated high in the atmosphere rather than impacting the ground.

At this point you may well be wondering how often such events are likely to occur. From our discussion so far, it would appear that sizeable impacts are fairly common events on Earth, even at this late stage in solar system history. One estimate suggests that a collision of the magnitude of the Tunguska event might occur on average once every 300 years. Another estimate puts it in a slightly different way. Craters 10 km (6 mi) or more across should be blasted out somewhere on a continent on the average once every 300,000 years. These figures do not put extraterrestrial impacts in quite the same worry category as unemployment or nuclear war, but viewed from a geological perspective they become events with the capacity for producing profound changes in the environment.

STUDY QUESTION

X-8. What is the principal difference between the Tungska event of 1908 and the Barringer crater event?

---

2. The Great Dyings

Paleontologists have been intrigued since the earliest days of their science by striking evidence in the geological record of what have been called "the great dyings" -- sudden shifts in Earth's biological populations marked by mass extinctions of species and their replacement by new ones.

A number of these extinction events have been identified in the fossil record. The most fascinating of these to the nonscientist occurred at the end of the Mesozoic Era some 65 million years ago at the boundary between the Cretaceous and Tertiary periods, for this marked the end of the Age of the Dinosaurs. Few questions have received so many hypotheses in answer, and the debate over the demise of the dinosaurs has continued for over a century.

As it turns out, the dinosaurs are only a part of the question. Many workers feel that they were already in decline as the Cretaceous period came to a close. On land, nearly all the large reptiles disappeared, leaving only turtles, crocodiles, snakes and lizards as survivors. Mammals, which began to proliferate during the Mesozoic, were mostly small in size and seemed to fare better during the crisis. Remarkably, no land animal weighing more than 25 kilograms (55 pounds) survived, indicating that, whatever happened, size was a liability.

Extinction visited the seas as well, taking many different types of creatures into oblivion, from microscopic plankton to large shellfish. In all, about three-quarters of all the known species of plants and animals disappeared forever.

What could have done this? Because most of the large dinosaurs are regarded as having been cold-blooded, climatic change has been a favored hypothesis. A sudden cooling over a long period of time would favor the warm-blooded mammals and work against the reptiles. Changes in ocean temperatures might have catastrophic consequences for marine life. But what could trigger such a climatic shift?

As we have noted already, when observational data are few, hypotheses flourish and multiply, unconstrained by the necessity to explain hard, cold facts. Among the hypotheses offered were: the effects of a relatively nearby supernova; changes in the output of the Sun; a sudden drop in sea level, causing a withdrawal of the shallow seas that had covered much of the continents during the Cretaceous, in turn causing a harsher climate with more pronounced seasons; a sudden influx of cosmic radiation on an unprecedented scale; and a period of increased planet-wide volcanism. The idea that the wave of extinctions might have been related to an asteroidal impact had been considered along with these others, but suffered for lack of evidence that such a cosmic encounter might actually have occurred.

In 1979, evidence of a fairly dramatic nature appeared. A thin layer of clay, only about a centimeter or two in thickness, separates Cretaceous from Tertiary limestones near Gubbio, Italy. This clay layer was found to contain the element iridium in concentrations thirty times those found in other nearby clay layers. Soon, iridium anomalies had been found in correlative layers at the end of the Cretaceous in other parts of the world and in deep-sea sediment cores as well. In each site, only one layer showed the increase in iridium.

As you learned earlier, iridium is very scarce on Earth, but is associated with meteoritic material. The iridium-rich layer was interpreted as being the fallout from an asteroidal impact. More recently, tiny grains of quartz have been found in the clay layer that are of types that are only formed under conditions of extreme shock. Taken together with the iridium enhancement, it constitutes powerful evidence in support of a major extraterrestrial impact. If the clay layer is a worldwide phenomenon, it is possible to calculate the size of the impacting body from the amount of iridium found in the clay. The answer: a walloping 10 km (6 mi) in diameter!

Could the resulting dust-cloud, "Nuclear Winter", or Greenhouse Effects have produced the pattern of extinctions observed at the end of the Cretaceous? If the effect was to obscure the Sun and produce sufficient darkness to halt photosynthesis worldwide for on the order of a year or so, the answer may well be "yes". Even so, some scientists are unconvinced and have posed alternative explanations.

In any case, speculation, so long as it is recognized as such, is considered by scientists to be good clean intellectual fun. It has been noted that during the Cretaceous certain smaller dinosaurs had begun to evolve rather large brains, with the ratio of brain weight to body weight being similar to that of early mammals. With the demise of the dinosaurs, the mammals took off in development and variety. It would seem that the dinosaurs had been suppressing the development of the mammals.

And so it is amusing to speculate that, but for a chance cosmic encounter with a 10 kilometer asteroid, the dominant intelligent species on Earth today might have evolved from the dinosaurs, and that the organism reading these notes might have been reptilian and not mammalian in form.

As profound as it was, the extinction at the end of the Cretaceous was not the greatest. An unprecedented dying occurred at the close of the Paleozoic, 230 million years ago. Nearly half of the known families of animals vanished. The trilobites disappeared, as did most corals and a significant proportion of the invertebrate marine population. The trilobites appear to have been in decline as the Paleozoic Era advanced, and so their disappearance may not be so remarkable, but other families that vanished appeared to be in robust health, having adapted to a wide variety of environments. Among these were the fusulinids, a tiny single-celled creature whose shell is preserved readily as a fossil. It was clearly a very successful organism in the late Paleozoic, with over 500 species represented from all continents. Not a single survivor is found in the overlying Triassic rock layers.

The question naturally arises: Could the great dying at the end of the Paleozoic have been precipitated by another collision? Geologists are busy today examining the rocks in which the Paleozoic - Mesozoic boundary is recorded, searching for iridium layers and other evidence of an extraterrestrial event. Already, a Chinese group of scientists has published a report of an iridium anomaly at this boundary. Additional but unconfirmed reports of iridium have been made for other layers associated with major extinctions: one at the end of the Devonian period and another that occurred about 37 million years ago.

Meanwhile, another approach has proved to be provocative. The occurrence of extinctions throughout the last 250 million years appears not to be random, but has a periodicity in the range of one event every 26 to 28 million years. Furthermore, a study of large well-dated impact craters on Earth suggests that they, too, have a similar periodic origin and seem to have occurred at about the same times as the extinctions. It seems highly unlikely that chance encounters with asteroids would fall into so regular a pattern.

One way out of this problem is to postulate that the number of objects in the vicinity of the Earth dramatically rises every 26 million years or so. Comets are the only bodies with extremely elliptical orbits that would bring them near the Earth in large numbers, and so the suggestion has been made that something may periodically perturb or disrupt a portion of the Oort cloud of comets. (See Chapter IX-C on page 313.) A massive body such as a star passing near the cloud could cause such a disruption, tearing large numbers of comets out of their more-or-less circular orbits and flinging them in all directions into highly elliptical orbits that would bring many into the Earth's neighborhood.

What could cause such periodic visitations by other stars? Once again hypotheses are little constrained by evidence, but two speculations currently are popular subjects of debate. One is that the Sun does not remain fixed in the disk of our galaxy, but oscillates above and below the galactic plane with an estimated period of about 66 million years. When the solar system is far from the galactic plane, nearby stars are few, but as it passes every 33 million years through the galaxy's spiral arm within the plane, encounters with other stars are more likely.

Still another scenario would provide our friendly Sun with a sinister companion. Called Nemesis, this faint dwarf star would spend most of its time very far from the Sun in a highly elliptical orbit. At its closest approach, Nemesis would not actually enter the solar system but would come close enough to influence the Oort comet cloud that encircles the Sun far beyond the orbit of Pluto.

In any case, you doubtless will be relieved to know that another disturbance of the Oort cloud is not due for another ten to thirteen million years.

Either of these mechanisms could provide an apparent cyclicity of extinctions and cratering events, but both have difficulties. The galactic plane argument seems to be out of step with the cycles in that the Sun is currently passing through the plane and is in a region of higher stellar density than average, yet according to the observed extinction cycle, we should be at a time of low risk. As for the Nemesis argument, some astronomers are skeptical about the existence of a solar companion and wonder, since it would be the nearest star other than the Sun, why it has never been observed. On the chance that it is a very faint star indeed, specific searches for it have begun. Perhaps Nemesis has been observed, but simply not recognized as being very nearby.

The search will concentrate on stars that are known to be of a type that are inherently dim and will look for a large degree of parallax in the star's position. This is an apparent slight wobbling motion of the star as seen from the Earth as it circles the Sun in its orbit. Parallax is an effect that is easily demonstrated from where you sit. Simply sway your head from side to side as you look at different objects, some near and some farther off, against a distant background such as the wall of your room. As you sway, nearer objects appear to move back and forth against the distant background to a greater extent than the farther objects. The motion of the Earth in its orbit about the Sun is the astronomical equivalent of your "head swaying". Astronomers have used parallax to measure the distance to nearby stars, and if Nemesis exists and can be seen from Earth, then it should reveal its presence by displaying a larger parallax than any other star.

This entire section has dealt with subjects that are regarded by many scientists with skepticism. Data are few and there are too many alternative hypotheses to be certain about any of them. Perhaps even the cyclicity of extinctions is open to question. In the unit **A Sense of Time** we noted that statistical predictions are not reliable when applied to small numbers, and fewer than ten extinction events are involved here. Even so, the feeling is growing that the evidence supporting a large extraterrestrial collision at the close of the Mesozoic era has become very convincing.

The effects of such a collision on the development of life on Earth is even less clear. Was it the cause of the dinosaur's demise or merely the final blow to a species already far into decline? Either view has its proponents. Already, however, biologists are looking afresh at evolutionary theory. The introduction of sudden, violent events such as a collision with a comet or asteroid poses new environmental stresses in addition to the pressures of natural selection, as possible driving forces for evolutionary change.

And so our studies of the solar system have brought us right back to where we started. But just as a tourist returns from extensive travels more worldly-wise, we have explored the solar system and returned home with a better understanding of the origin and development of our own planet and of our own species.

STUDY QUESTIONS

X-9. Why do the two great extinction events that occurred 230 and 65 million years ago fall right at the boundaries separating the Paleozoic, Mesozoic, and Cenozoic Eras?

X-10. If it has never been observed, why was the hypothesis of Nemesis, the Sun's companion, put forward?

RECOMMENDED READING

J. Kelly Beaty, B. O'Leary, A. Chaikin, eds., The New Solar System, Cambridge University Press, 1981.

UNIT XI GIFTS FROM THE EARTH:

MINERAL RESOURCES

BLACK SMOKER VENT ON THE SEA FLOOR

A. INTRODUCTION

1. Overview

Products made from the mineral resources of the Earth are so familiar in our industrialized society that we tend to take them for granted. In fact, most of the industrialized nations owe their present positions to an abundance of a broad range of mineral resources. In this unit you will examine the various uses that we make of important minerals, their worldwide distribution, and various mechanisms by which minerals are generally believed to be concentrated in some localities and not in others.

---

2. Objectives

Upon completion of this unit you should be able to:

1. distinguish between rocks and minerals
2. describe ways in which our civilization uses the minerals
3. distinguish between mineral reserves and mineral resources
4. relate fluctuations in mineral reserves to technical and economic factors
5. recognize that mineral reserves are not equably distributed among the nations of the world
6. discuss several mechanisms by which minerals become concentrated into ores and the conditions that favor this process
7. relate certain mineral concentration processes to volcanism associated with plate tectonic activity
8. relate mineral-rich localities visited in the corresponding television sequence to their geological settings and to the mineral concentration processes that produced them
9. recognize the finite nature of mineral resources and the difficulty in predicting shortages or gluts in particular markets
10. discuss some potential sources for future mineral supplies and strategies for dealing with finite supplies.

3. Key Terms and Concepts

mineral

rock   *3B*

ore   *END*

iron] *4A*
       000

aluminum

manganese

magnesium

titanium

copper

lead

zinc

nickel

molybdenum

mercury

chromium

tin

tungsten

uranium

gold

silver

platinum

industrial rocks and minerals

construction materials

fertilizers

gemstones

economic deposits (= ore)

mineral reserves

mineral resources

mineral concentration mechanisms:

    magmatic concentration

    hydrothermal processes

    mineral replacement

    sedimentary processes

    evaporation

    residual concentration

    mechanical concentration

pegmatite dikes

salt domes

bauxite

placer deposits

ophiolites

porphyry copper deposits

Kuroko-type massive-sulfide deposits

Bushveld igneous complex, South Africa

Troodos Massif, Cyprus

Noranda District, Quebec

mineral consumption and mineral

    production patterns

deep-sea mining

manganese nodules

mineral substitution

mineral conservation and recycling

4. Corresponding Video

Mankind is becoming painfully aware that Earth's resources are finite. Yet, new advances in science and technology now suggest that parts of Earth remain unexplored. In this program, views of ancient mines in Cyprus help to underscore our long dependence on minerals. Animated graphics will depict the widespread distribution of resources, and Landsat will give a global satellite view of mineral exploration. You will also visit South Africa and

Newfoundland, among other mining locations for the world's copper, gold, tin, iron, and silver.

B. THE TYPES AND USES OF MINERALS

1. Metals

The eminent philosopher Bertrand Russell once wrote, "It is to steel and oil and uranium, not to martial ardor, that modern nations must look for victory in war." And, he might have added, for prosperity in peace. The modern industrial nation makes prodigious use of mineral resources from the Earth. Our buildings and highways are made from concrete (cement from limestone, sand and gravel from stream deposits) and steel (iron and a variety of alloying metals); our automobiles and other machines are made of iron, specialty steels, and aluminum, with sizeable amounts of copper devoted to wiring. Much of our food is prepared in metal containers and packaged in cans of aluminum or tin-plated steel, or in metal foils. Even the pen with which you are (hopefully!) taking notes may be made of a half-dozen different metals encased in a variety of plastics produced from another resource from the Earth -- petroleum.

There is no question that the search for minerals has profoundly affected human history. In their classic textbook on economic mineral deposits, Mead L. Jensen and Alan M. Bateman give the following perspective:

> The quest for the wealth of minerals wrested from the Earth for man's vanity, necessities, or comforts has ever been a powerful incentive to discovery, exploration, and trade. Their search has given rise to voyages of discovery and settlement of new lands. Their ownership has resulted in industrial development and in commercial or political supremacy, and has also caused strife and war. In the ancient country of Saomes, the winter torrents brought down gravels containing gold, which the barbarians passed through inclined troughs lined with sheeps' fleeces to catch the gold. The fleeces that were hung on trees to dry, so that the fine gold could be beaten out of them, spurred Jason and the Argonauts in the ship Argo to seek the Golden Fleece near the shore of the Euxine. This is the earliest record of a placer gold rush and a poetic expression of an early mining venture. It was tin that drew the Phoenicians and Romans to Britain; it was gold and silver that lured the Spanish Conquistadores to the settlement of the New World. The gold rush of 1849 led to the settlement of California and then to the acquisition of the western part of the United States from Mexico and Spain.

Let us look at some of the most important materials that we take from the Earth and how we use them. Many of these, but by no means all, are metals. They seldom come from the ground as pure metals, but rather are found as chemical compounds -- metallic elements bound to other elements such as sulfur (to form sulfides) or oxygen (to form oxides). These compounds, along with the relatively few elements that occur in pure form, are referred to as minerals, in that they have definite chemical compositions. Rocks are heterogeneous mixtures of

346

minerals that can have a wide variety of different compositions, and <u>ores</u> are rocks in which certain minerals are concentrated by natural processes to such an extent that it is profitable to mine them.

A number of the metals are relatively abundant: iron, aluminum, manganese, magnesium, and titanium. Of these, iron is the most important, accounting for more than 95% of all the metals used in our civilization. You need only look around you to see the myriad uses for this versatile, strong, and inexpensive metal. Nowadays, iron is seldom used in its pure form, but is usually alloyed with other metals: nickel, chromium, tungsten, vanadium, cobalt, and manganese. Thus, the steelmaking industry is dependent upon supplies of all of these metals as well.

Aluminum is light and strong, and in recent years has taken over many uses formerly reserved for iron. Many beverage cans, for instance, are now made from aluminum rather than from tin-plated steel. Aluminum is replacing iron in many automobile parts such as engines and bumpers, as automakers strive to increase the fuel efficiency of their cars by making them lighter. It has good resistance to corrosion and so finds many uses in building construction and window frames. Its light weight makes it ideal for use in the construction of aircraft.

Manganese is seldom used by itself, but it is an indispensible ingredient in the production of many kinds of steel, including all carbon steels. It is also used in specialty steels that are extremely hard and tough and serve well in such uses as structural steel, gears, armor plate, and safes. Magnesium, not to be confused with manganese, oxidizes at low temperatures, burning with a bright white light. Fine powders and filaments of this metal are used in fireworks and flashbulbs. Magnesium is the lightest metal known, and so finds use in aircraft, automobiles, and instrument parts.

Titanium is not quite as light as aluminum but is stronger and resists corrosion better. It is difficult to separate from its compounds, however, and so has not seen widespread use as a pure metal in spite of its abundance. At the present time, it is used mostly in its form of titanium oxide (TiO_2), as a white pigment for paints.

The other metals are much more scarce in the crust: copper, lead, zinc, nickel, molybdenum, mercury, chromium, tin, tungsten, and uranium. The last, though a metal, is better treated as an energy resource, and we shall defer most discussion of uranium until the next unit.

Copper is an excellent conductor of electricity and can be drawn easily into flexible wires. As a result it is essential to our electrified civilization, carrying electrical power and communication signals throughout a web of wires that form the technological nervous system of the developed nations. It is also used in brass (copper and zinc) and bronze (copper, tin, and zinc).

Lead is used principally in storage batteries, but it finds other uses in bullets, solder, and, because of its extreme resistance to corrosion, as protective sheathing for electrical cables. Lead is toxic and long since has been removed from paints for this reason. For the same reason, a common use of lead today, as tetraethyl lead -- an antiknock compound in gasoline -- is gradually being phased out in several countries in order to protect the environment from local lead contamination. Nonetheless, there are many uses of lead that are environmentally sound, and its high salvage value means that with care it can be recycled through such uses with a minimum amount finding its way into the general environment.

Zinc has excellent corrosion resistance and is electrolytically deposited on sheet metal to form galvanized iron. Its greatest use, however, is in diecasting alloys. Dies are the tough metal stamps and forms used to punch out and shape complex sheet metal parts such as automobile fenders. We have already mentioned the use of zinc in brass and bronze. Nickel lends toughness, strength and anticorrosion properties to steel, and its principal use is in alloys such as stainless steel. It is also used for nickel-plating and in coinage.

Molybdenum is used almost entirely as an alloying agent in steels, to which it imparts strength and ductility or the ability to be drawn into rods. Tungsten has similar properties in a steel alloy, also imparting great hardness. As a result, it is used in high-speed cutting tools that are used to shape other steels in lathes and mills. It is also used as tungsten carbide, the hardest known cutting agent after diamond. A minor, though familiar use, is as a filament in ordinary light bulbs.

Chromium is used as an ingredient of stainless steel and other steel alloys. Chromium-plating was formerly used widely on automotive bumpers and trim, though this use is decreasing. It continues in many uses as a tough, attractive coating for other metals. The mineral ore chromite, which is an oxide of chromium and other metals, is an excellent refractory used in furnace linings.

Mercury is the only metal that is liquid at room temperature. It was known to the ancients and in the Middle Ages, fascinated alchemists with its ability to alloy (amalgamate) with many other metals, including gold and silver. For hundreds of years it has been used in the recovery of these two precious metals. Mercury, also known as quicksilver, has myriad other uses in thermometers, electrical switches, pharmaceuticals, insecticides and fungicides, explosives, and antifouling paints. Its use that comes closest to home, however, is in dental amalgam the metallic substance used to fill a cavity. In spite of its toxicity, the mercury remains safely trapped within the amalgam and only very minute amounts escape into our bodies.

Tin may have been one of the first metals used by mankind. Its use has been traced back to 3700 BC in Egypt, and its use in bronze affected an entire period of human history -- the Bronze Age. In more recent times, its uses in pewter and tin roofing have given way to other metals, and today its principal use is in tinplate, solder, bearing alloys, and bronze.

The precious metals gold, silver, and platinum have captured the fancies -- and greed -- of men and women for thousands of years, but in addition to jewelry, coinage, and bullion, they have many practical uses as well. Gold is used in electrical contacts, in plating, and in dentistry. Silver finds use in silverware utensils, other plated objects, and in the photographic industry. New uses are in the production of printed circuits for computers and other electronic devices. Platinum sees far more industrial than monetary or jewelry use, and finds employment in electrodes, crucibles, electrical thermometers, and medical and dental devices. A recent use that has spurred demand is as the catalyst in catalytic converters installed in the exhaust systems of automobiles that are designed to convert harmful emissions to more benign gases.

Because of its persistent value and the fact that it is nearly indestructible, most of the gold mined throughout all of history is still in use or in stockpiles of various sorts. It offers us the ultimate example of a recycled resource, in which part of an Egyptian king's adornment may now reside in one of your teeth or on one of your fingers.

Certain elements are of vital importance to agriculture: nitrogen, phosphorus, and potassium. Nitrogen is abundant in the atmosphere, but plants are not able to use nitrogen gas until it has been "fixed" or incorporated into a soluble form as a nitrate (KNO_3 or $NaNO_3$) or as an ammonia compound such as ammonium sulfate ($(NH_4)_2SO_4$). Bacterial action in soil and certain plants can fix nitrogen, but modern agriculture relies heavily on chemical fertilizers. The nitrogen in these is obtained from the atmosphere and combined with hydrogen to form ammonia. Phosphorus is obtained from the mineral apatite ($Ca_5(PO_4)_3OH$), which occurs in certain phosphate rocks or in marine sedimentary deposits called phosphorites. Potassium is obtained from salts left behind in the evaporation of seawater.

An "edible" mineral is common table salt ($NaCl$), which is found in thick sedimentary beds, salt domes, or may be obtained from evaporated seawater. Most salt is used not for eating but in the chemical industry and for road salt to melt snow and ice in northern climates.

Finally, there are "wearable" minerals: diamonds and gemstones. Diamonds are a high-pressure form of carbon, formed deep within the Earth. Rubies and sapphires are composed of aluminum oxide; emeralds are a beryllium aluminum silicate; amethyst is quartz (silicon dioxide). The distinctive colors of each gem are determined in some cases by minute amounts of impurities and in others by slight defects in their crystal structure.

Although man-made diamonds are making inroads on the natural stone for industrial use, natural diamonds are a critical industrial mineral widely used for cutting, grinding, and polishing. The drilling of many holes for petroleum and mineral exploration would be impossible without diamond drill bits.

2. Industrial Minerals [26]

In addition to the minerals discussed so far, there is a large group of mineral resources known collectively as industrial rocks and minerals. These comprise a myriad of diverse materials such as asbestos, clays, graphite, lithium, talc, and vermiculite, and although less well-known and glamorous than the metals, modern industrial society cannot function without them. Moreover, their total value far outstrips that of the metals.

Most important of these materials are those that make up concrete: sand, gravel, and cement which in turn is made by mixing and heating limestone and shale, the products of mines and quarries.

As concrete is literally our most important heavy construction material, so sulfuric acid is perhaps the most important industrial chemical. Sulfur, used to make sulfuric acid, is used somewhere in the production chain of nearly every important item used by modern man from oil refining products to plastics to steel. However, the single most important use of sulfuric acid is to combine it with ammonia and phosphate to form ammonium sulfate and super and triple

---

[26]This section was written solely for use in this text by Siegfried Muessig, Getty Mining Company

super phosphate fertilizers. We derive our sulfur from the flanks of volcanoes, from salt domes, sulfide minerals such as pyrites, and sour gas.

Various clays are used extensively for pottery, chinaware and other ceramics. Less well-known are specialty clays which are used to filter and clarify many liquids, including beer and vegetable oils; kaolin, used as a filler and coating on papers, in refractories, and in rubber; bentonite, used as a drilling mud in the petroleum industry; and fire clays used metallurgically.

Fluorspar (CaF_2), another widely used mineral, is the raw material for hydrofluoric acid, essential in the manufacture of the fluorocarbon compounds, so vital to our consumer economy. Borax, whose most advertised use is as a cleanser, finds its major use as an indispensable ingredient of high-temperature glasses and fiber glass. Another important element is lithium, which comes to us from natural brines and pegmatites, is a critical material in the manufacture of aluminum and finds extensive use as an ingredient in low-temperature greases.

A number of materials contain what are known as the rare earths, comprising such elements as europium, gadolinium, cerium, and zirconium. The United States is the largest producer and consumer of the rare earths which find such diverse uses as petroleum catalysts, super alloys, color TV tube phosphors and X-ray screen intensifiers, and specialty magnets.

Saline minerals, such as those containing sodium carbonate, impact our lives daily. Nearly half the mined product goes into the manufacture of glass and the chemical industry consumed much of the rest of such products as detergents and soaps.

STUDY QUESTIONS

XI-1. How was the Golden Fleece related to an early mining venture?

XI-2. What is the distinction between a rock and a mineral?

XI-3. What are some abundant metals?

XI-4. What are some important scarce metals?

XI-5. What industry is a heavy user of manganese?

XI-6. What are the principal uses of copper?

XI-7. What elements are of importance to agriculture as fertilizers?

C. WORLDWIDE DISTRIBUTION AND ABUNDANCE OF METALS

The Earth's crust contains all of the stable elements and some of the radioactive ones as well. Figure XI-1 shows the relative abundances of the most prevalent elements in continental crust and in seawater (excluding the hydrogen and oxygen of the water). In the crust, silicon and oxygen are the most abundant elements, reflecting the fact that the most common rock-forming minerals are silicates. The abundant metallic elements shown in the diagram -- aluminum, iron, calcium, magnesium, sodium, potassium, and titanium -- are for the most part bound up in the various silicates, carbonates, oxides, and other rocky materials that make up the crust.

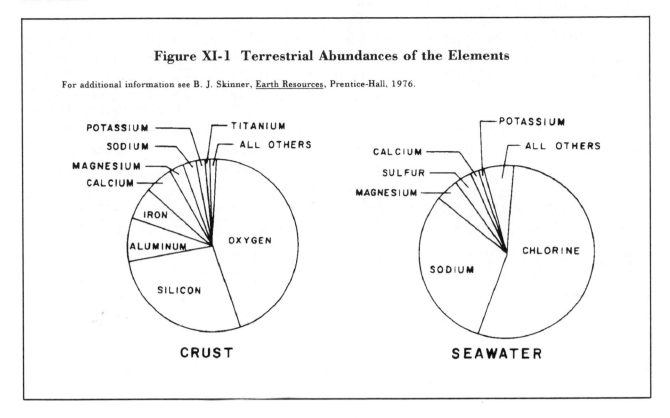

Figure XI-1 Terrestrial Abundances of the Elements

For additional information see B. J. Skinner, Earth Resources, Prentice-Hall, 1976.

Silicates are extremely stable in the chemical sense, and it requires prodigious amounts of energy to dissociate ordinary rocks in order to retrieve pure metals from them. While it is possible to obtain mineral resources in this way, until we find cheaper energy sources, it is not profitable to do so. Instead, as has been the practice since the dawn of civilization, it is easier to mine deposits in which nature has produced local concentrations of one or more minerals far in excess of what is found in ordinary rocks.

Perhaps the most striking characteristic of ore deposits is that they are far from being equably distributed throughout the world. Many countries may have few or no economic deposits, while others may be blessed with fabulous mineral wealth. Even on a very local scale, ore deposits may be found in one county or parish but not in its neighbor.

You probably have gold in your backyard. The real question, though, may be put as follows: Is there enough of it in a high enough concentration so that it is profitable to mine? Perhaps you noted that the word "profit" appeared in the last section when we defined what we meant by the term "ore". It is quite impossible to separate economics from any discussion of resources, and in fact the study of mineral resources is often called "Economic Geology".

The price of gold strongly influences the number of operating gold mines. When the cost of producing gold from ore at a mine rises above the value of the gold produced, the operation becomes unprofitable and either higher gold prices, higher grade ores, or improved extraction technology are needed for the mine to stay in business.

The use of iron ore in the United States serves as an illustration of how both economic and technical factors can affect the availability of mineral resources. Three essential ingredients are necessary to the making of iron: iron ore, coal, and limestone which is used as a flux to remove impurities. Iron ore contains iron oxides, which are very stable, and only by the application of high temperatures in an atmosphere that is deficient in oxygen can metallic iron be separated in pure form. A blast furnace is filled with a charge of the three ingredients and ignited. Air or oxygen is forced through the charge in order to bring the temperature up to the point where metallic iron is reduced by some of the coal, melted, and can be tapped off from the bottom of the furnace. The oxygen from the air is consumed in the burning process, producing the required oxygen deficiency.

Early centers of steelmaking such as Pittsburgh in the United States, Birmingham in England, and in Germany and Sweden, arose where all three ingredients were found together or in close proximity, resulting in low transportation costs. In the Pittsburgh area, coal, iron ore, and limestone were all found in the same immediate area. The local supplies of iron ore soon ran out, however, and ore had to be imported from other regions -- in this case, from the rich deposits of the Mesabi Range in Minnesota. These deposits contained more than 50% iron and were profitably utilized even though the ore had to be transported long distances.

As early as 1908, however, steelmaker Andrew Carnegie warned that these high-grade deposits were in danger of running out. Imminent shortage following World War II was overcome when it was discovered that lower-grade deposits, called taconite ores, could be treated in such a way as to increase their iron content from as low as 20% to more than 60%, higher than that of high-grade ores. These upgraded ores so improved the efficiency of the steelmaking process, that the costs of treating the ores were more than recovered. Because of the development of the concentration process, iron ore reserves in the United States have been vastly increased.

It is necessary to make a distinction between reserves and resources. Reserves are usually defined as known deposits from which minerals can be extracted profitably using existing technology and under present economic and political conditions. Because economic, technology, and legal/political approvals are essential parts of this definition, world reserves of some minerals could increase or decrease significantly with no significant change in the actual amount of that mineral that is known to be in the ground. There are a number of deposits that can be economically mined, but are precluded from production for legal/political reasons, in many cases related to environmental concerns.

Resources, on the other hand, are known potential sources of extractable minerals that might be used in the future if changes in technology or economic and legal conditions allow. For this reason, the term often appears as potential resources. In a sense, reserves are birds in hand, while resources are birds in a bush -- we may or may not someday actually mine resources. Also, you should realize that the two categories are not fixed and immutable. If mineral prices fall, as they have in recent years, some marginal reserves may slip into the category of potential resources; when prices improve, they may shift back into the active reserves.

With this distinction in mind, let us look at the distribution of reserves for a number of important metals among the countries and regions of the world. At first thought, you might

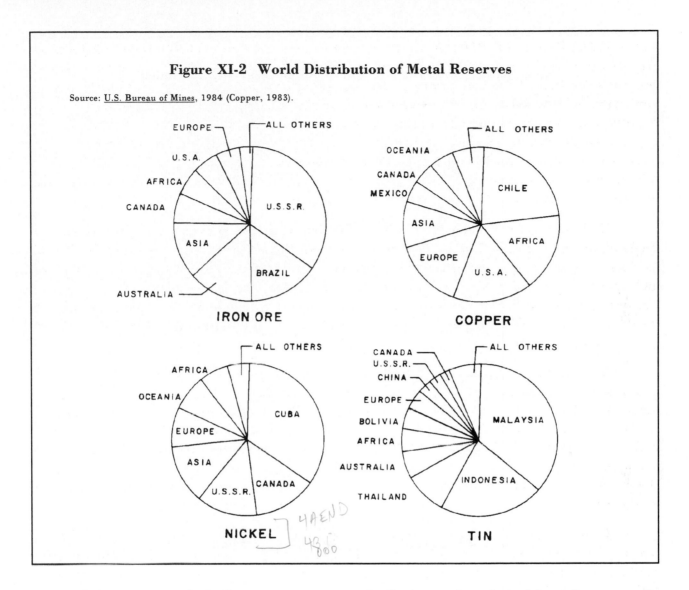

Figure XI-2 World Distribution of Metal Reserves

Source: U.S. Bureau of Mines, 1984 (Copper, 1983).

expect the countries with the largest areas to contain the most mineral wealth, and to a certain extent this is true. The U.S.S.R. and the United States certainly are among the leaders in mineral reserves. But there are some surprises as well. Figure XI-2 and Figure XI-3 show the distribution of reserves for eight important metal ores. Iron ore is distributed more or less as we might expect on the basis of area, though Africa has clearly been slighted. On the other hand, the United States and South America have a disproportionate share of copper resources. Figure XI-2 as reserves, recent economic conditions have shifted sizeable amounts of United States copper into the category of resources, resulting in a domestic copper industry that is extremely depressed. This is an excellent but unfortunate example of the extreme volatility of reserve estimates due to economic conditions, and of the high-risk situation that is a continual part of life in the mineral industry. The United States holds more than half the total world's reserves of molybdenum, but has almost no reserves of chromium, manganese, and tin. Alaska, however, holds ample tin resources.

Even more surprising is the dominance of some small countries for certain metals:. Cuba accounts for nearly a third of all the nickel reserves in the world. Four small countries,

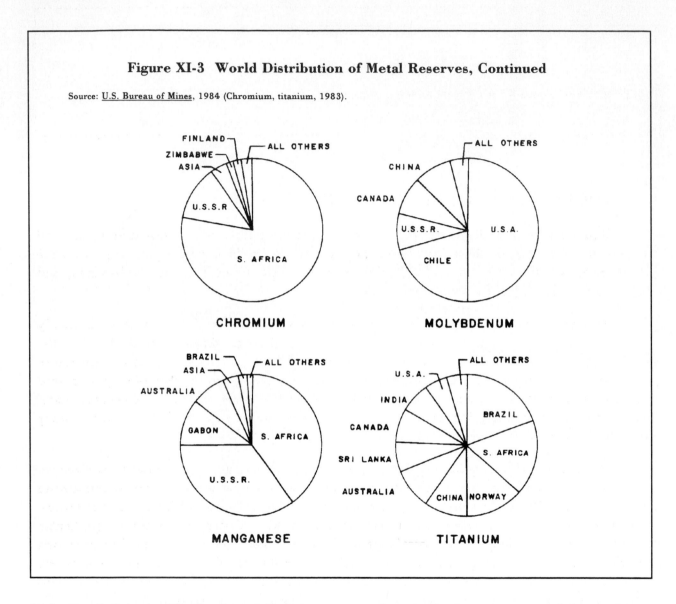

Figure XI-3 World Distribution of Metal Reserves, Continued

Source: U.S. Bureau of Mines, 1984 (Chromium, titanium, 1983).

Malaysia, Indonesia, Thailand, and Bolivia, account for nearly three-quarters of the world tin reserves, and South Africa, in addition to its gold and diamond mines, has a third of world manganese reserves and more than three-quarters of world chromium reserves.

Why should mineral reserves be so unequally divided? And what is the mechanism that concentrates minerals in ore deposits, sometimes achieving concentrations that are many thousands of times richer than those found in ordinary rocks? We shall take the two questions in reverse order and treat them in the next two sections.

STUDY QUESTIONS

XI-8. What are the two most abundant elements in the crust of the Earth?

XI-9. What is the distinction between mineral reserves and mineral resources?

D. PROCESSES OF MINERAL CONCENTRATION

Suppose I were to mix thoroughly a pinch of ground pepper with a spoonful of salt, and then ask you to separate the pepper from the mixture. It would not take you long to realize how to do it: simply place the mixture in warm water, mix it until the salt is dissolved, and pour off the solution, leaving the pepper grains behind.

Or perhaps you might have thought of a different method. If the mixture is slowly poured in a thin stream from a container while you gently blow across it from the side, the lighter pepper grains will be deflected more than the salt grains and will tend to concentrate in the downwind direction. In this case, however, the pepper will not be completely separated from the salt, but will become concentrated considerably over its earlier dispersed state. Ancient methods of winnowing grain work in the same way, with the heavier wheat being separated from the lighter chaff by the effects of an air current.

The process of concentration of minerals can proceed in nature in a number of different ways, just as in our experiment with salt and pepper. Different processes of ore concentration may work with different minerals, or with the same mineral under different circumstances. Sometimes it is difficult to determine just which process has acted to concentrate a particular ore, but general knowledge of the workings of these concentration mechanisms has improved over the years. The principal mechanisms that have produced valuable mineral deposits are magmatic concentration, hydrothermal processes, sedimentation, evaporation, residual concentration, and mechanical concentration into placer deposits. Let us examine each in turn.

Magmatic concentration refers to the process by which molten rock at depth (magma) segregates into different constituents as it cools and solidifies. An analogous situation in the kitchen is the process of clarifying butter. In this procedure, butter is melted and allowed to stand for a few minutes. The milk solids precipitate out and settle to the bottom of the container, allowing the clear (clarified) butterfat to be skimmed off the top and separated. One might look at it as a process for concentrating milk solids at the bottom of the container that previously had been dispersed throughout the butter.

In a similar manner, when a body of magma far underground begins to cool, certain minerals (often the more mafic ones) will crystallize first and, being denser than the surrounding fluid, will sink to the bottom of the magma chamber. The residual magma will become increasingly felsic, as was described in the unit **Plate Tectonics** in relation to the chemical differentiation processes that occur at oceanic ridges and in subduction zones. Ore minerals frequently have different chemical affinities for the different silicates that are involved in this differentiation process, and so will be concentrated either among the early crystals or in the portion of the magma that is late in solidifying.

In some cases, the last remaining fluids, together with a slurry of crystals already solidified, may be injected into cracks or fissures in the surrounding rock to form pegmatite dikes, coarsely crystalline masses that are often rich in metals and, sometimes, gemstones. The term dike refers to a tabular sheet of rock that has been injected into older, already existing rocks.

Magmatic concentration or segregation appears to be important in some deposits of chromium, iron, titanium, platinum, nickel, and copper.

Among the most important concentrating mechanisms are the hydrothermal processes. "Hydrothermal" simply means "hot water". We may draw upon another kitchen analogy to explain how these processes work. You probably know that if you boil vegetables too long, many of the vitamins, which are soluble, will be lost to the water. This is fine if you are making soup, but not good if you plan to discard the water and eat only the vegetables.

In the geological situation, hot water circulating among cracks will dissolve minerals from the rock and carry them away from the source rock to where the minerals can be deposited. Solution may take place at depth where temperature and pressure are high. Under these circumstances, the ability of water to dissolve minerals is enhanced, especially if the fluid contains chlorine or fluorine in solution. The metal atoms are transported in the hydrothermal fluid as metal chlorides or fluorides until they meet an external source of sulfur, perhaps in the form of hydrogen sulfide that may be derived from magma. A chemical reaction between the fluid and the sulfur creates metal sulfides that are highly insoluble in water. These precipitate out, taking the metal out of solution and concentrating it into an ore deposit. This may deposit mineral veins in fissures or pores of the surrounding rock, or perhaps the hydrothermal fluid may react chemically with the rock, altering it and producing a deposit via the process of replacement.

In replacement deposits, rocks are altered when they come into contact with hot fluids. These fluids are commonly rich in hot water, gases exhaled from the magma, or rarely magma itself. Some elements in the rock become dissolved in the fluid and other elements carried in the fluid replace them, chemically altering the rocks in the process. In this way, mineral ores can be formed, especially in the hot region immediately surrounding a body of magma underground.

Replacement can form very rich deposits. Although some of these are large and important, others are often very local and may be hard to find. They can also be exasperating to mine, because they may look very promising at first, but then may peter out abruptly as the mine is extended. Ores formed by replacement often contain iron, copper, zinc, lead, tin, tungsten, molybdenum, graphite, gold, silver, manganese, and corundum (an abrasive).

In the units on the oceans, we saw that the process of hydrothermal activity is taking place in the oceanic spreading ridges. Seawater, filtering through cracks in the rock, reacts with magma from the mantle, dissolves minerals from it, and then is ejected from hot springs in the ocean floor. On encountering the cold water of the ocean bottom, minerals are precipitated from solution, producing in some cases the black smokers referred to in the unit **Dynamics of** the Oceans. Local mineral concentrations are created in this way, and it is now realized that some mineral deposits on the continents may have had their origins in hydrothermal activity associated with the ocean ridges. Other deposits formed from groundwater (mostly derived from rainwater) acting hydrothermally in the vicinity of cooling magma bodies. Many natural hot springs, such as those found in health spas, are noted for their "mineral waters", and the colorful deposits that often form around hot springs in Yellowstone Park or in Iceland or Japan attest to the mineral concentrating mechanisms at work in hydrothermal fluids.

Hydrothermal activity accounts for a number of important deposits of gold, silver, copper, lead, zinc, tin, tungsten, mercury, antimony, cobalt, and germanium.

The processes of sedimentation can also act to concentrate minerals. Exposed to weathering -- the effects of rainfall, freezing and thawing, chemical reactions with air and water, and mechanical breakdown during floods or landsliding events -- rocks eventually become reduced to sedimentary detritus. This takes the form of small grains of sediment with widely varying composition. Transported by water, sediment is deposited on river banks, floodplains, in the deltas of major river systems, or on the sea floor. Throughout the process there is ample opportunity for minerals to become dissolved in the transporting water. As an example, when river water encounters the colder ocean there is a tendency for dissolved minerals to precipitate out of solution, helping to cement the deposited sediments into rock and sometimes forming widespread layered mineral deposits. Bacterial action may also come into play in aiding the mineral concentration process during and after deposition.

Hydrothermal fluids described above are also discharged directly onto the sea floor where chemical precipitation of their dissolved metals forms metal-enriched chemical sedimentary strata known as "exhalites". Sedimentary processes produce deposits of iron ore, manganese, phosphorus, sulfur, copper, cobalt, lead, zinc, silver, gold, uranium, limestone, and clay.

Evaporation is a familiar process for concentrating minerals. A glassful of tap water that is allowed to evaporate will leave behind a slight film of minerals on the walls and bottom of the glass. When ocean water is evaporated, salt deposits reflecting the concentration of elements shown in Figure XI-1 on page 351 will form. Lakes that have no outlet to the ocean, such as the Great Salt Lake of Utah or the Dead Sea in the Middle East, collect and concentrate minerals from the river water draining into them. If such a lake completely dries up, an enormous store of mineral deposits is left behind to be incorporated into the local rock strata. In some cases, basins with a weak connection to the ocean can dry up from time to time, creating extensive deposits. Recall from the unit Circulation of the Oceans that the Mediterranean Sea once dried up totally, producing immense deposits of salts that became buried in beds of sediment beneath its floor. The sediment layers now protect the salt beds from being redissolved in the present waters of that sea.

Evaporite deposits include gypsum, common salt (NaCl), potash (potassium ore), from the evaporation of seawater, and borax from the evaporation of saline lakes on the continents. When thick beds of salt are buried by overlying sediments, as in the Mediterranean and in the Gulf of Mexico, the extremely low density of the salt produces a curious and important effect. Since the overlying sediment is more dense, it tends to sink, displacing the salt upwards in rising columns called salt domes. These features (see Figure XII-5 on page 378) are of considerable importance to the petroleum industry, as we shall see in the next unit, and as sites for the deposition of sulfur.

Processes of weathering can operate in more than one way. In addition to carrying away desirable minerals for deposit elsewhere, weathering can remove common rock minerals, leaving concentrations of less easily-weathered minerals behind. This is called residual concentration of minerals, and can result in deposits of iron ore, manganese, bauxite, nickel, and clay.

Bauxite is the principal ore of aluminum, and it tends to form as a red aluminum-rich soil in tropical climates with high average temperatures and abundant rainfall. Source rocks rich in aluminum are also necessary for the creation of bauxite.

We conclude our list of mineral enrichment mechanisms with mechanical concentration. Heavy metals such as gold that do not react with oxygen or water may be released from deposit by weathering of the rocks and may then be transported in streams. Because of their density, the metal grains will not be carried as readily as the silicate sediment grains and will tend to become deposited in the sands and gravels of the stream bed that drains the source rock, or "mother lode". Concentrations of this kind are called placer deposits. The word placer, by the way, is pronounced with a short "a", as in the word "act".

Placer deposits may contain gold, platinum, tin, titanium, rare earths, diamonds, and other gemstones. The stereotyped grizzled prospector with his mule and pan is long since gone in the United States, though in some parts of the world he remains an important part of today's exploration team. Many people today follow in his footsteps, panning stream deposits for recreation and occasional modest profits. The process of panning essentially duplicates the mechanical concentration that nature used to produce the placer deposit. Sediment and water from the stream bed are placed in the pan and swirled around. The water current and the centrifugal force of the swirling motion separates the lighter sediment from the gold, forcing the former over the rim of the pan and leaving any flecks of gold behind. More serious operations use dredges and water sluices to accomplish the same thing, continuing the ancient tradition that began with the Golden Fleece.

STUDY QUESTIONS

XI-10. What condition is necessary to both magmatic and hydrothermal concentration?

XI-11. What minerals can be concentrated into important deposits by evaporation?

XI-12. What is bauxite and in what climatic condition is it most likely to form?

E. PLATE TECTONICS AND OCCURRENCE OF MINERAL DEPOSITS

Many of the processes for mineral concentration that were described in the previous section depend upon the presence of magma bodies underground. In the unit **Plate Tectonics** you learned that magma often results from tectonic activity near plate boundaries, and so it should seem reasonable that there are links between the occurrence of mineral deposits and the plate tectonic history of a region. In particular, magma generation is associated with oceanic spreading ridges and with subduction zones. Let us examine the deposits that are likely to result from each of these settings.

1. Mineral Deposits at the Oceanic Ridges

We have already discussed the process by which hydrothermal fluids can concentrate metal sulfide deposits at the oceanic ridges. Ocean water circulating through cracks in the new basaltic sea floor created at spreading ridges is heated by contact with hot rocks. Metals are dissolved in the saline water and precipitate near the discharge points of the hot springs when they combine with hydrogen sulfide to form metal sulfides. In this manner copper, iron, lead, and zinc sulfide minerals precipitate from solution. When the mineralized water encounters and mixes with cold ocean water, oxides of iron and manganese may precipitate.

At the sites along the East Pacific Rise where the submersible <u>Alvin</u> was used to investigate the spectacular hot spring vents described in the unit **Dynamics of the Ocean**, the black smoker chimneys were found to consist largely of metal sulfides. Chimneys have been discovered 15 meters (50 ft) tall resting on a metal sulfide mound of possibly equal height and up to 30 meters (100 ft) across. Such chimney-mound structures may weigh several thousand tons and one that was sampled was found to contain 14% iron, 0.7% copper, and 31% zinc, along with some cobalt, silver, and gold.

Only a small portion of the oceanic ridge system has been investigated in detail to date, but on the basis of present evidence, it seems that the extremely active chimney-vent type of activity may be restricted to fast-spreading ridges such as the East Pacific Rise, where new sea floor is being added at a rate of about 18 cm (8 in) per year. This type of vigorous sea floor spreading activity is reflected in the hydrothermal regime present at the ridge crests.

Sea-floor spreading rates are much slower in other parts of the world, and this seems to influence the style of hydrothermal activity as well. For instance, at sites investigated on the **Mid-Atlantic Ridge** where the spreading rate is on the order of two to three centimeters (one inch) per year, surface deposits consist of nearly pure manganese oxide encrusting the ocean floor. Where deeper layers of the sea floor are exposed, metallic sulfide deposits are found, indicating that the hydrothermal fluids have had a chance to mix thoroughly with ocean water before emerging from cracks in the walls of the central rift valley at the crest of the ridge.

Still another type of spreading boundary environment is encountered in the Red Sea mineral deposits. The Red Sea is a new ocean, formed by the rifting of the Arabian peninsula away from Africa. Its deposits were discovered in the 1960's during an international oceanographic expedition when echo sounders on the research vessels recorded an unusual reflection within the ocean water some distance above the seafloor. When the water was sampled at that depth, very high salinities were found, indicating that the sonar reflection marked a boundary between normal seawater and denser salt brine that was warm and rich in metals. The brine was collected in a series of pools located along the axis of the Red Sea. When cores were taken of the sea floor sediments, it was found that they contained layers of metal-rich sediment ten meters or more thick.

The Red Sea deposits are rich in iron, copper, zinc, and small amounts of silver and gold. 17513 It is estimated that the largest of these pools, the Atlantis II Deep, contains three million tons of metals, not counting the iron minerals. The two countries flanking that part of the Red Sea, Saudi Arabia and Sudan, have formed a joint commission to study the feasibility of mining this deposit. If the venture appears to be profitable, the Red Sea may be the site for the first commercial application of deep-sea mining. The technique most likely to be used is a dredge combined with a powerful suction device controlled from a sea-surface vessel. The device would scoop up loose sediment from the ocean floor and pump it up as a slurry to a ship on the surface.

In a few places, bits of oceanic crust have been scraped off the lithosphere at a subduction zone and added to continental crust, making them easily accessible. These hunks of displaced ocean floor are called ophiolites and have been mined for their stores of copper and other metals since antiquity. The best-known example is the Troodos Massif on the island of Cyprus in the Mediterranean Sea.

2. Mineral Deposits Related to Subduction Zones

Important occurrences of copper ore are found in what are called porphyry copper deposits. These form by magmatic concentration with hydrothermal alteration and replacement in magma intrusions into continental crust. Figure XI-4 shows the worldwide distribution of regions containing major porphyry copper deposits as shaded zones. Note that many occur in close association with subduction zones, such as those in South America, the Philippines, and the Middle East. In other cases, such as in western North America, eastern Australia, and in the Ural Mountains of the U.S.S.R., the deposits possibly were associated with subduction that took place in the past.

Recall that in a subduction zone, oceanic crust on the downgoing slab is heated, melted, and chemically differentiated, producing relatively felsic magma that is supplied to the volcanic arc above the subduction zone. The subduction process not only supplies the magma intrusions to fuel the hydrothermal process, but because the subducted ocean floor was already enriched in metal deposits by the processes described previously, it taps a source that has an abundant supply of metals.

Porphyry copper deposits tend to be large and relatively low grade, containing from 0.2% to 2% copper. Fortunately, the refining process is relatively easy, and because of the size and grade of the deposits, they are often profitable to mine. A single porphyry deposit may contain up to several million tons of copper, though most are considerably smaller.

In South America, the subduction zone associated with the Peru-Chile Trench is hard against the continental coastline, while in many places along the northwest margin of the Pacific Ocean, subduction takes place some distance offshore from the continent of Asia, forming a volcanic island arc separated from the main continental landmass by an oceanic basin

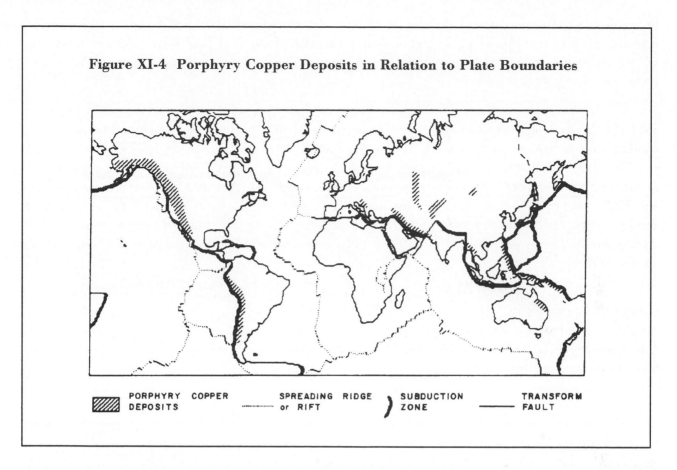

Figure XI-4 Porphyry Copper Deposits in Relation to Plate Boundaries

PORPHYRY COPPER DEPOSITS   SPREADING RIDGE or RIFT   SUBDUCTION ZONE   TRANSFORM FAULT

in which slow sea-floor spreading may be taking place. The Japan Sea is an example in which spreading has widened the separation of the Japanese islands from China. Figure XI-5 shows the relation between this spreading basin and the subduction zone.

A number of different mineral deposits are associated with back-arc spreading, resulting from similar concentrating mechanisms to those active on the other oceanic spreading ridges. Chief among these are the Kuroko-type massive sulfide deposits found in northern Japan and elsewhere that contain copper, zinc, lead, gold, and silver concentrations. Although smaller than the porphyry copper deposits, these are often significantly higher-grade ores and are of economic value.

In some places, the extensional or pulling-apart forces that generate back-arc basins can operate on a continent as well, so long as active subduction takes place nearby. This forms continental rift systems much like the oceanic rifts, but within the continent. Where these rifts are flooded, either with inflow of seawater or as lakes, like the great lake system of the East African Rift, subaqueous hydrothermal activity may form other types of lead-zinc-silver-rich deposits in a manner like that described for the Red Sea.

The western part of the United States was the scene of continuous subduction of Pacific Ocean floor along the west coast until about 26 million years ago. Continental rifting ocurred inland from the subduction, similar to back-arc spreading and, where this was accompanied by certain types of igneous intrusions, resulted in hydrothermal emplacement of large molybdenum

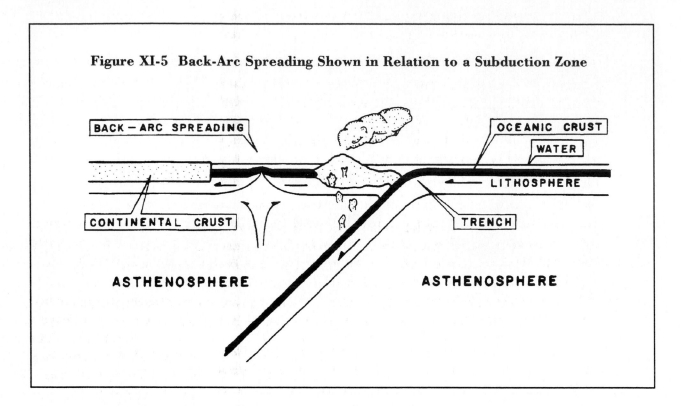

Figure XI-5 Back-Arc Spreading Shown in Relation to a Subduction Zone

deposits in the front ranges of Colorado. These give the United States its dominant position in reserves of this important steel-making element, as shown in Figure XI-3 on page 355.

STUDY QUESTIONS

XI-13. Are metallic oxide deposits more likely to be found in association with fast or slow sea-floor spreading rates?

XI-14. What is an ophiolite?

XI-15. What process is responsible for the formation of copper porphyry deposits?

XI-16. What is a back-arc basin?

F. SOME NOTABLE MINERAL DEPOSITS

The television film corresponding to this unit will take you to a number of sites where important deposits are being mined. This is a good opportunity for us to put together the various concepts discussed so far in terms of the origin and setting for different kinds of deposits.

1. The Bushveld Igneous Complex, South Africa

The Bushveld igneous complex in South Africa is magmatic in origin and consists of a series of layered igneous rocks that contain vast reserves of chromium along with platinum, nickel, and iron. It is the largest single deposit of chromium ore found anywhere in the world and has been estimated to contain reserves in the range of 6,000 million tons of chromite.

The Bushveld complex is a good example of magmatic concentration as described earlier, in which chromite (chromium ore) accumulates toward the bottom of a layered magma intrusion due to its high density and tendency to crystallize before the rest of the magma. The metals are concentrated in a series of relatively thin layers ranging in thickness from centimeters to as much as one meter. They extend over an area of many thousands of square kilometers, however, accounting for large reserves.

Radioisotope age dating of the Bushveld complex yields ages on the order of 2 billion years, placing its origin nearly half way between the origin of Earth and the present. The complex occurs as a series of nearly circular igneous intrusions that are relatively isolated in that the surrounding rock is much older and essentially undisturbed. It does not appear to be associated with either subduction or sea-floor spreading and so its origin is still a bit of a mystery, although an ancient form of contintental rifting may have been involved. Two interesting hypotheses have been advanced -- one is that the intrusion is due to hot spot activity, and the other is that the circular igneous complexes resulted from the impacts of fragments of an asteroid.

2. The Troodos Massif in Cyprus

The Troodos Massif in the western part of the Mediterranean island of Cyprus contains the Troodos ophiolite segment of ocean floor produced by sea-floor spreading in Cretaceous time. It contains massive sulfide deposits of iron, copper, zinc, and cobalt that were concentrated by sea-floor hydrothermal activity at an ocean ridge, as well as magmatic nickel sulfide and chromium deposits that were formed in deeper intrusive rocks at the ridge. The copper deposits on Cyprus have been worked since antiquity, and indeed the metal derives its name from this locality, having been called the "Cyprian metal".

Ophiolites around the world have their origins in the ocean floor and in addition to their mineral deposits are of interest to geologists because they provide dry-land exposures where the structure of what was once oceanic crust may be studied with relative ease.

3. Kuroko Massive Sulfide Deposits, Japan

We have already mentioned the Kuroko-type deposits in the previous section, relating their origin to back-arc sea-floor spreading. The Kuroko massive sulfide deposits are found in the northern portion of the main Japanese island of Honshu and contain zinc, copper, and lead along with minor amounts of precious metals. The concentrating mechanism was hydrothermal activity on the sea floor approximately 12 to 15 million years ago, making this a relatively young deposit, but there are many similar, much older deposits in more ancient rocks in other parts of the world.

4. The Noranda District of Quebec, Canada

The Horne Mine and others at Noranda in northern Quebec is an example of a layered massive sulfide deposit of considerably greater age, dating back to the early Precambrian. It is a classic example of a submarine volcanic process forming layered sulfide deposits over the hot springs on the ocean floor, and is similar to the Kuroko deposits of Japan. It produces copper, zinc, gold, and silver.

5. Uranium Deposits in Saskatchwan, Canada

Uranium deposits are generally found in sedimentary environments and are of several types, but their exact origin is still a matter of debate. Some geologists believe that the concentration mechanism is chemical in nature, due to circulating groundwater that collects dispersed uranium from the sediments and deposits it in favorable settings.

STUDY QUESTIONS

XI-17. What metal was mined in ancient Cyprus?

XI-18. What mineral concentration process seems to be responsible for the massive sulfide deposits of the Kuroko, Japan and Noranda, Quebec mining districts?

G. PRESENT SUPPLIES AND PROSPECTS FOR THE FUTURE

In discussing the availability or scarcity of various minerals, a distinction should be made between taking a local or worldwide view of the situation. You should recall the situation presented in Figure XI-2 and Figure XI-3 on pages 354 and 355. As an example, chromium has worldwide reserves adequate for the immediate future, but an industrialized country like the United States must import virtually all of this metal to meet its needs. No satisfactory substitute for chromium has been found, and so this makes the industrialized world dependent on only a few exporting countries such as South Africa, Zimbabwe, and Turkey. The U.S.S.R. and Albania are also significant producers.

Assessing the future supplies for any of these minerals is a hazardous business at best. Projections of exponentially escalating demand and imminent shortages in a number of minerals were common in the middle 1970's, but worldwide economic slowdown turned shortage into glut and today many mineral companies are struggling to make a profit. If economic improvement continues, the situation is likely to turn around again, and perhaps critical shortages will once again be a concern.

Another uncertainty in forecasting mineral supply is the near-impossibility of predicting major new discoveries. However, many knowledgeable observers are pessimistic about truly radical changes in known mineral reserves, and feel that the principal changes will be driven by economics and not by real changes in the resource base. Exploration techniques have developed to the point where they feel that most of the major surface deposits of high-quality mineral ores have already been found in most parts of the world.

At present rates of consumption, most scarce metal reserves should last another 20 to 100 years. Metals that are most likely to become scarce during your lifetime are silver, mercury, tin, and tungsten. The common metals, iron, aluminum, manganese and, to a lesser extent, copper and nickel, are relatively abundant, but many of the higher-grade ore deposits are becoming exhausted. The result is that lower-grade ores will have to be mined and refined, at greater cost. The prognosis in the long run, at least, seems to be gradually rising prices for these metals.

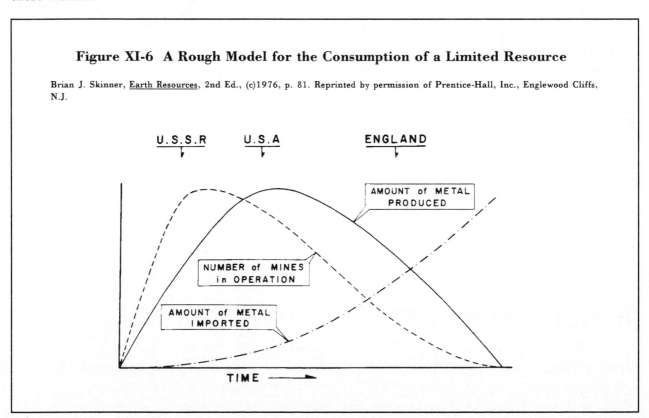

Figure XI-6 A Rough Model for the Consumption of a Limited Resource

Brian J. Skinner, Earth Resources, 2nd Ed., (c)1976, p. 81. Reprinted by permission of Prentice-Hall, Inc., Englewood Cliffs, N.J.

Figure XI-6 presents a model for the consumption of a limited resource within an industrialized country. Initially, the number of operating mines and the production of metal increase together. Eventually, as more deposits are worked out, the number of operating mines decreases, followed by the production. If demand continues, a greater and greater proportion of the metal must be imported from other countries.

If we consider total mineral production, the U.S.S.R., with its vast mineral reserves, is still in the stage of expansion at the left of the diagram; the United States is somewhere near peak production, with imports rising; and England has long since exhausted many of its resources and is heavily dependent on imports of raw materials. Japan joins England on the right side of the curve, having a smaller resource base to begin with and having passed through the process much more rapidly, partly due to its large population and partly due to its prodigious use of resources during World War II.

There is a significant disparity in resource use between the developed and the less-developed nations. At the present time 5% of the world's population consumes 90% of world mineral production. This divides the world into two camps: the highly industrialized consuming nations and the less-developed exporting nations. To a considerable extent, the industrialized nations are those that had a broad and rich resource base to begin with. The nations of Europe built the industrial revolution on domestic resources of minerals and energy, and even today total mineral production is dominated by only five nations: the U.S.S.R., the United States, Canada, Australia, and South Africa.

As the developed nations exhaust their domestic reserves, they are turning increasingly to mineral imports from their less-developed neighbors. To some, this smacks of exploitation of Third World Countries, but to many of those countries, the exporting of minerals is their principal means of obtaining capital for development, jobs, taxes, and a favorable influence on their balance of trade.

STUDY QUESTIONS

XI-19. Why did mineral shortages predicted ten or so years ago fail to come about?

XI-20. Why do you think the U.S.S.R. has used less of its mineral reserves than has the United States and Europe?

H. MEETING FUTURE MINERAL DEMANDS

There is sharp dispute even when mineral forecasts are made for 20 years into the future; attempting to see farther into the future can be little more than guesswork. We have seen that economics are a strong part of estimating reserves, and it turns out that the cost of energy is a strong part of the economic picture in the mineral industry. Refining metals from ore requires energy, and lower-grade ores require more energy. The combination of rising energy prices and the need to mine lower grade ore deposits can have a dramatic effect on metal prices. This effect is partly offset by technological breakthroughs that can increase the efficiency of extraction and refining, however.

Finding new sources for minerals is another distinct possibility for the future. We have already seen that steps are being taken toward the development of mineral deposits on the floor of the Red Sea, and it is clear that the sea floors hold large deposits of a number of minerals. The question is, will it ever be economical to mine them?

Serious consideration has been given to mining a mineral resource that is reasonably concentrated and accessible on the ocean floors. Large parts of the Pacific Ocean floor are covered with manganese nodules, concretions of manganese and iron oxides along with copper, nickel, and other scarce metals that were discovered during the voyage of the research ship H.M.S. Challenger in the 1870's. Already trial dredging operations have been conducted, and total reserves are estimated to be huge. In the cases of manganese and copper, the resources present in manganese nodules may surpass present continental reserves. Development of these resources has a number of problems associated with it, however. Resource analyst Brian J. Skinner[27] points out:

> What bottom dredging of the ocean will do to sea life, who really has the right to recover the material, and how mining is to be monitored and policed are vast problems for the future. An entire new field of legal expertise will apparently have to develop.

There are still a number of regions of the world that have seen little if any mineral development. This is true especially for the polar regions. Technical advances are slowly making it possible to operate in the frigid reaches of northern Canada, Siberia, and Antarctica, where significant mineral deposits have been found and more are likely to turn up with intensive exploration. Antarctica, however, is presently off limits for commercial development of any kind under the terms of the Antarctic Treaty, which dedicates that continent solely to scientific research. Political pressure is already rising to modify the treaty, however.

If prudence makes it unwise to predict the mineral situation for no more than a few decades into the future, it is only natural to speculate on the longer-term needs of humankind. We would like to think of our civilization as enduring for hundreds or thousands of years, yet the finite nature of mineral deposits casts doubt on our ability to maintain high consumption patterns for very long.

Some have envisioned the mining of extraterrestrial sources for future mineral supplies, and serious thought has been given to mining techniques that might be used on the Moon to supply any colonies that might be established there. Space stations might be constructed from minerals mined from small asteroids, making it unnecessary to supply the materials from Earth, and even more importantly, saving the energy necessary to lift large masses of metal into orbit.

Whether or not such sources are eventually developed, the most likely scenario for insuring near-future mineral supplies is through substitution and conservation. At present, the scarce metals are being used far out of proportion to their abundances, and efforts are constantly being made to substitute cheaper and more abundant materials for the expensive and scarce. Conservation is a reliable means of reducing demand for raw materials. The design of products that are more durable and more easily repaired, and implementation of effective

[27]B. J. Skinner, Earth Resources, Prentice-Hall, 1976

recycling of materials are approaches that are likely to become more desirable as raw material prices rise. The recycling of lead, gold, and aluminum is a substantial industry, and the recovery of many metals from junked automobiles is already profitable. Eventually, we may turn to our garbage dumps as high-grade "ores" of the next century. In the distant future, it may prove necessary to extract minerals from seawater itself, which probably contains more total metals in all the oceans than may be found on the continents. At present, the cost of extracting these very low concentrations makes this impractical.

STUDY QUESTIONS

XI-21. Where are manganese nodules found?

XI-22. Why is there no mining activity in Antarctica today?

XI-23. What is the difference between substitution and conservation in regard to mineral resources?

RECOMMENDED READING

Robert L. Bates, Geology of the Industrial Rocks and Minerals, Dover, 1969.

Brian J. Skinner, Earth Resources, Prentice Hall, 1976.

George F. Kunz, Gems and Precious Stones of North America, Dover, 1967.

UNIT XII GIFTS FROM THE EARTH:

ENERGY RESOURCES

WORKERS ON A RIG
Included with permission of Mark C. Van Veen, American Association of Petroleum Geologists.

A. INTRODUCTION

1. Overview

One of the most striking differences between the developed and less-developed nations is found in their patterns of energy use, with industrialized nations using far more energy per capita, of very different types. Oil and gas constitute the most important energy fuels in the United States today, and you will have a chance to examine their nature and origin along with the techniques being employed in the search for new oil and gas fields. World reserves of oil are dominated by those of the Persian Gulf region, but even with these huge fields the total supply is finite and some estimates put peak production only a decade or so away. Production in the United States already appears to have peaked. Tar sands and oil shales may extend petroleum reserves considerably if technological advances make them more profitable to work.

Coal, on the other hand, is present in large quantities, and its utilization is more likely to be impeded by environmental factors than those related to supply. You will consider the origin of coal, and relate its worldwide distribution to conditions that persisted in the Carboniferous world.

Uranium for use in nuclear power reactors is a finite resource in the type of reactor currently in use, but the development of breeder reactors could greatly extend the usefulness of existing fuel supplies. Public fears concerning reactor safety and security of nuclear fuel have virtually crippled the nuclear industry in the United States, however. Research in controlled thermonuclear fusion reactors continues, offering hopes of an alternative source of energy that may become available in the next century.

Renewable energy sources such as geothermal and solar energy, hydropower, and energy derived from wind, tides, and the thermal stratification of the oceans are all being investigated or developed as supplemental energy sources. As oil and gas production diminish in the future, alternative energy sources must be developed to take their place.

2. Objectives

Upon completion of this unit you should be able to:

1. distinguish between patterns of energy use in the developed and less-developed nations

2. describe different products derived from petroleum

3. describe the conditions that are required to produce oil fields

4. relate methods used in the search for oil and gas to the geological settings in which these resources are found

5. describe the worldwide distribution of oil and gas reserves

6. describe the nature of tar sands and oil shales and relate their properties to their present low level of utilization

7. relate the origin and occurrence of coal to geographical, climatological, and geological environments that favor its formation

8. relate the different types of coal to stages in coal formation

9. compare and contrast the origin of coal with that of oil and natural gas

10. describe the worldwide distribution of coal reserves and relate it to the geography of the Carboniferous world

11. describe environmental problems associated with coal use

12. distinguish between the operation of light water and breeder nuclear reactors and contrast their efficiency in utilizing world uranium reserves

13. discuss methods that have been proposed for the disposal of radioactive waste products

14. describe alternative energy sources that are being developed: geothermal, solar, hydropower, wind, biomass, tidal, and ocean thermal gradient

15. relate the production history of a finite resource to a model based on the increasing difficulty of extracting that resource.

3. Key Terms and Concepts

oil seeps

hydrocarbons

"cracking" of organic molecules

anaerobic environment

source rocks

migration of oil and gas

structural trap

anticline

gravity and magnetic surveys

seismic reflection surveys

oil fields

heavy oil

tar sands

oil shales

peat

lignite

bituminous coal

anthracite

sulfur in coal and acid rain

Uranium-238 and Uranium-235

light water reactor

breeder reactor

Plutonium-139

tidal power plants

geothermal energy

hydrothermal convection *p. 390*

passive and active solar heating

photovoltaic solar cells

wind energy

biomass energy

ocean thermal gradient power plant

satellite solar energy

production history of
   a limited resource

thermonuclear fusion

4. Corresponding Video

In this program you will see the formation and discovery of the riches contained within Planet Earth. Film footage and animation will help unravel the mysteries of coal, oil and mineral formation. Time-lapse microphotography explores the ecosystem of living soil, and new frontiers in exploration are highlighted.

B. THE CONSUMPTION OF ENERGY

One of the principal ways in which the developed countries are distinguished from the less-developed ones is in their use of energy. A poor farmer in a less-developed country must rely on humanpower and beastpower. Oxen or some equivalent are still used in many parts of the world to draw plows and provide motive power beyond the strength of men. In contrast, developed countries consume large quantities of energy for transportation, for industrial uses, and for heating or cooling of building space.

Stop for a moment and look around you, and think of how many energy-consuming devices are in your home or apartment building. Then consider how you would have lived if you had been born a thousand years ago -- or even two hundred years ago. Much of your present lifestyle is dependent upon abundant and affordable energy along with the devices that utilize it.

Figure XII-1 shows how energy is used in the United States. Three broad areas -- industry, buildings, and transportion -- have comparable shares of total energy use. Two of the most dominant uses of energy within the latter two areas are automotive and space heating and cooling.

The energy used in the United States is supplied by a number of fuels, as shown in Figure XII-2. Most of it is supplied by oil, natural gas, and coal. These are the fossil fuels, so-called because they are derived from energy that in most cases was emplaced in the ground millions of years ago. The heavy dependence on natural gas, by the way, is a peculiarity of North America, due to extensive domestic gas fields linked to many parts of the continent by a truly remarkable grid of gas lines. Other developed regions generally show a heavier reliance on oil and coal and a lesser use of natural gas.

This pattern of fuel use is quite recent; only a hundred years ago, the dominant fuel by far was wood, with coal contributing a much smaller proportion of the total. By 1900, coal became the most heavily-used fuel, but in the 1940's, oil and gas use began an explosive growth.

The quantities of fossil fuels that are consumed today are so great that even minor imbalances between supply and demand can cause considerable societal disruption. The 1973 oil embargo by OPEC (Organization of Petroleum Exporting Countries) caused considerable distress in the United States even though total oil consumption was cut by less than five percent. As it turned out, much of the problem was in the political and public response to the apparent shortage. Many people tried to keep their automobile gas tanks as full as possible, making more visits to service stations and so creating long lines, and creating temporary shortages by increasing the amount of gasoline stored in individual gas tanks. Political attempts to ration and allocate supplies also contributed to local product shortages. Even without these multiplying effects, though, small disturbances in energy supplies can have profound effects on a society so dependent upon them.

Worldwide use of energy for several decades appeared to be increasing dramatically, but in recent years it has leveled off and even dropped somewhat, as shown in Figure XII-3. All forms of energy use are represented in the diagram in terms of the amount of coal that would provide the equivalent energy.

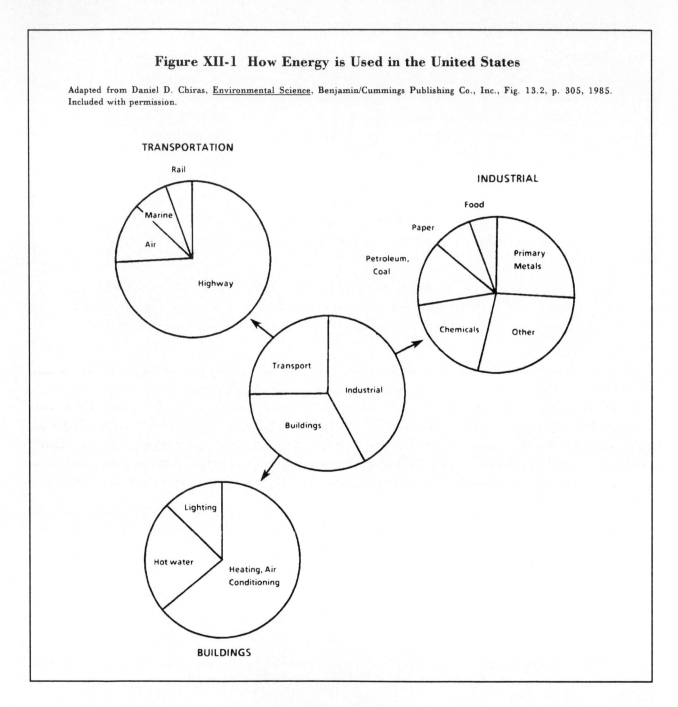

Figure XII-1 How Energy is Used in the United States

Adapted from Daniel D. Chiras, Environmental Science, Benjamin/Cummings Publishing Co., Inc., Fig. 13.2, p. 305, 1985. Included with permission.

This turnaround in energy demand was virtually unforeseen in the 1970's; most predictions made at that time assumed that demand would continue to accelerate, causing severe energy shortages. Instead, today we find a surplus of energy on the worldwide market, which has resulted from economic downturn coupled with a tenfold increase in the price of oil during the past two decades. In the United States, for example, new automobiles are smaller and more fuel-efficient, and many people have cut back on energy use. Whether energy consumption will remain depressed or will resume something more like its historical rise is quite unknowable at this point. Many observers, however, expect some resumption of increased demand due, at least, to the pressures of an expanding population.

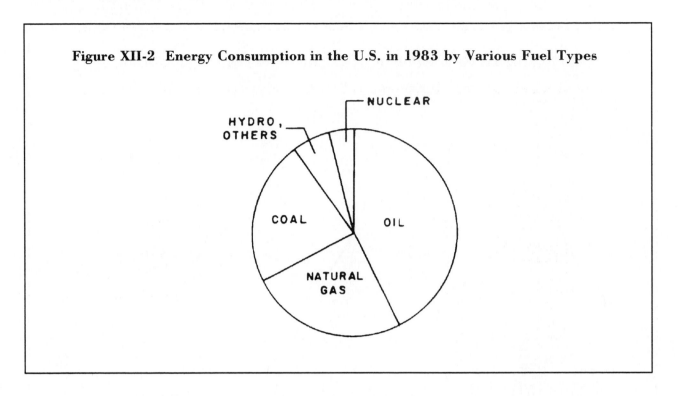

Figure XII-2 Energy Consumption in the U.S. in 1983 by Various Fuel Types

Figure XII-4 shows the effect for the United States. Energy use is plotted on a per capita basis, and it can be seen that individual energy use began to decline between 1970 and 1980 in the United States. The figure also shows the enormous discrepancy in per capita energy use between the United States, with its high state of industrialization, and less developed countries, represented by an average for the whole African continent and by India.

Let us examine each of the major energy resources, and some minor ones as well. We shall be most concerned with how they are formed, how they are distributed, and how they are found.

STUDY QUESTIONS

XII-1. Approximately what fraction of total energy use in the United States is taken by highway transport (automobile and trucks)?

XII-2. What are some of the factors that have affected the recent downturn in world energy use?

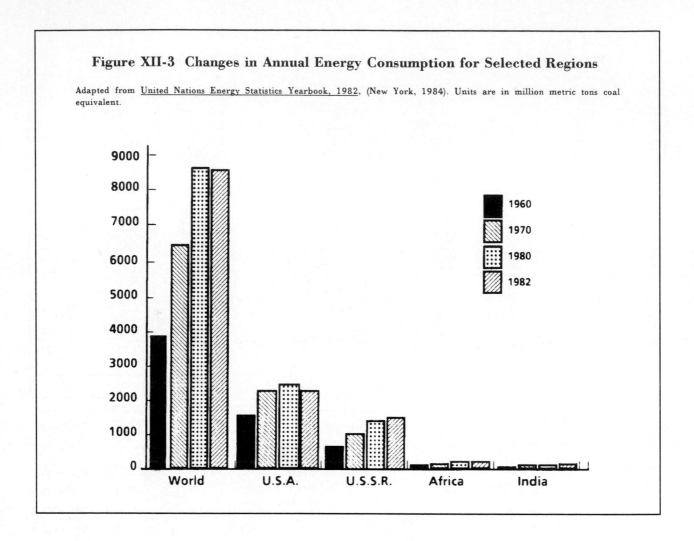

Figure XII-3 Changes in Annual Energy Consumption for Selected Regions

Adapted from United Nations Energy Statistics Yearbook, 1982, (New York, 1984). Units are in million metric tons coal equivalent.

C. OIL AND GAS

Today we must expend millions of dollars to prospect and drill for oil that is buried deep below ground, but there are a number of places where small quantities of crude oil find their way to the surface and are exposed as oil seeps. The famous La Brea Tar Pits in Los Angeles formed a natural trap for unwary creatures of past ages. Mired in the tarry pools, they died and their bones were preserved in the bitumen that still can be seen welling up from the ground.

Natural seeps occur in many parts of the world, and were used by Stone Age people in construction, by ancient Egyptians as an aid to the preservation of their mummies, and by seafaring people everywhere for caulking. Kerosene became popular for use in lamps, and in 1857 commercial oil production began in Romania. Two years later, Edwin L. Drake sank the first oil well at Titusville, Pennsylvania, to a depth of 21 meters (69 feet), beginning our modern age of oil consumption. That well is still producing. In the intervening century and a quarter, the availability of oil as a concentrated and portable source of energy has become a major influence in determining the life styles of people living in the industrialized nations.

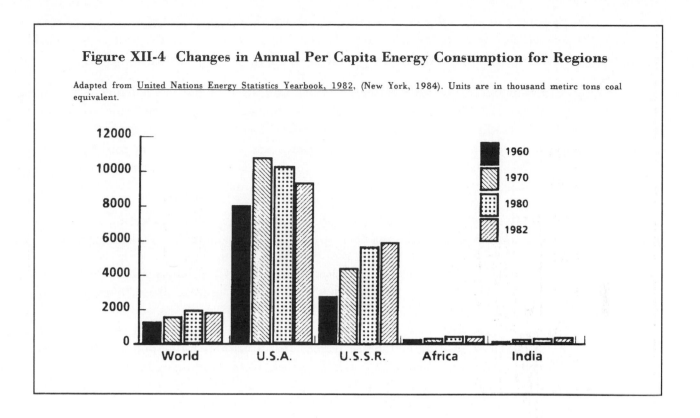

Figure XII-4 Changes in Annual Per Capita Energy Consumption for Regions

Adapted from United Nations Energy Statistics Yearbook, 1982, (New York, 1984). Units are in thousand metirc tons coal equivalent.

Petroleum is composed primarily of organic compounds made from carbon and hydrogen (hydrocarbons) that occur in a variety of mixes. Hydrocarbons can take a number of different forms, occurring as gases (natural gas, mostly methane), liquids (gasoline, kerosene), and solids (paraffin). Crude oil is "cracked" in refineries, a process of breaking complex organic molecules into simpler ones, to produce a wide variety of products. These include gasoline, kerosene, fuel oil, lubricants, paraffin, and raw material for the chemical industry. Referring to the latter uses, an oil minister of Saudi Arabia was once quoted as saying, "Oil is too precious to burn." Indeed, the chemical industry is heavily dependent on petroleum in the production of plastics and pharmaceuticals. Differing hydrocarbon mixes produce different grades of crude oil. Some, like the light oils of Pennsylvania, are suited for the production of lubricants, while heavier grades are better suited for the production of fuels.

1. Origin of Oil and Natural Gas

Oil and natural gas are found almost exclusively in sedimentary rocks. Oil begins to form with the burial of organic remains from plants and animals (usually microscopic in size) in an oxygen-deficient or anaerobic environment. Where oxygen is plentiful and supplies of organic material are slow in accumulating, decay, which is largely oxidation, sets in and returns the hydrogen to water and the carbon to carbon dioxide. In these environments, potential sources for production of oil and gas are destroyed. In certain geological settings, however, accumulations of organic materials in sediments rapidly deplete the oxygen supply, and if no external sources of oxygen are present, the formation of hydrocarbons can begin. This appears to happen most often for oil and gas in marine sediments. River deltas and other near-shore environments are rich sources of both organic debris and sediment, and the source rocks of many important oil deposits are associated with them.

Bacterial action on the organic material during and after burial is probably a part of the oil-forming process. As the organic material becomes buried deeper beneath the accumulating sediments, pressure and temperature rise and a natural form of the cracking process begins to occur, reducing the complexity of the organic remains and converting them into hydrocarbons. The deeper the burial and the longer the process goes on, the lighter the crude oil becomes, accounting for some of the different types that are found.

Natural gas is formed in much the same way, with methane being formed almost as soon as organic material is buried and continuing to be produced throughout the cracking process.

Both liquid crude oil and natural gas are less dense than water and are much less dense than rock. As a result, they have a strong tendency to migrate upward from their points of origin, provided that they can find permeable rock in which to do so. Some sedimentary rocks are relatively porous in that there are open spaces between sediment grains and small channels connecting them. A coarse sandstone may be so porous that a drop of water placed on it can be seen to spread out and be absorbed as though the rock were a sponge. It may be a slow process, but given enough time, oil can migrate through porous or fractured rock layers. Other sedimentary rocks, shales for instance, may be relatively solid and impermeable to oil and gas. These can form barriers to further upward migration of oil. In this way, pools of oil can accumulate beneath impermeable rocks and form economically important deposits.

Most of the oil that has been produced throughout geologic time has probably succeeded in migrating to the surface where it has oxidized to carbon dioxide and water. This is seen in the relative scarcity of oil in very old rocks. The majority of large oil pools are found in rocks less than 200 million years old.

2. Exploration for Oil and Gas Deposits

A wide variety of techniques have been developed to aid in the search for oil and gas. Because they are fluids and are contained within the pore spaces and fractures of rock, their presence deep underground is very difficult to detect. As a result, most exploration techniques are aimed at detecting the underground geological structures that serve to trap oil and gas in their upward journey. Figure XII-5 shows two such structural traps that are common sites for oil and gas deposits -- anticlines and salt domes.

An anticline is formed when rock strata are folded into an arch. If permeable rock is present beneath impermeable layers and the latter have not been so fractured that the oil or gas can leak through, then a concentration might form if hydrocarbon fluids are working their way upwards through the permeable rock. The odds are against a significant oil accumulation, however. In order for oil to form, there must be present a suitable source rock, an appropriate maturation history, a migration pathway, a sufficiently porous rock to form a reservoir, a structural trap, and the top of that trap must be sealed with an impermeable rock to prevent the oil's escape.

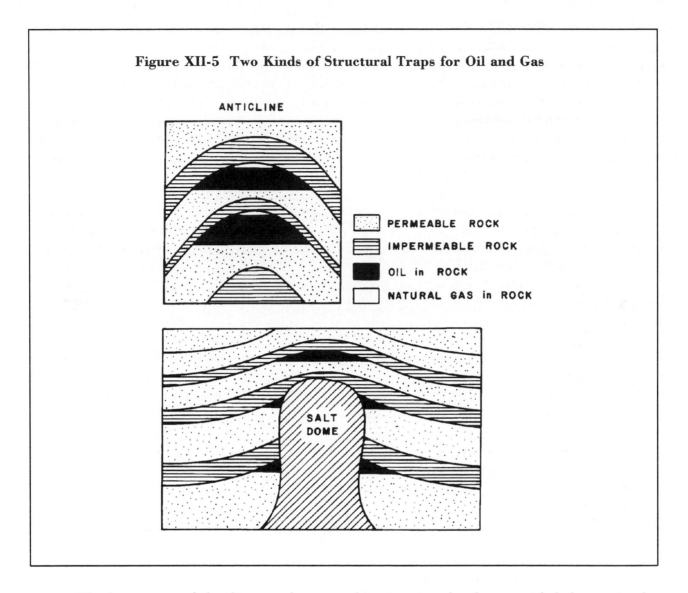

Figure XII-5 Two Kinds of Structural Traps for Oil and Gas

ANTICLINE

PERMEABLE ROCK

IMPERMEABLE ROCK

OIL in ROCK

NATURAL GAS in ROCK

SALT DOME

The lower part of the diagram shows another situation that has provided the setting for significant oil and gas finds. In the previous unit we referred to the formation of salt domes when evaporite beds become deeply buried. Because of its low density, the salt may force its way through the overlying strata, buckling them upwards and forming structural traps in which oil and gas may accumulate. Large oil deposits near the mouth of the Mississippi River and in the Gulf of Mexico and elsewhere have been found in association with salt domes.

The presence of a structural trap does not by any means guarantee that it contains oil. Most do not, but these are the most likely places to begin looking in a region that appears to have potential for oil production. In order to locate them, the structure and kind of rock layers must be worked out in considerable detail. Petroleum geologists apply a wide variety of techniques to this task. Precise surveys of the Earth's gravity and magnetic fields may serve to locate large subsurface features of interest, while seismic surveys provide more detailed views of the underground structure. In the Seismic Reflection Method, sound waves generated by a small explosion or other artificial source near the surface travel down into the ground and are reflected from rock layers. The recorded echoes from these layers are then processed by computers to provide diagrams of structures such as those shown in Figure XII-5. Test holes are drilled, to look for traces of oil or gas, and to obtain rock cores or chips that allow the geologist to identify the different strata. In addition, sensitive instruments are lowered down the drill holes to aid in the identification of rock layers.

Oil and gas exploration is extremely expensive. The drilling of a single oil well may cost many millions of dollars, and more millions may be expended in the careful scientific work that precedes any substantial drilling.

STUDY QUESTIONS

XII-3. What is the meaning of the term "cracking" as used in the context of petroleum formation and refining?

XII-4. Why is petroleum seldom found in igneous rocks?

XII-5. Why does oil tend to migrate toward the surface?

XII-6. Why is an anticline a favored structure for the trapping of oil?

D. WORLDWIDE DISTRIBUTION AND RESERVES

Sedimentary rocks are widely distributed over the Earth and so one might expect that the same would hold true for petroleum resources. To a certain extent this is true in that every continent contains major oil fields, but distribution on a finer scale turns out to more capricious. Of the 600 or so sedimentary basins known to exist, only about three dozen contain giant oil fields, which account for a sizeable portion of worldwide reserves. These giant fields may occur in clusters and cover only a fairly small geographical area within an individual basin.

The distribution of estimated reserves, shown in Figure XII-6, contains one very great anomaly -- the producing areas of the Middle East. This region, measuring only about 1,500 by 700 kilometers (900 by 400 miles) in and around the Persian Gulf, accounts for nearly 57% of total world reserves. Throughout long stretches of geologic time, all the factors involved in the generation of oil and gas -- proximity to sources of sediment and organic material, nearly enclosed oxygen-deficient seas, and the production of structural traps -- have been optimized to produce a series of giant oil fields that even today are not fully explored. Other large oil fields are found in the arctic regions of the U.S.S.R. and Alaska, in the Gulf of Mexico region, and in the North Sea.

The major portion of the continental shelves have not yet been fully explored and some may have considerable potential for substantial new discoveries. Even though offshore drilling brings with it a host of new difficulties, not the least of which is coping with severe weather conditions at sea and preventing oil spills during drilling and during transportation of oil to its destination, drilling has taken place offshore from more than 100 countries and in water depths as great as 7,000 feet (2,100 m) off the east coast of the United States.

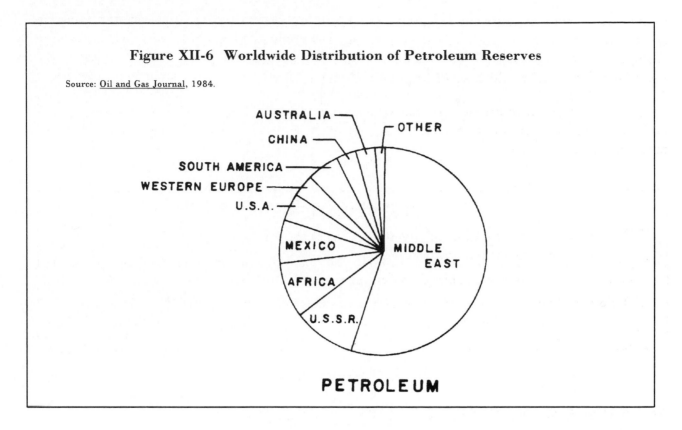

Figure XII-6 Worldwide Distribution of Petroleum Reserves

Source: Oil and Gas Journal, 1984.

PETROLEUM

How long will the oil last? This is a question frequently asked, but one whose answer depends on so many unknowable factors such as future oil demand, future exploratory success, future technological developments, and future economic conditions, that no definitive answer can be given. For example, worldwide proved reserves at current rates of consumption equates to a 32-year lifetime, but if estimates for undiscovered recoverable resources are included, a value of 70 years is obtained. However, both these estimates are based on a recovery of about 35% of the original oil in place, and new technology will certainly increase this factor to 50% or more. With this addition, the number becomes 100 years. If improved efficiency of usage and fuel-switching possibilities are considered, the time period becomes even longer. Furthermore, these observations don't include the potential additions from tar sands and heavy oils discussed below.

Clearly, oil is a finite resource and if viewed with respect to the totality of human history, the "oil age" may cover only a relatively brief period. There are many known sources of energy that will be developed to supplant oil in response to our ever-changing technologies and economic conditions.

In addition to crude oil pumped from oil fields, there are also substantial amounts of hydrocarbons found in heavy oils and tar sands. Tar sands contain heavy oil in the form of tar or very viscous oil that are incapable of migrating from the reservoir rock. As a result, these deposits cannot be pumped from the ground like crude oil. To make the oil fluid enough to flow, it may be heated with steam, and once recovered, it can be refined to recover the lighter oil components. Tar sands may also be utilized by mining tar sand rocks, which are then crushed and treated to recover the tar which is then refined.

Two large tar sand and heavy oil deposits are known: the Athabasca tar sands in Alberta, Canada, and the Orinoco deposit in Venezuela. Their total resources have been estimated at around one trillion barrels, or about half as much as worldwide crude oil resources. Commercial mining operations of tar sand deposits in Canada are ongoing. Recovery of heavy oil deposits by using in site methods such as steam drive is becoming common, especially in the United States.

Another potential petroleum resource is oil shale. These are fine-grained sediments containing organic matter that can be distilled to yield oil. They must be mined and then heat treated on the surface, requiring that the energy equivalent of about 40 liters of oil must be expended to recover the oil from a ton of shale. As a result, only shales that will yield in excess of 75 liters of oil-per-ton are worthwhile for development. Particularly rich oil shales have been used for commercial production of oil in the U.S.S.R. and in China for many years, but the largest oil shale deposits known are found in the United States. Worldwide reserves from oil shales are estimated to be comparable to those of tar sands. Whether these petroleum resources are ever utilized to any great extent will probably depend on the development of new lower-cost technology for mining and refining. At present, large-scale surface mining and disposal of residues left over from retorting the shale can pose environmental difficulties. These will need to be addressed if the cost of oil recovery from oil shale becomes competitive with that of crude oil.

STUDY QUESTIONS

XII-7. In what region of the world are the greatest oil reserves to be found?

XII-8. What is the distinction between tar sand and oil shale?

E. COAL

During the Middle Ages, wood was the dominant energy fuel, and in areas of high population density, the forests began to suffer from overcutting. In the twelfth century AD, so-called "sea coles" were found on the beaches of northeast England, and with the discovery that they were flammable, widespread use for home heating began. These were chunks of coal that had weathered from coastal cliffs and, traced to their source, were later mined directly from coal seams underground.

Coal use expanded rapidly with the invention of the steam engine in the eighteenth century. As the principal energy source for this new form of motive power, it was the essential fuel that ignited the industrial revolution in Europe. It can hardly be a coincidence that two nations that led this revolution, Great Britain and Germany, were the European nations most endowed with coal deposits.

1. The Origin of Coal

Of all mineral deposits, the origin of coal is among the best understood. Coal is a rock composed largely of carbon that is organic in origin. Within coal may be found fossil imprints of thousands of species of plants that flourished at the time that coal formed. These tell us that, unlike oil, which is largely marine in origin, coal formed entirely on the continents from a wide variety of plant life. In order for this organic matter to be preserved, however, a particular environment is necessary.

When a tree or other plant dies in a forest environment, it falls to the ground and immediately begins to decay, attacked by insects and micro-organisms that aid in the processes of decomposition. In this process, the complex carbohydrates that make up plant tissues are broken down into, primarily, water and carbon dioxide, both of which are returned to the environment. The carbon is recycled to carbon dioxide and not trapped, and so the usual forest environment is not likely to result in the formation of coal.

The preservation of carbon requires an oxygen-deficient setting, such as that found in stagnant waters. Swamps are a good environment for carbon preservation, as may be demonstrated by turning up a few inches of the muck at the bottom of a stagnant swampy pool. If you do this, you will generally find that old leaves, twigs, and stems are matted in layers and that the sediment is dark in color, indicating high levels of carbonaceous material.

Even at such shallow depths, the coal-making process has begun. Bacteria attack the wood and plant tissues, reducing the hydrogen and oxygen content and concentrating the carbon. The result is a carbon-rich humus that is called peat, generally regarded as the first stage in the making of coal. Though peat is not yet coal, it is in fact a low-grade fuel. For centuries, peat has been cut from bogs and dried for domestic use in cooking and heating.

Many coal deposits extend over large areas, indicating that the original coal swamps were quite extensive. Today, we might go to a place like the Everglades of Florida for a similar environment. Because running water brings in new supplies of oxygen, swamps in low, flat terrain with sluggish circulation of water serve the purpose best. In addition, incorporation of inorganic sediments such as sand or mud are minimized in flat regions, leading to the generation of high-purity coal.

Another environment in which peat may form is that of the tidewater region of coastal zones. A long, thin barrier island of sand takes the brunt of the sea's waves and storms, protecting shallow tidal pools behind it. If the coastal terrain is flat, freshwater marshes and swamps are commonly built between the streams that feed the tidewaters. Many of today's major coal seams appear to have originated in such an environment.

With an increasing overburden of sediment, the peat becomes progressively more compacted and dehydrated. Biological action largely ceases, and further changes are brought about by increasing pressure, temperature, and chemical changes. In this step, peat is changed into lignite, sometimes called "brown coal". It is woody in character and brown or brown-black in color. Because of its high moisture content, it tends to disintegrate when it dries. Lignite is often burned as a power-plant fuel.

With further heat, compaction, dehydration, and chemical action, lignite turns into sub-bituminous coal and then into bituminous coal. These are the most abundant and commonly used forms of coal. They are dense and stable when dry and produce a large amount of heat when they burn.

A further step may be taken when coal seams are exposed to metamorphic processes that occur in tectonic regions. Coal seams that have been subjected to intense folding and heating from nearby igneous activity are often found in the form of anthracite, a dark and brittle jet-black coal that burns with a blue flame, has high heating value, and contains little sulfur, a major contaminant of many bituminous coals.

The earliest coals appear in Devonian rocks because, prior to this time, the land plants that form coal had not yet evolved. It was, however, in the Carboniferous period that the coal swamps reached their greatest development. In tropical regions of the Carboniferous world, conditions must have optimized in the luxuriance of vegetation, climate, rainfall, and the geographical setting, that produced enormous peat bogs that would later be preserved as coal. When oil and gas are produced, they have a high probability of escaping to the surface, but coal seams are destroyed only when they are uncovered by erosion. It has been estimated that most of the coal produced in the past 350 million years is still in existence.

The origin of coal in broad, shallow swamps and bogs produces coal seams that are widespread but thin. They appear sandwiched between strata of other sedimentary rocks, frequently in groups one above another. Because of this, coal seams are not hard to find, and it is likely that most of the world's major coal deposits are known at this time. For this reason, the major involvement of the coal scientist is not in finding new coal seams to develop, but in determining the value and properties of known deposits and in aiding the coal mining industry in carrying out its task with safety and efficiency.

2. Worldwide Distribution and Reserves

On a global basis, coal is the most abundant fossil fuel, and known recoverable reserves would last for over 200 years at present rates of consumption. Estimated resources are truly huge, and if it should become necessary and profitable to use these as well, present consumption rates could be maintained for more than 1,500 years.

The distribution of coal reserves is shown in Figure XII-7. The most striking feature of this distribution is the relative scarcity of coal in the continents of the southern hemisphere. Keeping in mind that the most favorable climate for coal-swamp formation is temperate to subtropical, turn to Figure III-27 on page 85 and look again at the geography of the Carboniferous world. The large coal deposits of Appalachia in the eastern United States, of England and Wales, and of Germany in Europe were all located very near the equator at that time. Now look at the maps for succeeding geological periods and note how these regions have moved from the tropics into the northern temperate zone, carrying their coal deposits with them. Indeed, this pattern of distribution of Paleozoic coal was used as evidence by Alfred Wegener in his reconstruction of the supercontinent that we call Pangea.

In the Carboniferous period, much of the African and South American portions of Gondwana were situated at high southern latitudes, and the rest of Gondwana covered the South Pole. In fact, Permian-Carboniferous times marked one of the prominent periods of

384

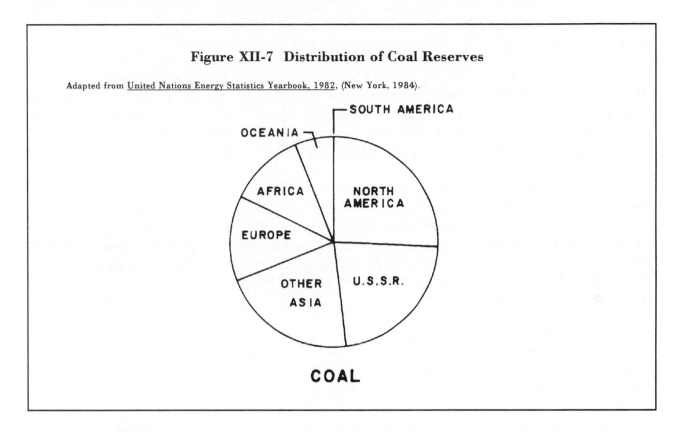

Figure XII-7 Distribution of Coal Reserves

Adapted from United Nations Energy Statistics Yearbook, 1982, (New York, 1984).

glaciation in world history, but with only the south polar region covered by continental ice sheets. The southern portions of South America and Africa were thus covered in ice and devoid of vegetation.

It is fascinating to speculate that the emergence of the Industrial Revolution in the northern hemisphere and the present asymmetric division between the developed and affluent northern nations and the less-developed and generally poorer southern nations, may have been influenced by the geography of the Carboniferous world that existed 300 million years ago.

3. Environmental Problems Associated with Coal Use

Bituminous coal constitutes an obvious source of energy that could last for the next century or more. Unfortunately, there are severe problems associated with increased dependence on the burning of coal. We have already investigated the climatic implications of increasing the carbon dioxide concentration of the atmosphere in the unit **Climates of Earth.** The fossil fuel reserves of the world constitute an enormous store of carbon that is safely locked away in the rocks below ground. Burning them combines carbon with oxygen to form carbon dioxide that is released to the atmosphere. Even at present rates of fossil fuel consumption, the carbon dioxide level of the atmosphere is increasing at rates that show that the natural processes that would normally dissolve excess carbon dioxide in the waters of the oceans cannot maintain equilibrium.

The carbon dioxide problem is common to all the fossil fuels, of course, and so as oil and natural gas resources are used up, simply replacing their usage with coal burning may not aggravate the situation. If we are to avoid adding to the buildup of CO_2 in the atmosphere, we must avoid an increase in the total production of energy from fossil fuels.

A common contaminant of bituminous coal is sulfur, and when coal is burned it is released into the environment as sulfur oxides (SO_2 and SO_3), which are major air pollutants. Once in the atmosphere, the sulfur oxides (along with nitrogen oxides from other sources) may react with water to form acids. Many observers believe this to be a contributor to the phenomenon of acid rain, which seems to be particularly prevalent in regions downwind of concentrations of coal-fired power generation plants. In some of these places rainfall has been measured to have an acidity greater than that of vinegar, and damaging acidification of lakes and soil has been blamed on coal burning. At the present time, the relative importance of the different contributors to acid rain is not known with certainty.

Air-pollution devices installed in power plants can cut back substantially on sulfur emissions (but not on carbon dioxide emissions). These devices are costly to operate and maintain, and they produce a waste-disposal problem in the sludges that remain from the process of scrubbing the sulfur gases from smokestack emissions. Taken together with the fact that underground mining of coal is dangerous and unhealthy, these problems and the regulations that have resulted from attempts to mitigate them have worked to hold coal consumption down in recent decades. Perhaps even more important as a constraint on coal production has been the relatively low price of oil and gas, which are competing sources of energy. Nonetheless, coal production in the United States in 1984 reached an all-time high, and the sheer abundance of coal will make it more attractive as time goes on and the limited reserves of oil and gas become more depleted.

STUDY QUESTIONS

XII-9. What are the principal differences between the environments in which coal and oil form?

XII-10. What is a common factor in sedimentary environments that are capable of producing either coal or oil?

XII-11. Why are world coal reserves so much greater than world oil reserves?

XII-12. Why are continents like South America and Africa deficient in coal reserves?

XII-13. What are two environmentally-troubling substances released in the burning of coal? Which one cannot be removed by pollution control equipment?

F. NUCLEAR FUELS

Although most uranium ore concentrations occur in a variety of sedimentary rocks, there is one important type of uranium deposit also occurring in igneous rocks. Deposits in Ontario, Canada and in Namibia in Africa are examples. As mentioned in the previous unit, there is still some controversy over the exact concentration mechanism, but it is thought that some are deposited at the same time as the enclosing sediments and some are deposited later by circulating uranium-bearing groundwater.

Figure XII-8 depicts the distribution of uranium reserves, recoverable at current prices. North America holds the greatest reserves, with significant additional reserves in Europe, Africa, and Australia.

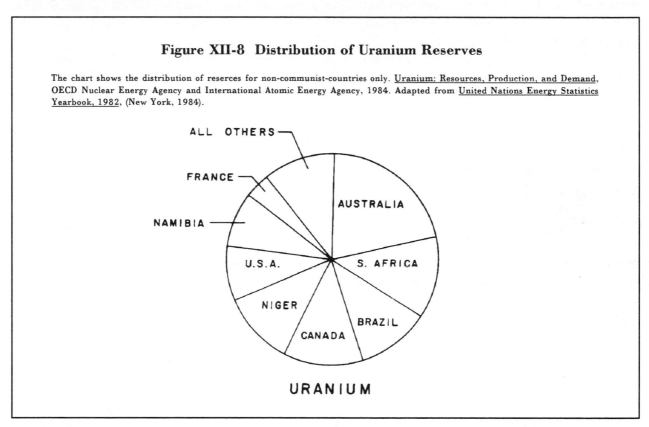

Figure XII-8 Distribution of Uranium Reserves

The chart shows the distribution of reserces for non-communist-countries only. Uranium: Resources, Production, and Demand, OECD Nuclear Energy Agency and International Atomic Energy Agency, 1984. Adapted from United Nations Energy Statistics Yearbook, 1982, (New York, 1984).

URANIUM

Once separated from its ore, uranium consists of three isotopes: U-238, which accounts for 99.3%, U-235, which accounts for only 0.7%, and U-234, which is very short-lived and accounts for less than 0.006%. U-235 is fissionable and hence is usable in nuclear reactors, while U-238 is not. Before being used as fuel, uranium is processed to enrich its U-235 concentration. It is then made into pellets, ready for use in reactors.

While the fuel pellets are in a light-water reactor, the type commonly in use today, the U-235 nuclei are bombarded with neutrons that cause fissioning, in which the U-235 nuclei are split into two smaller nuclei, releasing additional neutrons and a large amount of energy. The neutrons just produced can then strike other U-235 nuclei, causing them to fission in a chain reaction that can be controlled and sustained. The heat energy released is used to create steam that drives turbines attached to electric generators.

Because U-235 is such a small part of naturally occurring uranium, most of the uranium goes to waste, severely limiting the total amount of energy that can be derived from existing uranium reserves. However, U-238 can be converted into Plutonium-239 by absorption of neutrons. Plutonium-239 is fissionable and so may also be used as a fuel in reactors.

This circumstance has led to the development of the breeder reactor. In this reactor, U-235 or Plutonium-239 is fissioned in the usual way, but some of the neutrons produced are used to convert U-238 to Plutonium-239. With proper design, the breeder reactor using Plutonium-239 as fuel will convert 130 U-238 nuclei into Plutonium-239, while using up only 100 Plutonium-239 nuclei. This opens up the possibility for converting much of the existing U-238 reserves into fissionable Plutonium-239.

The tremendous multiplying effect that the breeder reactor would bring about means that existing reserves of uranium in the United States could supply all electricity needed by the country for several hundred years. In fact, the U-238 presently stored in uranium waste stockpiles could meet electrical needs for nearly a century.

In spite of such promise, the nuclear industry in the United States is in a state of severe depression. In the wake of the Three-Mile Island accident in Pennsylvania in 1979, public confidence in the safety of nuclear power plants plunged. Five years after the accident, extremely expensive cleanup operations were still in progress. While nuclear reactors are incapable of exploding like a nuclear bomb, a severe accident resulting in a total meltdown of the fuel assembly could release dangerous amounts of radiaoactivity into the surrounding countryside. Public fear of accidents and possible terrorist attacks on nuclear plants have combined with extremely high construction costs and delays caused by litigation to create an unfavorable environment for the nuclear power industry at the present time. As a result, new orders for nuclear power plants in the United States have effectively ceased during the past five years, though new plant construction continues in other countries.

Many scientists feel that the hazards of nuclear power generation have been exaggerated to the point of hysteria in the popular press. They point out that the safety record of the industry in general has been excellent and that reliance instead on coal burning carries with it far greater hazards in mining accidents and black-lung deaths, and in environmental problems such as air pollution, water pollution, and CO_2 generation. Others feel that problems unique to the nuclear industry make it undesirable for major expansion. Of considerable concern is possible proliferation of nuclear weapons as a result of reprocessing of spent fuel to recover weapons-grade plutonium.

The disposal of radioactive wastes that have already accumulated or that will result from future reactor operation is a major problem that has not yet been solved. In the past, wastes have been stored in tanks and barrels below ground, encased in concrete and dumped into the sea, or stockpiled above ground. The question of how to safely dispose of these wastes without contaminating the environment has been given a great deal of study in recent years, and experimental sites are now being chosen in the United States for this purpose.

Among the most likely disposal sites are salt formations and stable igneous rocks deep underground. Salt formations have the advantage that groundwater does not move through them rapidly (if it did, they would have dissolved a long time ago), but salt is mobile and the effects of heat generated by the stored wastes on the salt are not fully understood. Igneous rocks are very stable, but it is hard to guarantee that groundwater would not invade the storage area during the hundreds or even thousands of years that radioactivity in the wastes would persist. A major related problem is the design of waste containers that will not degrade over long periods of time.

Oceanic sediments far from spreading ridges or other tectonic zones on the sea floor contain very little circulating water, and it has been suggested that burial in these sediments would be safe, especially in areas of rapid sediment accumulation. An interesting suggestion has been made to bury radioactive wastes in deep ocean trenches where they eventually would be subducted. In both of these schemes, though, it is very hard to guarantee that the waste containers will remain intact until they are sufficiently buried, or that disturbances to the sea-floor sediments will not occur that might expose the wastes to circulating ocean waters.

At this point in time, it is very hard to separate the scientific from the social and economic difficulties that now beset the further development of nuclear power. In a democratic society, the ultimate decision must be made by the citizenry, who must take the time and trouble to listen carefully to arguments on both sides and then come to a reasoned conclusion.

STUDY QUESTIONS

XII-14. On what continent are the largest reserves of uranium found?

XII-15. Newspaper reports often say that breeder reactors "create more nuclear fuel than they burn". Does this mean that the breeder would produce an unending supply of Plutonium-239?

G. RENEWABLE ENERGY SOURCES

With recognition that oil and gas supplies are finite, increasing attention is going to sources of energy that are renewable in the sense that they can be used without exhausting the source of the energy. Among these are hydropower, geothermal, and solar energy in its various forms.

1. Hydropower

The traditional forms of hydropower make use of the energy of the Sun that goes to power the hydrologic cycle. Low-lying water from the sea is evaporated, carried to high elevations in the atmosphere, and then falls as rain onto the continental highlands. Energy is released by the water as it makes its way to lower elevations and, ultimately, back to the sea. Hydropower is generally utilized by damming rivers and streams and using the energy of falling water to turn electric generator turbines. This is a well-established technology that has been in use throughout the past century and indeed, water power has been used directly, without the intermediary of electric power generation, for thousands of years.

The potential for expansion of hydropower in the industrialized countries is limited, since most of the best sites have already been developed. In energy-poor continents of South America and Africa, however, hydroelectric projects may be able to provide substantial amounts of economical electrical energy. However, even if all potential sites were developed, hydropower would supply no more than 13% of current world energy demands. In addition, dams and reservoirs eventually fill with sediment, losing their storage capacity and becoming useless after lifetimes that extend typically for 50 to 100 or 200 years. Finally, dams alter streamflow and destroy habitats, landscape, and agricultural land along rivers and streams.

Another form of hydropower uses tidal energy instead of solar energy. In some coastal areas, tides are exceptionally large in amplitude, with the sea surface rising and falling five to ten meters (16 to 33 ft) twice each day. Two small tidal power generating plants have been constructed in France and in the U.S.S.R., but because in most places tides are much lower in amplitude, tidal power is unlikely ever to be more than a useful local option.

2. Geothermal Energy

A promising source of energy is produced by the Earth itself, in the slow decay of radioactive isotopes within the body of the planet. As such, geothermal energy is a natural form of nuclear energy. It can be exploited in places where hot rocks are found relatively close to the surface. These are generally found in regions of Quaternary volcanism.

Natural hot springs have been used for thousands of years to provide free sources of heated water for bathing, space heating, and even cooking, and a geothermal power plant has been operating in Italy since 1904. Today, some two dozen geothermal power plants are in operation, with about 4,000 megawatts of electric installed capacity. This is roughly equivalent to the output of four large coal-fired power plants. These make use of natural hydrothermal convection systems in which heat from hot rock at depth is conveyed to the surface by naturally circulating groundwater that emerges at the surface as steam or as hot water. Steam can be used to turn electric generator turbines, while hot water can be used for low-grade purposes such as providing inexpensive space heating for home use and commercial applications such as greenhouses.

In Iceland, sitting astride the volcanically active Mid-Atlantic Ridge, nearly 80% of all homes are heated geothermally and large greenhouses heated with geothermal hot water and steam provide nearly all the island's fresh vegetables. A television sequence will take you to New Zealand, situated on a major plate boundary in the South Pacific, where geothermal energy has been utilized for power generation for more than a quarter century.

At present, geothermal power has been developed from naturally existing hydrothermal convection systems. An experimental approach is to sink a well into a fractured dry hot-rock zone and pump water down to where steam can be produced. Returning to the surface via the same or another well, the steam can then be used to generate electricity. In order to make the rock sufficiently permeable to water, small conventional or nuclear explosions might be used to fracture the rock at depth. There are numerous difficulties with this scheme, and to date there are no commercial geothermal plants using it. The potential for future application may be high, however, since it would open extensive regions for development that have hot rock at depth but poorly developed or absent natural hydrothermal convection.

For the present, localities on plate boundaries or near hot spots offer the best hope for geothermal development. The Philippines, Mexico, Japan, and Italy are rapidly developing their potential. It has been estimated that by the year 2000, California could derive as much as 25% of its total electrical supply from geothermal energy. The Geysers area of northern California is presently the world's largest geothermal development, with about 1,300 megawatts of electric installed capacity.

3. Solar Energy

Among the renewable energy sources, solar energy has particular appeal because it is probably the most environmentally benign. A substantial technology has already developed, but actual employment has been hampered by inefficiencies and high initial costs.

The Greenhouse Effect, this time using actual glass instead of gases in the atmosphere, is the basis for operation of passive solar heating systems, so-called because they have no moving parts. Large, south-facing windows allow sunlight to enter an enclosed space where suitable surfaces are heated by sunlight. Heat so produced is stored in walls, floors, or water containers and is then circulated to the rest of the space to be heated. Many houses in sunny climes can derive all of their heating requirements from such a system, but in cloudier regions a supplemental heat source may be required. Careful attention must be paid to insulation and reducing air leaks in solar-heated homes.

Active solar heating uses a collecting system, usually panels mounted on roofs or in fields. A circulating fluid is heated by sunlight within the panels and heat is transferred by it to a storage tank. The system is well-suited to production of hot water for domestic use, and millions of these units are in use throughout the world. Note in Figure XII-1 on page 373 that hot-water heating accounts for a significant part of total energy use in the United States.

Solar energy use for space heating will almost certainly increase dramatically as oil and gas supplies become more scarce and expensive, but the future for large-scale power generation using solar energy is less clear. Steam may be produced by collecting sunlight over a large area with mirrors and focusing it on a small boiler, but the costs of construction and maintenance for this type of installation compared to its power production are still very high. In addition, energy produced during the day must be stored for use at night, and electrical energy storage technology is extremely expensive at present.

Direct conversion of sunlight into electricity is possible using photovoltaic solar cells, but at the present time these are inefficient and expensive to manufacture. New technological developments in this area promise to increase efficiency dramatically, but a fundamental limitation is the rate of energy input from the Sun to a given area of the Earth's surface. As an example, even if a 30% energy conversion efficiency might be obtained, it would still be necessary to cover eight square kilometers (2,000 acres) of ground in order to produce as much energy as a single modern nuclear power plant. Solar energy may best be used in small, distributed power systems provided that technological advances can increase efficiencies and lower costs. If this should occur, solar energy may prove to be a boon especially in underdeveloped tropical and subtropical countries that lack other resources.

Indirect solar energy is being explored in a variety of ways. Wind energy is being developed using new efficient designs for windmills in regions of persistent air currents. The same problems of efficiency and cost are the principal drawbacks to more widespread use.

The burning of wood has remained a prime energy source within undeveloped regions, but in all too many cases, the results have been disastrous in terms of deforestation and its resultant effects that will be explored in the final unit of this course. Utilization of other forms of biomass (literally all organic matter contained within plants and animals) for energy production is another field of active research, particularly in the production of synthetic fuels. Certain kinds of desert shrubs and even sunflower seeds may eventually be used to provide synthetic oil.

A more speculative use of indirect solar energy employs temperature differences between the different layers of the ocean. Temperature differences across the thermocline between warm surface currents and colder intermediate or bottom waters may amount to as much as 22°C (40°F), and large floating power plants have been envisioned that would use this oceanic thermal gradient to produce electric power.

Finally, a proposal has been made to establish large satellites in Earth orbit that would collect solar energy unhindered by clouds or atmosphere, convert it directly to electrical energy and beam it to Earth receivers via microwaves. Huge costs and low efficiencies are the principal difficulties with this scheme at present.

STUDY QUESTIONS

XII-16. What form of hydropower does not ultimately draw on radiant power from the Sun?

XII-17. What kind of geological setting is necessary at the present time for the development of geothermal energy?

XII-18. What is the distinction between passive and active solar heating?

XII-19. Why are there not more solar heating systems in use in the United States today?

H. PROSPECTS FOR THE FUTURE

We have already raised the question of how long our present supplies of fossil fuels will last. To understand the kinds of problems that we face, it is necessary to understand the manner in which finite resources are used up. Look back at Figure XI-6 on page 366, in which a rough model was established for the production of metal from a limited resource. The same model may be applied to oil produced within the United States, simply replacing the words "metal" and "mines" with "oil" and "oil wells".

Before the peak of production is reached, the number of oil wells drilled annually increases rapidly to meet the mounting demand. At about the time that the resource is half gone, the peak production rate is reached, and production declines thereafter. The reason for this decline is simply that the easy deposits to find and extract are the ones that are first exploited. As time goes on, smaller and less economical deposits are being developed. In the case of oil, the average depth of drilling has increased steadily, greatly adding to increased costs. Exploration costs are also increasing rapidly as smaller and more deeply buried deposits are sought. The result is that fewer wells are drilled and production falls. If the resource is still abundant elsewhere and domestic demand continues, the percentage of the resource that is imported will rise.

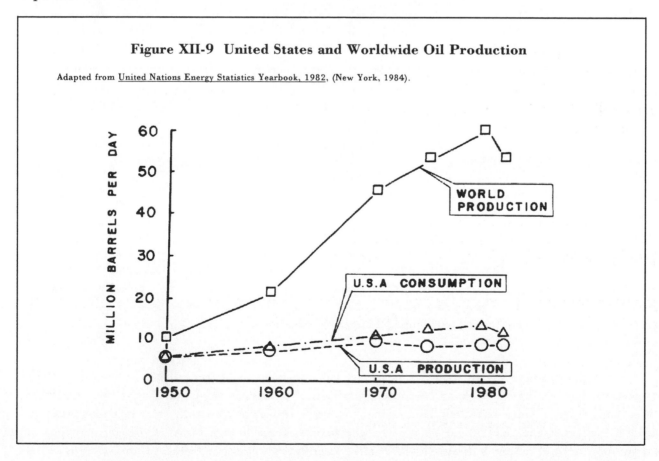

Figure XII-9 United States and Worldwide Oil Production

Adapted from United Nations Energy Statistics Yearbook, 1982, (New York, 1984).

Figure XII-9 shows world and United States oil production over a thirty-year period. Note that United States production peaked in about 1970, in good accord with predictions that had been made a decade earlier on the basis of estimates of the recoverable domestic oil resource. In 1970, it was felt, the United States would have used approximately half of its supply of recoverable oil. More recent estimates have not revised this figure significantly. In spite of an intense exploration effort, domestic production has declined somewhat since then, and until about 1980, imports of foreign oil grew. Economic downturn and the shift to smaller and more fuel-efficient automobiles has produced a lessened demand, as shown by the curve of United States consumption of oil. In addition, the number of wells drilled recently has increased again as oil companies have intensified their search for domestic oil. Importation of oil has decreased as a result.

The recent downturn of world oil production is due to economic factors rather than a peak reached due to passing the halfway mark in resource supplies. If the world economy should improve and the upward trend in production is resumed, it is widely felt that the natural peak in world production will be reached during the 1990's, and that total production will begin to fall irreversibly. The picture for natural gas supplies is not so clear, and world supplies of it may last longer. Even so, we can see that the oil and gas burning age in which we live will be a very short episode in human history.

Because of this situation, pressure is growing to switch to alternative energy sources such as the ones that we have already discussed. At the present time, many of our energy needs in the developed countries are being met by the elaborate electrical power distribution system. The only immediate replacements that are available for oil and gas use within this system are nuclear and coal, both of which have their attendant difficulties. Energy for transportation is an even more difficult situation, because no alternative technology to the gasoline or diesel engine exists in an acceptable form. Switching from gasoline to alcohol produced from fermented biomass has been suggested as perhaps the least disruptive alternative, but many observers are concerned that fuel needs will compete with food production for valuable farmlands, decreasing food supplies to an increasingly hungry world population.

The problems of finding sufficient energy resources to fuel the needs of a modern industrialized society are legion, especially as the populations of the developing nations improve their standards of living, with an attendant increase in per capita consumption of energy. A glance at Figure XII-4 on page 376 shows that it may be impossible for the rest of the world to increase their energy consumption to that found today in the United States, given the present energy supplies and the environmental problems that are already becoming manifest. But who would deny the people of the Third World the right to as high a standard of living as that enjoyed in the developed countries? And what implications does this have for world stability?

A development with the potential to substantially alter this scenario would be the successful harnessing of thermonuclear fusion. This, of course, is the process that fuels the Sun. We shall defer a more detailed discussion of it until the unit **The Sun**, but you may wish to take a look at Figure XIII-3 on page 404. The process fuses four protons (hydrogren nuclei) to form one helium nucleus along with a large amount of energy. Rather than starting with hydrogen atoms, the process could use deuterium (heavy hydrogen) as a fuel. Deuterium is a minor but recoverable constituent of seawater, present in such huge total quantities that world energy needs could be supplied for perhaps a million years at present rates of use.

Unfortunately, the technology of fusion is still in a very primitive state and, unlike fission reactors currently in use, nothing approaching a commercially feasible fusion reactor has yet been produced. Even if good progress continues to be made in this line of research, it is widely assumed that commercial-scale fusion reactors would not become available for another 30 to 40 years. In the meantime, we will have to use existing alternatives.

Certainly one means of stretching our energy resources lies in more conservative use of what we already have. Upgrading the fuel efficiency of American automobiles and reduction in oil use because of increased prices have had a dramatic effect on oil consumption in the United States in a remarkably short period of time. Effecting even greater energy savings will require incentives and an increased awareness on the part of the public of the many ways in which energy is wasted or is used in an inefficient manner.

STUDY QUESTIONS

XII-20. Why is it that energy production from fossil fuels will begin to decrease long before the reserves are actually exhausted?

XII-21. Where would we obtain the supplies of deuterium that might be used as a fuel for thermonuclear fusion reactors?

RECOMMENDED READING

Richard C. Selley, Elements of Petroleum Geology, W.H. Freeman, 1985.

UNIT XIII   THE SOLAR SEA:
THE SUN

CORONA

SUNSPOT MINIMUM   **SUNSPOT MAXIMUM**

A. INTRODUCTION

1. Overview

In spite of the Sun's centrality to all life in the solar system, it is only in recent years that the details of its features and workings have become known to us. Sunspots were the first to be observed and recorded on a regular basis, and studies of the solar spectrum later told us of the composition of the Sun. More recently, our ability to carry instruments above the obscuring atmosphere has provided us with a new view of the Sun, extending observations into the ultraviolet and X-ray portions of the spectrum. The Sun is a great gaseous ball, fueled by nuclear reactions in its core that convert hydrogen into helium. The energy produced there travels by radiation and convection to its surface, where it is radiated into space as sunlight and as a flood of particles known as the solar wind, pouring out into the surrounding solar system. The visible parts of the Sun -- photosphere, chromosphere, and corona -- have been shaped by the solar magnetic field, an influence so pervasive that the Sun may truly be regarded as a magnetic star.

1A
63

---

2. Objectives

Upon completion of this unit you should be able to:

1. relate the use of Fraunhofer lines in the solar spectrum to determining the composition of the Sun

2. recognize the need to make observations from above the atmosphere in order to complete our picture of the Sun

3. relate the prodigious energy production by the Sun to nuclear processes that convert hydrogen into helium in the solar core

4. discuss the apparent deficit of neutrinos produced in the Sun and relate it to our understanding of nuclear processes

5. relate the structure of the Sun's interior to the physical processes of energy transport that carry energy from the core to the surface

6. describe the nature of granulation and spicules found in the photosphere and chromosphere of the Sun

7. describe the function and operation of coronagraphs, spectroheliographs, and magnetographs

1B
21

8. relate the study of oscillations of the solar surface to the internal constitution and rotation of the Sun

9. describe the nature and origin of the solar wind

10. describe the shape of the solar magnetic field and compare and contrast it to that of a simple dipole field

11. describe the nature of coronal holes and relate them to the production of the solar wind

12. describe the sunspot cycle and relate it to cyclic changes in the solar magnetic field

13. describe solar prominences and flares and relate their behavior to the solar magnetic field.

3. Key Terms and Concepts

sunspots

spectroscopy

Fraunhofer lines

emission and absorption lines

Skylab

spicules

transition zone

solar corona

coronagraph

spectroheliograph

ultraviolet and X-ray imaging

magnetograms

solar oscillations

differential rotation of the solar interior

solar wind

interactions between electrical conductors
    and magnetic fields

spiral shape of the solar magnetic field

coronal holes

umbra

penumbra

sunspot cycle

Maunder Minimum

polarities of sunspots

reversals of the Sun's magnetic field

solar prominences

thermonuclear fusion

deuterium

gamma ray

neutrinos and the neutrino deficit

solar core

radiation zone

convection zone

photosphere

chromosphere

granulation

solar flares

white-light flare

ions

Doppler Effect

4. Corresponding Video

The Sun powers our planet and stimulates myth. This program will help to bring scientific clarity to our centuries-old study of the Sun. It will explore historical sun study -- from the Egyptians and Mayans, Galileo and Newton, to Kitt Peak, home of one of the most sophisticated land-based telescopes in the world. Correlations between solar activity and drought in the western states is stimulating scientists to explore solar variability and its effects on weather and climate. Dramatic spacecraft views of the Sun will reveal new and previously unseen features of its surface. Scientists will also explore the mystery of sunspots and sunspot cycles including their strange disappearance in the 17th century.

B. HUMANITY AND THE SUN

The effect of the Sun on our lives is so pervasive that it is easy to overlook. It provides light, warmth, fuel, and life itself to our planet. The daily passage of the Sun governs our activities and our life rhythms.

Because of the tilt of Earth's axis, we have distinct seasons except in the tropical zones. Between the Tropics of Cancer and Capricorn, each day of the year is much like all others, with the Sun riding high into the sky each noon, and days and nights dividing the 24-hour allotment of time. As we travel closer to the poles, seasons become pronounced, with distinct winter and summer climates, and the length of the sunlit day varies accordingly.

Above the Arctic Circle (or below the Antarctic Circle) the seasons take on a dramatic form: in the summer the Sun never sets, and in the winter it never rises (see Figure VII-4 on page 222). The year is divided into one huge "day" and "night" separated by spring and autumn seasons of fluctuating light and dark. It is little wonder that Soldag -- Sun Day -- is a time of celebration for the people of Tromso in Norway, 200 miles north of the Arctic Circle. January 29th is the day that the Sun returns to Tromso after the long winter's night.

Annual rituals celebrating the return of the Sun are as old as recorded history and as young as collegiate spring fever. Worship of Sun gods has figured prominently in many world cultures, acknowledging our dependence on the star that stands in the center of the solar system.

In recent years, our understanding of the operation of that star has taken significant strides forward, but the beginnings of that understanding may be traced back to Galileo. His monumental curiosity led him to turn his new instruments on every heavenly object, and the observation of dark spots on the surface of the Sun caused him consternation. Almost 2,000 years earlier, the great Aristotle had taught that the Sun was perfect and unblemished, and his teachings were not easily overturned. Perhaps for this reason, Galileo hesitated in publishing his solar observations, and so priority of publication went to Johannes Fabricius, a German scientist who observed sunspots only a few months later and published his notes in 1611.

Actually, scattered records exist from Aristotle's time onward of observations of sunspots, though many of the early observers thought them to be objects passing between the Sun and Earth. Some of these earlier observations used the naked eye, while others used a camera obscura -- a pinhole camera. Light from the Sun passes through a pinhole in a sheet of opaque material and projects an image of the Sun's disk on the opposite wall of a darkened room.

This, by the way, is the only safe way for you to view the Sun. <u>Never look directly at the Sun, either with the naked eye, through a telescope, or through binoculars!</u> Permanent eye damage can result before you are aware what is happening.

After Galileo's time, sunspot observations became frequent. From about 1850 on, they were recorded systematically. But little more was learned of the Sun until the development of spectroscopy -- the study of the spectrum of wavelengths in light. In the unit **The Atmosphere** we studied the spectrum of electromagnetic radiation, noting that Earth's atmosphere is transparent to visible light from the Sun, and that radiation due to the Sun's surface temperature of 5700°C peaks at wavelengths in the visible spectrum. If we use a glass prism to separate sunlight into its individual colors or wavelengths, we would find that all the colors of the spectrum are present. Indeed, the colors of the rainbow are nothing more than a display of the solar spectrum caused by the prism-like action of water droplets.

In 1814 Joseph von Fraunhofer studied the solar spectrum using primitive spectroscopes and discovered the presence of dark lines superposed on the continuous colors that make up sunlight. He catalogued the wavelengths of the lines and labeled them with letters of the alphabet, but had no explanation for them. Figure XIII-1 shows some of the stronger lines that he found. The horizontal bar in the diagram represents sunlight passing through a narrow vertical slit that has been spread out into its spectrum of colors -- short wavelengths on the left and long wavelengths on the right. The dark Fraunhofer lines show as an absence of light of particular wavelengths against a background of a continuous spectrum of colors.

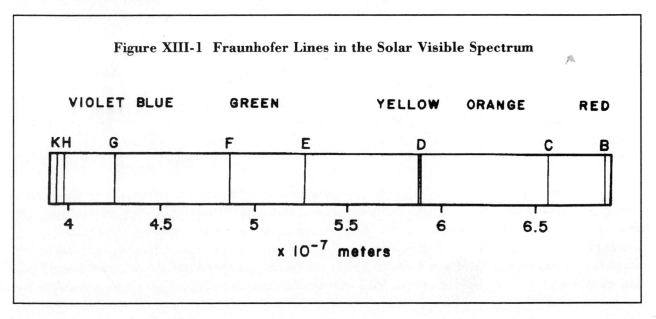

Figure XIII-1 Fraunhofer Lines in the Solar Visible Spectrum

VIOLET BLUE   GREEN   YELLOW   ORANGE   RED

KH   G   F   E   D   C   B

4   4.5   5   5.5   6   6.5

x 10⁻⁷ meters

Fifty years later the German chemist, Gustav Kirchoff provided an exciting explanation for the lines. He found that when chemical compounds were heated in a flame and the light given off was analyzed with a spectroscope, the light was not a continuous spectrum, but consisted of very narrow bright lines. Some of these lines were in the same positions as the Fraunhofer lines. On the other hand, he found that the light from a carbon arc lamp gave a continuous spectrum, without lines. In another test, he placed the flame-heated chemicals between the carbon arc lamp and the spectroscope and found that dark lines now appeared superposed on the continuous spectrum in the same positions that the bright lines previously had occupied. Kirchhoff recognized that the particular lines that appeared were characteristic of the chemical elements that vaporized in the hot flame, and that this method would enable him to determine the composition of the Sun.

The Fraunhofer lines arise from the nature of absorption and emission of light by atoms. A <u>photon</u> is a packet of radiation energy that has a particular wavelength. When a photon with an appropriate amount of energy is put into an atom, one of its electrons will absorb it and jump to what is called an <u>excited state</u>. This is a higher energy state than the <u>ground state</u> that the electron normally occupies. At some later time, the electron falls back down to the ground state, releasing a precise amount of energy in the form of another photon. The wavelength of the light emitted is determined by the energy difference between the excited state and the ground state. The higher the energy of a photon, the shorter its wavelength, and vice versa.

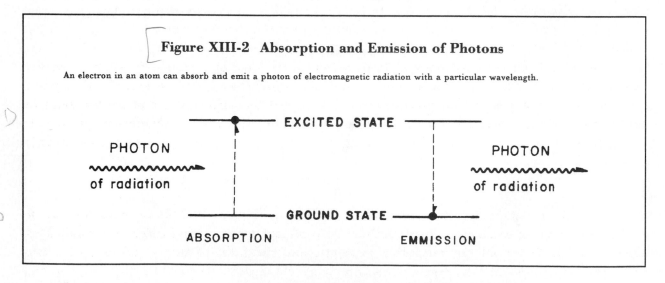

Figure XIII-2 Absorption and Emission of Photons

An electron in an atom can absorb and emit a photon of electromagnetic radiation with a particular wavelength.

Figure XIII-2 shows how this works. The energy difference between the excited state and the ground state of an electron depends upon the type of atom. On the left side of the diagram, an electron is initially in the ground state. An incoming photon is absorbed if it possesses just the right amount of energy to allow the electron to jump up to the excited state; if the incoming photon does not have the right energy, it is not absorbed. At some later time, usually very short, the electron pops back down to its ground state, most often giving up its energy as a photon identical to the one previously absorbed. The emitted photon may be sent out in quite a different direction from that of the one absorbed, however. In many cases, the electron may return to the ground state via one or more intermediate states, with the result that photons emitted have wavelengths different from that of the absorbed photon.

Now consider the behavior of a gas in the atmosphere of the Sun. Radiation with a wide range of visible wavelengths is being sent out toward Earth. A certain atom present in the upper layers of the Sun selectively absorbs just those wavelengths that possess the proper energy to allow its electron to jump to the excited state. When it jumps back to the ground state, the emitted photon is sent out in random directions, probably not toward Earth. As a result, a dark line appears in the solar spectrum, indicating that photons of this particular wavelength have been diverted from their journey to Earth.

More importantly, because each element absorbs wavelengths that are characteristic to itself, the dark lines may be used to positively identify which elements are present in the outer layers of the Sun. All elements have more than one excited state, and so produce more than one absorption line. As an example, the two closely spaced Fraunhofer D lines in Figure XIII-1 have wavelengths of precisely 5.88998×10^{-7} and 5.89594×10^{-7} meters and identify the presence of the element sodium. These are the only strong sodium lines in the visible spectrum, and this accounts for the strange visual effects that you may have noticed under the golden-orange light of sodium vapor lamps, sometimes used for street lighting. Virtually all color except for that of the lamp disappears, rendering everything monochromatic. The reason is that there is, in fact, only one color emitted by the lamp -- the yellow D lines of sodium.

Modern spectrographs permit the measurement of thousands of absorption lines in the solar spectrum. Which lines appear depend upon the temperature, density, and the abundance of elements present. Abundant elements produce especially dark lines, and so the relative darkness of each line may be used to calculate the relative amounts of each element present in the Sun. As a result, we know the composition of the Sun's outer layers to a considerable degree of accuracy.

The most abundant element in the Sun by far is hydrogen. It alone accounts for 92.1% of the total number of atoms present. Helium is in second place, with 7.8% of the total. This allows only 0.1% for all the remaining elements, and the bulk of that is taken up by two elements, oxygen and carbon. Nonetheless, nearly all other elements have been detected in the solar atmosphere, though many measure only in the parts per billion range.

The Sun is an immense gaseous body, made up almost entirely of the two lightest elements, hydrogen and helium. Helium, by the way, was discovered on the Sun before it was found on Earth. Its spectral lines showed its presence clearly, but no known element could be found whose absorption lines matched. Named in 1868 for the Greek word for the Sun, helios, helium was not discovered on Earth until seventeen years later.

STUDY QUESTIONS

XIII-1. If observations of sunspots were made from the time of Aristotle on, why do we credit Galileo with their discovery?

XIII-2. Which has more energy, a photon of yellow light or a photon of green light?

XIII-3. Why are the Fraunhofer lines called <u>absorption lines</u>?

XIII-4. How can the spectrograph be used to identify the composition of the solar atmosphere?

C. A NEW VIEW OF THE SUN

You may recall from the unit **The Atmosphere** that ultraviolet rays are absorbed in the Earth's atmosphere and for the most part do not reach the surface. X-rays have great penetration powers, but they too are absorbed in Earth's atmosphere. Spectrographic observations of the outermost region of the Sun -- the corona -- indicated that the temperature there must be very high, perhaps as much as two million degrees Celsius. Using Rule (1) in Chapter VII, we find that the peak wavelength emitted at such a temperature should be around 1.4 x 10^{-9} meter -- in the region of X-rays. It became clear that there were new views of the Sun to be had, in the X-ray and ultraviolet regions of the spectrum.

Attempts were made early in this century to carry cameras and spectrographs above the densest part of the atmosphere in balloons, but it was not until instruments were carried totally above the atmosphere in 1946 by V-2 rockets left over from the war, that significant ultraviolet and X-ray observations could be made. Later, orbiting scientific satellites extended these observations. It was Skylab, however, that provided us with the best and most spectacular views of the Sun in short wavelengths. Launched in 1973, Skylab was an orbiting space station that, during the next year, was visited by three teams of astronaut-scientists who spent a cumulative total of nearly six months in orbit. A battery of eight large telescopes were on board, including instruments designed for observations in the ultraviolet and X-ray regions of the spectrum.

The success of the Skylab mission exceeded all expectations. Many of the spectacular images that you will see on the television program corresponding to this unit were taken during this mission that at long last lifted solar astronomy above the murky atmosphere and gave it new short-wavelength "eyes". But before you can understand what it is that you are looking at, you need to learn something about the inner workings and structure of our star.

D. THE NUCLEAR FURNACE

As we have seen, the Sun is a large gas ball made up almost entirely of hydrogen and helium. It seems reasonable to assume that its source of energy must be derived from these two elements. The fusion of hydrogen atoms to form helium is almost certainly that source of energy, and is similar to the process that liberates vast amounts of energy in a hydrogen bomb explosion. In the Sun, however, the process occurs continuously and steadily and, because of the great distance involved and various shielding effects of Earth's atmosphere, benignly.

In thermonuclear fusion, four hydrogen nuclei, each consisting of a single proton, are fused in a series of steps to form one helium nucleus, which consists of two protons and two neutrons. The mass of the helium nucleus turns out to be some 0.7% lighter than the sum of the masses of the four hydrogen nuclei, and this is the source of the prodigious amount of energy released in the process. Einstein's famous equation, $E = mc^2$, comes into play here, giving the relation for the conversion of mass (m) to energy (E), with the speed of light (c) squared determining the proportionality between them. In the fusion process, the missing mass is converted into energy and sent out as electromagnetic radiation.

The first step in this process is to bring two protons together with sufficient energy so that they can fuse. The result is a nucleus that contains one proton and one neutron, called deuterium or heavy hydrogen and represented by the symbol H^2 (see Figure XIII-3). In this step the positive charge of the proton is carried off by a small particle called a positron. Additional energy is carried away by another small particle called a neutrino. Once this stage is reached, the deuterium readily absorbs another proton and yields a nucleus of helium that is deficient by one neutron, represented by the symbol He^3. More energy is carried off by a photon of electromagnetic radiation -- a gamma ray. The final step combines two He^3 nuclei to form one stable helium nucleus (He^4) with two leftover protons that are now free to participate in a repetition of the whole process.

Note in the diagram that six protons are consumed and two are released, for a net consumption of four protons. The difference in mass between the four protons and the resulting helium nucleus is converted almost totally into energy. Even the positron that is released will soon encounter an electron and the two will annihilate each other, releasing another gamma ray. Thus, in the process of nuclear fusion hydrogen is converted into helium. Excess energy is released in the form of neutrinos and high-energy photons of electromagnetic radiation.

The greatest impediment to this process is the fact that all of the particles that need to be brought together have positive charges. Two positively charged particles repel one another (like charges repel, unlike charges attract), and so the protons must be hurled together very forcefully for the fusion to take place. This can occur only at extremely high temperatures, on the order of ten million degrees Celsius. At such high temperatures, the velocities of individual particles become very high, and when the density is high enough, random collisions between them can initiate the fusion process. It is for this reason that a hydrogen bomb must be detonated by an ordinary uranium bomb -- only in this way can the necessary temperatures be reached. It is also the reason that the achievement of controlled thermonuclear fusion in a reactor for economic power generation so far has proved beyond practical application. Any

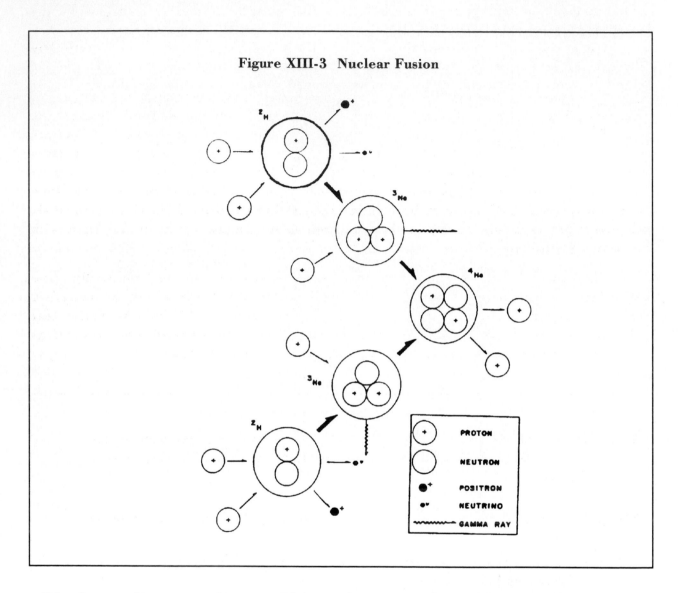

Figure XIII-3 Nuclear Fusion

solid substance known to science would instantly vaporize if it came into direct contact with temperatures of this order.

We have every reason to believe, however, that temperatures as high as 15 million degrees exist in the core of the Sun. We know the mass of the Sun from astronomical measurements, and the pressures that must exist at its center would produce such high temperatures by the familiar effect of heating with compression of a gas. And so it is in the core at the center of the Sun that we must look for the great nuclear furnace.

There is a significant problem with this scenario, however. The fusion process outlined above produces neutrinos along with radiation. The neutrino is a very strange particle: it has no electrical charge, it has a very small mass, and it travels nearly at the speed of light. This combination of properties results in a particle that hardly ever interacts with anything else. The vast majority of the neutrinos produced in the core of the Sun immediately escape, passing right through the immense mass of the Sun and spraying out in all directions into space. They pass, essentially unhindered, right through Earth -- and you. For every square centimeter of surface that you present to the Sun, there is a flux of some 70 billion neutrinos passing through you every second! The fact that they don't interact with our tissues means that they have no effect upon us -- fortunately. In fact, detecting their presence at all is extremely difficult.

One apparatus for the detection of neutrinos has been constructed one and a half kilometers (about one mile) underground in the Homestake Gold Mine in South Dakota. The depth of burial effectively screens out the effects of cosmic rays that would otherwise interfere with the experiment. A large tank contains 100,000 gallons of carbon tetrachloride (ordinary cleaning fluid). If a neutrino should happen to interact with a chlorine atom in the cleaning fluid, it would change it into an atom of radioactive argon, which then could be flushed from the tank periodically and measured by sensitive analytical instruments. Even so, only about six such events are expected per day in the tank, necessitating measurement of the argon with extraordinary sensitivity.

The experiment has been running now for more than a decade, and the results show that the neutrino flux is only about 30% of what is expected from the models of thermonuclear fusion. The explanation for this discrepancy is still unknown, nor is it known whether that explanation will be trivial or revolutionary in its impact on our understanding of the workings of the Sun. It remains one of the great outstanding questions in all of science.

STUDY QUESTIONS

XIII-5. In the hydrogen fusion process, four protons are needed to create one helium nucleus. But a helium nucleus has only two protons in it. What happened to the other two?

XIII-6. If nuclear fusion can release so much energy, why is it so hard to get started?

XIII-7. In the Homestake Gold Mine neutrino detector, how many argon atoms arc actually being produced each day?

E. JOURNEY THROUGH THE SUN

The Sun has been shining for at least 4,500 million years, slowly converting the abundant hydrogen into helium. The mass that is converted into energy is lost from the Sun and radiated into space. This lost mass alone amounts to some five million tons per second. Even so, the Sun is so huge that over its lifetime so far, it has managed to convert only about 4% of its available hydrogen into helium.

Energy produced in the core of the Sun must work its way to the surface by two different mechanisms. In the deep interior the gases, though dense, are fairly transparent to gamma and X-radiation and so the principal mechanism of heat transport is via radiation. Even so, the gamma rays produced in the fusion reaction are quickly absorbed by ions and

2A
? 341 ?

re-emitted as X-rays. These in turn are absorbed and re-emitted countless times by the process of Figure XIII-2 on page 400 as the energy slowly makes its way outward.

At a depth of about 200,000 km below the surface, the gases have cooled and become more opaque, giving rise to a zone in which convection is the dominant means of energy transport. Here the gases are in violent motion, churning in ever-changing cells and transporting the energy to the visible surface of the Sun. Even so, energy produced in the core probably requires anywhere from one to ten million years to make its way through the radiation and convection zones to the surface. From that point, it travels by direct radiation through the near-vacuum of space, requiring only eight minutes to complete its journey to Earth.

As we proceed outward from the thermonuclear core, the density of the gases drops dramatically. From a value of 160 times that of water at the center, solar density reaches that of water within the convection zone. By the time we arrive at the luminous surface of the Sun, the density is equivalent to that found in the Earth's atmosphere at an elevation of 50 km -- the top of the stratosphere. Throughout all of those portions of the Sun that are directly visible to us and our instruments, densities are such as can only be produced on the surface of Earth with vacuum pumps.

Figure XIII-4 diagrams the various divisions of the Sun. You should be aware, however, that these divisions do not indicate changes in the chemical composition of the Sun or in transitions in states of matter in the way that Earth is divided into crust, mantle, and core. The Sun is entirely gaseous. These zones denote different behavior, temperature, and density regions. It is unlikely that the composition of the Sun's gases changes drastically at any of the boundaries.

The visible parts of the Sun are contained in a narrow region shown in the blow-up at the top of the diagram. At the base is the photosphere -- the luminous surface that we see as the Sun's disk. The thickness of the photosphere is only about 400 km -- approximately the width of the state of Ohio. This is as far as we can see down into the convection zone. The dark portion of sunspots that are seen in visible light are found within the photosphere. In addition, a mottling of the photosphere is seen in many photographs, giving the appearance of the skin of an orange. This structure, called granulation, shifts constantly and is due to the presence of the convection cells below.

Above the photosphere is the chromosphere, in which the temperature drops to a minimum of about 4,000 degrees Celsius in its lower part and the density drops to values attainable only in the finest laboratory vacuum systems. The chromosphere appears pink or magenta when the photosphere is obscured by the Moon during a total eclipse of the Sun.

The upper boundary of the chromosphere is not smooth, but is distorted into sharp protruberances called spicules. Giving a somewhat furry appearance to the edge of the Sun, they are numerous and appear small only in comparison with the Sun's disk. In fact, they can extend to heights greater than Earth's diameter. Measurements have shown that spicules can grow to their full height and fade away in a space of only ten minutes, like waves that rise and fall on the surface of the ocean.

Separating the corona from the chromosphere is a very narrow layer called the transition zone. In this thin region, only a few tens of kilometers thick, the temperature takes a

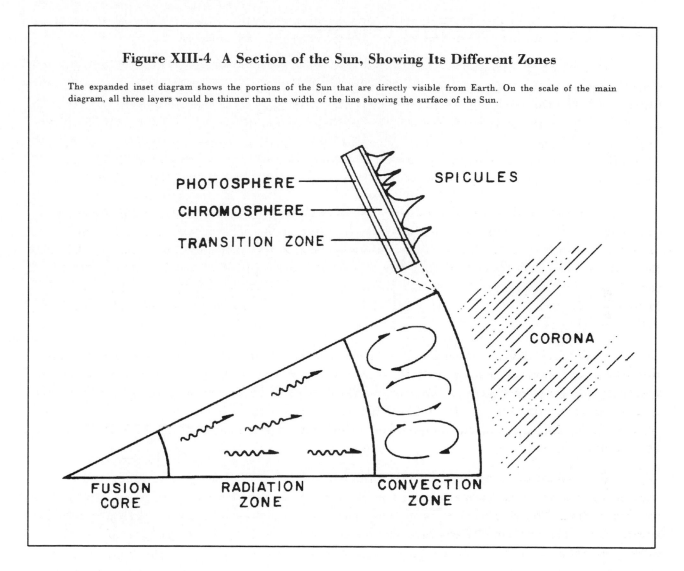

Figure XIII-4 A Section of the Sun, Showing Its Different Zones

The expanded inset diagram shows the portions of the Sun that are directly visible from Earth. On the scale of the main diagram, all three layers would be thinner than the width of the line showing the surface of the Sun.

tremendous jump, from 8,500 degrees in the upper chromosphere to half a million degrees or more at the base of the corona.

The corona (or "crown") presents a grand spectacle during a total eclipse of the Sun. Its opalescent glow is shaped into streamers that extend far out into space with a total brightness comparable to that of the full moon. Because of the extreme temperatures reached in the rarefied coronal gases -- up to a million degrees -- it emits largely in the X-ray portion of the spectrum. X-ray telescopes carried into space have provided remarkable views of the structure and behavior of the corona.

<u>STUDY QUESTIONS</u>

XIII-8. What is the coolest portion of the Sun?

XIII-9. On what basis is the Sun divided into different layers or regions?

F. TYPES OF SOLAR OBSERVATIONS

The desire to investigate in detail the wealth of phenomena present on the Sun has led to the development of a variety of specialized instruments, each designed to permit the observation of particular regions or features. One of them is the coronagraph, in which a disk is placed over the image of the Sun at the focus of a solar telescope. This cuts out the bright glare from the photosphere and permits observation of the corona, in effect producing an artificial solar eclipse. For ground-based telescopes, the instrument is unable to cut out the bright skylight of Earth's atmosphere, and so it is limited to observing the coronal structure near the Sun where it is bright. In space, however, there is no atmosphere and so the coronagraph sees the corona against the black of space and can trace it to great distances away from the Sun, routinely as far as five solar diameters.

The spectroheliograph is another instrument of considerable utility that combines an image-forming telescope with a spectroscope. With it, an image of the Sun can be formed using only the light that is emitted at a single wavelength. As an example of its use, as seen in an ordinary telescope, light from the chromosphere is completely overwhelmed by the much brighter light from the photosphere below it. However, the light given off by the chromosphere in the visible spectrum is dominated by a single spectral emission line -- the red hydrogen-alpha line at 6.563×10^{-7} meters. It is this spectral line that gives the chromosphere its characteristic color and name. Unlike the dark absorption lines, emission lines are bright and can be used to form images. If a spectroheliograph is set to look only at the hydrogen-alpha emission, the image of the Sun that it produces will show the chromosphere and ignore the photosphere below it.

Other emission lines are found to be characteristic of certain temperature and density ranges and so can form images in which bright areas represent the presence of material having these conditions. In this way, it is possible to use the spectroheliograph to look at particular layers or features within the chromosphere and photosphere, much as changing the focus of a set of binoculars allows one to see distant objects more closely.

Ultraviolet and X-ray imaging works in a very similar manner. Emissions in the ultraviolet region, for instance, comes largely from the hot, low density transition zone. Detailed views of the structure of that important zone came largely from ultraviolet views taken during the Skylab and subsequent missions. X-ray emission from the intensely hot corona makes it possible to view the corona in three dimensions, not only as seen in profile outside the disk of the Sun, but everywhere on the hemisphere facing Earth. In an X-ray view, the most active and hottest regions of the corona blaze like fiery coals against the dark disk of the Sun, whose surface is far too cool to produce such short-wavelength energetic radiation.

Another effect that may be observed with a spectrograph provided the key to unraveling one of the key influences on solar behavior. Magnetic fields have proved to be an intrinsic feature of sunspots, the structure of the corona, prominences, and flares, and a host of other solar phenomena. Magnetic fields, of course, are invisible, but they affect emission and absorption lines in a way that allows us to determine the strength and configuration of the magnetic field and its polarity. When atoms are subjected to strong magnetic fields, some of their spectral lines will split into two or more evenly-spaced lines whose separation is proportional to the strength of the magnetic field. Furthermore, the light making up the individual lines is polarized in such a way that the polarity or direction of the field may be determined as well. Using polarizing filters, similar in operation to those used in polarized sunglasses, light selected from the individual magnetically split lines may be used to construct images called <u>magnetograms</u> that effectively map the magnetic fields present on the Sun.

As we have seen before, advances in technology and instrumentation often lead to major breakthroughs in basic understanding. Those technological advances, of course, are based on previous generations of basic research, providing a symbiotic relation between science and technology that permits both to move forward, building upon each other's successes.

STUDY QUESTION

XIII-10. How can the spectroheliograph be used to form images of the chromosphere?

G. SOLAR OSCILLATIONS

Studies of the motions of granulations in the photosphere led to the discovery that the surface of the Sun is oscillating. The Doppler Effect applies to spectral lines, and permits detection of the motion of gases along the line of sight direction by measuring the shift of the lines toward longer or shorter wavelengths. Just as with a train whistle that is approaching, the spectral lines of gases that move toward us are shifted to higher frequency -- that is, shorter wavelengths. Those that are moving away from us are shifted to lower frequencies, or longer wavelengths. The "red shift" of receding distant galaxies may be familiar to you if you have done some reading in astronomy.

The Doppler Effect has been used to explore the motions associated with solar granulations, showing that the bright centers of the granules are associated with hot rising convection currents and the darker furrows or lanes separating them are caused by cooler descending currents. Using this technique, it was discovered some 25 years ago that superposed on the motion of the granules was an up-and-down motion of the whole Sun with a period of about five minutes and an amplitude of about 25 km (16 mi). Ten years later a mechanism was suggested that would provide an explanation for these oscillations. They were due to sound waves traveling below the surface of the Sun, alternately being refracted upwards by the increasing velocity of propagation at depth and being reflected downwards again by encounters with the solar surface, in effect, being trapped between two reflecting media.

You may recall from your studies of seismology in the unit **Continental Tectonics** that seismic waves in the Earth behave in a similar fashion, being continually bent upwards as they travel through the mantle and then reflecting from the surface to continue their travels. You may find it useful to look at the PPP or SSS waves shown in Figure IV-9 on page 116 as examples of waves similar to those causing solar oscillations.

The solar oscillations are not caused by sunquakes -- the Sun, after all, is a totally fluid body and cannot behave like brittle rock. Instead, the solar sound waves are generated in the

Within a closed region, it is much more difficult for particles to escape into space, because in order for them to do so, they must cross the looped magnetic field lines. On the other hand, in the open regions the particles are free to coast along the field lines and head outward with little impedance. These open regions are referred to as coronal holes, and at times when the number of sunspots is at a minimum, there is generally a coronal hole in each polar region of the Sun. When sunspots are present in large numbers, however, the field of the Sun is greatly complicated, and coronal holes may occur anywhere, taking on complex shapes. When a coronal hole occurs near the Sun's equator, it funnels the solar wind outward and is easily detected at Earth as it sweeps around the face of the Sun.

STUDY QUESTIONS

XIII-12. Why is the solar wind more intense above coronal holes?

XIII-13. Why is the magnetic field of the Sun spiral in shape when viewed from above the plane of the solar system?

I. SUNSPOTS

Sunspots are the most easily observed transitory features on the Sun. A long-focus telescope can be made to project a sizeable image of the solar disk on a screen. This image may then be traced onto a sheet of paper, recording the number and shapes of spots that are visible at that time. They range considerably in size, from tiny, barely visible pores to fully developed spots that are several times larger than the Earth.

The visible structure of a sunspot consists of a central dark region called the umbra surrounded by a mottled or fibrous appearing ring called the penumbra. The overall appearance is remarkably like that of a human eye, with the umbra representing the pupil and the penumbra the iris. Measurements have shown that the umbra is a cooler, slightly depressed region, several hundred kilometers below the surrounding photospheric surface.

The number of sunspots has been shown to fluctuate with a cycle lasting about eleven years. Figure XIII-6 shows a compilation of sunspot observations spanning the past 270 years. During a typical cycle, the number of spots visible at one time ranges from zero at sunspot minimum to a hundred or more at sunspot maximum. Note that from about 1650 to 1700 very few spots were observed on the Sun and it would appear that the usual sunspot cycle was suppressed. This period has been called the Maunder Minimum after the British astronomer who wrote about it in the late 1800's. You may recall from earlier units that this was also the period of the Little Ice Age, and the coincidence has spurred speculation that this climatically cooler period was spawned by changes in solar activity. We shall return to this idea in the next unit.

Most sunspots are found in a band within 30° of the solar equator. Near the equator, they, along with the entire solar atmosphere, rotate once around the Sun in about 27 days, but those found at higher latitudes take a longer time, ranging up to 37 days near the poles. We

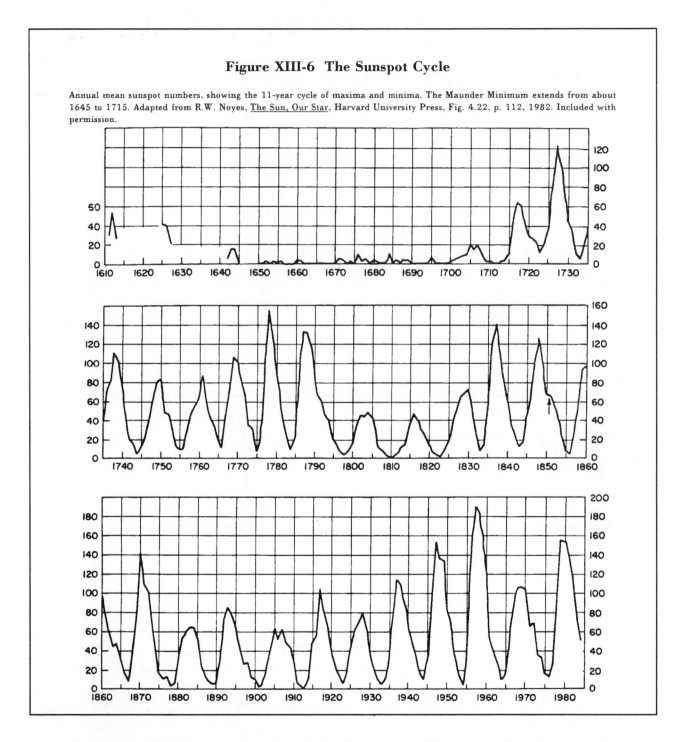

Figure XIII-6 The Sunspot Cycle

Annual mean sunspot numbers, showing the 11-year cycle of maxima and minima. The Maunder Minimum extends from about 1645 to 1715. Adapted from R.W. Noyes, <u>The Sun, Our Star</u>, Harvard University Press, Fig. 4.22, p. 112, 1982. Included with permission.

have already seen that the interior parts of the Sun do not all rotate at the same rate; the same is true of the surface as well.

An extremely important observation was made when the spectroscope was turned on sunspots. It was found that spectral lines showed a splitting into multiple lines that is characteristic of strong magnetic fields within the spots. These fields can attain values on the order of 1,000 times the strength of the Earth's magnetic field -- comparable to the magnetic fields produced by large electromagnets. This is also many times the average fields found elsewhere on the Sun. Sunspots, then, represent tight bundles of magnetic field lines emerging from beneath the solar surface. They are cooler than surrounding material because the gases within the magnetic bundles are confined there by the magnetic lines of force, and so convection is locally suppressed. As a result, the material making up a sunspot does not conduct heat from the interior to the surface as readily and the surface of the spot is locally cooler.

Sunspots are often (but not always) found in groups, with two large spots dominating the group. Magnetograms show that the leading spot has the opposite magnetic polarity from the trailing spot -- just as though two powerful bar magnets had been placed just below the photosphere, one with its North Pole at one spot and the other with its South Pole at the other. In contrast to the Earth's regular dipole field, with a North Magnetic Pole at one end and a South Magnetic Pole at the other, at sunspot maximum the Sun can possess dozens or even hundreds of north and south magnetic poles, as localized pairs.

There are further magnetic peculiarities on the Sun. In the northern hemisphere, the leading spots of a pair typically have a north polarity and all the trailing spots have a south polarity, but in the southern hemisphere the situation is reversed: the leading spots will have a south polarity. If we then wait through sunspot miminum until the next maximum, we would find that the entire situation has reversed: now the leading spots in the northern hemisphere have a south polarity while the leading spots in the southern hemisphere have a north polarity. In fact, the 11-year cycle is really a 22-year cycle, because it takes 22 years for the process to repeat itself from the same starting point.

Figure XIII-7 shows a model that may explain the sunspot cycle along with the peculiar magnetic behavior associated with it. We begin at the time of sunspot minimum in (a), with the solar magnetic field being essentially dipolar, as shown in Figure XIII-5a. The magnetic lines of force continue into the Sun's interior, where they are constrained to follow the motion of the ionized gases. Because of differential rotation, with the interior rotating more rapidly than the surface, the lines of force inside the Sun are distorted in such a way that they are stretched out and start to wrap around the Sun (b). This can continue (c), wrapping the lines of force several times around the Sun and stretching them out into thin, tight bundles. Where the bundles break the surface (d), the lines of force emerge with one polarity, loop around, and reenter the surface with the opposite polarity, forming two spots of opposite polarity. Note that the leading spot (the ones on the right) have opposite polarities in opposite hemispheres, as is observed.

The exact cause of the reversal of the solar field is not known, but it has been observed that as sunspot groups dissipate, the following spot tends to drift toward the poles, leaving the leading spot behind. The following spots have the opposite polarity from the polar regions, however, and tend to cancel out the prevailing field and initiate a field reversal.

So we find that the solar field, though far more complex than that of Earth, also reverses its polarity. The Earth's field reverses polarity on the order of once or a few times per million years, with no clear repeating pattern, while the solar field reverses itself in a fairly regular pattern every 11 years or so.

Figure XIII-7 A Possible Model of the Sunspot Cycle

Adapted R.W. Noyes, The Sun, Our Star, Harvard University Press, Fig. 4.22, p. 112, 1982. Included with permission.

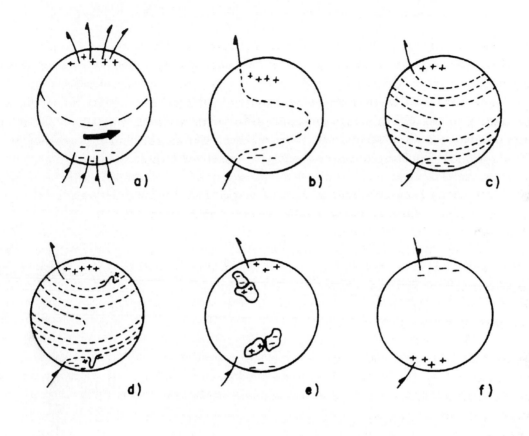

STUDY QUESTIONS

XIII-14. Are sunspots hotter or cooler than the surrounding photosphere?

XIII-15. Why are they hotter or cooler?

XIII-16. What is the difference between the 11-year and the 22-year sunspot cycles?

J. PROMINENCES AND FLARES

Among the television views of the Sun that you will see, some of the most spectacular are those of solar prominences -- bright arches and streamers extending into the corona far beyond the Sun's disk, sometimes being blown outward from the Sun at tremendous speeds. The arched or curved shapes of prominences give a strong hint that the solar magnetic field is once again involved.

It may surprise you to learn that the prominences are much cooler than the surrounding corona. While the corona reaches 1,000,000°C or more, the prominences are only about 10,000°C -- not much hotter than the photosphere. Prominences can last for weeks or more, demonstrating how cool ionized gases trapped within magnetic lines of force can be refrigerated and kept separate from the surrounding much hotter coronal gases. The prominences are also much denser than the corona, and are apparently suspended in space by the magnetic fields. Just how that material gets into the corona is not understood, nor is the mechanism that occasionally causes the magnetic field to act like a spring and eject prominences into surrounding space at speeds that sometimes exceed 200 km/sec (450,000 mph).

This kind of magnetic violence is often associated with the occurrence of solar flares. Flares are among the most remarkable, and most intensely studied, phenomena in the solar system. When a flare occurs, a prodigious amount of energy is released, sometimes attaining levels that would be equivalent to the simultaneous explosion of a thousand million one-megaton hydrogen bombs! Originating in the corona, flares may be accompanied by temperatures soaring to as high as a hundred million degrees -- considerably hotter than the center of the Sun. At temperatures so high, nuclear reactions can occur in the tenuous gases of the flare, and gamma rays produced in such reactions have been detected. The densities are so much lower, however, that flares contribute only negligibly to the total energy produced by the Sun.

Solar flares appear to be a kind of magnetic explosion, in which energy stored in the magnetic fields linking active regions on the solar surface is suddenly released. Flares are usually associated with sunspot groups that have both polarities present. Perhaps the magnetic field lines connecting the opposite polarity regions become twisted and stretched by the convective churning of the surface layers, storing magnetic energy in the field loops that extend into the corona above. At some point the magnetic field configuration becomes unstable and reconnects its lines of force into a simpler and more stable configuration. The immense energy that had been stored in the previous configuration of magnetic field lines is released as thermal energy into the corona, producing a flare event.

Even though flares may be initiated in the corona, much of the observed activity takes place in the chromosphere, and the larger flares can also cause heating in the photosphere, producing a "white-light flare". Usually most of the radiation released is in the ultraviolet and X-ray.

With each flare a burst of X-ray and ultraviolet radiation is released, and large numbers of high-energy charged particles are ejected from the Sun. The electromagnetic radiation travels at the speed of light, reaching Earth within minutes; the slower-moving particles arrive during the next few days. As we shall see in the next unit, solar flares can manifest themselves in a variety of ways when their energy and particles reach our own environment.

STUDY QUESTIONS

XIII-17. Are all solar nuclear reactions confined to the Sun's core?

XIII-18. In which region of the Sun do prominences and flares originate?

RECOMMENDED READING

John A. Eddy, A New Sun: The Solar Results from Skylab, NASA, Washington, D.C. (1979). Spectacular illustrations and accompanying narrative emphasizing the new short-wavelength views of the Sun -- highly recommended.

Fire of Life: The Smithsonian Book of the Sun, Smithsonian Exposition Books, Norton, 1981. Lavish illustrations.

Robert W. Noyes, The Sun, Our Star, Harvard University Press (1982). A comprehensive and well-written text on the Sun, suitable for non-scientists.

Ronald G. Giovanelli, Secrets of the Sun, Cambridge University Press (1984). Another recent illustrated discussion of the Sun.

UNIT XIV THE SOLAR SEA:

INTERACTIONS BETWEEN SUN AND EARTH

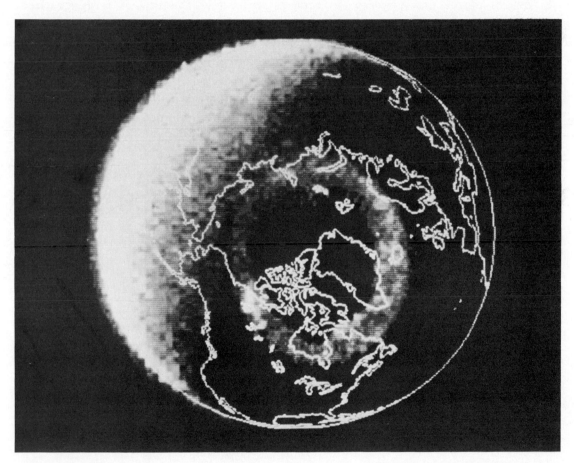

AURORA FROM SPACE

A. INTRODUCTION

1. Overview

Electromagnetic radiation and the solar wind exert substantial influences on Earth's upper atmosphere. As they arrive in the vicinity of the Earth, they cause a cascading series of interactions with the gases of the upper atmosphere and with Earth's magnetic field. These effects, and changes in them due to solar activity cycles, are most prominent in the highest and less dense regions of the atmosphere, but tend to die out as the surface is approached. The energetic particles that reach Earth can cause magnetic storms that produce such diverse phenomena as the aurora, changes in atmospheric drag on orbiting satellites, disruption of short-wave radio communications, and power grid blackouts in our cities. For more than a century, scientists have sought connections between changes on the Sun (especially the sunspot cycle) and climate changes on Earth. Though nothing conclusive has yet been found, there are a number of intriguing coincidences that point to possible links, including the Maunder sunspot minimum, recurring droughts in the western United States, and a number of other effects.

2. Objectives

Upon completion of this unit you should be able to:

1. describe the effects of solar radiation on the different layers of the atmosphere

2. relate the shape of the magnetosphere to the interaction between the solar wind and Earth's dipole magnetic field

3. describe how the magnetosphere shields Earth from most of the particles arriving in the solar wind

4. relate the existence of layers of charged particles in the ionosphere to long-distance radio communications on Earth

5. relate changes on the Sun that affect the solar wind to the occurrence of magnetic storms in the magnetosphere

6. describe several effects of magnetic storms

7. describe the aurora and its geographical occurrence

8. relate auroral displays to emissions from excited atoms in the ionosphere

9. relate the occurrence of aurora to the occurrence of magnetospheric substorms

10.   describe the phenomenon of airglow

11.   describe several climatic phenomena that have been purported to be related to the sunspot cycle and other changes on the Sun.

---

3. Key Terms and Concepts

plasma	magnetospheric substorms
cometary dust and gas tails	the fall of Skylab
magnetosphere	Aurora Borealis
magnetopause	Aurora Australis
magnetosheath	auroral oval
magnetotail	airglow
bow shock	solar constant
Van Allen Radiation Belts	Maunder Minimum
plasma sheet	Carbon-14 index of solar activity
ionosphere	Precambrian lake sediments and
magnetic storms	the sunspot cycle

---

4. Corresponding Video

In this program, you will focus on the interactions between the Sun and Earth. You will see incredible pictures of the aurora, one of the visible effects of that interaction, broadcast from the dynamic Explorer I satellite. You will also see spectacular film of the Aurora Borealis shot on location in Alaska. Scientists are studying auroras to find out what happens when trillions of watts of electrical energy are pumped into our atmosphere by the Sun. You will also be treated to a tour in animation through the Sun's interior from the core to the photosphere, then its chromosphere and outer corona. Scientists will also explore solar oscillations and the solar wind.

B. HOW THE SUN INFLUENCES THE EARTH

You have seen how the Sun pours out its energy in a variety of forms, both as electromagnetic radiation and as a stream of charged particles called the solar wind. When these reach Earth, they interact with our atmosphere and with our own magnetic fields to produce complex effects. Most of these effects are confined to the upper reaches of the atmosphere, but some are experienced on Earth's surface in the form of disruptions in communications, the visual spectacle of the aurora, and magnetic field changes that can affect power systems.

In the atmosphere and climate units the effects of solar heating on the troposphere and stratosphere were seen to be the cause of weather in the lower atmosphere. Not all of the energy reaching Earth makes it down to these lower levels, however. It is primarily the visible light (see Figure VII-3 on page 218) that is able to penetrate to the surface; the shorter wavelengths of the ultraviolet are to a considerable extent intercepted and absorbed at higher elevations. You may recall that the ozone layer results from the interaction of ultraviolet rays from the Sun with molecular oxygen (O_2) to form ozone (O_3).

Above the stratosphere with its ozone layer is the mesosphere, at the top of which air temperature reaches its lowest value (see Figure VII-2 on page 216). Above 100 km (62 mi) elevation, the temperature once again begins a steady rise with increasing height, giving the thermosphere its name. This region of extremely thin gases has received a great deal of study in recent years, since the Space Shuttle and many satellites orbit the Earth within it.

The thermosphere derives its high temperature from its absorption of the more energetic radiation from the Sun - the extreme ultraviolet. Thus there is a progression in the altitude at which the different wavelengths of radiation are absorbed: the shortest wavelengths are absorbed at the highest elevations and the longer wavelengths penetrate deeper into the atmosphere before being absorbed. Most of the visible and infrared radiation are able to penetrate all the way to the surface.

Because of this effect, solar variability, which is most extreme in the short wavelengths and much more slight in the visible and infrared, has its greatest effects in the uppermost atmosphere. We, on the surface, feel very few direct effects of the great short wavelength radiation outbursts from the Sun during the occurrence of solar flares. That turns out to be very fortunate for us, since ultraviolet rays can be damaging to eyesight and are a leading cause of skin cancer. They are also capable of disrupting or tearing apart and ionizing certain molecules, and this is one of the mechanisms that produces heating in the thermosphere.

One of the principal results of absorption of ultraviolet rays in the thermosphere, in addition to heating, is the production of atomic oxygen (O) from normal molecular oxygen (O_2). The two atoms of oxygen in molecular oxygen are torn away from one another to yield two oxygen atoms that then travel independently of one another. Recombination of atomic oxygen back to molecular oxygen at these altitudes is very slow, however, since it turns out that the two oxygen atoms must also collide with a third atom or molecule in order to come back together in a stable arrangement, and this is unlikely at the very low densities in the thermosphere. As a result, atomic oxygen produced in the upper atmosphere must migrate downward into denser regions before recombination into O_2] becomes a dominant process.

Partly for this reason, and partly because the upper atmosphere tends to segregate its constituents by atomic weight, the denser molecular oxygen (O_2) and nitrogen (N_2) settles to the lower portions of the thermosphere and the lighter constituents (mostly atomic oxygen and helium) become dominant at higher elevations. Atomic oxygen becomes the most abundant constituent of the atmosphere at a height of about 150 km (93 mi).

Throughout the remainder of this unit, you will see how this downward percolation of solar energy to the lower levels of the atmosphere produces different effects in the different layers. Because the greatest variation in solar output is contained in the X-rays and extreme ultraviolet rays, the greatest influences of solar variation are to be found high in the thermosphere. On the other hand, solar variations in the visible wavelengths are very small and their effects on Earth's climate are still largely unknown, since they are the ones that directly affect the troposphere in which weather and climate are produced.

In addition to the effects of the Sun's radiation, we must also consider what happens when the charged particles of the solar wind encounter Earth's environment. We shall do this in the next section and then return later on to examine the consequences of these interactions with the thermosphere and with the atmospheric regions below.

STUDY QUESTIONS

XIV-1. What is an important source of heating in the thermosphere?

XIV-2. Why is atomic oxygen (O) more abundant in the upper thermosphere than molecular oxygen (O_2) or nitrogen (N_2)?

XIV-3. In which layer of the atmosphere do we find the greatest variability due to changes on the Sun? In which do we find the least?

C. THE SOLAR WIND AND THE MAGNETOSPHERE

In the previous unit we saw that the solar wind is an extension of the ever-expanding corona, a blast of charged particles traveling outward from the Sun on a journey through the solar system, carrying lines of force from the solar magnetic field trapped within it. All of this occurs within the near-vacuum of space and forms a new kind of entity that is not familiar on the surface of Earth -- a plasma or rarefied ionized gas.

The presence of the solar wind had been deduced even before the space age by observations of the tails of comets. It was noted that some comets had not one but two tails, and spectrographic examination soon showed that each tail had a different composition. One consisted of dust particles and the other of ionized gases. The uncharged dust particles form a tail that is directed radially outward from the Sun. The ionized gas tail may be at some other angle because it follows the magnetic field spiral of the solar wind. The presence of these gas tails argued persuasively for a steady, outward-flowing solar wind.

Rather than wait passively for the occasional comet that comes by, scientists more recently have turned to an active type of experiment: that of "painting" the sky with an artificial comet tail. During 1984 a rocket was launched at Cape Canaveral that placed three satellites in orbit around Earth. One of these contained a cannister of powdered barium. At the end of the year, as scientists watched with Earth-based telescopes and the companion satellites rendezvoused to observe from space, the cannister released its cloud of barium. When hit by the solar wind it vaporized and began to glow, first appearing as a dot as bright as the North Star. It expanded to a reddish-yellow ball with a green halo, turned to a magenta tint and grew a shimmering tail that appeared from Earth to be several times longer than the diameter of the Moon. In this experiment, the barium cloud would help to study interactions between particles and the solar wind, permitting a more detailed understanding of phenomena in the near-space environment of our planet.

The solar wind carries to us the influence of the solar magnetic field, those spiral lines of force emanating from coronal holes on the Sun. But Earth, too, has its own magnetic field. What happens to it when it interacts with the solar wind?

Rocket probes in the early years of the space program began to provide answers by mapping the particles and magnetic fields in the space surrounding Earth and in the nearby solar system. First came discovery of the Van Allen Radiation Belts -- bands of charged particles trapped within the lines of force of Earth's magnetic field. Next was the first direct observation of the solar wind. It soon became clear that beyond the thermosphere above 500 km was a region dominated by magnetic fields and their interactions with the solar wind. This region received a name: the magnetosphere.

Figure XIV-1 shows a simplified view of the structure of this complex region. Just as the solar wind distorts the magnetic field of the Sun, stretching it out into a spiral shape, so it impinges upon and distorts Earth's field. Now, however, it is not moving outward from the Earth, but is coming inward from the Sun. As you saw earlier, magnetic field lines tend to become trapped or frozen in to moving electrically conducting gases such as the solar wind. At the same time, charged particles find it easy to move along magnetic lines of force, but hard to cross them.

When the solar wind encounters the terrestrial magnetic field, the two fields collide rather than mix. Because of the generally looped nature of Earth's field, it acts effectively to exclude most of the particles arriving from the Sun. The barrier between the terrestrial and solar magnetic fields is called the magnetopause. Outside the magnetopause we find ourselves immersed within the solar wind, but there is a compressed, hot region called the magnetosheath that is a zone of turbulence in the solar wind.

Like a boat moving through water, the magnetosphere parts the solar wind and forces it to flow around the magnetosphere. In this analogy, the magnetosheath is similar to the bow wave of the boat, and the leading edge of the bow wave is called the bow shock. Just as a water skier encounters bumpy water in the towing boat's wake but smooth water once he or she moves sideways out of the wake, a spacecraft leaving Earth measures turbulent magnetic fields within the magnetosheath, but on crossing the bow shock, enters a region of magnetic calm within the surrounding solar wind.

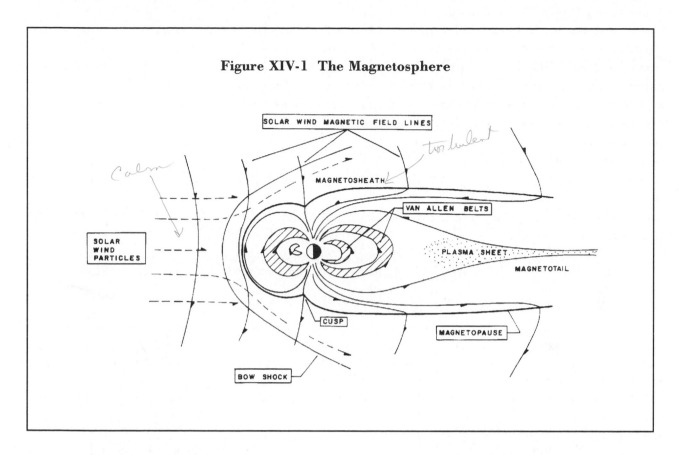

Figure XIV-1 The Magnetosphere

Under the influence of the relentless solar wind, the magnetosphere is blown out and away from the Sun into a long tail-like appendage, appropriately called the magnetotail. Though only the auroral zones are actually visible, in many respects a diagrammatic view of the Earth and its surrounding magnetosphere has the appearance of a comet.

In the picture we have drawn so far, it would appear that the magnetosphere is totally shielded from the particles of the solar wind, but we know that this is not the case. There are the charged particles that make up the Van Allen Belts, and it has been known for years that in the wake of solar flares, charged particles race into the thermosphere along lines of magnetic force to produce the aurora. How do they get there?

As seen in Figure XIV-1, there are several places that appear to be good candidate zones for leakage of solar wind particles into the magnetosphere. One of these is at the cusps in the magnetopause, where lines of force from the closed loops facing the Sun meet those going to or coming from the magnetotail. Here is one place where the solar magnetic field may connect with Earth's, allowing particles to flow into the magnetosphere along lines of force. The magnetotail presents another possibility, reaching 1000 or more Earth radii above the midnight zone on Earth's surface. The far reaches of the magnetotail are still poorly mapped, but it may be that some particles work their way into the magnetosphere following a circuitous route along the length of the magnetotail. Still other particles may simply diffuse their way slowly across the magnetopause and then follow nearby lines of force into the upper reaches of Earth's atmosphere.

Particles from the solar wind that make it past the magnetopause accumulate in two main regions within the magnetosphere. The first consists of the Van Allen Radiation Belts that are contained within the closed loops of magnetic field shown as shaded in Figure XIV-1. The other is a plasma sheet extending the length of the magnetotail on the midnight side of Earth. This plasma sheet is essential to the maintenance of the magnetotail, serving as the source of electric currents that stretch out the lines of force into the tail's elongated shape.

More recently discovered are reservoirs of particles just inside the magnetopause, undoubtedly resulting from diffusion of particles from the magnetosheath across this boundary.

STUDY QUESTIONS

XIV-4. What influence acted on the cloud of barium atoms released into space to create an artificial comet tail?

XIV-5. How would you define the magnetopause?

XIV-6. What are two major reservoirs of charged particles within the magnetosphere?

D. THE IONOSPHERE

High in Earth's atmosphere is another region containing for charged particles: the ionosphere. Solar radiation in the extreme ultraviolet and charged particles entering the upper atmosphere have sufficient energy to ionize atoms and molecules in the thermosphere. The ionosphere is a weakly ionized plasma within the thermosphere maintained by ionizing radiation and charged particle bombardment. Beginning at a height of 50 km, the ionosphere overlaps the mesosphere and thermosphere as defined in Figure VII-2 on page 216 and extends beyond them into the magnetosphere. Though a direct demonstration of its existence would have to wait until well into the present century, several eighteenth-century scientists (Benjamin Franklin was one of them) noted the similarity between auroras and electrical discharges in evacuated jars and suggested that the upper atmosphere might be electrically conductive.

The existence of the ionosphere was discovered in 1901, shortly after the invention of the radio, when Marconi successfully sent a message from England to Newfoundland using short-wave radio. It was not understood how the radio waves, which were known to be absorbed by the ground, could make their way around the curvature of Earth's surface over such a distance. Short-wave radio, of course, would become a major instrument of long-distance communications, and within a year it was suggested that these radio waves could travel around the world by being reflected from layers of ionized particles in the upper reaches of the atmosphere. This conjecture was confirmed in 1924 when radio waves directed vertically upward were found to generate an echo from the charged regions of the ionosphere.

Several distinct layers or regions of charged particles make up the ionosphere. Ionization in each layer is accomplished by radiation in the extreme ultraviolet coming from the Sun, and in high-latitude regions by high energy particles that travel down the near-vertical lines of force. The lowest is called the D layer, with E, F1, and F2 layers occurring at higher elevations. During the day, the direct radiation from the Sun builds up these layers and at night the D, E, and F1 layers tend to dissipate.

Particles from the magnetosphere spiral down the magnetic lines of force into the ionosphere and collide with atmospheric atoms and molecules and produce heating and ionization. These particles produce electric currents that stream into the ionosphere, amounting to a million amps or so. These contribute to resistive heating of the ionosphere and thermosphere, much as an overloaded wire becomes hot from the current flowing through it. In addition, heating by the Sun's radiation during daytime causes the thermosphere to expand and sets up convection within the ionosphere. The motion of this conductor through the lines of force of Earth's magnetic field generates additional electric currents that flow around the world through the atmosphere. This dynamo process operates in the atmosphere at an elevation of around 100 km. These currents can cause changes in the the magnetic field measured at the surface on the order of a few percent. Changes of this magnitude are readily observed, and serve to provide a means for measuring and mapping the currents that cause them.

Circulation in the thermosphere operates somewhat differently from the weather of the troposphere. As is the case in the stratosphere, temperature increases with height in the thermosphere, resulting in a fairly stable arrangement. As a result, there are no cyclonic storms in the thermosphere. There is, however, substantial circulation caused by heating differences between day and night and between the seasons, when one polar region or the other is heated almost continually by sunlight. Fairly steady winds amounting to 100 meters per second (225 miles per hour) may result from this circulation of the thin gases of the thermosphere.

From day to night, circulation in the neutral (uncharged) atmosphere causes the charged-particle layers to move up and down considerably in elevation, and these, together with the changes in the number of particles in each layer, contribute to substantially different radio propagation between day and night. If you have a short-wave radio, you know that distant stations are often best received at night. At times, however, these regular variations are upset by events that can cause a variety of freakish problems: it may be possible to pick up stations from nearly halfway around the world but not those from a city relatively nearby. Some of these effects are caused by distortions of Earth's ionospheric layers that respond to solar disturbances such as electromagnetic radiation from the Sun during solar flares and enhanced auroral particle bombardment during what are called magnetic storms.

During these storms that intimately involve the magnetosphere, electric currents and energetic particles entering the thermosphere may deposit as much energy in it as that from solar ultraviolet radiation. Air temperatures in the thermosphere may rise by hundreds of degrees, causing it to expand and increasing the density of the atmosphere in its upper reaches, where a number of satellites have been placed in orbit. The increased density of air causes greater drag on the satellites and accelerates the decay of their orbits. For this reason, it is not always possible to predict just when and where a satellite will eventually fall to Earth, because magnetic storms are not predictable in detail.

STUDY QUESTION

XIV-7. Would you expect the drag on a satellite to be increased or decreased during the maximum of the solar sunspot cycle?

E. MAGNETIC STORMS

Events on the Sun can have a profound effect on the magnetosphere and ionosphere. The 11-year cycle of sunspots alternates between periods of low magnetic activity and peaks of high activity when coronal holes spray an additional cloud of particles into the solar wind. Solar flares vastly increase the density and speed of the solar wind that then collides with the magnetosphere, compressing it on the Sun side and distorting its shape and structure.

On Earth, this is seen as a small but sudden change in the strength of the magnetic field all over the world, lasting an hour or more. This may be followed by slow changes in field strength occurring over the next day or two, with superposed irregular fluctuations of strength and direction of the field. These changes are generally too small to be recorded by a simple compass needle, but are clearly shown in the traces of sensitive magnetometers that are capable of recording changes in Earth's field to one part in 10,000 or so. Magnetic storms tend to occur on the average once or twice a month, but recent discoveries have shown that there are much more frequent disturbances whose principal low-altitude effects are confined to circular bands surrounding the poles in which auroras are frequently seen. Called magnetospheric substorms, these disturbances originate on the night side of the magnetosphere and occur on average several times a day. They may be due to instabilities in the plasma sheet that involve dumping of large numbers of particles into the ionosphere, resulting in spectacular displays within the auroral band.

Intense currents flowing in the upper atmosphere near 100 km produce magnetic fields that extend down to the ground. These variations in the magnetic field can induce electrical current surges in long conductors such as pipelines and power transmission lines. The effects of magnetic storms were largely responsible for the pursuit of research into the behavior of the magnetosphere. Consider some of these effects:

- A taxi driver in New York City found that he was being radio dispatched not by his local office but by an office in Miami.

- On February 10, 1958, users of the undersea telephone cable from Newfoundland to Scotland noted that transmissions from North America were garbled, while voices traveling in the opposite direction were received clearly.

- In September of 1957, planes of the U. S. Navy early-warning network found that all communications with their home bases were cut for several hours. During that time, the United States may have been open to attack without warning.

- On March 24, 1940, nearly all long distance telephone lines serving Minneapolis, Minnesota, went dead for several hours. On the same day a major power failure blacked out a sizeable portion of the northeastern United States.

All of these events, mundane and major, have been attributed to magnetic storms. Radio communication problems, of course, are related to changes induced in the ionospheric layers, but problems in power grids and wire-based communications are caused by the fact that electric currents can be induced in a conductor by moving or changing magnetic fields. Magnetic storms can induce large surges of electric current in long wires or conductors on Earth and so can overload transformers and circuit breakers in power networks. When the Alaska pipeline was constructed it was found that large currents were being induced along the length of the pipe, causing unexpected corrosion of valves and interference with electrical monitoring devices.

The influence of magnetic storms has proved to be pervasive. When line-of-sight communications via geostationary satellites was instituted, it was assumed that these would be immune to interruption during magnetic storms. Though the situation was certainly improved, it was found that transmissions to and from the satellites were being distorted or absorbed as they passed through the disturbed ionosphere.

At the Space Environment Services Center of the National Oceanic and Atmospheric Administration in Boulder, Colorado, activity on the Sun is monitored continuously. When flares are detected, alerts are sent out to scientific laboratories, telecommunications agencies and firms, and power companies that a magnetic storm is probable during the next few days. Short-term prediction of solar activity has proved to be difficult, and at present major flares can be predicted with some reliability only about one to three days before they actually develop in magnetically active regions. Because of the inherent complexity of the behavior of these regions, it is not clear just how much further in advance flare predictions can eventually be pushed.

An event that received considerable press coverage involved the fall of Skylab in 1979. As reported in the last unit, Skylab was a manned space station whose measurements of the Sun from above the atmosphere generated important new insights into the operation of the Sun. It had not been intended to allow Skylab to fall, and when it did, it was not clear just where this massive object would come down. A report of the National Academy of Sciences describes what happened: [28]

> It became a media event with all the ingredients of suspense and potential for catastrophe to frighten a confused public. When the last astronaut left Skylab in 1974, it was thought that the spacecraft was in a safe parking-orbit, where it could await a visit by an early Space Shuttle flight, which would push it to a higher orbit for safekeeping until it could be refurbished and reactivated. Unfortunately, the plan was frustrated by a delay in the Space Shuttle schedule and by the rapid rise of solar activity toward sunspot maximum. With high sunspot activity came a hotter and denser atmosphere at Skylab altitude, which increased the drag

---

[28]Solar-Terrestrial Research for the 1980's, pp23-24, National Academy Press, Washington, D. C., 1981

on Skylab and caused the orbit to decay much faster than anticipated. Skylab thus fell victim to solar activity. Our inability to predict where it would fall exemplifies our current lack of instruments to observe with adequate precision the solar output of extreme ultraviolet and X-rays, which control the density of the atmosphere at satellite altitudes.

And so Skylab was destroyed by the activity of the Sun that it had been designed, in part, to study. Fortunately, after a fiery reentry, its remains fell in a remote area of Australia and harmed no one. Scientists and politicians, along with everyone else that had been following the story, breathed a sigh of relief.

STUDY QUESTIONS

XIV-8. How can power blackouts be caused by magnetic storms?

XIV-9. Why did Skylab fall earlier than had been expected?

XIV-10. At what time of day would you expect the plasma sheet to be more or less directly overhead?

XIV-11. Why is it that short-wave radio communications can travel around the curve of the Earth's surface?

F. THE AURORA

The most familiar of solar-terrestrial interactions are manifested in the auroras, and yet the majority of dwellers of temperate latitudes are probably not aware of ever having witnessed them. Popularly known as "The Northern Lights", the auroras were named in the 17th century for the Roman goddess of the dawn. In an auroral display, the nighttime sky may be splashed with shifting lights of several colors, providing a display that leaves few observers unmoved. Writing in The Alaska Sportsman,[29] Dorothy Jean Ray describes one such:

> In the northern latitudes a display of the aurora begins "inauspiciously, with not even a hint of the drama to come in the steady iridescent haze that forms on the horizon early in the evening. The sky, dark and so far away, gradually begins to open up as the haze settles into a comfortable glow, warming the sky above and gradually widening its expanse to immeasurable depths and widths. Then, almost without prelude, the sky moves.

> Streaks of light toss about with abandon. Gargantuan, ghostlike arms, chasing and darting, appearing and disappearing spontaneously, writhe across the upper sky. Suddenly, for a second, all light melts away and the sky is almost dead with darkness. But just as quickly the lights blossom again in pulsating waves and arcs, and then, as if to test the credulity of man, giant draperies of light wash in

---

[29]April 1958

quickly undulating movements across the whole heavens, sometimes stabbing the ends of their folds toward the Earth, dripping with the green of grass or the red of blood.

Native people of the Northlands were accustomed to such displays, and the Inuit (Eskimos) incorporated them into their legends. A common legend describes the aurora as lights placed by spirits to light the paths of the dead who are coming to join them or as reflections from a walrus skull that is being used as a football by recreating spirits. Most Inuit did not fear the lights because they were so familiar -- in many places of the North the aurora is visible on more than half the nights of the year.

In more temperate climates, however, auroral displays are seen much less frequently. When they do appear at lower latitudes it is often the red form that is seen, and this bloody appearance has more often than not been taken as an evil omen. In this form the aurora might be mistaken for reflections from a distant fire, and even in recent years, fire engines have been called out in response to an unusual auroral display.

Contrary to popular opinion, the frequency of auroral displays does not increase continuously as you approach the poles, but peaks in a roughly circular band centered on the geomagnetic poles. In the western hemisphere the band cuts through central Alaska and across the southern shores of Hudson Bay. Crossing the southern tip of Greenland and Iceland, it continues past the northern tip of Norway and skirts the Arctic Ocean shore on the northern boundary of the U.S.S.R.

Within this zone there is a shifting band of activity at any one time called the auroral oval. (See Figure XIV-2) It is fixed with respect to the Sun and not quite centered on the magnetic poles, with the result that the display appears to advance and retreat as the night goes on. In Alaska, the auroral oval shifts from as far south as Anchorage at 8 AM to off the north coast in the Beaufort Sea by 2 PM. Auroras do occur during the day, though they are obscured from view during daylight hours. The greatest concentration of aurora is found on the night side, however, because the cause of the aurora is found within the plasma sheet in the magnetosphere.

There is an auroral zone in the southern hemisphere as well. The English explorer James Cook was the first European to describe them during his famous voyage around the world in 1773. He named the Southern Lights the Aurora Australis to distinguish them from the Northern Lights, the Aurora Borealis. In all respects except for geographical location, the two are identical.

Displays are also seen outside the auroral zone, with frequency of occurrence decreasing with distance from the zone. From five to ten auroral displays may be seen each year as far south as Washington, D.C. and London. Because of the tilt of the geomagnetic axis, the auroral zone extends farther south in the western than in the eastern hemisphere. Paris, Switzerland, and Texas can expect to see only one auroral display per year. The frequency increases to more than 200 per year within the auroral oval, then decreases to less than 50 per year at the North and South Poles.

The aurora appears to observers in a variety of different shapes, but all are variations on a single type as seen from different viewing angles. The basic type is that of the curtain or

Figure XIV-2 The Auroral Oval

Adapted from S.-I. Akasofu, <u>Aurora Borealis: The Aurazing Northern Lights</u>, Alaska Geographic Vol.6, No.2, 1979

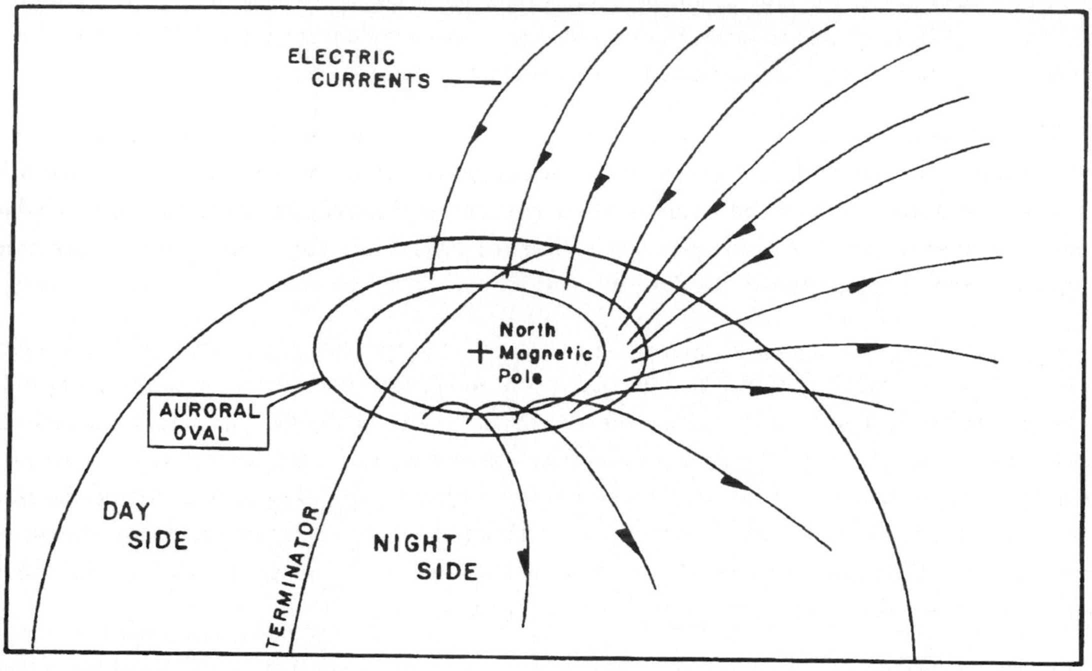

drapery -- a band of light stretching across the sky containing rays or folds that are aligned with Earth's magnetic field. In polar regions, the field lines are only slightly inclined to the vertical, giving the impression of hanging drapes or curtains, especially when seen from some distance to the north or south of the display. When seen edge on at lower latitudes, the curtains take the shape of arcs or folds rising from the horizon, and when seen from directly below they take the form of streamers radiating outward from the zenith. In the latter form, the aurora is said to have a coronal form, referring to its crownlike shape and not to the solar corona. Watch for the different types in the television views, and see if you can detect the parallel strands of Earth's magnetic field lines made visible in the aurora. This is the only circumstance in which we can actually see the magnetic field lines in nature.

Many observers have reported hearing whistling or swishing sounds in association with the aurora. Inuit legends insist that if you whistle back at them, they will approach closer out of curiosity. In spite of widespread popular belief in these sounds, scientists have not succeeded in recording any associated sounds that would be audible to the human ear.

The aurora are composed of a somewhat limited number of basic colors, though combinations provide a satisfying artistic palette for nature's use. When examined by the spectrograph, auroral lights are found to consist of only a few strong emission lines, without any continuous spectrum such as sunlight contains. Most prominent is a green line at 5.577 x 10^{-7} m whose origin was a mystery for many years, since it did not seem to correspond to the spectra of familiar gases. It was finally discovered that this line arose from atomic oxygen (O) rather than from the more familiar molecular oxygen (O_2). Recall that high in the ionosphere, molecular oxygen tends to be split into its individual atoms by the effects of solar radiation.

Atomic oxygen is also responsible for the blood-red color often seen at lower latitudes with a single emission line seen at 6.300×10^{-7} m. Atmospheric nitrogen (N_2) contributes a band of lines that appear crimson red, a color often seen at the base of auroral curtains during especially active displays. Ionized nitrogen molecules also contribute a band in the ultraviolet, but these are at wavelengths too short to be seen by the human eye.

These emission lines arise from particles traveling down the magnetic lines of force. When they strike atoms and molecules in the ionosphere, they ionize some and cause others to jump to excited states. When the excited atom returns to its ground state, it emits a photon of characteristic energy (see the right portion of Figure XIII-2 on page 400), in just the same way that emission lines are produced on the Sun.

From the color of the aurora it is possible to determine the energy of the bombarding particles and the height of the peak ionization. A red aurora is excited by low energy electrons with maximum ionization high in the thermosphere near 250 km. A green aurora is more energetic and is caused by particles that penetrate down to 120-100 km. A green aurora with a red lower border is from even more energetic particles that produce maximum ionization near 90 km. Thus, the color of the aurora is a rough guide to determine the energy of the bombarding particles and the height of the aurora. In addition, the brightness of the aurora provides a measure of the particle fluxes into the atmosphere.

The process is very similar to that produced within a neon tube used in advertising signs. The glass tube is filled with rarefied gases that are bombarded by electrons that travel the length of the tube between electrodes placed at either end. The colors emitted by the tubes depend on the gases placed within them, and a spectral study of light from such a tube would show emission lines characteristic of each gas. Neon is only one of the gases used in the tubes and produces a bright orange-red color.

Early auroral scientists studied the spectrum and determined the height of the aurora by comparing photographs taken simultaneously from widely separated locations. Triangulation allowed them to calculate the heights of various prominent features of a display. They found that auroras reach no lower than about 90 km, and extend to elevations greater than 500 km, placing them within the ionosphere and thermosphere.

Modern observations of the aurora are increasingly being made from space. Orbiting satellites have provided wide-ranging views that clearly define the auroral oval and can provide almost continuous coverage of auroral activity. From such observations has come a better understanding of the origin of the particles that give rise to the aurora.

Auroral displays are intimately linked to the phenomenon of the magnetospheric substorm that was mentioned earlier. As the solar wind streams past the magnetopause that surrounds the magnetotail, positively and negatively charged particles from the Sun are diverted to opposite sides of the tail, producing a positively charged region on the morning side of the tail (to the east, as seen from an observer on the night side of Earth) and a negatively charged region on the evening side of the tail. This accumulation of charges feeds the plasma

435

sheet that runs down the middle of the magnetotail and helps to maintain the tail's shape. Every so often, the plasma sheet becomes overloaded and dumps particles down the streaming lines of force that reach into the polar latitudes. This is the essence of a magnetospheric substorm.

Look at Figure XIV-1 on page 425 and make sure you see why this should give rise to auroral displays only in the polar regions. It is less clear why an auroral oval should result -- to see this it is necessary to be able to extend the two-dimensional slice of the magnetosphere in Figure XIV-1 into three dimensions as seen by Figure XIV-2. The magnetic lines of force that connect the plasma sheet with Earth would then be found to converge in two circular bands surrounding the magnetic poles.

Magnetic storms caused by disturbances on the Sun can severely distort the fields within the magnetosphere, causing the auroral oval to shift to more southerly latitudes and allowing those of us who live in the temperate zones an occasional glimpse of the auroral spectacle. Scientists study the aurora not only to gain an understanding of the origin of these fascinating displays, but also to provide a powerful tool for probing the nature of the complex interactions between events on the Sun and resulting effects on Earth.

STUDY QUESTIONS

XIV-12. Where on Earth would you expect the likelihood of observing auroral displays to be highest?

XIV-13. Why is the "curtain" or "drape" the basic shape for auroral displays?

XIV-14. Emissions from what two gases in the atmosphere are most responsible for the lights of the aurora?

G. INTERACTIONS WITH THE MIDDLE ATMOSPHERE

At the beginning of this unit we stressed the fact that the effects of solar variations diminish in the atmosphere as we approach the surface. There are, however, several important interactions with the middle atmosphere. One of these occurs near the top of the mesosphere and the bottom of the ionosphere and literally causes the air to glow.

Absorption of solar radiation and particles in the upper atmosphere not only produces the ions found in the ionosphere but also excites many atoms to higher energy levels in the manner shown in Figure XIII-2 on page 400. In most cases emission occurs almost immediately after absorption, producing an airglow in the daytime sky. This is faint compared to scattered sunlight, however, and is not generally noticeable to the naked eye. Photographs taken from space in the visible and ultraviolet clearly show the daytime atmosphere glowing brightly from airglow emissions.

The main airglow layer seen extending above Earth's surface in photographs made from space is caused by the chemical recombination of the atomic oxygen (O), created in the thermosphere, that is recombining near 100 km to form molecular oxygen (O_2). Energy released in the process excites remaining oxygen atoms to a higher energy state that radiates at 5.577×10^{-7} in the green line of atomic oxygen. This is the nighttime (and daytime) airglow layer near 100 km.

The sky is not totally dark, even on a moonless night far from artificial lights. If you were to hold an opaque object above you, you could still see its outline clearly against the night sky, and not just by the blocking out of stars. There is a general glow to the sky caused by emission phenomena that is faintly visible on most nights.

Study of the airglow is helping to determine the chemical and physical processes that dominate the middle and upper reaches of the atmosphere. In addition, there is a practical turn to airglow research in that the detection of nuclear weapon explosions and of rocket or missile liftoffs must be able to distinguish between electromagnetic radiation produced by these man-made events and by those of natural origin, such as airglow and auroral displays.

Another effect that concerns the mesosphere and stratosphere is in the production of nitric oxide (NO) in the thermosphere by solar ultraviolet radiation and by energetic particles associated with the aurora. Nitric oxide may migrate downward through the mesosphere in the polar night region where it can escape destruction by solar ultraviolet rays, and may ultimately find its way to the stratosphere where it serves as an effective catalyst in the destruction of O_3 in the ozone layer. This causes fluctuations in ozone concentration within that layer that depend on magnetic activity, making it difficult to observe changes that have been predicted on the basis of man-made introduction of nitric oxide and chlorofluorocarbons into the atmosphere.

Following certain solar flares, extremely energetic solar protons have been observed to bombard the middle atmosphere in the northern and southern magnetic polar caps down to about 60° latitude. The solar proton event of August 4, 1972 produced the largest flux of high energy protons that has ever been observed. It produced considerable amounts of nitric oxide near 50 km in elevation that was observed to catalytically destroy ozone in the upper stratosphere. These kinds of events are relatively rare, but when they do occur, they provide a unique natural experiment that allows us to test theories concerning ozone in the middle atmosphere.

This is as far down in the atmosphere as it is presently possible to trace effects of solar variability with any certainty. But that does not prevent scientists from searching for circumstantial evidence linking tropospheric weather and climate to changes on the Sun. A great deal of effort has gone in that direction, as you will learn in the next section, but you should keep in mind that most of these intriguing ideas are still considered to be very speculative in nature.

STUDY QUESTIONS

XIV-15. What is the principal source of energy that produces airglow?

XIV-16. Would you expect an increase or a decrease in ozone concentration within the stratosphere during a time of sunspot maximum?

H. SOLAR VARIABILITY: IS THERE A CLIMATE CONNECTION?

That the Sun should control Earth's weather goes without saying -- it is the source of the heat energy that drives the weather engine. But how constant is that energy supply? And how much of a change in the Sun's output would be necessary to upset Earth's climate systems?

You are already familiar with the Milankovitch Theory that purports to explain the comings and goings of the Pleistocene Ice Ages (see Figure VIII-5 on page 256). In this explanation, the changes in the distribution of solar energy received by the Earth are due to alterations in Earth's orbit and axis tilt, not due to inherent changes in the levels of radiation emitted by the Sun. In addition, these are very slow changes, with periods ranging from 19,000 to 100,000 years. Are there shorter changes due to the Sun itself that are climatically important? Are there significant changes in solar output associated with the 11- and 22-year sunspot cycles?

If there is such a relationship, then the regularity of the sunspot cycle holds out the hope of predictive power that is not currently available to climatologists. If droughts or other major changes in the climate system are tied to the sunspot cycle, then it should be possible to predict them years in advance and to prepare for them. Such a discovery would be of great practical value and would constitute a major scientific advance. Small wonder, then, that attempts to prove such a connection go back more than a century.

In 1801, William Herschel, perhaps the greatest astronomer of his day, thought that he saw a correlation between times of few sunspots and the price of grain on the London market. Higher prices, he figured, reflected reduced crop yield due to adverse climate that might result from a reduction in solar output. In this surmise, however, the man of science was bested by a man of letters -- Jonathan Swift. In his Gulliver's Travels, written in 1726, Swiftian astronomers feared that too many sunspots would eventually obscrue the Sun, blocking out its light. Some 250 years later, Swift was proved the more correct of the two by a satellite called Solar Max -- the Solar Maximum Mission launched in 1980 to observe the Sun during a time of maximum sunspot activity. The passage of a large group of sunspots across the face of the Sun was found to decrease the total solar output by something on the order of 0.1%, simply due to the lesser amount of light being radiated by the dark area of the spots. But then, Swift seemed to have a way with astronomy. Others among his fictional astronomers correctly predicted the discovery and general characteristics of the two moons of Mars, Phobos and Deimos, more than 150 years before their actual telescopic discovery.

The quantity of solar energy that reaches Earth amounts to 1.35 kilowatts per square meter. This quantity is called the solar constant, and attempts have been made for nearly a century to determine just how constant it truly is. Earth-based measurements have to account for the variable nature of the atmosphere, however, and the best that could be done until Solar Max went into orbit was to limit any changes in the solar constant to less than one percent. Solar Max is still in orbit and no changes greater than that have yet been observed. Is this change great enough to affect day-to-day weather on Earth? Probably not significantly, since the individual reductions due to sunspots last only a few days and this period of time is too short to have much influence on the temperature of land and ocean. They may have an influence in the longer term, where the effect is hard to find. But the workings of the climate system are not yet fully understood. To date, the customary approach has been to look for simple correlations between solar and climatic phenomena. This, of course, is just what Herschel did, and to avoid repeating his error, it is necessary to exercize extreme caution in assigning a cause-and-effect relationship to something that may be merely a coincidence.

A problem here is that reliable weather records do not extend back in time far enough to provide a convincing basis for comparison. The tree ring record from the American west, on the other hand, provides an almost unbroken rainfall record for that region extending over some 300 years. Trees in semiarid regions will respond to wet years with vigorous growth and wide rings, while dry years produce narrow rings. At the University of Arizona, scientists at the Laboratory for Tree-Ring Research have compiled records from long-lived trees and sampled throughout the western United States, searching for evidence of climatic cycles. In fact, there does appear to be a distinct cyclicity of drought conditions with a period of 18 to 22 years, suggesting an effect due to the solar magnetic cycle. Recall that the solar magnetic field reverses itself once each 11-year sunspot cycle, requiring two complete cycles to begin again with the same polarity.

This calls to mind the Maunder sunspot minimum shown in Figure XIII-6 on page 413 that has been used as a possible cause for cold periods during the Little Ice Age that lasted from approximately 1400 - 1850 AD. But this single coincidence could easily be pure happenstance, and it even seems to repeat Hershel's apparent error of attempting to correlate few sunspots with colder climate. What is needed is a longer record of solar activity than that afforded by the observations of sunspots, as well as better histories of world climate.

It may seem surprising that the same trees that provide evidence for the drought cycle would also have recorded the level of solar activity, but this indeed seems to be the case. Radioactive Carbon-14 is present in the wood of the tree rings and Carbon-14 provides the necessary link. It turns out that Carbon-14 is produced when cosmic rays entering the upper atmosphere collide with nitrogen atoms. The link to solar activity is via the solar magnetic field. Magnetic fields of both Sun and Earth provide a shield against cosmic rays that tends to deflect them and prevent them from entering the atmosphere at middle latitudes. When interplanetary magnetic fields are turbulent or when Earth's magnetic field is strong, fewer cosmic rays make it through and less Carbon-14 is produced. During those times less Carbon-14 enters the trees.

There are two effects here -- one from Earth's magnetic field and one from the solar wind magnetic field. Paleomagnetic studies have shown that Earth's magnetic field changes with time, having reached a peak around 200 AD and having decreased ever since. Investigators checking the accuracy of the Carbon-14 age-dating method against tree rings first found this effect and suggested that significant deviations in Carbon-14 ages could be explained by a change in production that could be attributed to changes in the field strength. The smooth sinusoidal curve in Figure XIV-3 shows how the changing strength of the Earth's magnetic field appears to have affected the Carbon-14 record in tree rings.

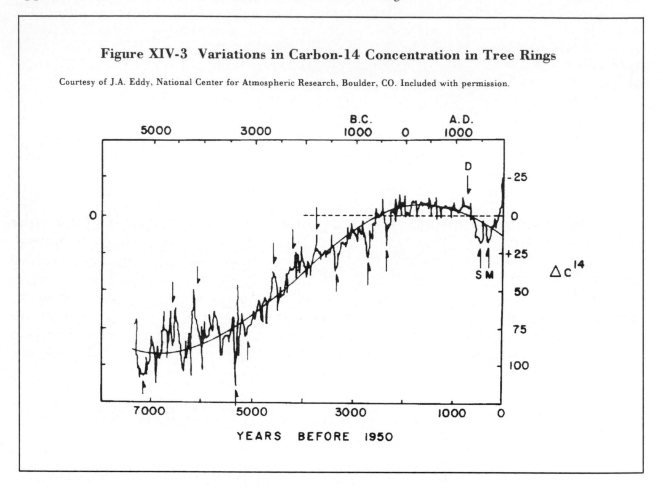

Figure XIV-3 Variations in Carbon-14 Concentration in Tree Rings

Courtesy of J.A. Eddy, National Center for Atmospheric Research, Boulder, CO. Included with permission.

The jagged line in the figure shows the measured deviations of Carbon-14 in the tree rings, and while they generally follow the smooth line for the Earth's field, there appear to be significant smaller deviations as well. For instance, the deviation marked "M" corresponds to the Maunder sunspot minimum. With fewer sunspots there should have been increased Carbon-14 production (a downward deflection in the diagram.) Just prior to that is a similar deflection marked "S". Together, they bracket the entire lifetime of the Little Ice Age. Furthermore, the upward deflection marked "D" corresponds in greater solar activity and in climate to the Medieval Optimum, a warm period that extended from about 900 to 1200 AD

If this interpretation is correct, then the data in Figure XIV-3 represents a more than 7,000-year record of solar activity, with significant departures from the norm marked with arrows. The climate connection is more tenuous, but solar activity in this data does seem to correlate with advances and retreats of Alpine glaciers for the past 2,500 years or so.

If trees can record the coming and the going of sunspots, could there exist some other terrestrial record of the long-term behavior of the Sun? Very ancient sediments that accumulated in the bed of a Precambrian lake in Australia may provide just such a record. Many lake sediments are found to accumulate in layers, one being deposited each year. These geological equivalents to tree rings provide some of the most detailed records available in geology, and with possible year-by-year resolution.

The Australian lake sediments show a very distinct grouping of layers into sets of 11, with pronounced, darker layers separating the sets from one another. There also appears to be a less pronounced 22-band cyclicity in the record. It is tempting to regard this as an indication of the solar cycle some 680 million years ago during the Precambrian Era, when these sediments were laid down. A difficulty with this interpretation is that more modern lake sediments in Australia or elsewhere do not appear to reflect the solar cycle, implying either that the solar interpretation is wrong or that the effect of solar activity on climate was more intense in the Precambrian.

Several other changes in the climate system have been reported to follow solar phenomena. Russian astronomers have suggested that atmospheric pressure changes systematically a few days after the beginning of major magnetic storms, increasing in Alaska and northeastern Europe while decreasing in central North America and in China. Another recent study attempts to link temperature changes in the northern hemisphere to the ratio of the average size of sunspot umbras (the dark inner portion of the spot) to the size of the penumbras (the mottled outer portion). This ratio, like northern hemisphere temperatures, has been observed to increase from the middle nineteenth century until about 1940 and to decrease since.

A 1982 study[30] released by the National Academy of Sciences had this to say:

The obvious terrestrial importance of the Sun, coupled with hopes of simplicity and predictability, have made solar variations a popular subject for investigation by generations of scientists who have sought the causes of changes in weather and climate. Despite much research, no connection between solar variations and weather has ever been unequivocally established. Apparent correlations have almost always faltered when put to critical statistical examination or have failed when tested with different data sets. As a result the subject has been one of continual controversy and debate.

There is no question that the constant Sun directly influences world climate. What is still debated is whether there are also changes in solar output great enough to be detected in the natural variability of the climate system. So far, the largest changes detected are about .01% and at the margin of significance. The debate will probably not be settled until someone can demonstrate the physical mechanisms by which such small solar effects can be translated into measurable climate changes.

---

[30]Solar Variability, Weather, and Climate, National Academy Press, Washington, D. C.

STUDY QUESTIONS

XIV-17. What is the maximum change that has been observed in the energy output of the Sun?

XIV-18. What is the evidence indicating recurrence of droughts with a 22-year period in the American west?

XIV-19. The Maunder Minimum and other small deviations shown in Figure XIV-3 on page 438 appear to be due to variations on the Sun. What is the large, slowly varying change in Carbon-14 concentrations due to?

XIV-20. Why are scientists so reluctant to ascribe climatic changes to solar causes when it is obvious that the Sun is the principal energy source for the climate machine?

RECOMMENDED READING

S.-I. Akasofu, Aurora Borealis: the Amazing Northern Lights, Alaska Geographic magazine, Vol. 6, No. 2 (1979). Beautifully illustrated and interesting discussion of the aurora, including historical aspects, legends, and a brief description of magnetospheric phenomena. Highly recommended.

Robert H. Eather, Majestic Lights - The Aurora in Science, History, and the Arts. American Geophysical Union, 1980. Illustrated, with emphasis on history.

James A. Van Allen, Magnetospheres and the Interplanetary Medium, in The New Solar System, Cambridge University Press (1981). A good discussion not only of Earth's magnetosphere, but that of the other planets as well.

Edward Peary Stafford, Sun, Earth, and Man, Washington, DC, NASA, 1982. A well-illustrated tour of space missions designed to explore the upper atmosphere and magnetosphere.

UNIT XV FATE OF THE EARTH:

THE BALANCE OF NATURE

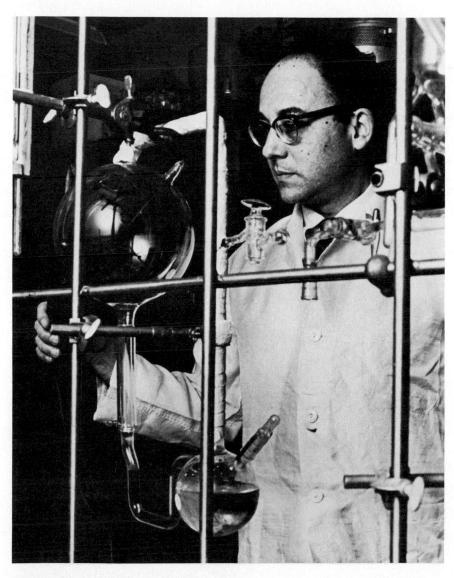

STANLEY MILLER STUDIES THE ORIGIN OF LIFE

A. INTRODUCTION

1. Overview

The establishment of life, early in its history, was an event that would totally transform our planet. The atmosphere originally was virtually oxygen-free, and in this environment the basic building blocks of life, the amino acids, appeared. We don't know how the next step, the production of nucleic acids to enable reproduction of cells, was achieved, though interesting hypotheses have been proposed. Once this step was achieved, life and the environment evolved together, and the development of life proceeded to ever higher levels of sophistication. With photosynthesis, life acquired practical method for obtaining energy from the Sun, and the subsequent rise of oxygen in the atmosphere had profound consequences for both the geosphere and the biosphere. With the arrival of mankind, a new force for change appeared -- one that acts with great rapidity and force, though by no means always benignly. An examination of the energy budget of the Earth and of three important biogeochemical cycles -- those of carbon, nitrogen, and phosphorus -- illustrates the interconnectedness of the planet-wide systems that comprise the biosphere and geosphere. Before we can predict the consequences of our actions as stewards of Planet Earth, we must have a reasonable understanding of this complex system of interacting cycles that serve to keep the environmental house in order. This system is what is popularly referred to as the "balance of nature". You will see, however, that natural "balances" are not static but are dynamic and have undergone dramatic shifts throughout geologic time.

2. Objectives

Upon completion of this unit you should be able to:

1. describe the nature of the chemical environment that may have existed when life first formed on Earth

2. describe experiments designed to show how the first amino acids may have been produced

3. relate certain self-replicating processes in clay to a possible explanation for the development of self-replicating organisms

4. describe some of the oldest fossil life forms found in rocks and relate them to the early environment of Earth

5. relate the development of photosynthesis to a change in the atmosphere from reducing to oxidizing conditions

6. trace the flow of nutrients and energy through the food chain within an ecosystem

7. compare the energy production of mankind with that of natural processes

8. describe the operation of the biogeochemical cycles for carbon, nitrogen and phosphorus

9. relate the distribution of these three elements within the biosphere and geosphere to the operation of the biogeochemical cycles

10. describe the effects of fossil fuel burning and the use of chemical fertilizers upon the operation of these cycles.

3. Key Terms and Concepts

geosphere

biosphere

prokaryote

amino acids

nucleic acids

reducing atmosphere

oxidizing atmosphere

stromatolites

eukaryote

photosynthesis

producers

consumers

herbivores

carnivores

omnivores

decomposers

energy cycle in the biosphere

nutrients

biogeochemical cycles:

    carbon cycle

    nitrogen cycle

    phosphorus cycle

biomass

nitrogen fixation

eutrophication

Greenhouse Effect

Gaia Hypothesis

4. Corresponding Video

Here, you will explore the role of life in shaping our planet and its future. Through animation, the program will recreate Earth as it may have been 4.6 billion years ago. You will visit the outback of Australia to find the oldest record of life on Earth. In the tidepools along the California coast, the early chemistry of life and the formation of the first cells will be recreated. A visit to the Pacific Islands of Palau will show how life creates coral reefs and how it may then literally eat them away. The program will explore the Gaia Hypothesis, the suggestion that life manipulates the Earth to make the environment hospitable for itself.

B. GEOSPHERE AND BIOSPHERE

Of all the thousands of photographs taken by the Apollo astronauts, the one that has most captured the attention of the public, non-scientists and scientists alike, is the remarkable picture taken from a command capsule in orbit around the Moon that serves as the symbol for this telecourse. Above the bleak and cratered terrain of the Moon's horizon rises the blue, white, and brown orb of Planet Earth. When this picture first appeared in magazines and on television it seemed to proclaim, "We have truly cut loose from Earth and traveled in space", showing both the destination and starting point of our travels. But as time went on, the picture seemed to take on new meaning: the connotation of spaceship Earth. It is hard to deny the finite nature of a world that can be viewed in a single glance.

The concept of that small glittering sphere as home to the entire mass of humanity, in all its forms, customs, tribes, and nations is humbling and sobering, reminding us of the overwhelming need for all humanity to coexist in peace and cooperation as we share this small planet. The sense of finiteness conveyed by the picture also seemed to echo environmental concerns that had grown and matured in preceding decades. These concerns had earlier roots in thinkers like St. Francis, Darwin, Emerson, Thoreau, and Muir, who marveled at the finely tuned balance of natural processes and worried about the impact that intruding mankind would have on such delicate machinery.

But now a new interpretation of this photograph became possible, using a line of reasoning followed by a growing band of scientists with eclectic backgrounds and wide-ranging interests. This was to be a holistic view of the planet, in which mankind, life, oceans, atmosphere, and the rocky tectonic engines of the physical planet itself were not to be regarded as separate and compartmented, but one vast, intricate, interconnected living machine. With its immediate roots in the relatively new science of ecology, this view also stretches out tendrils to gain support and understanding from such diverse sciences as astronomy, climatology, oceanography, geology, geophysics, paleobiology, and even planetology -- in short, from all the fields mentioned in this telecourse.

In this view, life is inextricably part of the Earth and part of the balance of nature. To separate studies of the geosphere (the physical world of geology) from those of the biosphere (that part of the world inhabited by life) or vice versa is to break so many connections as to render the smaller view incomplete and often misleading. It is not too hard to accept the influence of the geosphere upon the biosphere -- the adaptation of life to its physical surroundings -- but more difficult is the notion that the biosphere has an important counter-influence upon the geosphere. Earth itself has been shaped by life, and for that reason is unique in so many ways among the planets of our solar system. Consider the remaking of the atmosphere by life on Earth. Only on this one planet do we find an atmosphere rich in oxygen and nitrogen. Even so distant, separate, and physical a phenomenon as the aurora has been shaped by life. The principal colors that we observe in these high-altitude light shows are due to the presence of oxygen and nitrogen in the atmosphere.

C. THE ORIGINS OF LIFE

Let us take a closer look at the original events that so changed Earth. Recall from the unit **Origins** that Earth's first 1,000 million years were extremely turbulent ones, with a rain of infalling debris left over from the formation of the solar system dominating the geological processes. We don't know exactly when or how life first made its appearance on Earth, but we do know that life had already been around for some time when Earth was only about 1,100 million years old, 3,500 million years ago.

The earliest fossils discovered so far date from that time and are found in Australian rocks. These are fossils of very primitive single-celled organisms of a group called prokaryotes. Today this group is represented by blue-green algae and bacteria, which reproduce asexually by cell division. As a result, the offspring are genetically similar to the parents. Just how long life had existed before these oldest fossils is not known.

By the time these microscopic messengers became sealed in their rocky time capsules, the final stages of volcanic flooding of the maria were drawing to a close on the Moon. On Earth, the landmasses were barren; the sea was to be the mother of all life.

1. The First Living Cells [31]

How did this remarkable phenomenon come about? What was the origin of life on Earth? There is hardly any other question that holds such a fascination for our species. Every religion and every primitive folklore has its own particular answer. Science can only look at the available evidence and construct logical models that predict the eventual appearance of tiny fossils in the rocks of Australia.

One way to go about answering this question is to reason back from what we know today about the chemical basis of life. The unit of life is the cell, and at some point cellular organisms must have assembled from the components available on the early Earth. If we could somehow transport ourselves 4,000 million years back in time, what would the first living cells be like? In order to be called alive, they would certainly have the primary characteristics of the living state, which include growth, reproduction and the ability to evolve. Let's examine these properties individually.

Growth in cells today occurs through a process by which energy and nutrients are extracted from the environment, the energy being used to cause the nutrients to be assembled into new cell components. The assembly depends on metabolism, particularly enzyme-catalysed biosynthetic reactions, and is directed by a kind of molecular blueprint containing the information needed for the assembly process. The information itself is copied during cell reproduction by replicating the molecules in which it is stored. In even the simplest modern cell, vast amounts of information are present, and this is used to direct the synthesis of hundreds of enzymes and other proteins involved in metabolism and structure.

In the earliest life forms, probably nothing more was present than a small, slowly replicating molecule, together with an equally small catalytic molecule which happened to be

---

[31]This section was written solely for use in this text by David W. Deamer, University of California, Davis

able to enhance the rate of replication. Most important, a mechanism must also have been present which permitted the replicating molecule to direct the synthesis of the catalytic molecule. Only when this loop between information and catalytic activity is closed can evolution occur, because only then can random changes in the replicating molecule occasionally produce a favorable change in the catalyst that in turn increases the ability of the system to extract nutrients and energy from the environment. Finally, for an actual cell to exist, by definition there would need to be a boundary membrane of some sort which could encapsulate the system and provide a microenvironment conducive to its overall function.

Even at this relatively simplified level, the first cell seems fairly complicated. Is it conceivable that such a system could have assembled by chance on the early Earth? First we can ask what sorts of molecules might have been available. The major molecular components of living cells today include nucleic acids like DNA and RNA for the replicating molecules which transmit information, and proteins called enzymes for the catalytic molecules. The genetic code carried in DNA directs the synthesis of proteins from amino acids, while lipid bilayers[32] form the boundary that separates the cell interior from the external environment. Could any of these general classes of molecules have been present 4,000 million years ago?

The answer to this question is a cautious yes, and this is where our first step toward a plausible scheme for the origin of life begins. First we can ask whether any of the simpler molecules of the life process might have been available. The early solar system involved not only geological processes, but also chemical processes, with vast clouds of gas and dust being acted upon by violent energies, including ultraviolet light and heat from the sun, electricity (lightning), and heat from volcanic activity. A historic experiment was conducted in 1953 which attempted to simulate these conditions in order to observe the chemical reactions that might occur. At that time, it was believed that the composition of the early atmosphere was what chemists call "reducing", in contrast to today's oxidizing atmosphere. That is, no free oxygen was present, but instead the mixture was dominated by hydrogen, the most abundant element in the solar system, with smaller quantities of the reduced forms of carbon (methane), nitrogen (ammonia), and oxygen (water). In the experiment, a mixture of these gases was subjected to an electrical discharge to simulate lightning, and the composition was analyzed over a period of time.

The results were truly astonishing. The mixture turned reddish brown after only a day or so, and a week later, when analyzed, was found to contain numerous organic substances including several amino acids, the building blocks of proteins. This clearly suggested that there was nothing special about amino acids, but instead they could have been readily available as dilute solutions in the lakes and oceans of the early Earth. Similar experiments have been repeated under a variety of conditions, and it is now clear that all of the main components needed to form the major kinds of biomolecules could have been synthesized by the chemical processes occurring on the prebiotic Earth.

---

[32] Lipids are biochemical compounds that have "split personalities." That is, one portion of the molecule is oil-like, and does not readily mix with water, while the other portion is water soluble. Some examples of lipids include fat, vegetable oils, cholesterol, and phospholipids such as the lechithin found in egg yolk. Biological membranes contain phospholipid arranged in two layers (the lipid bilayer) and this sheet-like structure provides a relatively impermeable barrier that separates the cell interior from the external environment. Other membranes compose various sub-cellular structures inside the cell. Essentially all biological membranes have a protein component as well, often present as embedded enzymes that mediate fundamental cell activities such as light-trapping, protein synthesis and transport of nutrients.

Despite the satisfaction of discovering a chemical basis for early chemical evolution, simulation experiments still leave something to be desired. They only represent our best guesses at the actual conditions 4,000 million years ago, and in fact, we already are reasonably certain that the guess about a reducing atmosphere was off the mark. Instead, the Earth was probably at volcanic heat early in its history, due to the energy content of the solids that accreted to form the planet. It follows that the reducing atmosphere would have been rapidly replaced by a "volcanic" atmosphere of carbon dioxide, water, and nitrogen, which would have boiled off from the accreting dust as it heated up. However, recent experiments have shown that significant synthesis of organic molecules can occur under these conditions as well.

Is there any way to test the concepts brought out by the simulation experiments? Probably the best test comes in the form of certain meteorites that crash to Earth every few decades. Most meteorites are made of stony material or metallic alloys of iron and nickel. However, on rare occasions a meteorite falls that contains small amounts -- a few percent -- of carbon. Furthermore, when this carbon content is analyzed, a substantial portion of it turns out to be organic, composed of amino acids, hydrocarbons, lipid-like molecules, and even traces of purines and pyrimidines, building blocks of nucleic acids. All of this organic material was synthesized by abiotic processes, probably in planetoids that never got large enough to become true planets. Since similar chemical processes must have occurred on the early Earth, this discovery represents a remarkable confirmation that the simulation experiments are on the right track.

We now come to the most important question of all, and yet the one we know the least about: assuming that the simpler organic compounds of the life process were present on the prebiotic Earth, how could they assemble into the first cells? An important concept related to this question is that many biologically relevant molecules have an extraordinary ability to fit together into larger aggregates called supramolecular structures. This process of self-assembly is still not understood in detail, but some important examples include the manner in which purine and pyrimidine bases form hydrogen bonds to stabilize the double helical structure of DNA. Another example is the self-assembly of lipid molecules into the lipid bilayer structure of membranes. Self-assembly is fundamental to the architecture of all modern cells, and certainly must have been involved in early life forms.

The question can now be stated more clearly: what kinds of self-assembling molecules could have provided the first replicating information carriers, the first catalysts and the first membrane boundaries? If we can answer these questions, perhaps the assembly of a whole cell will be less mysterious. First, what sort of replicating molecules might have been present? The only one we know of is nucleic acid, which has the ability to make copies of itself with the help of enzymes called polymerases. Recently, an experiment has been performed which in its way is as important as the original finding that amino acids are produced under prebiotic conditions. In the experiment, chemically active forms of nucleotides were mixed with a small amount of nucleic acid which could act as a template. When the mixture was analyzed a week later, it was found to contain polymers of the nucleotides, with chains up to 40 nucleotides long. This is the first indication that polymerization and replication can occur in the absence of enzymatic catalysis, as it must have in the earliest chemical systems on the evolutionary pathway to the first true cells.

What about catalysts? A catalyst is anything that speeds up a chemical reaction without actually taking part in it and becoming changed itself. Biological catalysts today are all proteins which have active sites with catalytic activity. Could amino acids have assembled into protein-like compounds on the early Earth? This turns out to be relatively simple. When mixtures of amino acids are simply heated to 60° C or higher, after a few weeks numerous chemical linkages have formed and polymeric compounds can be isolated. There is even evidence that some of these polymers have modest catalytic ability.

The last question is related to the boundary membrane. Again, this has turned out to be relatively straightforward because of the self-assembly properties of lipid-like molecules. It has been found that fatty acids, glycerol and phosphate, when dried under the same conditions described above for the formation of protein-like polymers, are able to produce simple phospholipids. These in turn readily form membranous sacs when exposed to water, and the sacs can encapsulate other molecules even as large as DNA. Recently it has been discovered that some of the organic components of carbonaceous meteorites are able to form membrane sacs, again showing that we are not limited to simulation experiments.

We can finally return to our original question. How might the first cell have been assembled? Imagine an environment such as a tide pool which undergoes daily cycles of drying and heating by the Sun, followed by rehydration as the tide comes in. Under these conditions, the dilute solutions of organic molecules described above would be concentrated to thin films on the surfaces of rocks and sand particles, and the heating would drive polymerization reactions so that larger molecules would be continually forming. Lipid-like molecules present would tend to encapsulate the polymerized products, and on rehydration membranous sacs containing the polymers would float away to land somewhere else for another drying cycle. Now imagine this process of natural experimentation occurring along tens of thousands of miles of coastline, and over millions of years of time. It seems almost inevitable that just once, a simple replicating molecule would find a way to interact with a simple catalyst in a membranous sac, and life would begin.

2. Did Clay Provide the Framework for Life?

While the basic building blocks of a cell can be constructed by random events such as in the spark experiment, it seems unlikely to some scientists that a fully-functional cell could result from random processes. As one scientist puts it, if you take a pile of aluminum, copper, and insulation, grind it all up and throw it down a flight of stairs, you don't expect a jetliner to fly away at the bottom.

Recent observations, however, have suggested a surprising possibility. Ordinary clay is made up in part of very small mineral grains that are constructed of flat layers like sheets, piled one atop another. These layers are only lightly bonded together, resulting in the slipperiness characteristic of clay. Like cards in a deck, clay mineral layers can pack a lot of surface area into a small volume. In addition, the surfaces of each layer are chemically active and can bond organic molecules to them. It has been suggested that clay surfaces might act as a kind of template, governing the structure of bound molecules and aiding in their assembly into more complex and regular forms.

It is even possible that the clay itself is able to replicate its own structure, including variations that might exist within that structure. Water causes clay layers to expand and separate slightly. New layers can then form in the spaces opened up between the old layers, and the new layers are imprinted with the structure of the old layers. In this way, distinctive clay layers have the potential to "breed true" in generation after generation.

The combination of large surface area, possible template activity, and a crude kind of inorganic self-replicating ability might have introduced the necessary degree of order and reproducibility to have allowed a primitive type of natural selection to begin its task. As an example of natural selection in such a system, certain chance clay-organic combinations may have proved especially successful at self-replication and so they would have become more numerous.

Finally, if a molecule like a nucleic acid were to result from large numbers of repetitions of this process, the newly-formed organism would be able to replicate itself in a wholly organic manner and would no longer require the clay, but could cast it aside like a crutch that is no longer needed. It is a highly speculative but captivating idea, that clay might have provided the catalyst that enabled the quantum leap from random organic molecules to the first living, reproducing cell.

3. The Next Steps

However that momentous first assembly was accomplished, our only evidence is in those 3,500 million year old single-celled fossils. Already, these ancestors of modern blue-green algae were fairly complex and may have been capable of photosynthesis, the conversion of light into chemically useful energy. An important byproduct of this process is the conversion of carbon dioxide into oxygen. Before the ancient algae must have come more primitive forms yet, tiny chemical factories that scavenged the abundant organic molecules that had been produced by lightning and other nonbiological processes. Next came single-celled organisms that could manufacture their own food from readily available inorganic compounds such as carbon dioxide, hydrogen sulfide, and ammonia (NH_3).

Some structures built by blue-green algae today bear striking resemblance to early fossils found in western Australia. They form dome-shaped masses about the size of cabbages, and are similarly layered. The structures, commonly made of carbonate rock, are called stromatolites. Today they form mostly in intertidal zones, alternately being exposed to air and covered by water as the tides come and go, but also extend into the subtidal. The first Precambrian fossil stromatolites very much resemble the structures formed by their present-day counterparts.

Colonies of microscopic blue-green algae trap and bind sediment and can bring about the precipitation of mineral material, which in time builds up the layered structures characteristic of stromatolites. The video corresponding to this unit will take you to the shores of Australia where you can witness these living remembrances of things long past.

With the development of photosynthesis, the release of oxygen by chlorophyll began a slow but profound change in Earth's atmosphere. The first half of Earth history, until about 2,000 million years ago, was characterized by anaerobic conditions (deficient in molecular oxygen). However, the rock record contains abundant rocks, called <u>banded iron-formations</u>, that are rich in oxidized iron minerals. How can this apparent enigma be explained? Under anaerobic conditions in the ancient oceans, the reduced variety of iron, ferrous iron (Fe^{++}), is soluble in water. If oxygen is introduced, the ferrous iron will oxidize, combining with oxygen to form the ferric oxide mineral hematite (Fe_2O_3), which will precipitate out of the water and be deposited. Stromatolites are constructed by blue-green algae (called cyanobacteria) and can produce oxygen. The oxygen produced by the stromatolitic blue-green algae and phytoplankton reacted with the ferrous iron dissolved in the seas, resulting in the accumulation of beds rich in iron oxide. Indeed, most of our iron ore comes from these ancient banded iron-formations.

The seas had many hundreds of millions of years in which to build up considerable quantities of ferrous iron and this had to be swept from the oceans before oxygen could begin to accumulate in the atmosphere. With the buildup of atmospheric oxygen, ozone (O_3) could be created by the action of sunlight in the upper atmosphere, establishing a filter for harmful ultraviolet radiation. With the establishment of the ozone layer, the level of ultraviolet radiation from the Sun at the Earth's surface was reduced to the point where organisms could live in very shallow water environments and later could even live full-time out of the water.

The change from a reducing to an oxidizing atmosphere has been described as the greatest air pollution event in the history of Earth. For it almost certainly brought disaster to the anaerobic prokaryotes that had flourished until then. Oxygen is a poison to all life. We and other aerobic organisms have evolved special enzymes to neutralize the toxic effects of oxygen yet permit the oxygen to function in metabolism. The evolution of oxygen-mediating enzymes must have co-occurred with the first release of oxygen from blue-green algae.

Once oxygen appeared, the dominant anaerobic biota had three choices: 1) retreat to permanently oxygen-deficient places such as muds; 2) evolve oxygen-coping abilities; or 3) become extinct. No doubt all three occurred. The remaking of planet Earth by its own life forms had begun.

The next major step was the development of eukaryotes. Unlike the prokaryotes, these organisms are far more organized and complex. Their genetic material is organized into a well-defined nucleus that is surrounded by a nuclear wall membrane within the cell. In addition, eukaryotes contain small specialized units called <u>organelles</u>. Among these are plastids such as <u>chloroplasts</u>, which carry out photosynthesis, and <u>mitochondria</u>, which combine carbohydrates and fatty acids with oxygen to release energy. Working together, the chloroplasts and mitochondria act as a means of gathering and utilizing solar energy, storing it in the form of carbohydrates and drawing on these reserves when there is no sunlight.

The first eukaryotes may have formed as a symbiotic association of specialized pro-karyotic individuals, each providing a different function and gradually evolving into organelles within a single eukaryotic cell. This new complexity and degree of organization opened the way to additional developments: the appearance of multicelled organisms and, most importantly, sexual reproduction. The latter opened up tremendous variability within a species because offspring now derived their genetic makeup from a mixture of genes from both parents. The processes of evolution could now advance at a far more rapid pace.

Unlike the early prokaryotes, which were adapted to living in an anaerobic environment, eukaryotes were aerobic, adapted to deal with an environment rich in oxygen. By the time that they appeared, oxygen must have begun to accumulate in the atmosphere and oceans of Earth. The first eukaryotic fossils appear in rocks about 1,300 million years old.

With the remaking of the atmosphere, life placed its stamp on the third planet, the only planet with liquid water. Eventually, a species would evolve that would begin anew the process of alteration of the natural environment. Homo sapiens had arrived, and even from space, the dark side of Earth could be seen to be spangled with the lights of his cities.

STUDY QUESTIONS

XV-1. What distinguishes a eukaryote from a prokaryote?

XV-2. What is the significance of Precambrian banded iron-formations?

XV-3. What role might clay minerals have had in the early development of life?

D. THE ENERGY BUDGET OF THE BIOSPHERE

Embedded in the environmental ethic is the concept that nature is governed by many interacting parts that, prior to the influence of mankind, had evolved into a state of balance and internal harmony. This concept has become familiar in the term the balance of nature. To many, it seems that the introduction of man's influences have hopelessly upset that balance. Let us look at this situation in a little more detail.

Taken literally, the idea of a balance in nature can be misleading, especially if it is considered to be a static balance. Rather, what we are really talking about is a complex, internally balanced set of interactions which essentially produce the distribution of species and the physical and chemical environment. As the species have changed with evolution, so have the interactions, the balances, and the environment as well. To change any of the latter is certainly going to change the species distributions; some more, some less. Some balances are very delicate; others are not. As one of the species inhabiting the biosphere, we affect those balances. For the remainder of this course, we shall use this rather more complex and dynamic meaning for "the balance of nature."

From an environmental point of view, the principal issue of science is to identify those points at which our interventions most seriously change the distributions presently found in nature, and those which are not so likely to produce significant change. Once potential changes and their likely causes have been identified, then society as a whole can make intelligent value judgements as to whether those changes are desirable or undesirable.

Before we can properly assess the impact of mankind on the biosphere and on the biota that inhabit it, we should gain a better understanding of the processes that maintain ecological conditions. In the limited time and space remaining to this course, we cannot cover all the appropriate principles of ecology, but we should examine at least those that are most important to interactions with the geosphere, which has been the focal point of our interests so far. Then, perhaps, we will be in a better position to determine which among mankind's activities are most hazardous in the long run to us and to all the other inhabitants of our planet.

You have already seen the prodigious increase in our use of energy, much of it in the form of fossil fuels drawn from the Earth. The biota of Earth are also consumers of energy on a grand scale, though nearly all of the energy consumed in nature is solar. Only about half of incoming solar radiation makes it through the atmosphere to the ground or water surface; about 20-30% is reflected back into space; and the rest is absorbed by the atmosphere. Much of what strikes the surface is immediately converted into heat, but a small portion -- less than one percent -- is utilized in the process of photosynthesis to produce complex carbohydrates from water, carbon dioxide, and nutrients. This process adds to the biomass -- the total mass of living matter.

The process is not efficient for a number of reasons. For instance, the annual net photosynthesis in the higher plants requires that a significant portion (probably 10 to 20%) of the net heat available at Earth's surface must be used to drive transpiration (the direct transfer of water in vapor form from the plant to the atmosphere). This is more than ten times greater than the biochemical energy stored in the carbohydrates.

The process of photosynthesis is a highly complex process that may be represented by a simple chemical equation:

$$6\ CO_2 + 12\ H_2O \underset{\text{respiration}}{\overset{\text{photosynthesis}}{\rightleftharpoons}} C_6H_{12}O_6 + 6\ O_2 + 6\ H_2O$$

The equation does no more than state that six molecules of carbon dioxide along with twelve molecules of water are changed in the process of photosynthesis into one molecule of the carbohydrate sugar ($C_6H_{12}O_6$), six molecules of oxygen, and six molecules of water. The process requires energy from sunlight and the intervention of chlorophyll, an organic substance that is itself constructed of carbohydrates and nutrient minerals. You may check for yourself to see that there are the same number of carbon, hydrogen, and nitrogen atoms present on either side of the equation. The individual atoms are conserved; it is only their arrangement into molecules that is changed.

Sugar is one substance in which energy may be stored by organisms for use when needed. The equation may be reversed, oxidizing sugar to produce carbon dioxide and water, which are released into the atmosphere in the process of respiration. The energy formerly needed to produce the sugar is now released. This, of course, is just the process that occurs within your body when you eat a candy bar and then "work it off" by jogging or by some other physical activity. You "burn up" (oxidize) the sugar to provide needed energy.

Photosynthetic plants are called producers, in that they produce carbohydrates and form the base of the food chain. Figure XV-1 shows how energy and nutrients move through the biosphere starting with the producers.

Herbivores are animals that eat plants, and carnivores are animals that eat other animals. Some carnivores eat only (or primarily) herbivores, while some eat other carnivores. There are, of course, omnivorous animals like ourselves who will eat almost anything.

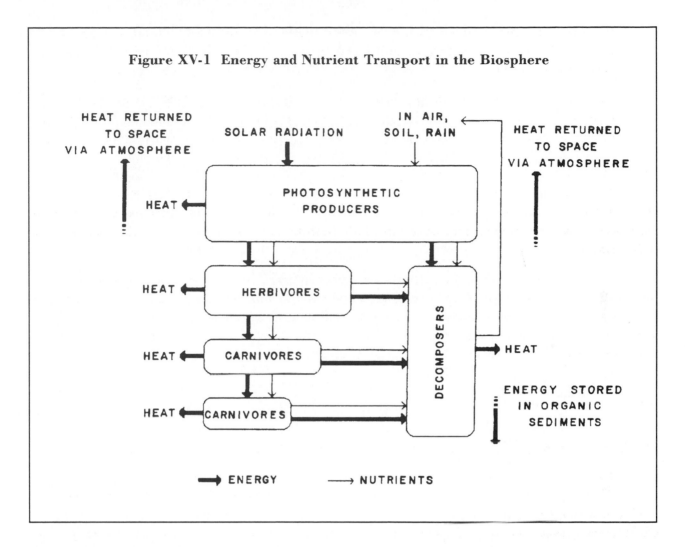

Figure XV-1 Energy and Nutrient Transport in the Biosphere

Collectively, we are all called <u>consumers</u>. In the process of consuming other organisms, we who are farther along the food chain, gain energy and nutrients from earlier members of the chain.

In the end, though, all energy is derived from the photosynthetic producers. At each step of the way, heat is lost via body heat, heat produced by friction in movement, and so on. In fact, most of the energy that is being passed along the chain is lost to heat. This results in smaller and smaller populations that can be supported at each stage of the food chain.

Producers are most abundant, but only about 10% of their stored energy is passed along to herbivores. Only about 10% of the stored energy of herbivores is passed on to primary carnivores, and the secondary carnivores get only 10% of that. This is why there are huge numbers of edible plants, lots of rabbits, and relatively few foxes. It also explains why there is a great advantage to being omnivorous, a condition that allows us to break the restrictions of being at the end of the food chain by picking and choosing from whatever part of the chain we wish. Many of the more highly developed mammals share this trait with us. Our own specialty, however, is technology. We are able to intervene in the natural cycles, adding energy wherever it suits our purposes in hopes of achieving a simplification of the cycle and a resulting greater efficiency of energy use.

In addition to this somewhat linear food chain, there are also the decomposers (see Figure XV-1). These are the agents of decay and decomposition, the bacteria and the fungi. They take the wastes and the remains from all the other stages and complete the process of oxidation, returning water and carbon dioxide to the atmosphere and nutrients to the soil. In effect, they close the cycle, reclaiming these substances for use again by the producers. You may recall encountering this concept earlier, in the unit **Dynamics of the Oceans**. There it was used to explain the high level of nutrients found in ocean bottom water due to a steady rain of decomposing organic material from the surface layers.

Energy, however, is not fully recycled. It passes in only one direction along the food chain and is eventually lost to heat. This heat contributes to raising the temperature of the environment and is finally lost to space as infrared radiation, along with heat from sunlight that never participated in the activity of life (which accounts for greater than 99% of the solar energy flux).

Where does modern, industrialized mankind fit into this scheme of things? First, as omnivores, we have an enormous variety of food choices available to us, giving us considerable adaptability to changing conditions. The diet of a Chinese farmer is vastly different from that of the average American, and yet both manage to survive. Second, through agriculture and animal husbandry we can manipulate the food chain itself. Cattle, which we like to eat, are encouraged to multiply, while mice, which most of us do not like to eat and which tend to compete with us for food are discouraged. Third, and most important, we are able to go outside the food chain entirely and manipulate energy directly. In this way we can intervene in the natural energy cycle and manipulate populations, such as in the case of domesticated cattle.

Our heavy dependence on fossil fuels is merely a way of drawing energy from long-vanished ecosystems, but our manipulation of direct solar energy, geothermal, tidal, and nuclear energy is a radical departure for life on Earth. Because the machinery by which we accomplish this is relatively compact, the principal effects are often localized in nature. A large power plant that discharges waste heat into a river can raise the temperature of river water for kilometers downstream to beyond the tolerance limits for some species of fish, resulting in massive kills. To avoid this, heat is more often vented in huge cooling towers that use the evaporation of water to carry waste heat off into the surrounding atmosphere. Even this, though, may result in local weather modification, leading to increased precipitation and the creation of "heat islands". The effect has been called thermal pollution, and is one of the most difficult kinds of pollution to avoid, since all our activities result in the production of waste heat.

How does our energy production compare to that of nature? On an annual basis, mankind's total world energy consumption is roughly 300 times the energy released in all the earthquakes that occur on Earth and about three times the total energy dissipated by the tides. It is about one-fifth as great as the radioactive heat generated within the entire planet. Because most geothermal heat escapes quietly by conduction through Earth's crust, this means that our rate of energy production is probably at least comparable to that of all the volcanoes on Earth. Clearly, our activities are not to be dismissed as inconsequential.

On the other hand, the net radiation energy from the Sun at Earth's surface is some 6,000 times our own heat production. If photosynthesis utilizes about 0.2% of the solar energy flux, then the energy stored in biomass on a worldwide basis is only about 12 times our total energy production. While this may seem small on a worldwide basis, much of our energy consumption is carried out in a few fairly restricted geographical regions, mostly in the northern hemisphere. This should (and does) give rise to noticeable effects on a local scale.

A 1977 report from the National Academy of Sciences[33] concluded that direct energy release from mankind's activities would probably cause no more than a 0.5°C (0.9°F) increase in global temperature over the next century or so, but it also warned that much more severe effects might be seen due to increasing carbon dioxide levels in the atmosphere, acting via the Greenhouse Effect.

To what extent can the balances of nature deal with our increased production of carbon dioxide? In the next section we shall look at how these balances work for carbon and two elements important to agriculture: nitrogen and phosphorus.

<u>STUDY QUESTIONS</u>

XV-4. Why does the population decrease with each step along the natural food chain?

XV-5. Is the industrial heat that we are generating likely to drastically increase global temperatures?

E. BIOGEOCHEMICAL CYCLES

Many elements essential to life are cycled and recycled through both the geosphere and the biosphere. As a result, their concentrations in the atmosphere, oceans, and soil are dependent on the balances that are struck within the cycles that involve them. Though we shall treat each cycle separately, you should not assume that they operate independently from one another. In each of them, you will see common components, particularly the photosynthetic <u>producers</u>, the carbohydrate <u>consumers</u> (herbivores, carnivores, omnivores), and the <u>decomposers</u>. Changes in population and makeup of any of these components cause changes in all of the biogeochemical cycles and generate interactions between them.

1. The Carbon Cycle

Carbon is the fundamental building block of life. The reason is that carbon is one of the few common elements whose chemistry is sufficiently complex to permit the construction of the tremendous variety of compounds required in life processes -- the organic molecules. Something as complex as the genetic materials DNA and RNA are only possible because of the large number of ways in which carbon can form chemical bonds with other elements.

---

[33]<u>Energy and Climate</u>, National Academy of Sciences, Washington, D. C. 1977

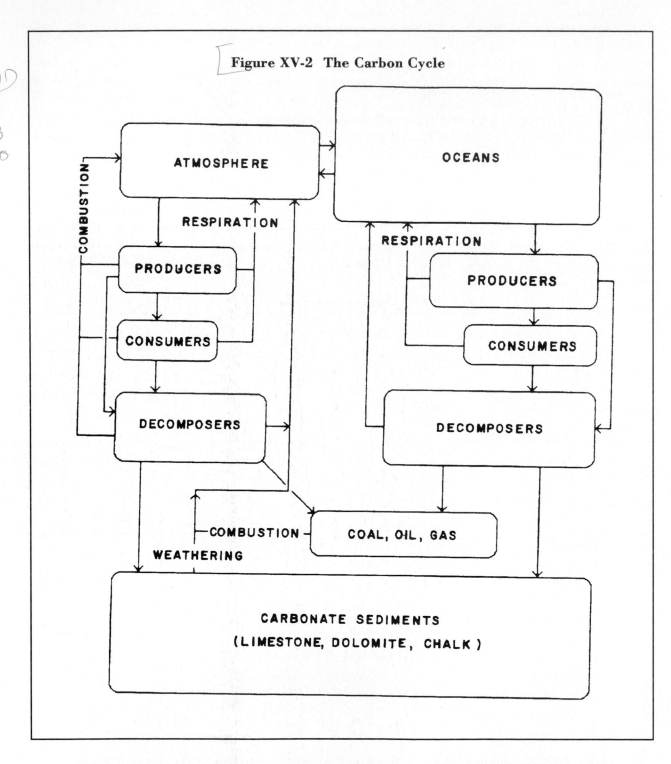

Figure XV-2 The Carbon Cycle

The carbon cycle is diagramed in a simplified way in Figure XV-2. Each box represents a reservoir or storage component for carbon; in its cycling through the process, carbon may reside in any of these reservoirs for a greater or lesser length of time before continuing its journey through the cycle. The arrows connecting the boxes represent the paths that carbon may take in moving from one reservoir to another.

Consider the atmospheric portion of the cycle, in the upper left part of the diagram. Photosynthetic producers (plants) draw carbon from the atmosphere in the process of converting carbon dioxide and water to carbohydrates. Plants do not only breathe out (respire) oxygen, but parts of their tissues also respire carbon dioxide, just as we do. These processes provide the plant with energy for growing and maintaining its life support systems, and go on at all times. During the sunlit day, more carbon dioxide is consumed than is released in respiration, but at night photosynthesis ceases and the plant respires pure carbon dioxide, returning a portion of its carbon to the atmosphere.

The consumers (herbivores, carnivores, omnivores) eat plants and each other and pass carbohydrates along the food chain. Each of them respires carbon dioxide, returning a portion of the carbon to the atmosphere. Waste products contain carbon, and when the organisms die, their bodies are attacked by decomposers who oxidize the carbohydrates and return the carbon to the atmosphere as carbon dioxide. A forest fire might consume the entire local food chain at some point, oxidizing producers, consumers, and even decomposers, and returning at one stroke most of the carbon to the atmosphere in the process of combustion.

Not all terrestrial carbon is recycled back into the atmosphere, however. As you may recall from the unit **Energy Resources**, coal is formed from decaying vegetation that is prevented from oxidizing in an anaerobic environment such as a swamp. The carbon in coal is effectively withdrawn from the cycle for very long periods of time and remains trapped within sedimentary rocks until it is exposed by erosion or by mining.

The right side of the diagram is concerned with the oceanic part of the carbon cycle. Producers (phytoplankton), consumers (zooplankton, fish, crabs, etc.), and decomposers operate much as their terrestrial counterparts, except that outright combustion does not occur. Again, carbon is withdrawn from the cycle in the process of oil and gas production. In the sea, however, a far more important withdrawal mechanism exists: the production of calcium carbonate, the principal constituent of limestone. Calcium carbonate has the chemical formula $CaCO_3$ and is the principal geological repository for carbon. Other carbonate rocks, like dolomite which is composed of magnesium carbonate $(Ca,MgCO_3)$, also serve in this capacity. Freshwater limestones also form in lakes, but marine carbonates are far more abundant.

TABLE XV-1

Amount of Carbon Residing in Different Reservoirs
Quantities in Gigatons (10^9 metric tons)[34]

Atmosphere	710
Continents	
Biomass	590
Litter	60

[34]Data from U.S. Department of Energy, 1980, <u>Carbon Dioxide Effects Research and Assessment Program</u>, Washington, Carbon Dioxide and Climate Division, Report 008, and from B. Bolin, <u>Scientific American</u>, September 1970.

Soil	1,670
Fossil fuels	5,000
Carbonate sediments	20,000,000
Oceans	
Biomass	4
Seawater	
Surface layers	680
Intermdiate	8,200
Deep waters	26,000
Sediments	4,900

Table XV-1 lists the approximate sizes of the different reservoirs in the carbon cycle. Note that the combined terrestrial and oceanic biomass reservoir is comparable to that of the atmosphere, followed in size by soil carbon, deep ocean sediments, seawater, and, dominating the total, the carbonate sediments. Here, then, is where the vast bulk of the carbon dioxide that may at one time have dominated Earth's atmosphere has gone, placed in the Earth's crust by biological action. Some aquatic plants are so efficient in converting dissolved carbon dioxide to calcium carbonate that they can precipitate it at the rate of two percent of their own weight for every ten hours of exposure to sunlight. Perhaps it is appropriate that many of our finest buildings -- the great cathedrals among them -- are often constructed from limestone, one of ouur most useful rocks. These stones contain the vast stores of carbon that have been collected from the atmosphere over the aeons and trapped safely in a benign solid form.

Notice also that the atmosphere and the oceans are able to exchange carbon dioxide between them. This is accomplished at the sea-air interface, where carbon dioxide is readily dissolved in seawater. Because the oceans hold in seawater 50 times the carbon dioxide found in the atmosphere, they provide not only a large reservoir but also a mechanism for regulating the carbon content of the atmosphere. On short time scales, however, only the surface layer of the ocean exchanges carbon dioxide freely with the atmosphere, and this constitutes a reservoir comparable to that of the atmosphere.

Recall from the unit **Climates of Earth** that the carbon dioxide content of the atmosphere has increased by some eight percent between 1958 and 1982. This effect is due to the burning of fossil fuel, but the total amount of carbon dioxide released to the atmosphere is actually twice the amount of the observed increase. The oceans (and perhaps increased biomass production as well) have apparently absorbed half of the total excess carbon dioxide released in that time.

2. The Nitrogen Cycle

Nitrogen is the most abundant gas in the atmosphere, accounting for 78% of the total. It is also biologically critical to organisms because of its role in the structure of proteins and nucleic acids (RNA and DNA). Like carbon, nitrogen supports a complex chemistry that allows it to participate in many and diverse organic molecules.

With its great abundance, you might expect that a nitrogen shortage would be difficult to produce in the environment. In fact, molecular nitrogen in the atmosphere cannot be used by most organisms, but must first be "fixed" or converted into the highly soluble forms of ammonia (NH_3), or the nitrate (NO_3^-) or nitrite (NO_2^-) ions.

Few inorganic processes can muster sufficient energy to accomplish the task. Meteors, cosmic rays, and lightning fix some nitrogen, but the amount is negligible. The primary natural agents for this task are nitrogen-fixing bacteria in the soil, many of which exist in symbiotic relationship with plants. These plants -- alfalfa, clover, peas, beans, and others -- have long been used to "build up" soil that has become depleted in nitrogen. When depletion occurs, leaves turn yellow and plant growth and productivity are stunted. Blue-green algae are also able to fix nitrogen.

Other bacteria in the soil derive their energy by converting fixed nitrogen into nitric oxide (NO), nitrous oxide (N_2O), and molecular nitrogen (N_2), which are then returned to the atmosphere. Providing the principal return path to the atmosphere, these denitrifying bacteria serve to keep the overall nitrogen cycle in balance.

Plants take up fixed nitrogen through their roots and pass it on in the form of proteins and amino acids to the rest of the food chain. The nitrogen is returned to the soil in the form of wastes and the products of decay. Note that the atmosphere is not directly involved at this stage. So far as your body is concerned, nitrogen in the atmosphere is simply an inert gas that carries vital oxygen to your lungs. The reason is that dinitrogen (N_2) is inert, while odd nitrogen (N) is chemically active.

Soil erosion carries fixed nitrogen into streams, lakes, and the oceans, often leaving behind barren and infertile soils. In addition, intensive farming practices do not allow fields to lie fallow long enough to regain their depleted fixed nitrogen supplies by natural processes. For this reason, large quantities of industrially fixed nitrogen are added to the soil in the form of chemical fertilizers. It is produced by passing atmospheric nitrogen and hydrogen over a catalyst at high temperatures (about 500°C or 900°F). It is interesting to note that we must take such extreme and energy-consuming measures, while nitrogen-fixing bacteria can do the same thing very efficiently at low temperatures by processes that are not yet understood. They are slower to act, however, and modern farming cannot always wait for natural replenishment of the soil to occur. Critics of modern farming practices have sardonically labeled them a means by which oil and gas are converted into food. Indeed, when the energy requirements of fertilizer production, plowing, tilling, and harvesting are added up, the energy yield of the crop may be only a little more. Some highly processed foods actually yield less energy than that consumed in their production.

There is some question as to whether the denitrifying bacteria can keep up with this added influx of fixed nitrogen from chemical fertilizers. Increased nutrient levels in stream

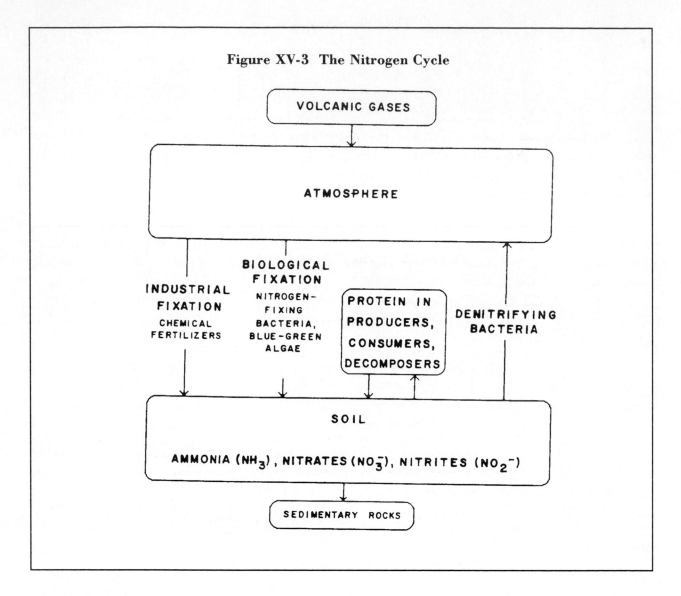

Figure XV-3 The Nitrogen Cycle

runoff and in the lakes and bays that they feed have led in some cases to the condition of
eutrophication -- an oversupply of nutrients in water that can cause algal blooms -- rapid
growth of algae and bacteria that can use up all the oxygen in the water and choke out other
life. In addition, excess nitrous oxide and ammonia may be released into the atmosphere,
perhaps contributing to the dissociation of ozone in the stratosphere, a problem to which we
shall return in the next unit.

Volcanic gases contain nitrogen that is added to the atmospheric supply, and incorporation of fixed nitrogen into sedimentary rocks serves to remove it from the biosphere for
periods of geological time. The overall nitrogen supply within the biosphere is maintained at
high levels because of the critical need for nitrogen in complex organic molecules. And so,
along with oxygen, virtually the entire composition of our atmosphere has been influenced by
the presence of life on Earth.

3. The Phosphorus Cycle

Though phosphorus is found in fairly small quantities in organic matter, it is nonetheless essential to life. It is necessary to the manufacture of proteins and is found in nucleic acids. It is found to be concentrated in the hard parts of our bodies, in our teeth and bones. Occurring as the phosphate ion (PO_4^-), it is leached from phosphate-containing rocks by water and deposited in soil, where it is taken up by the roots of producers and introduced into the usual food chain. No chemical compound of phosphate is gaseous, however, and so the atmosphere does not come into play in this cycle. The return path to the soil is via waste products and decompositon (see Figure XV-4).

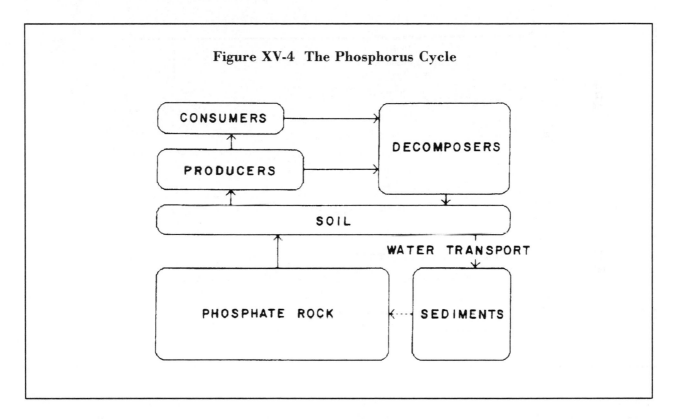

Figure XV-4 The Phosphorus Cycle

Phosphates leached from soil and rocks find their way into streams and eventually to the continental shelves where they are deposited in sediments as the mineral phosphorite. Eventually, these sedimentary rocks may be uplifted and eroded to provide new sources of phosphate rocks.

It is often found that the lack of phosphorus as a nutrient is to limit the growth of the primary producers in an ecosystem. As a result, phosphates are mined extensively for incorporation in chemical fertilizers along with nitrogen. Excessive use of fertilizer can result in phosphate overload in streams and rivers, a condition made worse by the extensive use of phosphate in detergents. As in the case with excess nitrogen, excess phosphate can lead to eutrophication, explosive growth of algae and bacteria, and depletion of oxygen. The result is most visible as a massive fish kill, but in the process, most of the river's life can be destroyed. Far downstream it may recover as the water becomes oxygenated once again.

STUDY QUESTIONS

XV-6. Which of the reservoirs in the carbon cycle holds the greatest amount of carbon in storage?

XV-7. Where do you fit in the carbon cycle shown Figure XV-2, and what are the paths of carbon flow leading to you and from you?

XV-8. Why is there no arrow connecting the atmosphere box in Figure XV-3 to the box labeled "Protein in producers, consumers, decomposers"?

XV-9. Why does the atmosphere not appear in Figure XV-4?

F. THE GAIA HYPOTHESIS

Our discussion so far of the balances of nature as exemplified in Earth's energy budget and in the biogeochemical cycles has shown that the biota interact closely with the geosphere, and this over Earth history has brought about a state of affairs quite different from what would be expected if biota were not present. What is remarkable, though, is that in most cases the changes wrought are ones that have made Earth into a more secure environment for present-day life.

So far we have treated these changes as the natural consequences of the physical environment and of the evolution of life, without providing any real insight into how such a situation would come about. That living organisms can flourish on Earth is not merely due to chance, according to an English scientist, James Lovelock, and Lynn Margulis, of Boston University. In their view, life is not simply a passenger on the planet, passively adapting itself to the existing physical environment. Rather, it actively manipulates Earth to make the environment hospitable for itself.

In fact, the biota of Planet Earth may be viewed as a single giant organism, with elaborate mechanisms for regulating itself. Lovelock summarizes the hypothesis as follows: "... the biosphere is a self-regulating entity with the capacity to keep our planet healthy by controlling the chemical and physical environment." [35] Lovelock calls this idea the Gaia Hypothesis, after the Greek goddess of Earth, and traces its roots back to ancient conceptions of a protective and nurturing Mother Earth.

What follows is a synopsis of the Gaia Hypothesis, and of Lovelock's arguments in support of it. It should be emphasized that many scientists are highly skeptical of it, and that it is by no means the only plausible explanation for the present Earth environment. In the next section, Dr. Stephen H. Schneider of the National Center for Atmospheric Research will present an alternative view. These two somewhat different views will provide you with an opportunity to explore the role of controversy in the development of scientific knowledge, for all competing hypotheses must be challenged and tested to determine which is the best rational

---

[35] J. E. Lovelock, GAIA: A New Look at Life on Earth, Oxford University Press, 1979

explanation for all available observations. At the present time, there is insufficient evidence to make a definitive choice between these two fascinating hypotheses.

Let us consider some of the evidence marshaled by Lovelock. You have already noted the remarkable transformation of Earth's atmosphere by life. Lovelock emphasizes the magnitude of these changes in the following table, in which the present atmosphere of Earth is contrasted with those of Mars and Venus, and also with a model Earth atmosphere that he believes might have resulted if the biosphere had never become active.

TABLE XV-2

Gas	Venus	Planet Earth without life	Mars	Earth as it is
Carbon dioxide	98%	98%	95%	0.03%
Nitrogen	1.9%	1.9%	2.7%	78%
Oxygen	trace	trace	0.13%	21%
Argon	0.1%	0.1%	2%	1%
Surface temperature°C	477	290	-53	13
Total pressure (bars)	90	60	0.0064	1.0

The striking differences in the composition of Earth's atmosphere from those of Mars and Venus are not what would be expected if an atmosphere were allowed to come into equilibrium on a purely physical or chemical basis, Lovelock argues. Many of these characteristics happen to fall within quite narrow ranges of tolerance for life.

Take Earth's surface temperature, for example. The range of surface and water temperatures in the tropical regions of Earth are precisely optimum for the existence of present-day life, yet the temperature of the planet under conditions of a carbon-dioxide-dominated atmosphere surely would be much higher. Fossil evidence indicates that the prevailing temperatures in the tropics have been stable enough to support life throughout the past 3,500 million years, even though solar output has increased by as much as 30% during this time, according to fairly reliable models for solar evolution. This has given rise to what is referred to as the Early Faint Sun Paradox. Lovelock maintains that an active controlling mechanism could have manipulated the workings of Earth's atmosphere in such a way as to keep the temperature nearly constant throughout the whole of this time. He points to the inherently unstable conditions that can be produced by positive feedback mechanisms in climatic systems and suggests that a controlling mechanism would have been necessary to prevent temperature extremes from occurring that might have destroyed all life on Earth.

The oxygen level in our atmosphere is clearly a result of biological activity, but what is not so obvious is that the percentage of oxygen is precisely at optimum levels for use by life.

Increasing oxygen levels provide greater energy conversion efficiency, but too much oxygen could bring about worldwide holocaust. The present level stands at 21%, but recent experiments have shown that the probability of a forest fire being started by a lightning stroke increases by 70% for each 1% rise in oxygen concentration above the present level. Ignition is strongly dependent upon the moisture content of combustible material, and so naturally-set fires are much more prevalent after prolonged drought. If the oxygen content should exceed 25%, however, ignition becomes highly probable even in the damp vegetation of a rain forest.

How is the oxygen concentration regulated? Lovelock thinks that methane (CH_4) released into the atmosphere by the operation of the carbon cycle provides the necessary mechanism since it is readily oxidized, converting molecular oxygen into water vapor. On the other hand, nitrous oxide produced in the nitrogen cycle eventually decomposes in the atmosphere into molecular nitrogen and oxygen. This source of oxygen may provide the opposite controlling mechanism by which Gaia keeps the oxygen level optimized.

In Lovelock's view, even the acidity of the atmosphere and oceans and the degree of saltiness of the oceans are under biological control. We live in an environment that is nearly neutral -- neither acidic nor alkaline. This is in fact the optimum condition for most life forms, though some have adapted to local extremes that may be found in certain environments. Most natural atmospheric processes tend toward the acidic, and the environments of Venus and Mars are in fact highly acidic. Ammonia produced in the nitrogen cycle, thinks Lovelock, serves to keep Earth's acidity in check.

In the unit **Physical and Chemical Makeup of the Oceans** you learned that the saltiness of the ocean has been maintained at a nearly constant level throughout geological time in spite of continual production of new salts from river input and from hydrothermal activity at the spreading ocean ridges. This level, too, is poised in the optimum range for present-day life, at 3.4%. Salt levels higher than 6% would disrupt cell walls, a fatal situation. It is not at all clear just what the biological regulation mechanism is in this case, however.

If the Gaia Hypothesis has any validity, then it becomes critically important that we understand the mechanisms by which Gaia exerts her control. Lovelock likens them to those of a cybernetic entity, in which sensors are linked through a feedback mechanism to controls.

A modern home heating and cooling system provides an example of such a system. A thermostat is linked to furnace and air-conditioning controls in such a way that if the temperature falls below the desired range, the furnace is turned up or the air-conditioner is turned down, whichever is appropriate. If the temperature rises, the furnace is turned down or the air-conditioner is turned up. When appropriate, control may be switched from furnace to air-conditioner or vice versa. Each individual part of the system has a simple and well-defined function, but taken together, the system functions as a highly efficient regulator of the house environment, protecting it from external changes.

Such a feedback mechanism provides an important element of resilience to disruption. For instance, if you leave the house on a cold winter's night and accidentally leave the front door wide open, your family sleeping inside will not freeze to death because the house's environmental controls will spring into action, turning up the furnace to compensate for the sudden heat loss. A price will be paid for this accident when the heating bill arrives, but the inhabitants will have survived.

Many organisms possess the ability for self-regulation in order to withstand stresses produced by a changing environment. This ability, called homeostasis, is familiar in the human body's ability to regulate its temperature to within one degree Celsius of an optimum level, even though the temperature of the surrounding environment may vary over a very wide range.

With the help of an ingenious computer model, Lovelock has shown how life might regulate the temperature of a planet. His simplified Daisy World model has just three kinds of creatures: black daisies that grow better when it is cool, white daisies that grow better when it is warm, and cows -- who eat the daisies.

Suppose that Daisy World is initially quite cold. Black daisies will grow faster and the planet's surface will darken. The darker surface absorbs more sunlight and the planet begins to warm. The cows notice the black daisies and begin eating them. Meanwhile, the rising temperature allows the white daisies to grow faster. Soon, there are more white daisies than black. The planet's surface lightens again, and it begins to cool. The cows turn their attention to the by now more numerous white daisies...

Whatever the starting conditions, warm or cool, the result of the computer simulation is always the same: the Daisy World model settles down to a temperature that is at least tolerable to both kinds of daisies and, presumably, to the cows as well.

Lovelock sees in this resilience to environmental change strong echoes in the actual biosphere. For decades, environmentalists have warned of the dangers of ocean pollution through oil spills and climatic change due to thermal and atmospheric pollution. Yet in almost every case, the actual changes observed have proved to be less than those predicted on the basis of the direct consequences of the effect itself. Oil spills can have severe local consequences, but the rebound of even local ecosystems has been more rapid than previously expected. You have already learned that half of the carbon dioxide released into the atmosphere from fossil fuel combustion has already been removed by the oceans.

Quick recovery from environmental disaster is clearly a trait that contributes high survival value, and the self-regulating mechanisms of Gaia may well have evolved in response to such Darwinian pressures. Lovelock cites the changeover from an anaerobic to an oxidizing atmosphere as an environmental disaster of the first rank for the biota that existed in the early history of life on Earth. In this case, however, Gaia's response was not to restore the status quo but to generate new forms of life that could take advantage of this new world in which energy was more readily available. Here is one of the weakest parts of the Gaia Hypothesis. The rise of oxygen in Earth's atmosphere was clearly a stimulus for immense change, not for stability. An effective self-regulating mechanism, one would think, would have prevented it from happening. Perhaps in this case, Gaia lost control of the situation and proved powerless to prevent the onset of an environmental disaster of her own making.

From the foregoing argument, the Gaia Hypothesis might be construed as going against the grain of the environmental movement that is presently active in many countries. To a certain extent it does in that it admits of greater resilience in the response of the biosphere to environmental changes caused by mankind than previously acknowledged. If, however, Gaia truly exists, then the hypothesis indicates those features of the environment that must be preserved at all cost. These are the essential workings of the biogeochemical cycles and the huge biomass contained in the photosynthetic producers that flourish in tropical rain forests and in shallow seas on the continental shelves.

Lovelock sees an important threat to the integrity of Gaia's operation in uncontrolled human population growth. A larger human population requires more and more energy consumption, competing with the rest of Gaia for available supplies. There is also competition for space, and large-scale deforestation of tropical regions could act to cripple substantial parts of the self-regulating mechanisms of the biosphere. He also sees danger in large-scale ocean farming on the continental shelves for the same reason.

Gaia, which is the biosphere with its biogeochemical cycles and other self-regulating mechanisms, is not regarded by Lovelock as a sentient being. In speaking of Gaia as though she were an embodiment of Mother Nature, he says,

> This is meant no more seriously than is the appellation "she" when given to a ship by those who sail in her, as a recognition that even pieces of wood and metal when specifically designed and assembled may achieve a composite identity with its own characteristic signature, as distinct from being the mere sum of its parts.

Even so, Lovelock suggests that Gaia has evolved a form of intelligence in the collective intellect of humankind. We are certainly a part of the biosphere, and what we choose to do as a species is clearly of a magnitude that can bring about vast changes in Gaia's ability to function. What seems to be imperative is that we recognize the consequences of our actions. But consequences cannot be foreseen reliably without a thorough understanding of the extremely complex interactions between biosphere and geosphere, and throughout the enire web of life. Gaining that understanding, it would seem, constitutes a vital agenda for the coming decades.

G. GAIA OR COEVOLUTION OF CLIMATE AND LIFE? [36]

It's hard for me to imagine a more original and profound concept than the Gaia Hypothesis, for which its popularizers James Lovelock and Lynn Margulis deserve significant recognition. The realization that climate and life mutually influence each other is profound, and provides an important counterpoint to the parochial view of the world as physical environment dominating life; this had been the predominant paradigm in the physical sciences for many decades. Nevertheless, the fact that climate and life "grew up together"[37] and mutually influenced each other -- the concept of coevolution -- is not the same thing as to say that life somehow self-optimizes its own environment. It is the latter idea, the most radical of Lovelock and Margulis, on which I feel some elaboration -- and caution -- is needed. Moreover, there is also competition at all levels of organization from cells to communities.

Lovelock wrote that the "climate and chemical properties of the Earth, now and throughout history, seem to have been optimal for life". But let's examine that a bit more.

---

[36]This section was written solely for use in this text by Stephen H. Schneider, National Center for Atmospheric Research, Boulder, CO
[37]S.H. Schneider, R. Londer, The Coevolution of Climate and Life, Sierra Club Books, 1984.

What does it mean? For example, it is pointed out that production of oxygen, a largely biological phenomenon, helped make the world fit for modern life. But at the same time, it made the world unfit for other forms of anaerobic life that had preceded it. How is this profound change an optimization? From the point of view of oxygen-loving life it is certainly welcome! From the point of view of anaerobic life it was a disaster that relegated these forms to small niches on Earth where oxygen is barely present. The early physical environment largely carved out ecological niches which early life forms had. And, life altered the physical environmental constraints on itself by changing the composition of the atmosphere. This changed the competitive balance of species, and forced evolutionary change -- indeed, coevolutionary change between organic and inorganic parts of the environment. Change, yes; but optimization, I don't quite see how. Only if one defines life in terms of the best current adaptations to the current environment might life be optimizing itself. What about the losers? No doubt they won't see the current environment as having been optimized.

One of my principal difficulties with the whole idea of self-regulation of life is what "life" is. Is "life" to be optimized by maintaining for the longest period of time the stability (i.e. the survival) of extant species? Is "life" the maintenance of the maximum biomass? Is life the maintenance of the maximum diversity of species? All of these seem to me to be legitimate definitions of life, yet to "optimize" each one is inconsistent. Consider a specific example. At the end of the Pleistocene, when the last Ice Age receded, the carbon dioxide content of the atmosphere was perhaps a third less than it is today. This implies that the weakened Greenhouse Effect would have made Ice Ages even colder then they otherwise would have been. A plausible explanation for the decrease in CO_2 during the height of the Ice Age has to do with the biochemistry of the oceans and the planktonic response to altered nutrient runoff associated with exposed continental shelves when sea levels were lower. In other words, life was altering the chemical composition of the atmosphere, the climate, and it's own environment. But it seems hard to imagine how making an Ice Age even colder could be "optimization of life". We now have considerable evidence that the biomass on Earth was some ten percent smaller during the Ice Age than now. It is reputed that Einstein once said when asked by a lay person how many experiments it would take to prove the theory of relativity: "No number of experiments can prove me right, but one can prove me wrong." Indeed, scientific hypotheses often fall when some nasty facts force alteration. If carbon dioxide content was diminished during an Ice Age by life, this certainly seems to throw a monkey wrench into any general law that somehow environmental conditions were being altered by life for its self-optimization.

For me, what Lovelock and Margulis have really done is to point out the role of feedback mechanisms between organic and inorganic components of Earth. This is a major contribution and a brilliant insight. But, feedback can be a two-way street, so to speak. Feedback processes are not just interactions which tend to stabilize, but also can be interactions which tend to destabilize -- like the Ice Age-life-CO_2 case. I believe that life and the environment have coevolved, but I also believe that those interactions have not always been optimum to all forms of life, but simply interactions which lead to mutual change -- some beneficial and some detrimental for some forms of life at some times. That alone is enough for me to sing the praises of those looking beyond the narrow disciplines of biology, climatology, geophysics, and so forth -- people who insist that the organic and inorganic parts of the planet must be viewed as coupled systems. However, it strikes me as speculation at best and environmental brinksmanship at worst to believe that somehow Gaia, through self-regulation, will protect the planet from the negative consequences of all human intervention.

Coevolution is not the same as homeostasis -- that is, self-regulation. To most scientists I know, the idea of planetary-scale homeostasis, which is the principal intellectual thrust behind the Gaia Hypothesis, is more religion than science. As religion I find Gaia deep, beautiful, and fascinating. As science, I find the hypothesis not very well formulated, hard to test empirically, and too full of contradictory examples to show much promise. Brilliant ideas are what drive future understanding, regardless of whether the initial hypotheses emerge intact. Let us then do the interdisciplinary work needed to help understand the coevolutionary path of our physical and biological worlds.

RECOMMENDED READING

Edward J. Kormondy, Concepts of Ecology, 3rd ed., Prentice-Hall Inc., pg. 19.

Daniel D. Chiras, Environmental Science: A Framework for Decision Making, Benjamin/Cummings Publishing Co., 1985.

Gonzalo Vidal, The Oldest Eukaryotic Cells, Scientific American, February 1984.

James E. Lovelock, GAIA -- A New Look at Life on Earth, Oxford University Press, 1979.

Stephen H. Schneider and Randi Londer, The Coevolution of Climate and Life, Sierra Club Books, 1984.

UNIT XVI FATE OF THE EARTH:

THE IMPACT OF MAN

THE WORLD'S INCREASING POPULATION

A. INTRODUCTION

1. Overview

"There is no medicine like hope, no incentive so great, and no tonic so powerful as expectation of something better tomorrow."

O. S. Marden

"In the absence of hope it is still necessary to strive."

attributed to William of Tell

Your previous studies in this course have suggested a number of areas of concern in terms of the impact of mankind on existing environmental and planetary balances. In this final unit we shall attempt to identify those that are most critical to the continued integrity of the workings of the geosphere and biosphere and their interactions. Identifying the potential problems and understanding the consequences of our actions are the first steps that must be taken before solutions may be found.

2. Objectives

Upon completion of this unit you should be able to:

1. describe environmental problems involving air and water pollution

2. describe the nature of the ozone layer depletion problem and relate the dangers of releasing nitrous oxide and chlorofluorocarbons into the atmosphere to what you have learned about the ozone layer in previous units

3. identify regions in the world that are most vulnerable to the spread of desert conditions

4. identify activities of mankind that encourage the spread of deserts

5. discuss the occurrence and causes of the Sahelian drought in Africa

6. relate the need for preservation of tropical rain forests to the operation of the biogeochemical cycles and to their roles as the habitat for a large proportion of Earth's species

7. describe the growth of world human population and relate it to the consumption of resources

8. describe the concept of Nuclear Winter and the manner in which it was postulated

9. relate major temperature drops in the aftermath of a nuclear war to injections of smoke and soot into the stratosphere, and describe the climatic response to such an event

10. recognize that the major environmental problems cannot be solved by technical means alone but may require difficult social choices as well.

3. Key Terms and Concepts

toxic substance pollution	exponential population growth
urban "heat islands"	doubling time for population
acid rain	demographic transition
ozone layer depletion	Nuclear Winter
desertification	Volcanic Dust Veil Effect
Sahelian drought	firestorms
deforestation	"technical fix"
tropical rain forests	Luddites
extinction of species	

4. Corresponding Video

Can mankind destroy the habitability of Planet Earth, or is Earth a self-regulating organism manipulated by life to insure its survival? In this program you will explore soil depletion and deforestation, and see how scientists are trying to save the most complex and threatening ecosystem on Earth -- the disappearing rain forests. Dramatic special effects will depict the global consequences of a "Nuclear Winter", with its projected devastation to life. The program will also recreate the Dust Bowl of the Depression and present it as a possible model of what is happening in Africa today. Using everything from microbiology to advanced satellite technology, scientists are seeking answers to the African drought. The program will ask you whether or not we have a real choice in the fate of the Earth.

B. THE IMPACT OF MANKIND

In the previous unit you learned something about the complex interactions that link biosphere and geosphere. Now perhaps it may be possible to look at these workings and attempt to determine where the natural systems are most vulnerable to interference and disruption from the activities of mankind.

There is no shortage of suggestions. In recent decades the public has been bombarded with warnings of imminent disaster from scientists and environmentalists alike to the point that many are weary of the constant doomsaying. In addition, for many who live in the industrialized countries, the quality of life does not appear to have suffered. Where it has, it is often perceived as a local problem.

The totality of these environmental problems is far too great for us to tackle in detail here and now. Our approach in this final unit of the telecourse will be to review some of them briefly and then to concentrate on those that have the most significant connections to the topics of the previous units. These turn out to be largely concerned with effects on global climate.

C. POLLUTION

Problems of water and air pollution were among the first to be discussed and combated in the industrialized countries. As urban centers grew, the necessity for sewage treatment became obvious, but problems associated with the proper disposal of industrial and other hazardous wastes remain with us today. Indeed, many less-developed regions have yet to deal effectively with the sewage problem, even though the technology of waste treatment facilities is quite advanced.

Toxic substances in the environment have caused severe difficulties in many local areas. These range from radioactive mine tailings to PCB (polychlorinated biphenyl) contamination to overuse or improper use of pesticides such as DDT. Even though DDT use is banned in the United States, heavy use continues in many less-developed countries where it is considered essential to public health as a means to control mosquito and other pest populations.

Lead and mercury have both found their way into the environment, leading to regulation of the use of lead as an additive to gasoline in the United States, and to disruption of Japanese coastal fisheries due to such high levels of mercury compounds as to cause illness and death among seafood consumers.

In semiarid regions where irrigation has allowed marginal lands to be cultivated, the flushing of fields with water that is then returned to local rivers, has caused such a buildup of salt that downriver users may be severely affected. While this may seem like a problem for local governmental action, it may be worthwhile pointing out that the majority of the major river systems of the world are shared between two or more nations.

Air pollution, which has important and undeniable health effects on urban and sometimes even rural populations, can also produce substantial local climatic change. So-called "heat islands" form in urban areas due to a variety of changes that include air pollution, replacement of vegetation by heat-absorbing and non-evaporating surfaces, intensive local

474

energy release, and a general interference in the energy and hydrologic cycles. In essence, large cities can create their own local climate systems.

Sulfur emissions due to the burning of coal already have been mentioned in the unit **Energy Resources**. In the atmosphere, these along with nitrogen oxides from the nitrogen cycle form acids that give rise to acid rain. Severe acid deposition in the Adirondack Mountains of New York and in adjacent regions of New England and Canada, has turned the water of many lakes so acidic that virtually all complex life in them has been destroyed. In Norway the spawning of salmon has come to a halt because of river acidification, crippling the inland fishing industry, and in Sweden thousands of lakes are threatened by the same effect. Many of the worst-hit areas are downwind from major coal-burning utilities and industries, indicating that these may be the principal sources of the acid. Nitrogen oxides released from automobile exhausts and from improper use of chemical fertilizers may also be implicated in some regions.

The ozone layer of the stratosphere has been referred to many times throughout this course; it functions as one of the "greenhouse" gases and serves to absorb dangerous ultraviolet radiation from the Sun. Throughout the past decade the ozone question has remained one of public debate and controversy. Briefly put, certain compounds released into the atmosphere have the potential for finding their way into the stratosphere and encouraging ozone (O_3) to revert to molecular oxygen (O_2), which does not behave at all like ozone. These substances include gaseous chlorinated compounds such as the chlorofluorocarbons (CFC's) and the nitrogen oxides. The former are used as foam-blowing agents, as working fluids in refrigeration systems, and as propellants in spray cans. The nitrogen oxides arise from the nitrogen cycle, from direct injection into the stratosphere from the exhausts of high-flying aircraft such as military jets and supersonic transports, and from nuclear explosions in the atmosphere.

A recent report by the National Academy of Sciences (NAS) [38] concludes on the basis of refined models of stratospheric processes that if production of CFC's continues at the rate prevalent in 1977, global ozone eventually would reach a steady-state reduction of from five to nine percent. The NAS report points out that increased releases of nitrogen oxides over present levels would result in an even higher depletion of ozone.

Ultraviolet rays from the Sun have been implicated in the production of skin cancers, and it has been estimated that for every one percent depletion of the ozone layer, basal and squadmous cell skin cancer rates might increase by five percent. Thus, at even a five percent total ozone decrease, basal and squamous skin cancers might increase by twenty-five percent. Basal and squamous cell skin cancers are usually (but not always) easily treatable, though they are not to be taken lightly; of greater concern is the melanoma type of skin cancer, which is far more dangerous. The connection between ultraviolet radiation and melanoma incidence is more tenuous, however. Though there appears to be some evidence for increased melanoma incidence among populations living at low latitudes (who presumably have greater exposure to sunlight), the National Academy report declined to make quantitative estimates of the effects of reduced concentrations of atmospheric ozone on the incidence of melanoma.

You may find it strange that so much fuss is being made in spite of the fact that direct measurements of ozone concentration show no striking systematic changes at all. Natural

---

[38]Causes and Effects of Stratospheric Ozone Reduction: An Update, National Academy Press, Washington, D. C., 1982

fluctuations of measured ozone occur on a scale that is large compared to effects predicted for the present time (see Figure XVI-1). Modeling indicates that depletion of the ozone layer should have reached no more than a percent or so at present, due to the slow rate of chemical reactions in the ozone layer. The natural fluctuations tend to mask any change of this magnitude, and the direct observational data to date are not regarded as indicating any significant trend.

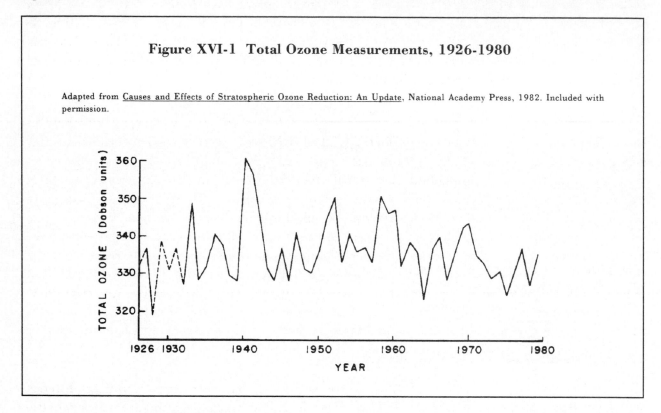

Figure XVI-1 Total Ozone Measurements, 1926-1980

Adapted from Causes and Effects of Stratospheric Ozone Reduction: An Update, National Academy Press, 1982. Included with permission.

The National Academy report points out:

A notable feature of the ozone issue is that a reduction due to increases in the tropospheric concentrations of CFC's or N_2O, once it has taken place, is expected to persist for more than 100 years even if the practices that caused it are stopped immediately. It is therefore important to detect an anthropogenic (man-caused) effect at the earliest possible time.

Much of the controversy stems from this feature -- at present we do not see any substantial change in ozone concentration, but if we wait until we do, it may be too late to take meaningful action to prevent further change. By then there will already be sizeable quantities of the offending substances within the ozone layer, ready to slowly destroy ozone throughout the next century. Recall from the unit **The Atmosphere** that the stratosphere is stratified and stable, with little mixing between it and the troposphere that could serve to remove accumulated CFC's or N_2O.

In a case like this, predictions are based on theoretical models that use all of our accumulated knowledge about how CFC's, for example, react with ozone in the presence of sunlight rich in ultraviolet rays. Careful laboratory experiments attempt to duplicate strato-spheric conditions, and the experience gained from them is incorporated into the equations that define the model. Such models can become very complex, and once again a computer is called in to handle the arithmetic. The essentials of the model, though, are contained in the equations that govern what the computer does, just as you saw in the model that you worked with in the unit **Climates of Earth**. There, as here, the goal is to simulate the operation of the natural system sufficiently well that it can serve as a substitute for the real thing in "experiments" that would take too long or are too hazardous to try in nature. For instance, the model might be asked what effect a certain release rate of CFC's might have on stratospheric ozone a hundred years from now. Clearly this is not an option available to us in actual measurements of the real thing, at least not without waiting a hundred years.

STUDY QUESTIONS

XVI-1. What are two possible sources of acidity that have been implicated in acid rain?

XVI-2. What substances seem to pose a danger to the integrity of the ozone layer?

XVI-3. By how much has the ozone layer decreased so far?

D. DESERTS AND FORESTS

The spread of deserts and the destruction of forests are two distinctly different problems that are nonetheless related. Both arise from the mounting pressures of population and both are likely to have unfortunate climatic repercussions on a local or global scale.

Deserts are regions largely barren of life and incapable of supporting habitation. Figure XVI-2 indicates existing desert regions of the world along with regions that are considered to be at high risk of becoming deserts in the near future. If all the lands at risk were actually to succumb, the total area given over to desert would triple.

The spread of deserts has sources that are both natural and man-made. There is ample historical and geological evidence for the existence of extensive desert areas long before the pressures of mankind could have exerted any influence. Deserts naturally tend to form in the rain shadow downwind of mountain chains that force air masses to rise, cool, and dump their moisture in the mountains. When they descend in the lee of the mountains, they warm and their relative humidity drops, sometimes to extremely low levels. In addition, as you saw in the unit **Circulation of the Atmosphere**, some deserts form downwind of persistent downwelling masses of air in the atmosphere that are influenced by the distribution of continental and oceanic areas.

At the same time, there is no question that regions with marginal rainfall and burgeoning populations have suffered severely and are being driven into a more arid state by

Figure XVI-2 The Potential Spread of Desert Regions of the World

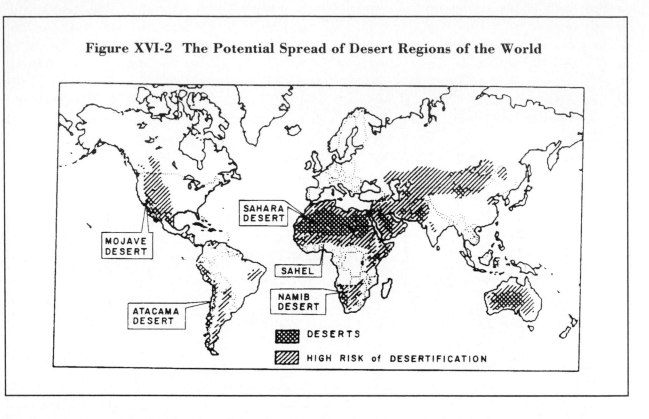

mankind's activities. Overgrazing, leading to denuding of the soil, and clearing of forests are the principal agents of desertification in these areas. Deforestation, first in response to increased pressures for more agricultural land, and later as a result of firewood gathering in less-developed societies, is proceeding with alarming speed in many areas. Where natural climate fluctuations have stressed already marginal lands, the combination can prove disastrous.

The Sahel region of Africa extends in a narrow band of wooded grassland and bushland across the entire continent, from Mauretania in the west to Ethiopia in the east (see Figure XVI-2). A long-standing drought has wreaked havoc in this region, bringing starvation and malnutrition to millions, along with physical and mental impairment in many children of the Sahel. Economies of a number of nations in the region have been undermined and long-term damage to the soil has been done that may take a long time to repair, even if the drought ends.

Figure XVI-3 shows averaged departures in rainfall from normal values for the past 45 years in the broader sub-Saharan region. Note that from 1950 to 1967, the region enjoyed above-average rainfall. Such temporary fluctuations often encourage changes in local agricultural practices and population distribution. When the abnormally wet period ends or a drought sets in, the agricultural system may collapse. The Sahelian drought and its causes are discussed in more detail in the accompanying essay by noted climatologist Stephen H. Schneider.

Even in developed countries, the spread of deserts can be a problem. Many marginal lands are made productive by use of irrigation schemes, but in many cases this has led to increasing salinity of the soil. Surprisingly, waterlogging of the soil is sometimes a problem because irrigation water is brought in by canals that may leak, raising the local water table to the point that the roots of plants may be drowned. The Global 2000 Report to the President of

478

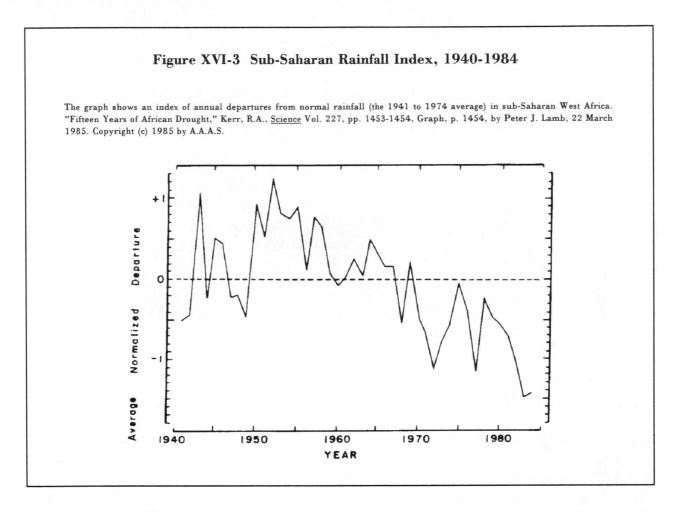

Figure XVI-3 Sub-Saharan Rainfall Index, 1940-1984

The graph shows an index of annual departures from normal rainfall (the 1941 to 1974 average) in sub-Saharan West Africa. "Fifteen Years of African Drought," Kerr, R.A., Science Vol. 227, pp. 1453-1454, Graph, p. 1454, by Peter J. Lamb, 22 March 1985. Copyright (c) 1985 by A.A.A.S.

the United States[39] points out that some of the richest farmlands in the United States are threatened. Irrigated farmland in the San Joaquin Valley of southern California is an important source of year-round produce and other foods to the entire nation, but high, brackish (salty) water tables pose an increasingly serious threat to productivity. The report goes on to say:

> The salting problems of the valley have been compared to those that resulted in the collapse of civilization in Mesopotamia and Egypt's upper Nile when early signs of agricultural overproduction were not heeded. Loss of the productive capacity of the San Joaquin Valley lands would be a serious loss to the people who work them, to the economic community of the valley, to the State of California, and to the country as a whole.

Forests are among our most valuable natural resources. They reduce soil erosion to almost negligible rates and act to even out the flow of water between times of flood and drought. They influence the biogeochemical cycles strongly, accounting for a large proportion of the biomass in the cycles, and, along with the oceans, are important regulators of the carbon dioxide content of the atmosphere. In addition, they supply us with wood for construction,

[39]Pergamon Press, New York, 1980

479

paper, and packing materials. Finally, forests provide the habitat for a disproportionate number of the total number of species of living things in the world.

In recent years, however, forests throughout the world have suffered from increasing population pressures. Today, some one-third of all land area on average is covered by forests, but it has been estimated that we have already lost some 30 - 50% of the forest lands that originally covered the continents. Most of this land has been converted to agricultural use. In high population density regions such as Europe, some 70% of the original forests are gone. Haiti, the most densely populated nation in the western hemisphere, has lost 91% of its forests. Today, only 6% of Africa is covered by forest land, in contradiction to popular notions in other lands of a continent covered by vast, teeming jungles. These exist, but mainly in the limited region of the Congo River Basin, and even there are increasingly threatened.

The tropical rain forests occupy a position of importance among world forests because of the sheer amount of biomass contained within them. They are believed to be important to carbon dioxide balance between biomass and atmosphere in the carbon cycle, and some experts for this reason feel that preservation of the rain forests is of concern.

Tropical rain forests may be contrasted with temperate hardwood forests in a number of ways. Most dramatic is the sheer diversity of species found within the rain forests. Fully two-thirds of the plant and animal species on Earth are found there; it has been estimated that only one-sixth of these have been described and named. The Global 2000 Report estimates that by the year 2000 between 1.5 and 2 million species are likely to be forced into extinction, largely due to the clearing or alteration of tropical rain forests. Such a loss, accounting for between 15 to 20% of all species now on Earth, would constitute an irreparable loss to the diversity of the genetic pool available to this planet. As it is, the rate of extinctions has been accelerating, largely due to habitat encroachment by the activities of mankind. Figure XVI-4 shows the increase in extinction rate for animal species only, which are for fewer but better known over the time period indicated than the vast number of plant species.

There are other good reasons for sparing the rain forests from clearing for agricultural purposes. Surprisingly, the soil beneath the lush tropical forests is often nearly sterile. The circulation of nutrients takes place almost entirely between the biomass itself and the atmosphere. As a result, when rain forests are cleared and burned, the nutrients tend to wash away along with the ash, leaving a barren and unproductive soil behind. Finally, the burning of forests produces a large amount of carbon dioxide that is released into the atmosphere, adding to the growing amount already there.

The problem of the rain forests is made more acute by the fact that they are concentrated in less-developed nations in which population growth rates are among the highest in the world.

Figure XVI-4 Extinction Rates for Animal Species Over Four Centuries

Adapted from Daniel D. Chiras, Environmental Science, Benjamin/Cummings Publishing Co., Fig. 10.2, p. 221, 1985. Included with permission.

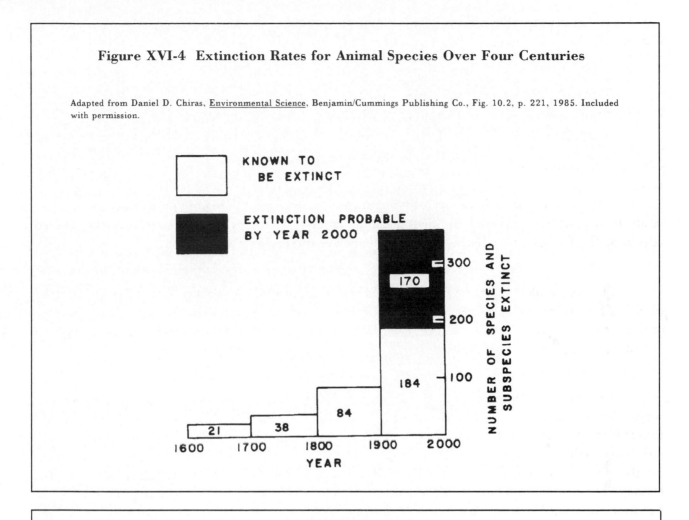

STUDY QUESTIONS

XVI-4. What are some natural causes of deserts?

XVI-5. What are some factors that can cause desertification in less-developed countries?

XVI-6. What conditions are threatening productivity in the rich farming land of California's San Joaquin Valley?

XVI-7. What are some of the ways in which tropical rain forests are important to ecological balances?

E. THE SAHELIAN DROUGHT: A CLIMATIC AND SOCIETAL TRAGEDY [40]

A modern example of a climate-related human disaster took place in the Sudano-Sahel region of north central Africa, just to the south of the Sahara Desert. Nearly everyone agrees that during the early 1970's in the sub-Saharan region of Africa, there were two predominant events: rainfall for several years running was generally well below long-term averages; and famine occurred along with considerable losses of livestock and crops. While most analysts of the Sahelian crisis believe that these two events were cause and effect, respectively, this view is increasingly being challenged. A number of conflicting interpretations of the Sahelian drought cloud the general view of what should have been done to offset its effects -- and what might currently be done as a matter of policy to prevent people from suffering in the future as a result of climatic variations. Identical sounding arguments can be -- and have been -- advanced with regard to the terrible famine in Ethiopia in the mid-1980s.

1. Climatic Trend as Cause

One theory, propounded by Reid Bryson, is that the suffering followed drought caused by a cooling trend in the northern hemisphere, creating a "Sahelian Effect" -- shifts in the monsoon rains in the Sahelian part of Africa and the northern parts of India. [41] If this theory is right, and such a Sahelian Effect continues to suppress the monsoon rains in inhabited areas, deserts will spread southward and hundreds of thousands or more will die -- or be forced to migrate -- in succeeding decades. This is a controversial position, particularly since Bryson attributes part of that cooling to atmospheric aerosols generated from activities outside of the Sahel. The inherently divisive nature of potentially human-induced climatic modification is obvious from this example, regardless of whether the assumed modifications were perceived, rather than proved, as true.

2. Climatic Fluctuation as Cause

Other researchers contend that droughts in the Sahel are a regular, hazardous feature of the climate, and not part of an evolving trend. The 1970's event, they say, was much like previous extremes.[42] University of Maryland climatologist Helmut Landsberg suggests that the recurrence of drought was "an entirely forseeable event, if not precisely predictable on a time scale as to when it would strike."[43] Both this view of climate as a hazard and Bryson's view of climate as a trend suggest that populations of humans and animals were too high to be sustained in bad climate years.

---

[40] This section was written solely for use in this text by Stephen H. Schneider, National Center for Atmospheric Research, Boulder, CO
[41] Bryson, R.A., and Murray, T.J., Climates of Hunger Madison: University of Wisconsin Press, 1977.
[42] Hare, F.K., Food, climate and man, in Biswas, A.K., and Biswas, M.R. (eds), Food, Climate and Man New York: Wiley, p.7, 1979.
[43] Landsberg, H.E., The effect of man's activities on climate, in Biswas and Biswas (eds.), Food, Climate and Man, pp. 187-236, 1979.

3. Inappropriate Technology and Foreign Aid as the Culprits

Political scientist Michael Glantz of the National Center for Atmospheric Research stresses another aspect of the problem. He says that western countries intervened with an inappropriate technology that exacerbated the drought's impact. Foreign aid financed the digging of wells, or boreholes, which upset the well-established nomadic way of life that was well adapted to periodic droughts while these wells provided a short-term solution to the perennial problem of water shortage in the Sahel, they encouraged nomads to increase their herd's size beyond the long-term carrying capacity of the land and to stay in areas close to the wells.[44] As a consequence, the herds overgrazed these areas and depleted and trampled the available vegetative cover.

4. Social Structure as Cause

Still others believe the roots of the Sahelian crisis to be even more deeply imbedded in social factors. In their book **Food First** Frances Moore Lappe and Joseph Collins argue that the governments that took over after independence in the Sahelian countries forced peasants to grow cotton for French export markets, even though the soil of this region was ill-suited to yearly cotton growing.[45] Over the centuries Sahelians had developed a way of coping with periodic drought. Farmers traditionally rotated millet and legumes to replenish soil nutrients -- and to nourish themselves. But the continual planting of cotton depleted the soil, forcing further expansion of cotton cropping and locking up land and other resources that would otherwise have been used for farming and grazing. Thus, in this view, modern technological, social, economic, and political influences actually increased the vulnerability of subsistence farmers and nomads to climate stress. Only by overthrowing "elites" who "put profits ahead of people" will the harm of periodic bad years be alleviated, they believe. In a similar vein, Argentinian meteorologist and social critic Rolando Garcia led a major study on the Sahel events and published his controversial views in a book with the title **Nature Pleads Not Guilty**. He describes "the official view" -- malevolent nature, overpopulation, and indigenous misman-agement -- of the Sahelian catastrophe, and instead argues that it was a "structural problem, the unavoidable consequence of a system" that combines the "prevailing international economic order," plus the "international division of labor," the application of "comparative advantages" plus the "prevailing ideas on international aid."[46] He suggests "structural adaptations" to prevent recurrence of catastrophes (based on reduction of the influence of market economics), redefinition of the value of "productivity" (to de-emphasize output and encourage resilience), and reformulation of productive organization to aid in agricultural work.

---

[44]Glantz, M.H., Nine fallacies of a natural disaster: The case of the Sahel, Climatic Change 1: 69-84, 1977.
[45]Lappe, F.M., Collins, Jr., and Fowler, C., 1978, Food First: Beyond the Myth of Scarcity New York: Ballantine. For example, see pages 89-92, 1978.
[46]Garcia, R.V., 1981, Nature Pleads Not Guilty (Oxford: Pergamon Press). See for example Chapter 7; our quote is from p. 195.

5. The Technology and Free Market Incentive View

Another interpretation, however, forwards the opposite view: that the only way to avoid chronic food shortages and to build food-growing capacity is to give farmers a bigger stake in a growing economy through greater cash sales. Producers need an incentive to purchase necessary technologies -- such as fertilizers -- in order to increase production, according to food analysts Sterling Wortman and Ralph Cummings.[47] Access to markets and the profit motive, they argue, would encourage farmers to grow more food -- whether cash or subsistence crops. More food and production, or course, are the best means of preventing famine.

6. Summary

The example of the Sahelian disaster shows that climate is only one element of societal vulnerability to climatic stresses. While nearly all analysts would concede this point, the differing emphasis placed on environmental, social and political factors often stems more from the differing ideological viewpoints of the analysts than from disagreements over the events themselves. Missing, incorrect, or distorted facts have been piled on top of the conceptual differences, providing even further cause for disagreement over interpretation of cause and effect.

F. NUCLEAR WINTER

The belated realization, almost 40 years after the first nuclear bomb was dropped on Hiroshima, that such weapons might cause global catastrophe reveals one of the more remarkable oversights of twentieth century science. While the consequences for the human survivors of a nuclear war were carefully documented, the fate of the Earth was not. In fact the discovery that Earth's climate could be altered drastically by a major nuclear exchange came as a surprise to the scientists involved. As in many discoveries, a strong element of serendipity was involved in that it was made as a result of work that was originally designed to investigate something else.

Prior to the space program, dark features had been observed moving across the surface of Mars, leading some to suggest they might be due to the presence of life. As it turned out, they were caused by gigantic dust storms that developed in the southern hemisphere and rapidly covered the planet. When the spacecraft Mariner reached Mars in 1971, the entire planet was obscured by a dust storm that had started several months before. While most of the scientists involved in the project were dismayed at being unable to see the Martian surface, a small group of workers began a study of the dust clouds themselves. Fortunately, there was an infrared spectrometer on-board the spacecraft that allowed measurement of the temperature of the dust clouds at different elevations.

---

[47]Wortman, S., and Cummings, R.W., Jr., 1978, To Feed This World: The Challenge and the Strategy Baltimore: Johns Hopkins Press. For example, on page 7 they argue that "one objective of agricultural development must be to allow individual families to produce a surplus for sale so that the total output of a locality exceeds total local requirements and permits sales to urban centers, other rural regions or international markets. Imports required for higher productivity must be purchased and markets for products must be established. In short, traditional farmers must be brought into the market economy."

The scientists found that the normal thermal structure of the Martian atmosphere had been inverted. While the clouds appeared to be warm, the Martian surface was much colder than usual. Dust, high in the atmosphere, had absorbed sunlight and prevented some of the normal flux of solar radiation energy from reaching the surface of Mars. Instabilities developed in the atmosphere as a result, causing winds that picked up even more dust, perpetuating the storm.

To account for what they found on Mars, the group developed a computer program capable of modeling the effects of dust clouds on the planetary energy budget. Since the surface atmospheric pressure on Mars is similar to that in Earth's stratosphere, the model could also be used to compare these effects with those observed on our own planet following volcanic eruptions. You may recall the Volcanic Dust Veil Effect in the unit **Climates of Earth**. When the computer models were given data about the amounts of material injected into the stratosphere by different volcanic eruptions, the results appeared to simulate the observed effects accurately.

The matter might have rested there, except that in 1980 one of the scientists involved with the project read a paper suggesting that the extinction of the dinosaurs had been caused by the impact of an asteroid or comet. Such an impact would inject an enormous amount of dust into the stratosphere. The scientists decided to apply the model that they had developed for the Martian study to this problem and found that while the dust cloud would persist for only a few months, it would cause a significant drop in temperature on Earth. When they presented their results at a scientific meeting, one of the participants suggested that there might be a similarity between the asteroid-generated dust cloud and the totality of those generated in a major nuclear war. In the free exchange of scientific information and ideas, the vital connection had been made.

Meanwhile, a scientific collaboration between a West Geman and an American had produced a critically important result. This study calculated the amount of particulate matter -- smoke and soot -- that might be sent into the atmosphere during a nuclear exchange. Vast forest fires would be set by the nuclear explosions, they figured, and large quantities of fossil fuels in stocks and disrupted gas wells would burn as well.

The optical properties of soot are quite different from those of dust. Dust particles tend to scatter sunlight in all directions, but soot particles efficiently absorb visible light and are much more effective in preventing the energy of sunlight from reaching the surface. Moreover, they allow the escape of radiant heat from Earth's surface and lower troposphere to space. The researchers assumed that a major nuclear exchange might involve the explosion of warheads equivalent to 5,000 megatons of TNT -- less than one-half of the present nuclear arsenal -- and calculated that beween 200 and 400 million metric tons of smoke would be generated from forest fires and combustible fossil fuels. Initially, less than one percent of the sunlight would reach the surface and only ten percent might filter through during the next thirty days.

This work was published and came to the attention of the group that had been working with dust. They added the important element of the effects of smoke and soot to their computer model, along with the results of a third important study that quantified the large amounts of combustible material found in cities. During the bombing of Hamburg and Dresden in World War II, huge firestorms were ignited. Under certain atmospheric conditions many small fires may coalesce into a single conflagration. Hurricane-force winds were generated, sending flames and smoke up into the atmosphere in a pillar of fire. Similar fires were caused in Hiroshima and Nagasaki. At Hiroshima, thirteen square kilometers were burned. Within the fire zones, most combustible materials were consumed. Towering smoke plumes rose into the sky and oily black rains fell downwind. Survivors shivered in the rain as the temperature rapidly fell.

When the effects of smoke and soot emissions from the burning of cities was added to the earlier computer model, the results were profoundly disturbing. For a wide range of nuclear scenarios the global climate would be severely disrupted. In a 5,000-megaton exchange, surface temperatures would drop rapidly to subfreezing levels throughout much of the northern hemisphere, remaining there for several months. Even a year later, temperatures would be several degrees below normal.

Now other researchers took up the problem and applied to it the specialized computer models that they had developed for other purposes. The global circulation models of the atmosphere used at the National Center for Atmospheric Research in Colorado, discussed in the unit **Climates of the Earth** were brought into play, allowing a more realistic simulation of atmospheric circulation. These simulations confirmed the essential findings, while providing significant refinements, as will be discussed in the next section. They also indicated that worldwide atmospheric circulation might be disrupted.

Note in Figure VII-9 on page 229 that the two Hadley Cells operating in the tropics tend to keep air from the two hemispheres from mixing except in the narrow intertropical convergence zone. If the nuclear exchange were to take place between the United States and the U.S.S.R., the immediate effects would take place where the northern boundary of the Hadley Cell normally produces downwelling air. Heating of the troposphere in this zone might reverse the direction of circulation in the northern Hadley Cell, forcing it to combine with the southern cell in one huge cell that would carry smoke and soot into the southern hemisphere as well.

In a severe-case scenario, if the exchange were to take place during the northern hemisphere summer, crops around the world would be destroyed in the subfreezing temperatures. There is another effect, too. Atmospheric nuclear explosions generate large quantities of nitrogen oxides, and their sudden transport into the stratospheric ozone layer could have severe effects upon it, perhaps in time depleting it by as much as 50%. Plants and animals that have survived the direct effects of the blasts and the prolonged periods of darkness and cold would find, as the clouds cleared, that now they had to face increasingly high levels of ultraviolet radiation from the Sun.

Nuclear war may well be the ultimate environmental disaster, comparable to those that marked the great extinction events of the geological record. Scientists hope that an improved understanding of its effects will make it less likely to occur.

XVI-8. What is the cause of low temperatures expected in the aftermath of a nuclear war?

XVI-9. How could the smoke clouds from the northern hemisphere spread into the southern hemisphere?

G. NUCLEAR WINTER: THE FIRST THREE GENERATIONS OF CLIMATE MODELS AND SOME OF THEIR IMPLICATIONS [48]

The first calculation of the possible climatic effects of a hypothetical nuclear war-induced smoke cloud was made by a team of scientists known as TTAPS (after the authors' initials: Turco, Toon, Ackerman, Pollack, and Sagan).[49] The TTAPS paper predicted 20 to 40°C or more land surface cooling for up to many months from such nuclear smoke clouds. Although the authors were aware of many of their model's deficiencies, TTAPS was criticized, nonetheless, for a number of reasons. First, they used a so-called "one dimensional radiative-convective" climate model, whereas the real world is, of course, three-dimensional. TTAPS represents the first generation of climatic assessments of the effects of nuclear war. A set of second generation studies followed shortly, and within another year a third generation of work became available. The rate at which new climatic studies are being done is several per year. As a result, a number of TTAPS conclusions have been sustained, some contradicted, and yet other altogether new conclusions have been uncovered -- as we'll explore soon.

The TTAPS model was applied to either an all-land or all-ocean planet, and neglected the important effects of heat transported from the oceans to land which might ameliorate the land surface cooling from the nuclear smoke scenarios. Also, TTAPS used annually-averaged solar radiative input, ignoring potential effects of seasonality of the results. There are other criticisms that could also be directed to TTAPS, such as their neglect of interactions between perturbed atmospheric conditions and smoke transport and removal, or their inability to treat important smaller scale effects, such as fire plumes and their role in creating thunderstorm-like clouds which might wash out some of the initial smoke before it got very far in the atmosphere. However, most of the latter criticisms apply as well to all climatic modeling of Nuclear Winter to date -- although interactive smoke transport and removal calculations (the "third generation" mentioned above) studies have now been conducted at several national laboratories. But the first three difficulties mentioned above have been examined in the context of more complex, so-called "three-dimensional, general circulation models". Two such second generation studies have been published as of 1985: a Russian study by V. Alexandrov and S. Stenchikov[50] (AS)

---

[48] This section was written solely for use in this text by Stephen H. Schneider, National Center for Atmospheric Research, Boulder, CO

[49] R.P. Turco, O.B. Toon, T. Ackerman, J.B. Pollack, and C. Sagan, Global atmospheric consequences of nuclear war, Science 222, pp. 1283-1292, 1983.

[50] V.V. Aleksandrov and G.L. Stenchikov, On the modeling of climatic consequences of nuclear war, The Proceedings on Applied Mathematics, The Computing Centre of the U.S.S.R. Academy of Sciences, 1983.

and an American effort by C. Covey, S.H. Schneider, and S.L. Thompson[51] (CST) at the National Center for Atmospheric Research (NCAR).

The CST three-dimensional model results significantly modified the one-dimensional TTAPS findings. It, along with TTAPS and the Russian work, helped provide the scientific support needed to conduct several major national and international assessments of the overall Nuclear Winter problem, including a second U.S. National Academy of Sciences study[52] of long-term atmospheric effects of nuclear explosions. CST used a scenario for a large-scale nuclear war-induced smoke cloud at the direct request of the National Academy of Sciences study panel then in progress.

Like TTAPS, CST found significantly reduced light and surface temperatures in mid-continental regions under the assumed high altitude sooty smoke cloud. CST also observed that the smoke could be lifted by its own heating into the stratosphere, where it then could more easily spread out of the war zone, at least in the summer or spring seasons. However, CST also found that the magnitude of surface cooling under the smoke cloud was comparable to TTAPS only in mid-continental areas and in the summer. Along west coasts, the cooling was perhaps a factor of ten less. Moreover, in the three-dimensional CST model there was a great seasonal dependence to the cooling -- TTAPS and AS only made a mean annual calculation -- with CST's land average temperature drops ranging from about half of TTAPS in the summer to relatively little change in winter.

But the most important CST result was the discovery of the possibility that even a few days of dense smoke high overhead could drop surface temperatures on land below freezing, depending on location and meteorological conditions. Moreover, such a transient "quick freeze" could be exported by the lottery of wind directions -- a sort of "weather roulette" -- to almost anyplace in the war hemisphere, even if the overall size of the nuclear smoke cloud was many times less than hypothesized by TTAPS or the National Academy study.

This possibility of "quick freeze" -- it is too soon to say how certain it is -- is both scientifically interesting and has major policy implications of its own; and it also cancels one implication of TTAPS. TTAPS postulated the existence of a "threshold" for Nuclear Winter -- at a "modest" 100-megaton war in which major city centers were struck by 1,000 100-kiloton bombs. This scenario represents use of about 1% of the global nuclear arsenal. The implication is that wars with total megatonnage above 100-megatons exploded on cities could "trigger" a Nuclear Winter.

This threshold concept has led to conflicting policy implications. Whereas Carl Sagan had said it implies that nuclear weapons stockpiles need to be reduced on both sides by a factor of twenty or more in order to drop below the TTAPS "threshold", a U.S. Department of Defense official countered that the TTAPS threshold suggests that one side could launch a 99-megaton first strike on the other side, but the victims couldn't fire back without crossing the threshold.

---

[51]C. Covey, S.H. Schneider, and S.L. Thompson, Global atmospheric effects of massive smoke injections from a nuclear war: Results from general circulation model simulations, Nature 308, pp. 21-25, 1984.
[52]U.S. National Academy of Sciences, The Effects on the Atmosphere of a Nuclear Exchange, Report of the Committee on Atmospheric Effects of Nuclear Explosions, National Academy Press, 1984.

Fortunately, planners can stop thinking up such pathological "threshold" scenarios, because they only arise in the context of a one-dimensional model such as used by TTAPS. The three-dimensional models with more realistic regional geography and dynamics included show that is is more likely that there would be a spectrum of probabilities and consequences, ranging from a higher probability of transient quick freezes to a lower chance of a more devastating global-scale Nuclear Winter lasting for months -- which was the scenario that so concerned the 22 biologists who published a companion study to TTAPS[53] Indeed, these second generation results of CST have been largely sustained by the third generation of studies -- in which smoke is interactively transported and/or removed by the model's internal processes -- at NCAR, Los Alamos and Livermore Laboratories, and by Alexandrov and colleagues in the U.S.S.R.

The largest uncertainties surrounding the long-term climatic effects of nuclear smoke clouds arise not from the climatic models themselves, but from the assumptions used to drive them. In particular, what no climatic model can explicitly calculate is the height to which a plume of smoke will rise over a city consumed by a firestorm. If the bulk of the smoke is injected above a few kilometers of height, then this smoke will reside above the bulk of the water vapor in the atmosphere. The lifetime of the smoke, in turn, depends primarily on how effective that water is in washing out the smoke. Thus, the biggest questions in the theory of Nuclear Winter occur at the local scale: namely, how much smoke will initially get how high. Then, how much smoke will be removed and spread by atmospheric motions in the next few days, as hundreds of plumes of smoke follow twisted and interlocking paths over distances of a few hundred kilometers -- the so-called "mesoscale" problem.

If, as Crutzen and Birks[54], TTAPS, the 1985 NAS Report, AS or CST assumed, there is still some few hundred million tons of smoke spread high in the atmosphere after a week or so "post-war", then the likelihood of severe climatic effects on a hemispheric to global scale is high. On the other hand, if there were a great deal more initial washout of smoke arising from either lower plume heights or more rainfall than is presently anticipated, then the likelihood for global-scale Nuclear Winter would be lower, although the possibility of regional transient "quick freezes" would remain. (The third generation of studies, which do explicitly include smoke transport and/or removal properties, confirm these second-generation conclusions.)

It is simply too soon to estimate these relative probabilities in any quantitative way, although there has been some theoretical and experimental evidence that can be brought to bear in order to estimate how much smoke might get how high in the first few hours following the exchange of many nuclear weapons on a variety of targets.

In addition to the few cases of World War II firestorms, there have been observations of large forest fires. The best documented one took place in Alberta, Canada in the summer of 1950. The smoke plume from this fire could be clearly traced for days, and was so dense that when it passed over Cleveland and Detroit on Sunday September 24, 1950, it darkened the sky to the point that lights were needed for afternoon baseball games. Temperatures were also several degrees below the expected levels in Washington, D.C. This same cloud was observed by a Royal Air Force pilot over England several days later at a height of 9-12 kilometers! Clearly,

---

[53]P.R. Ehrlich et al., Long-term biological consequences of nuclear war, Science 222, PP. 1293-1300, 1983.
[54]P.J. Crutzen and J.W. Birks, The atmosphere after a nuclear war: Twilight at noon, Ambio 11, pp. 114-125, 1982>

when meteorological conditions permit, smoke from massive fires can both get very high and last a long time in the atmosphere.

But Nuclear Winter is not a proved entity, only plausible theory. How then, some ask, can policy be made on the basis of all that uncertainty? For instance, although there are examples of lots of smoke getting into the atmosphere from historical fires, some of the most fundamental issues such as how much smoke will initially get how high and will spread how far from a thousand burning cities can never be definitively estimated -- even by our most sophisticated computer models -- until lots of real cities burn. In other words, some elements of the theory of Nuclear Winter may be unverifiable.

What Nuclear Winter issues add to the arms control debate, in my opinion, is an extension to the global scale of the already well-traveled conclusion that use of even a small fraction of strategic nuclear weapons as now deployed could quite likely be mutual suicide. Additionally if even as little as one third of the "baseline" smoke levels postulated by the National Academy of Science's recent study were deposited over the United States or the U.S.S.R. alone -- that is, if there were a "successful" unanswered first strike -- there would be drifting clouds of smoke in the atmosphere that could pass over the territory of the attacker.[55] The third generation of studies now becoming available suggests that such a "first-strike feedback scenario" could cause significant near-freezing outbreaks in the summer in the northern hemisphere for weeks, including the territories of the United States and U.S.S.R., regardless of whether the attacked country fired back. Finally, Nuclear Winter also warns any non-combatant nations or terrorist groups that any deliberate attempts to start a superpower nuclear war will not leave them immune from devastating consequences. The implications of these potential effects on non-combatants could have consequences on their attitudes toward the nuclear nonproliferation treaty renewal discussions that are ongoing[56].

For all the reasons given, I believe that it is not premature to consider the implications of Nuclear Winter research to date for a host of applications. To be sure, large uncertainties remain, but many fundamental concerns over the potential climatic consequences of nuclear war have been sustained over three generations of increasingly complex investigations. As long as the Nuclear Winter scenarios are viewed probabilistically, there is no "scientific" logic in delaying such policy-oriented inquiries.

H. THE CRISIS OF MANKIND -- THE POPULATION EXPLOSION

The growth of human population over the past thousand years is diagrammed in Figure XVI-5. This general type of curve, in which the values being plotted increase in an accelerating fashion, is often described in terms of exponential growth. In this type of growth, values double with every passage of a fixed amount of time, which is the doubling time. You are probably familiar with this type of growth in a savings account that is growing with compound interest. In fact, population growth has been even more rapid throughout this period, with the time

[55]S.L. Thompson, Global interactive transport simulations of nuclear war smoke, Nature (accepted for publication).
[56]S.H. Schneider, Nuclear Winter: State of science and observations of its implications, Testimony before the House of Representatives Subcommittee on Natural Resources, Agricultural Research, and Environment of the Committee on Science and Technology, and the Subcommittee on Energy and the Environment of the Committee on Interior and Insular Affairs, March 14, 1985.

needed for a doubling of the population becoming shorter as time goes on. It took 80 years (1850 - 1930) for the population to double from 1,000 million to 2,000 million, but only 45 years (1930 - 1975) to double again to 4,000 million. Some projections see the next doubling, to 8,000 million, being accomplished by the year 2017. This doubling would require only 42 years, and chances are that most of you will still be leading an active life by that time.

The recent rise in population is made even more dramatic when you realize that <u>per capita</u> consumption of resources is also rising, compounding the effects. Consider, for instance, that in the period 1950 - 1970, world population rose by 46%, an impressive figure. In the same period, however, world fertilizer consumption increased by 330%. Similar per capita increases have been registered for many resources, including energy, and minerals such as copper.

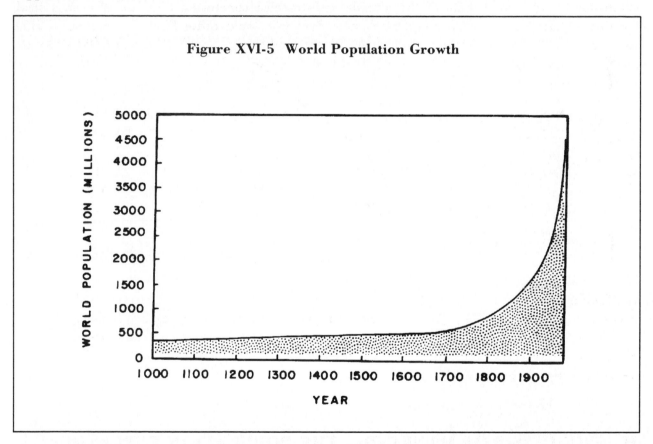

Figure XVI-5 World Population Growth

Much of the population increase is occurring in the less-developed nations. Today, annual population increase in western and northern Europe is only 0.1%, but it is 2.5% in South America, 2.9% in Africa, and 3.0% in Central America. Even so, it should be recognized that the highest per capita consumption of resources occurs in the developed nations that generally have low population growth rates. In 1982, a resident of the United States used the same quantity of energy resources as forty-seven residents of India.

The developed countries have achieved low population growth rates through a slow process of development in which a society evolves from initial high birth and death rates that tend to cancel one another for overall slow growth, to a condition of low birth and death rates, once again yielding slow growth. Between these two conditions, the society goes through a period of continuing high birth rates, but rapidly falling death rates due to improved medical and health practices. During this period, called the <u>demographic transition</u>, the society's population increases rapidly. Decreasing birth rates tend to come about with economic improvement, urbanization, and education. Unfortunately, many of the less-developed nations are stuck in the middle of the transition. They have succeeded in bringing down the death rate, but have been unable to advance their people's standard of living sufficiently to encourage comparably lower birth rates.

The population of the planet will not increase indefinitely, of course. If left totally to the processes of biological fertility, the population will ultimately become limited by the total energy supply, mostly in the form of food. It is obvious that a situation in which birth and death rates are set primarily by malnutrition is not a desirable one. Furthermore, the longer a populous nation waits to take corrective action, the less likely it is that a stable population will be achieved without severely altering the ecological situation.

The choices for developing countries are extremely difficult ones. There are, in fact, only two possibilities for reducing population growth: decreasing birth rates or increasing death rates. The former is greatly to be preferred, but the question is by what means it is to be accomplished. India has experimented briefly with forced sterilization and today forbids marriage for males under 21 years of age and for females under 18. This, together with extensive education in birth control methods and voluntary sterilization programs, has produced a decrease in population growth rates from 2.6% in 1969 to 2.0% in 1984. By encouraging families to have only one child, and placing penalties on families with more than two children, China has seen its growth rate drop from 2.5% in the 1960's to only 1.4% in 1983. Their goal is to reach 1.0% by the year 2000. Virtually all of these methods require impingement on personal freedom and are unacceptable to many people on moral and religious grounds.

Clearly the kind of increase shown in Figure XVI-5 cannot continue forever. The important question is, how many people can the world support, and how close are we to that limit? In 1969, the National Academy of Sciences released a report that concluded that a world population of 10,000 million "is close to (if not above) the maximum that an intensively managed world might hope to support with some degree of comfort and individual choice." A population of 10,000 million might well be reached by the year 2030.[57] The Academy also concluded that Planet Earth would appear to have an ultimate carrying capacity of around 30,000 million people, but only at the sacrifice of individual freedom and choice, and under conditions of chronic near-starvation for the vast majority.

> Indeed it is our judgment that a human population less than the present one would offer the best hope for comfortable living for our descendents, long duration of the species, and the preservation of environmental quality.

---

[57]<u>Resources and Man</u>, Committee on Resources and Man, National Research Council of the National Academy of Sciences, W.H. Freeman and G., 1969.

STUDY QUESTIONS

XVI-10. In what way does the recent human population growth rate differ from a true exponential growth?

XVI-11. Why do populations grow fast during the demographic transition?

I. THE FATE OF THE EARTH

You have probably noticed throughout the course that while scientific investigations are often initiated for reasons of pure intellectual inquiry, once they have been pursued sufficiently far and the proper connections established with related pursuits, then the insight gained usually has important social significance as well. Pure science, after all, is really nothing more than the business of learning the rules of nature. All of us are forced to abide by those rules, but they are still not very completely understood.

A baseball player that inadvertently wanders onto a cricket field may learn the appropriate rules by trial and error, but probably not before he has lost the game. A knowledge of the rulebook would help enormously, and that is true for us also as we attempt to steer the course of human destiny away from disaster and toward a fulfilling life.

At this point in our civilization there really can be no separation of scientific and social pursuits. Attempts to do so are still found in the popular press from time to time, using one or the other of two approaches.

The first false approach is blind reliance on the appearance of some future "technical fix". It is quite possible that some of our pressing problems -- we may even hope to say, many of them -- will at least be eased in some degree by future scientific advance. This has been true in the past, and we have every reason to believe that it will prove true in the future as well. But it would be a mistake to assume that all of them, or even any one of them, can be solved with a technical fix alone.

The development of practical, inexpensive, and safe power generation from a sustainable source would prove to be an immense boon for all of mankind. Even so, there would still be the need to instill the notion of frugal use of energy in all of Earth's citizens because of the inescapable problems of thermal pollution. Until such a development hopefully comes about, we are likely to take the easiest course, continuing to rely primarily on the fossil fuels: oil and gas, and when they run out, coal. Yet the carbon dioxide problem will remain. There may be no technical fix to this problem at all, other than shifting away from the fossil fuels entirely, and there are social and economic problems associated with that. To a remarkable extent, our lifestyle in the developed nations has been built around needs and opportunities afforded by the internal combustion engine. Eventually it may prove necessary to replace it with something else and a certain amount of social adaptivity will be called for.

The Global 2000 Report makes the point in this way:

The full effects of rising concentrations of carbon dioxide, depletion of stratospheric ozone, deterioration of soils, increasing introduction of complex persistent toxic chemicals into the environment, and massive extinction of species may not occur until well after 2000. Yet once such global environmental problems are in motion they are very difficult to reverse. In fact, few if any of the problems addressed in the Global 2000 Study are amenable to quick technological or policy fixes; rather, they are inextricably mixed with the world's most perplexing social and economic problems.

The second false approach takes the opposite view to a reliance on the technical fix; we may call it the Luddites' approach. In the early nineteenth century, English workingmen, worried that proliferating manufacturing machines would put them out of work, and perhaps sensing the immense changes that their use would bring about in the incipient Industrial Revolution, organized bands to destroy machines of every description that went beyond a certain degree of complexity. There are echoes of this kind of thinking today, though violence has not been its principal manifestation.

Most of us have entertained at times a yearning for the apparently simpler and more predictable lifestyles of the past, and some people have revoked a technologically-supported lifestyle entirely, preferring to adopt a more primitive existence that emphasizes self-reliance and close contact with nature -- the "back to nature" movement. There is no particular harm in this, as long as too many people do not attempt it. This style of life requires considerable acreage and would support a far lower population density than already exists on this planet.

The principal danger in this direction is more in the sense of a widening of the popularly perceived gap between science and society. Many scientists sense a kind of withdrawal on the part of their neighbors in the fields of business, arts, and humanities. In the United States the flow of bright young people into the basic sciences has slowed in recent years. But mostly, there is a perception on the part of many people that science is too abstract, too difficult for them to understand. Poor education in mathematics may be part of the problem. Mathematics is the natural language of science, and if you were too terrified of mathematics to even attempt the numerical model exercize in the **Climates of the Earth** unit, then you may have had difficulty grasping the power of this kind of approach in solving problems of extreme complexity.

In a technological world, it would be extremely unwise for the citizenry to remain ignorant of the workings of science. Nor is there any reason for it, even among the mathematically shy. Throughout the course we have stressed that the working rule of science is nothing more than a codified form of common sense: that theory is best which best explains all available observations. And common sense dictates that it is just as important for the artist or those in business to understand and even participate in science as it is for the scientist to understand and participate in economics and art.

In an earlier time, statesmen such as Thomas Jefferson and Benjamin Franklin delighted in their participation in the exploration of science and succeeded in making substantial contributions. While science is today far more specialized, amateurs in many fields continue to make significant discoveries. Even so, scientitst who are weekend painters or concert-goers participate in the arts not in order to compete with the professionals, but for personal satisfaction in having mastered to some extent a fuller range of mankind's endeavors. Science and technology stand among the highest achievements of mankind, and just as an understanding of arts, humanities, and business is vital to the scientist as a human being, so is some degree of understanding of science vital to the artist, the humanist, and the man or woman in the business world.

At this point in time, many scientists feel that it would be a great error to take a pessimistic point of view concerning the fate of the Earth. As an example, in 1984, 75 leaders of science, government, industry, and citizen's groups from 20 countries met near Washington, DC to participate in an international conference on "The Global Possible: Resources, Development, and the New Century." The conference was convened to address a fundamental question:

> Can the world reverse current resource and environmental deterioration while at the same time promoting a better quality of life for all and achieving a marked improvement in the living standards of the world's disadvantaged?

The participants came to a hopeful conclusion. In the summary statement of the conference[58], the participants outlined the resource and environmental problems and then concluded:

> The Conference is confident that these trends can be reversed. We can stabilize human populations, improve people's quality of life, provide more food, save tropical forests and disappearing species, and protect the environment. We can do these things using means that are within our grasp and in ways that further other critical goals, both economic and humanitarian. But these goals will be realized only if a concerted effort is made, with some urgency, to change many current policies and to strengthen and multiply the successful initiatives that already have been undertaken. Since the United Nations Conference on the Human Environment, held at Stockholm in 1972, many encouraging initiatives have pointed the way to a sustainable society. We have learned from both success and failure and have built many of the institutions needed for effective action internationally and locally. We must mobilize now to achieve the global possible. If we do, the future can be bright. We have sufficient knowledge, skill, and resources -- if we use them. If we remain inactive, whether through pessimism or complacency, we shall only make certain the darkness that many fear.

Your studies in this course have taken you through many worlds and realms of scientific thought. The sheer explosion of scientific knowledge that has occurred in the past few decades seems overwhelming, even sometimes to its practitioners. And yet we have gained powerful new tools to aid our scientific understanding, and there is considerable hope in that.

The tools that are being developed are complex in themselves, but their effect is to simplify and make more direct our understanding of the workings of nature. The numerical model for computers is a good example. Such a model may require millions of calculations, but it is not necessary for the human who is using it to follow all the details of these calculations. That is the computer's job. The human's job is to define the rules under which the computer

---

[58]The Global Possible: Resources, Development, and the New Century, World Resources Institute, 1984.

operates, and the protocol under which those rules are applied. The rules are generally well-understood and thoroughly reliable laws of nature. The scientist's job is to see that the model used by the computer is a good approximation to the situation in the real world that is being modeled.

The modern 35mm camera is a good analogy to how an increase in the complexity of a tool's construction can actually make it simpler to use. Many cameras now contain tiny microprocessors -- computers, actually -- that can take care of such things as exposure and even focusing that used to be the bane of many amateur photographers. As the sophistication of these cameras has increased, they have become easier to use, and their choices for exposure have come closer to what a highly trained photographer would have applied to a manual camera.

Throughout this course, you have seen many other kinds of tools that have become available to scientists. Some are conceptual tools: plate tectonics is a prime example. Their effect is almost always to simplify our understanding and at the same time to make that understanding far more powerful. Before the introduction of plate tectonics, geology seemed to be such a complex science that even many of the most basic questions, such as "why are there mountains?" could receive only the most circumstantial of answers. Plate tectonic theory provided so simple and understandable a framework for understanding that the science has been able to make enormous strides in a very short time.

Other recently-developed conceptual tools have included the recognition of different kinds of tectonic regimes on other planets, greater understanding of the mechanisms for mineral concentration, and the strong interconnectedness of the oceans with the atmosphere and the solid rocky parts of the geosphere. Indeed, the recognition of interconnections between what were formerly regarded as separate and isolated disciplines of study may be one of the most significant recent developments in Earth science. You have seen how recognition of these interconnections has developed: between geosphere and biosphere; between Sun, Earth, and the other planets; and between the activities of mankind and the physical environment in which we live -- an environment shaped to a considerable extent by the other organisms with thich we share this planet and by those that preceded us in the long evolution of life on Earth.

An increasingly sophisticated technology has added to the hardware in our toolbox as well: remote-sensing satellites, research submersibles, Doppler Radar, laser-ranging distance measuring devices, seismic sources and detectors, rigs for drilling deep boreholes into the Earth, and many others, including the computer itself.

Taken together, these tools not only provide help for the scientist but are making an increasing impact on decisions of societal importance. As depressing as the thought of a Nuclear Winter may be, there is a distinctly encouraging side to it as well: we now know a great deal about the environmental effects of nuclear war without having had to actually experience it. The problems that we face are greater than ever; but so are the tools that are available to help us deal with them.

What remains, then, is a broader understanding; an understanding of the problems from both a scientific and social point of view, and a careful working out of the available options. What science can offer in many cases, through its rule-finding and modeling of natural reality, is a somewhat imperfect ability to predict consequences of our actions, and to suggest possible alternatives. The actual choices will usually involve tough social and ethical choices, and scientists cannot (and do not wish to) make these for you. The value choices must be made cooperatively, by all segments of society.

In this course we have seen some consequences that suggest meaningful action. It appears that we should take steps to decrease the production of excess carbon dioxide; to control emissions of nitrogen oxides and chlorofluorocarbons; to prevent toxic wastes from entering the environment and in general to preserve the essential workings of the biosphere; to prevent extinctions of species; to control human population growth in a manner that is effective yet ethical and humane; and to avoid nuclear war. It is possible that as our scientific understanding increases, some of these concerns may prove unfounded. Other new ones are sure to emerge. For now it is the best we can do, and in most cases, some action is preferable to no action.

These are problems that will require all the imagination and energy of many different points of view. They are not the agenda of science alone; they are your agenda, for the rest of your life on Planet Earth.

RECOMMENDED READING

Gerald O. Barney, Dir., The Global 2000 Report to the President of the U.S., Pergamon Press, 1980.

National Academy of Sciences, Causes and Effects of Stratospheric Ozone Reduction: An Update, National Academy Press, 1982.

R.P. Turco et al., The Climatic Effects of Nuclear War, Scientific America, August 1984.

ANSWERS TO STUDY QUESTIONS

I-1.   Sediments are carried to the sea by water, and as they settle out on the sea bottom, they distribute over a wide area, forming a flat and mostly horizontal layer.

I-2.   a) Your answer could include any of the following: tidelands, deltas, large lakes, or shallow seas. b) in young active mountain ranges, such as the Alps or Himalayas. c) almost anywhere above in a desert or on a beach. e) almost anywhere above sea level.

I-3.   Both a) and b) have been observed to occur in recent years, and so are uniformitarian. The rate of erosion in the Grand Canyon is sufficient today to have carved out the canyon in a million years or so, and therefore d) is also uniformitarian.

        While floods are frequent occurrences, no worldwide flood has been observed in recent times, and so e), is catastrophist. An asteroidal collision has not been observed to happen on Earth or on other planets, though it probably did happen in the past. It, too, would be classed as catastrophist. The distinction between catastrophist and uniformitarian models is not always clear, because it depends on the definition of "recent times."

I-4.   According to the Principle of Superposition, the topmost layers are youngest. Therefore, the trilobites came first, and the dinosaurs appeared later.

I-5.   For the Principle of Original Horizontality, your answer might include something like: the thin layer of mud left at the bottom of a dried up puddle, mud and gravel washed out over a road <u>after</u> a storm, the mud and gravel left <u>after</u> a flood, or any other example of sediment spread out and deposited in a flat sheet by water or wind.

        For the Principle of Superposition, it is the sequence of deposition that is important. The mud on top of the road was deposited after the road was built. Mud and gravel covering grass came after the grass grew. In a pile of magazines in your living room or den, the oldest magazines are on the bottom of the pile, provided that no one has come along and reshuffled them.

I-6.   Each time a parent atom decays, it is replaced with one atom of the daughter. Hence the total number stays the same.

I-7.   a) There were zero Lead-207 atoms present when the zircon formed, since zircon excludes lead at the time of formation. b) There were 400 billion Uranium-235 atoms present at that time, since each Lead-207 atom present today formed by changing from a Uranium-235 atom. c) The parent Uranium-235 is down to 25% of its original number, hence the zircon is two half-lives old, or 1.4 billion years.

I-8.   Since Carbon-14 is the parent, you can read off the number of half-lives directly from Figure I-5. Locate 30% on the vertical axis and draw a horizontal line from it to the curve, then straight down to the horizontal axis. You should come up with something like 1-3/4 half-lives. Since one half-life for Carbon-14 is 5,730 years, the age of the wood is 1.75 times 5,730, or about 10,000 years.

I-9.   Ordovician, Carboniferous, Triassic, Jurassic, Tertiary.

I-10.   Top: Tertiary, Jurassic, Triassic, Carboniferous. Bottom: Ordovician

III-1.   The principal difference was in the assembly of all the continents into one supercontinent in the Late Carboniferous.

III-2.   According to Wegener's theory, Africa, South America, and Antarctica were still attached to one another when Mesosaur lived. Its original habitat split apart when the southern continents broke up.

III-3.   Your answer should include mention of thin, rigid plates that are moving about the surface of the Earth.

III-4.   Lines or bands of earthquakes are the most reliable evidence for a plate boundary. You should realize, though, that not all earthquakes occur on plate boundaries. Compare Figure III-2 to Figure III-3.

III-5.   Wegener's theory relied only on motion of the continents. In his theory, the ocean floor did not move. In plate tectonics it is the plates that move, and they may carry continent, ocean floor, or both.

III-6.   Either temperature or rigidity would be correct answers, since the asthenosphere is softer (less-rigid) than the lithosphere, because it is hotter and very close to its melting point.

III-7.   Either rock type or density would be satisfactory answers, though you should recognize that mantle rocks are much denser than crustal rocks.

III-8.   Lithosphere and asthenosphere refer to physical properties (temperature and rigidity) which can be different in different parts of the mantle.

III-9.   If every part of Mt. Everest were in perfect isostatic equilibrium with its root, the tip of that root would extend 4.5 times its height below the average crustal thickness, or 39,600 m (39.6 km). This is in addition to the normal continental crustal thickness of about 30 km (see Figure III-4).

III-10.   Since new oceanic crust is created at the ridge crest and is then carried away by sea-floor spreading, youngest ocean floor should be found near the ridge and oldest ocean floor far from the ridge.

III-11.   Deep oceanic trenches are formed where a subducting plate is drawn down under the overriding plate. This can only happen at a subduction zone.

III-12.   The volcanoes of the Andes are created on the overriding plate by magma rising from the subducting slab of oceanic plate below. This slab is the edge of the Nazca Plate which is diving under the edge of South America, west to east. In general, the volcanoes are found on that side of the trench where there are two thicknesses of plate, one above the other. See Figure III-7.

III-13.   Baja California is a strip of land that has split away from the coast of Mexico just to the southeast. You should be able to fit it back into the fairly obvious niche in the Mexican coastline by moving it almost vertically down in the figure to close the Gulf of California. In the future, Baja California together with everything west of the San Andreas, will move northwest along the fault. In ten million years or so, Los Angeles will be next to San Francisco.

III-14.   The Hawaiian hot spot must have been located in the vicinity of Oahu and Kauai, since the rocks on these islands formed about three million years ago. All the islands to the southeast with rocks younger than three million years had not yet been created.

III-15.   The Pacific Plate is moving to the northwest, along the Hawaiian chain, with respect to the hot spot. Remember the sewing machine analogy: the islands are the "stitches" moving away from the hot spot "needle", which is currently located at the big island of Hawaii.

III-16.   Fractionation refers to the separation of substances from a mixture on the basis of a physical property such as melting or boiling point. There are many commercial examples, including the recovery of pure metals from scrapped automobiles by heating scrap to progressively higher temperatures -- the metals with the lowest melting temperatures will melt first.

          Distillation works the same way using boiling, as in the separation of volatile gasoline from crude oil and the concentration of alcohol from a fermenting mash.

III-17.   Oceanic crust is thin because it is carried away from its place of origin at the ridge crest before it can accumulate to any great thickness. At subduction zones, however, lava and magma can accumulate for a very long time on the overriding plate because it is not being destroyed.

III-18.   Both the Eurasian and African Plates move toward the west as seen from the North American Plate, continuing the widening of the Atlantic Ocean.

III-19.   A collision between North America and Africa brought the Appalacians to their greatest height in the Late Paleozoic. The Alps are being formed by the slow northward motion of Africa against Europe, and a Late Paleozoic collision between Europe and Asia raised the Ural Mountains of Central Russia.

III-20.   The following approximate inclinations can be read directly from Figure III-18: New Orleans, 50° down into the ground; Oslo, 75° down; Porto Alegre, 50° up out of the ground.

III-21.   You would be somewhere on the magnetic equator, which in most places is not far from the true equator. If the magnetic field points north and horizontal the magnetic field points north and up out of the ground, then you would be somewhere in the southern hemisphere. On the basis of the information given in the question, you can't be more specific than that.

III-22.   We can determine the ancient latitude and orientation (that is, which way was north at that time) from the magnetic field direction recorded in rocks. We cannot determine the ancient longitude.

III-23.   After a magnetic polarity reversal takes place, the magnetic field at any point on Earth points in the opposite direction to that it had before the polarity flip.

III-24.   Using only the ocean floor age data in the exercise you cannot determine the ancient latitudes for South America and Africa, only their relative positions with respect to one another. Paleomagnetic data from each continent will allow the ancient latitudes to be determined, however.

III-25.   Gondwana consisted of the continents of South America, Africa, India, Antarctica, and Australia. It existed throughout the Paleozoic and began to break up in the Jurassic.

III-26.   Essentially all of the Tethys Sea has been destroyed by subduction. We do not know of any Paleozoic ocean floor that has survived to the present day.

III-27.   Your answer might include any of the following: he could not explain what were the driving forces that moved the plates; he could not explain how the continents could move through a stationary ocean floor (in fact they don't); alternate explanations were available to explain his data that were more in accord with prevailing views of the permanence of oceans and continents; he could not prove that the continents were actually moving at the present time. All of these contributed to the rejection of his theory.

III-28.   There is no certain answer to this question, but plate tectonic theory cuts across the lines separating all these disciplines to such an extent, that it seems likely its development entirely within one of the disciplines would at least have been delayed severely.

III-29.   Most scientists from around the world favor the free exchange of scientific data and publications among all nations. International conferences, multinational research projects, and the exchange of visiting scientists have frequently led to advances in all the sciences.

III-30.   California, the intermountain seismic belt (between the Rockies and California), the Seattle area, the New Madrid area of the midcontinent, and the Charleston, South Carolina area, all are assessed as having the potential for serious seismic hazard in Figure III-33.

III-31.   Two unit steps on the Richter magnitude scale corresponds roughly to a 30 x 30 = 900 times increase in energy release.

IV-1.   Intraplate seismicity is found more often on the continents than on the ocean floors. See especially North America, Asia, and Africa.

IV-2.   In the course of melting its way through the crust, the magma mixes to a greater or lesser extent with the crustal material. Since oceanic crust is basaltic, the resulting lava is usually basaltic. In the case of continental crust, the basaltic magma from the hot spot mixes with the andesitic and granitic crustal material to produce a wide range of composition.

IV-3.   The active volcanism marks the current location of the hot spot, which is at the southern end of the line of volcanoes. Volcanoes produced by the hot spot have been carried away to the north.

IV-4.   Volatile gases, such as water vapor, dissolved in the magma produce much of the violence of the eruption. Once these gases come out of solution, the effect is that of a steam explosion.

IV-5.   Basaltic magma tends to contain less volatile gas in solution than andesitic magma.

IV-6.   Aulacogens are associated with rifting, which is an extension of the crust.

IV-7.   The process is very similar to what happens at the oceanic ridges. The stretched crust becomes thinner, just as a stretched bar of taffy becomes thinner. Thin crust stands at a lower elevation than thick crust (see Figure III-5), so a valley results at the rift.

IV-8.   If a lithospheric block is approaching North America prior to accretion, subduction must be occurring in the region separating them in order to get rid of the intervening lithosphre. Subduction is the only process that can do this.

IV-9.   There has been no subduction off the United States east coast during the past 250 million years, hence no terranes could have arrived within the last 100 million years. Sea-floor spreading has been the dominant process in the North Atlantic during this time period.

IV-10. Ray C^1 penetrates the outer core, which has a much lower P-wave velocity than the mantle. Imagine a car traveling along the path of ray C^1. At the core-mantle boundary, its right wheels are in the slow-speed region of the core while its left wheels are in the high-speed region of the mantle. The result would be to pull the car around to the right, more sharply into the core. As the car emerges from the core, it is once again pulled to the right by the same effect. The rays behave in the same way.

IV-11. Ray G crosses each boundary almost head-on, and so it is not deflected much either way. In the car anology, the car drives off the end of a paved road onto a dirt road. It is slowed down, but because both left and right wheels see the change at the same time, the car continues straight ahead.

IV-12. Among other geophysical reasons, the outer core is believed to be liquid because S waves cannot travel through liquids and the large S-wave shadow zone indicates that S waves cannot penetrate the outer core.

IV-13. The core is metal while the mantle is rocky.

IV-14. Other than the liquid water of the oceans, the only major division of the Earth that is liquid is the outer core. However, there may be small local parts of the crust and asthenosphere that are at least partially melted beneath volcanic zones.

IV-15. On the average, we might expect crustal rocks to have a temperature of about 300°C at a depth of 10 km if we use the average rate of increase of 30°C per km.

IV-16. (a) is an example of convection, since heat is carried by a moving medium, in this case, air. A cool breeze simply means that warmer air is being displaced (moved away) by the motion of cooler air. (b) and (c) are examples of conduction, since heat must travel out through the stationary outside wall of the house through the metal bottom of the hot iron.

IV-17. Hot convecting material expands, becomes less dense, and rises, while cool convecting material contracts, becomes more dense, and sinks. The hot rising mantle material supplies heat to the volcanism of the oceanic ridges, and as it spreads apart (see Figure IV-7 on page 114) encourages rifting of the spreading ridge.

IV-18. Both X-rays and seismic rays can penetrate their respective subjects, and so the patterns that they produce after they have emerged allow us to deduce what is inside.

IV-19. In Figure IV-18b, the convection currents move in the same direction as the plates in the upper mantle, and so help to drive them. The effect is not so clear in Figure

IV-18a, where the upper mantle moves with the plates in some places, but in the opposite direction in others.

V-1.   Your answer should have been something close to 6,400 km.

V-2.   He noted that the shadow of the Earth cast on the Moon during a lunar eclipse is always circular. Note that if this observation is made at several different eclipses, when the Moon is in different parts of the sky for an observer, then the possibility that the Earth is a circular disk rather than spherical, is also eliminated.

V-3.   Only about the width of the present Atlantic Ocean.

V-4.   The merchant ship captains were familiar with the Gulf Stream and avoided it when traveling to the west. The mail packet captains did not know of it and sailed against its currents.

V-5.   The sounding line.

V-6.   The recovery of sediment and rock cores from the ocean floor.

V-7.   Soft sediments.

V-8.   To lower instruments, samplers, and corers into the deep ocean.

V-9.   Sea level is likely to be lower when seafloor spreading rates are low because the spreading ridges have less volume and displace less ocean water onto the continental shelves.

V-10.   Sediments and water.

V-11.   Near the continental shelf. This is older ocean floor and so has been accumulating sediments for a longer time. Also, there is a greater supply of terrigenous sediments near the continental shelf.

V-12.   At 2 mm per thousand years, it would take 500 thousand years for one meter of sediment to accumulate.

V-13.   The side-scanning sonar is able to map a swath of ocean floor in one pass, while the electronic depth sounder gives depths only directly beneath the track of the ship.

V-14.   A depression. The mass of rock on either side of the canyon gravitationally attracts the water away from the point over the canyon.

V-15.   Among the unusual properties of water are: its solid state (ice) is less dense than its liquid state; its high melting and boiling temperatures; its ability to dissolve so many other substances.

V-16.   Chlorine (Cl⁻), sodium (Na⁺), and sulfate (SO⁻ ⁻) ions are the most abundant   in seawater.

V-17.   Carbon dioxide (CO_2).

V-18.   The sodium may be supplied by river water, while the chlorine appears most likely to come from the mantle by way of the vents at the spreading ridges.

V-19.   A region of rapid change in salinity, generally dividing the surface from the   deep waters of the oceans.

V-20.   The Antarctic region, providing cold, saline water that is very dense.

VI-1.   Direct measurements track ocean currents   by observing how they move; indirect methods   rely on secondary effects like the generation of electrical currents or the relation between density changes and currents.

VI-2.   It would produce upwelling. The effect is reversed from that shown in Figure VI-15 because the Coriolis Effect is opposite in the southern hemisphere.

VI-3.   The ball will be deflected to the right as seen by you, since the merry-go-round is rotating in the same sense as the Earth as seen from above the northern hemisphere. The direction that you throw the ball does not matter.

VI-4.   The east coast of North America is washed by the northward flowing warm currents of the Gulf Stream, while the west coast is influenced by southward flowing cold currents.

VI-5.   Because water has such a high heat capacity, it can absorb a large amount of heat and not increase its temperture very much. Similarly, the oceans can give up a large amount of heat and cool only slightly. The effect is to produce cooler tropics and warmer poles than would otherside prevail.

VI-6.   A strong El Ni ño influences climate because the temperature of large areas of the sea surface is changed quite substantially. Interactions with the atmosphere produce the climatic effects.

VI-7.   Drought in some places, heavy rain and flooding in others; unusual storm activity.

VI-8.   By observing the color of the seawater.

VI-9.   The cores of ring currents consist of water pinched off from the other side of the Gulf Stream. Cold slope water is pinched off by a loop of the Gulf Stream and transported south into the warm Sargasso Sea.

VI-10.   They tend to be transparent and gelatinous, like jellyfish.

VI-11.   Chemical energy, probably derived from the oxidation of hydrogen sulfide.

VI-12.   The pressure is too great to allow steam to form. Boiling temperature increases with pressure.

VII-1.   The burning of coal and volcanic eruptions.

VII-2.   The atoms and molecules of a hot object are moving or vibrating more rapidly than those of a cool object.

VII-3.   In a mercury barometer, the level of mercury in the glass tube is controlled by the pressure of the atmosphere. When the atmospheric pressure drops, the level of mercury falls, hence the expression.

VII-4.   Ozone absorbs ultraviolet radiant energy from the sun, and this heat raises the temperature of the stratosphere.

VII-5.   Blue-violet light has a wavelength of 4×10^{-7} m. Using this in rule (2), we get a temperature of 6,927°C. Compare this to the much lower temperature of the red glowing coal.

VII-6.   The light-colored sand reflects much of the Sun's radiant energy, while the dark asphalt absorbs it, raising its temperature.

VII-7.   The bank of smoke or mist acts as a barrier to the escape of infrared radiation, and prevents the trees from losing as much heat during the night. In some cases, this can prevent the blossoms or fruit from freezing.

VII-8.   In September and April, when the Sun is directly overhead at noon. In December and June, the Sun reaches north or south of the overhead point, and its rays strike the ground at an angle less than 90°, resulting in slightly less heat. Nevertheless, the temperature would hardly change from season to season.

VII-9.   During the night, the land surface cools down and may become cooler than the ocean water. Air rises over the warmer water and causes a breeze blowing from land to sea. In the afternoon, sunshine has heated the land surface until it is now warmer than the ocean water and a sea breeze results.

VII-10.   It is being carried away by the latent heat contained in the steam. When the steam (water vapor) condenses into the fog droplets that you can see, the latent heat is released into the air.

VII-11.   North of the Hadley Cell, horizontal mixing of the atmosphere dominates due to the fact that the Coriolis Effect becomes stronger at higher latitudes. Rotary motion about low-pressure regions brings cold air to the south and warm air to the north, aiding the poleward flow of heat.

VII-12. In the southern hemisphere summer, that is, December to April. During this time the Australian continent is warmed and a low pressure zone forms over it, pulling moist air in from the ocean to the north.

VII-13. No answer necessary.

VII-14. Look at Figure VII-11 on page 232. Note that a cold front has a very blunt nose, with the boundary between cold and warm air rising very abruptly from the ground. The result is that the warm, moist air is forced up quickly, producing more localized cloudiness and storms. The warm front, however, is characterized by a shallow wedge of cold air overlain by warm air. The warm air is forced upward more gradually and over a larger area, producing widespread cloudiness.

VII-15. Winter air is generally drier than the air during the summer, and so the potent contributions of latent heat to the energy content of a storm are much less in the winter. The powerful updrafts that form thunderheads are largely due to the energy transported by latent heat.

VIII-1. Regions downwind of semipermanent high-pressure zones and those downwind of major mountain ranges are two types of areas that tend to be arid.

VIII-2. No, the ice in a glacier always moves downhill in response to gravity. But if more ice melts near the toe than comes down from above to replenish it, the glacier will get shorter and its end (toe) will move farther up the valley. Climate changes are reflected in the growth or shrinkage of the glacier.

VIII-3. Patterns of thick or thin growth rings reflect benign or harsh climatic conditions during the life of the tree.

VIII-4. The growth ring patterns in the logs can be matched to the same patterns found in trees that were alive then and are still living today. It is then just a matter of counting the rings to determine the age of the patterns.

VIII-5. Because O^{18} does not evaporate as readily as O^{16}, rainwater should be deficient in O^{18} and the O^{18}/O^{16} ratio should be lower in rainwater than in ocean water.

VIII-6. Fortunately the two effects work in the same sense, not against one another. Note that a clam will incorporate more O^{18} into its shell at lower temperatures, but that a greater volume of ice will increase the O^{18}/O^{16} ratio in the seawater that the clam draws upon. So a higher proportion of O^{18} in the clam's shell indicates either lower seawater temperature or a greater volume of ice, both of which are indicators of cooler climate. The only difficulty is precisely separating the two effects.

VIII-7. The glacier carries along with it boulders and rocks that act in an abrasive manner. During the passage of tens or hundreds of thousands of years, the glacier has ample opportunity to rework the landscape in a substantial way. In fact, glaciers are one of nature's most effective landscaping tools.

VIII-8.   The arrangement of continents that tends to cut off the polar regions from the circulation of major ocean currents seems likely as at least a partial explanation for the onset of Ice Ages in the Pleistocene.

VIII-9.   Regular and periodic changes in the Earth's orbit around the Sun.

VIII-10. If the initial conditions given to the model are always the same, then the predictions of the model in fact will not vary. But different results will be obtained if the initial conditions are changed.

The results that you would have obtained in steps 6 through 10 would have been exactly the same, because the results in any step depend only on the values that existed in the previous step. Notice that as you filled out the worksheet, you never had to look beyond the previous step.

VIII-11. Similarly, in climate and weather models, if the initial conditions are sufficiently well defined, then it is not necessary to know the entire previous history of the climate system. In order to model Cretaceous climate, the geographical conditions that existed during the Cretaceous are fed into the model and it should then reproduce Cretaceous climate, provided that all the mechanisms that were operating then have been incorporated into the model.

VIII-12. The burning of fossil fuels (coal, oil, natural gas) and wood, deforestation, fermentation processes, and breathing.

IX-1.   Venus moves more quickly in its orbit because it is closer to the Sun.

IX-2.   Neptune.

IX-3.   The twinkle of stars is caused by turbulence in the Earth's atmosphere. Astronomical observations are located on mountaintops in order to be above the densest part of the atmosphere. The Space Telescope will be placed in orbit so as to be above the atmosphere entirely.

IX-4.   Mercury and Venus are both closer to the Sun than Earth. Look at Figure IX-1b on page 291 and notice that from a point on Earth's orbit, all of Venus' orbit is within 45 degrees of the Sun and Mercury's orbit is within 23 degrees of the Sun. As a result, neither planet as seen from Earth wander farther from the Sun than that angle.

IX-5.   Your answer might include the following: Earth rocks contain water, while Moon rocks do not; sedimentary rocks are found on Earth, but not on the Moon; metallic iron is found in Moon rocks, but not in Earth rocks; Earth rocks are usually weathered to some degree, while Moon rocks are not.

IX-6.   The maria are some 1.5 billion years younger than the highlands, so this implies that the number of impacting bodies was much higher prior to the basaltic flooding of the mare plains than after.

IX-7.   There are two reasons -- Earth's oldest rocks date back to only 3.8 billion years. By this time the influx of meteorites had already slowed. Most of Earth's rocks are younger than 3 billion years, in fact. The second reason is that weathering and sedimentation on Earth has removed traces of all but a few of the craters that have formed in existing rocks.

IX-8.   Most of the maria are found on the hemisphere that faces us. This is probably due to a thinner crust in that hemisphere.

IX-9.   Comparison between the numbers of craters per unit area for the Moon and Mercury make rough dating possible, since we have age-dated lunar rocks. No samples have been returned from Mercury, so this is the only method available so far.

IX-10.   The dominant tectonic process formerly active on Mercury seems to have been shrinkage of the planet as it cooled early in its history.

IX-11.   The dense atmosphere with its clouds of sulfuric acid droplets makes it impossible to photograph its surface features. Detailed radar mapping of the surface has only just begun.

IX-12.   No, the temperature is far above the boiling point of water.

IX-13.   Channels eroded by running water are found in abundance on Mars. Today, all water on that planet must be frozen.

IX-14.   Carbon dioxide is involved in nearly all life forms found on Earth, either as a source of carbon (as in photosynthesis) or as a waste product.

IX-15.   The pressure in the lower atmosphere of Jupiter is so high that gaseous hydrogen changes gradually, and not abruptly, from gas to liquid.

IX-16.   A hurricane or other persistent cyclonic eddy in the atmosphere.

IX-17.   Io is close to Jupiter and derives much of its internal heat from tides raised by Jupiter. Callisto is far from Jupiter, and so its tides are very weak in comparison.

IX-18.   Jovian planets are large, gaseous, low in density, far from the Sun, and possess few or no satellites and have no rings.

IX-19. The rings are not solid, but are made up of millions of particles in orbit around Saturn. There is empty space between the particles.

IX-20. An atmosphere rich in nitrogen.

IX-21. Pluto differs from Uranus and Neptune in that it is very much smaller, and its surface is icy, not gaseous.

IX-22. Jupiter, Saturn, and Uranus. Perhaps Neptune also has a ring system, but it has not been observed from Earth, either, and is only visible from passing spacecraft. The question of Neptune's rings will have to wait until 1989.

IX-23. Many of the larger meteorites that fall to Earth have their source in the asteroid belt, while some of the smaller meteors, especially those associated with meteor showers, are probably derived from passing comets.

IX-24. The iron meteorites are most like the Earth's core.

IX-25. The solar wind blows the dust and gas from the solid body of the comet outward and away from the Sun. As a result, the tail can precede the comet, when the comet is moving farther from the Sun.

IX-26. Larger-sized craters tend to show central peaks. On the Moon, craters smaller than about 20 km (12 mi) in diamter usually do not have them.

IX-27. Plate tectonics is known to be active for certain only on the Earth. Venus is a possibility, but more detailed examination of its surface will be needed first.

IX-28. The ice worlds have smoother surfaces because ice can flow under the influence of gravity more readily than rock.

IX-29. Carbon dioxide.

IX-30. Biologic and oceanic processes have removed Carbon dioxide from the atmosphere and locked it up in sedimentary rocks, especially limestone and coal.

IX-31. The temperature on Venus is high because of a strong Greenhouse Effect caused by its dense carbon dioxide atmosphere that admits sunlight but prevents infrared (heat) radiation from escaping to space.

X-1. The oldest dated lunar rocks and meteorites are that age. In addition, studies of uranium and lead isotope abundances on Earth indicate the same age for our planet.

X-2.   These heavy elements were probably produced in one or more supernovae explosions that occurred in the vicinity of the primordial nebular cloud shortly before its collapse to form the Sun and solar system.

X-3.   The explosions helped to compress the original nebula to begin the accretionary process. Also, short-lived radioisotopes like Aluminum-26 were still active in the early solar system.

X-4.   Earth is small compared to the Jovian planets and its weaker gravity field cannot prevent the escape of these light elements. In addition, the intense early solar wind, when the Sun first ignited, probably swept away Earth's primitive atmosphere, which well might have been rich in hydrogen and helium. Finally, most hydrogen in Earth's present atmosphere would oxidize to (Hydrogen-20), water.

X-5.   Both are drastic but temporary effects on the worldwide climate caused by massive injection of explosion products (dust, soot, water vapor) into the atmosphere. In the case of a Nuclear Winter, the model depends largely on injection of soot and smoke particles from the burnings accompanying large numbers of separate nuclear explosions.

        In the case of an asteroidal collision, there is only one massive explosion that excavates far more dirt to form dust. This, together with soot from local forest fires, could produce a nuclear winter effect. On the other hand, if the impact occurred in the ocean, less rock would have been excavated and vast quantities of water vapor would have been injected into the atmosphere, possibly resulting in a temporary warming, not cooling, effect.

X-6.   The chondrites are a class of meteorites whose overall composition reflects that of the solar system as a whole (see page 311). They are felt to be representative of the primitive matter from which the solar system was made.

X-7.   The core probably formed early in Earth history, perhaps within the first twenty million years.

X-8.   The Barringer Meteorite hit the ground and produced a large impact crater, while the Tunguska body disintegrated completely in the atmosphere

X-9.   The extinction events define the boundaries between those three eras. It was precisely the patterns of dramatic change in the fossil record that prompted early geologists to these as the boundaries between the geological eras.

X-10.   If Nemesis has an elliptical orbit that brings it near the solar system once every 26 million years it would explain the apparent cyclicity of extinctions. It would spend much of its time far from the Sun and it may be a very dim star, which would account for the fact that is has not been observed, or at least noticed.

XI-1.   Sheeps' fleeces were used to line sluice troughs in order to separate gold from gravel.

XI-2.   A mineral has a definite chemical composition, while rocks are heterogeneous mixtures of minerals that may have varying compositions.

XI-3.   Your answer might have included: iron, aluminum, manganese, magnesium, and titanium.

XI-4.   Your answer might have included: copper, lead, zinc, nickel, molybdenum, mercury, chromium, tin, tungsten, and uranium.

XI-5.   The steel-making industry is a heavy user of manganese.

XI-6.   Copper is used in the production of electrical wires, brass, and bronze.

XI-7.   Nitrogen, phosphorus, and potassium, are important fertilizing elements.

XI-8.   Oxygen and silicon are the most abundant crustal elements.

XI-9.   Reserves are known deposits from which minerals can be extracted profitably using existing technology under present economic conditions. Resources, on the other hand, are potential resources that might become economically viable at some future time.

XI-10.   Volcanism to supply magma and heat.

XI-11.   Gypsum, salt, potash, and borax.

XI-12.   Bauxite is the principal ore of aluminum and it tends to form in tropical climates with abundant rainfall.

XI-13.   Metallic oxides are formed on slow-spreading ridges, where the hydrothermal fluid has had a chance to mix with cold seawater. Metallic sulfides are deposited directly by the hot hydrothermal waters and are more abundant on fast-spreading ridges.

XI-14.   An ophiolite is a piece of oceanic crust that has been scraped off the sea floor and added to continental crust.

XI-15.   Hydrothermal circulation in the vicinity of igneous intrusions produced by subduction.

XI-16.   A back-arc basin is spreading or rifting that tends to occur immediately behind a subduction zone and separates it from a main continental landmass.

XI-17.   Copper was mined in ancient Cyprus.

XI-18.   Hydrothermal processes were probably responsible for both deposits.

XI-19.   The worldwide economic downturn suppressed demand for minerals.

XI-20.   The U.S.S.R. has consumed few of its mineral reserves because of its relatively low population density, and because it began with substantial reserves.

XI-21.   Manganese nodules are found lying on the ocean floor.

XI-22.   Mining or other commercial activity in Antarctica is prohibited by the Antarctic Treaty.

XI-23.   Substitution involves replacing expensive and scarce metal uses with metals that are cheaper and more abundant. Conservation involves cutting back on demand for raw materials by reducing unnecessary consumption, designing products for durability, and recycling used materials.

XII-1.   Figure XII-1 shows that 1/4 of energy use goes to transport, and of that, 3/4 goes to highway use. We may then take 3/4 of 1/4 to get 3/16 of total energy use that goes to trucks and cars.

XII-2.   Worldwide economic downturn and increasing prices of oil have affected recent energy use.

XII-3.   The term "cracking" refers to the process of breaking complex organic molecules into simpler ones.

XII-4.   To create petroleum, the debris from organic matter must be sealed in an anaerobic environment. Sediments accumulating on the continental shelves or sea floor provide the best environment for this process.

XII-5.   Oil tends to migrate toward the surface because it has a lower density than either rock or water.

XII-6.   An anticline is a favored structure for the trapping of oil because upward-migrating oil beneath an impermeable rock layer will travel to the highest part of the anticline's arch and become trapped there.

XII-7.   In the Persian Gulf region of the Middle East.

XII-8.   Tar sands, at least in theory, could be pumped out of the ground once they have been made less viscous by heat treating with steam. The oil in oil shale is locked into the fine sediments and must be mined before the oil can be recovered by crushing and heating the rock.

XII-9.   Coal forms in a terrestrial environment that is swampy and characterized by a subtropical climate and abundant rainfall, with luxuriant growth of plant matter. Oil forms most often in a shallow-water marine environment that is near abundant sources of organic matter, often from microscopic plants and animals.

XII-10.   An oxygen-deficient (anaerobic) environment is necessary to the production of either coal or oil.

XII-11.   Only a small fraction of the oil produced in rocks remains, much of it having escaped to the surface and oxidized, while most of the coal that has formed remains in the ground.

XII-12.   During the Carboniferous period, when much of the world's coal reserves formed, these continents were at high southern latitudes and probably endured climates too cold for the encouragement of peat bogs.

XII-13.   Sulfur and carbon dioxide are released in the burning of most coal. Only the sulfur can be removed by pollution control equipment.

XII-14.   North America holds the largest reserves of uranium.

XII-15.   No. The breeder reactor simply converts Uranium-238, which is not fissionable, into fissionable Plutonium-239. The supply of Plutonium-239 would be limited by Uranium-238 supplies, but these are very large compared to currently-available supplies of the Uranium-235 presently used in most nuclear reactors.

XII-16.   Tidal energy draws on gravitational energy, not solar radiant energy.

XII-17.   A present or recently active source of volcanism is required for the development of geothermal energy.

XII-18.   Passive solar systems utilize direct heating of the air within the living space, while active solar systems use a heat-collecting system that is mounted externally to the building and usually uses a circulating fluid.

XII-19.   High initial cost is the principal reason.

XII-20.   The last reserves to be extracted are those that are most difficult and expensive to recover, resulting in higher prices and lesser production and consumption.

XII-21.   Deuterium can be obtained from seawater.

XIII-1.   Many of the early observers thought sunspots were objects passing between the Sun and Earth and not actually a part of the Sun. In addition, very few observations were made prior to the invention of the telescope.

XIII-2.   A photon of green light has a shorter wavelength, and hence higher energy than a photon of yellow light.

XIII-3.   Fraunhofer lines appear dark against the background of the Sun; they are due to atoms that have absorbed light that otherwise would have traveled toward Earth.

XIII-4.   Each element has its own characteristic absorption lines in the solar spectrum. The more prominent those lines appear, the more abundant that element must be on the Sun.

XIII-5.   They were converted into neutrons that are still in the helium nucleus. Their positive charges were carried off by two positrons. See Figure XIII-3 on page 404. Note in the figure that six protons are absorbed and two given off in the last stage for a net consumption of four protons.

XIII-6.   Nuclear fusion is hard to get started because all of the particles involved are positively charged, and like charges repel one another. The particles must be brought together with enough energy to overcome this repulsion.

XIII-7.   Only about two argon atoms are actually being produced each day. Six are expected from models of thermonuclear fusion, but only about 30% of that neutrino flux has actually been observed. Thirty percent of six is approximately two.

XIII-8.   The temperature minimum in the chromosphere, at about 4,000°C, is the coolest part of the Sun. Temperatures climb once again in the upper chromosphere and in the corona above the chromosphere.

XIII-9.   The Sun is divided into different layers on the basis of temperature, density, and of type of energy transport.

XIII-10.   By excluding all but the red hydrogen-alpha emission line given off by the chromosphere, for example, the spectroheliograph can form an image of that layer of the Sun.

XIII-11.   The depth of the convection zone and the differential rotation of the Sun's interior are two useful bits of information to come from studies of solar oscillations.

XIII-12.   The solar wind is more intense above coronal holes because the magnetic lines of force are open in these regions -- that is, they extend from the Sun's surface far out into the surrounding space. The particles of the solar wind can travel along these lines of force to escape easily from the Sun.

XIII-13.   Because the outer atmosphere of the Sun is a good conductor of electricity, the lines of force are rooted to the region in which they originated on the Sun. The rotation of the Sun twists the lines of force, which otherwise would have been radial, into a spiral shape.

XIII-14.   They are cooler, hence they appear darker than the surrounding photosphere.

XIII-15.   The strong magnetic fields in sunspots locally suppress convection, with the result that heat from the Sun's interior is not able to reach the surface so readily in the region of the spot. This results in a cooler temperature in the spot.

XIII-16.   Sunspot numbers increase and decrease with the 11-year cycle, but the magnetic fields not associated with them change polarity once in each cycle, requiring two such cycles, or 22 years, to return to the original polarity. Thus it is the magnetic cycle that lasts 22 years.

XIII-17.   No, sometimes temperatures in solar flares can reach high enough for nuclear reactions to take place.

XIII-18.   In the solar corona.

XIV-1.   Ultraviolet and extreme ultraviolet rays are absorbed by atoms and molecules in the thermosphere, causing heating. Later on, you will see that particle bombardment is an additional source of heating as well.

XIV-2.   Atomic oxygen is formed in the upper thermosphere by the action of X-rays and extreme ultraviolet rays. In addition, it is lighter than molecular oxygen and nitrogen and so tends to rise to the top of the atmosphere along with helium, which is a very light element.

XIV-3.   The greatest changes are found in the thermosphere; the least are found in the troposphere.

XIV-4.   The solar wind.

XIV-5.   The magnetopause is the boundary that separates the magnetic fields of Earth from that of the solar wind, preventing most solar wind particles from entering the near-Earth environment. The barrier is not perfect, however, and there is some leakage across the magnetopause.

XIV-6.   The plasma sheet and the Van Allen Belts. More particles are found within the ionosphere and in a band of particles that have diffused across the magnetopause.

XIV-7.   Increased magnetic activity produces greater heating in the thermosphere. This causes it to expand, increasing atmospheric drag on satellites.

XIV-8.   Magnetic storms cause rapid changes in Earth's magnetic field, and these changes can induce surges of electric current in long wires that may trip circuit breakers or overload transformers.

XIV-9.   Skylab fell sooner than expected because increasing solar activity heated and expanded the upper atmosphere, increasing atmospheric drag on the space station.

XIV-10.   At midnight. The plasma sheet is always on the side of Earth that is opposite the Sun.

XIV-11.   The magnetopause is the boundary that separates the magnetic fields of Earth from that of the solar wind, preventing most solar wind particles from entering the near-Earth environment. The barrier is not perfect, however, and there is some leakage across the the magnetopause.

XIV-12.   Within the auroral oval, which is a band centered on the poles. Auroral displays peak within this zone, then decrease as the poles are approached.

XIV-13.   The particles that excite emission from gases in the ionosphere are guided by Earth's magnetic field lines; their alignment along these lines of force cause the appearance of a hanging curtain or drapery.

XIV-14.   Oxygen and nitrogen.

XIV-15.   Recombination of atomic oxygen is the immediate source, but atomic oxygen is produced by the action of solar radiation and energetic particle bombardment.

XIV-16.   Increased solar activity would result in more energetic particles reaching the thermosphere, hence a greater production of nitric oxide. After a while, some of the nitric oxide would migrate down to the ozone layer where it would destroy some of the ozone present there.

XIV-17.   Less than one percent.

XIV-18.   Patterns of tree rings, with narrow rings indicating dry years.

XIV-19.   Probably due to changes in the strength of the Earth's magnetic field. These are thought to be unrelated to changes on the Sun.

XIV-20.   Solar changes are very small compared to the steady output -- less than one percent. This may be negligible compared to normal energies involved in climate variations on Earth. In addition, the mechanisms by which small solar changes can influence our climate are not sufficiently well understood to trace a cause-and-effect relation at this time.

XV-1.   Prokaryotes are primitive single-celled organisms that reproduce asexually by cell division. Eukaryotes are far more complex entities composed of organelles that are each as complex as a prokaryote, each performing a specific task within the cell.

XV-2.   The iron oxide bands indicate that Earth's atmosphere had shifted from a reducing to an oxidizing state.

XV-3.   They might have served as self-reproducing templates to which organic molecules could bind in regular but gradually changing patterns that might eventually develop into living organisms through the processes of natural selection.

XV-4.   There is only so much energy available from the carbohydrate producers, and most of it is lost to heat at each step of the food chain. The population that shares the available energy must therefore decrease with each step along the food chain.

XV-5.   Direct energy release from mankind's activities are likely to raise global temperatures by only about 0.5°C, but increases in "greenhouse" gases such as carbon dioxide are of much greater concern.

XV-6.   Carbonate sediments by far hold the greatest amount of carbon in storage.

XV-7.   As an omnivore, you are a consumer on the left side of the diagram. Virtually everything you eat contains carbon. Carbon leaves you in the form of bodily wastes and carbon dioxide exhaled in your breath. When you die the remaining carbon in your body will be passed on to the decomposers if you are buried or directly to the atmosphere if you are cremated.

XV-8.   Molecular nitrogen must be "fixed" before it can be used by the higher food chain. Nitrogen fixation occurs mostly in the soil.

XV-9.   Phosphates do not occur in gaseous form and so must be carried by water in nature.

XVI-1. Sulfur emission from coal burning and nitrogen oxides from chemical fertilizers.

XVI-2. Chlorofluorocarbons (CFC's) and nitrogen oxides.

XVI-3. No significant decrease has been detected yet, but the substances that can cause depletion are slow-acting and remain in the stratosphere for long periods of time, and can cause future depletion.

XVI-4. Rain shadows in the lee of mountains and persistent downwelling currents in the atmosphere that are influenced by the distribution of oceans on Earth.

XVI-5. Overgrazing and firewood gathering are two important factors in the spread of deserts in less-developed countries.

XVI-6. Increasing salinity of the soil and a high water table due to irrigation.

XVI-7.   Carbon dioxide regulation in the atmosphere and serving as the habitat for large numbers of species are two of the ways in which tropical rain forests are important to the biosphere.

XVI-8.   The large amount of smoke and soot injected into the atmosphere due to the burning of cities and forests.

XVI-9.   The rising columns of heated air over destroyed cities in the northern hemisphere might reverse the flow of the northern Hadley Cell, providing a direct circulation path between the two hemispheres.

XVI-10. True exponential growth doubles with every passage of a fixed length of time, called the doubling time. Recent human population growth is accelerating in such a way that successive doubling times are becoming ever shorter.

XVI-11. During the demographic transition, death rates fall due to increased public health measures, but birth rates reamin high.

GLOSSARY

Note: If you can't find a term in this glossary, try a standard dictionary. Whenever you are studying, you should always keep a good dictionary at hand.

Abyssal plain
A flat, sediment-covered region of the deep sea floor.

Accreted terrane
A region that formed apart from the landmass to which it is now attached, and that became attached or accreted to it at a later time.

Accretion
The process of growth of planets and planetesimals by the infalling of matter that is swept up from interplanetary space.

Achondrites
Stoney meteorites containing little or no metal that are igneous in origin and probably resemble Earth's mantle rocks.

Aerobic
Pertaining to life that requires free oxygen.

Airglow
A faint glow produced in the lower ionosphere and upper mesosphere due to the absorption of solar radiation and particles in that region.

Albedo
A measure of the extent to which a surface reflects incident light.

Algorithm
A procedure for obtaining a desired result, generally applied to mathematical procedures.

Amino acid
Simple organic compounds that are the fundamental building blocks for living organisms.

Anaerobic
pertaining to organisms that require or can live in the absence of free oxygen.

Andesite
A generally fine-grained igneous rock of intermediate chemical composition.

Anthracite
The highest grade of coal, with high heating value and generally low sulfur content.

Anticline
A fold in rock stata that is convex upward.

Asteroid
Small rocky bodies in the solar system, many of which are found between the orbits of Mars and Jupiter.

Asthenosphere
The weak, soft region of the mantle just below the lithosphere.

Aulacogen
The failed rift-valley arm of what began as a triple spreading ridge junction.

Aurora
The Northern Lights (Aurora Borealis) and the Southern Lights (Aurora Australis).

Auroral oval
A circular band roughly centered on the magnetic poles in which aurora are most frequently visible.

Baltica
Northeastern Europe, including Scandinavia.

Basalt
A fine-grained mafic igneous rock.

Basement
The igneous or metamorphic layer beneath sedimentary strata.

Bathymetric map
A map showing the depths to the ocean floor.

Bauxite
The principal ore of aluminum, occurring as an oxide.

Benthic storm
Turbulent transport of sediment along the deep-sea floor by rapidly-moving currents.

Biogenic
Of biological origin.

Biogeochemical cycle
The process in which elements such as carbon, oxygen, nitrogen, phosphorus, sulfur, and others are cycled through the biosphere and geosphere.

Biomass
Pertaining to all organic matter produced by presently (or very recently) living organisms.

Biosphere
That portion of the rocky and fluid world inhabited by life.

Bitumen
A general name for various solid and semisolid hydrocarbons.

Bituminous coal

A dense black coal of high grade that burns with a flame.

Black smokers

Rock chimneys on the sea floor venting hot water with dissolved minerals suspended in it, giving the appearance of smoke.

Brachiopods

Marine shelled animals with two unequal shells.

Bow shock

The boundary between the magnetosheath and the solar wind.

Breeder reactor

A reactor in which some of the neutrons produced are used to convert nonfissionable Uranium-238 into fissionable Plutonium-239.

Caldera

The crater or bowl-shaped depression formed at the summit of a volcano or by the collapse of a magma chamber.

Callisto

A moon of Jupiter.

Carnivore

A meat eater.

Catastrophism

The invocation of a general catastrophe at some time in the past in order to explain differences between fossils and fossil abundances found in different strata.

Centripetal acceleration

The acceleration of an object toward a fixed central point that produces circular motion.

Chemosynthesis

The use by organisms of energy from chemical reactions, rather than from light, as in photosynthesis.

Ceres

The largest asteroid.

Charon

The satellite of Pluto.

Chlorofluorocarbons

A class of relatively inert gases used as propellants in spray cans and as refrigerants in air conditioners and refrigerators.

Chondrites
Stoney meteorites whose compositions match that of the atmosphere of the Sun. They contain small grains of nickel-iron metal.

Chloroplasts
Organelles that carry out photosynthesis within cells.

Chromite
The principal ore of chromium, an oxide.

Chromosphere
The region of the Sun's surface that lies between the photosphere and the transition zone. It appear magenta or pink during total eclipses of the Sun.

Climate proxy
An indicator of past climates.

Comets
Small bodies in orbit around the Sun that are composed of rock and ice. When they approach the Sun, vapor is produced from the ice that is swept by the solar wind away from the Sun, producing a tail.

Conduction
The process by which heat travels through solid bodies from a warm region to a cooler region.

Continental margin
The edge of the continent consisting of the continental shelf, slope, and rise.

Continental rise
The gradual slope between the continental slope and the deep ocean floor.

Continental shelf
The relatively flat portion of continental crust that is covered by shallow seawater.

Continental slope
The edge of the continental shelf where it slopes down toward the deep ocean floor at an average angle of 4 degrees.

Convection
The process by which heat is carried by a moving fluid (gas or liquid) from a warm region to a cooler region. Convection is caused by density differences in the fluid caused by heat.

Coriolis Effect
The apparent tendency of moving objects, as seen from the rotating surface of a planet, to veer off to one side or the other, depending on whether the object is north or south of the equator.

Corona
The outermost portion of the Sun, visible during total solar eclipses.

Coronagraph

A telescopic device that blocks out the bright disk of the Sun, allowing the much dimmer corona to be photographed.

Coronal holes

Regions of the Sun in which magnetic field lines extend far out into interplanetary space before looping back to the Sun.

Cracking

The process of breaking large, complex molecules (especially hydrocarbons) into smaller, simpler ones.

Creationists

Those who believe that life originated in one or a series of specific creations, as opposed to evolutionists, who believe life evolved in a slow and relatively gradual process from simple to complex forms. Creationists generally believe that geological history should be consistent with events described in the Book of Genesis in the Bible.

Crust

The outermost layer of the Earth, above the Mohorovicic discontinuity.

Cyclone

A circulating air mass that rotates counterclockwise in the northern hemisphere and clockwise in the southern hemisphere. Also used in many parts of the eastern hemisphere as synonymous to "hurricane".

Declination

The angle between true or geographic north and magnetic north.

Deimos

A satellite of Mars.

Demographic transition

The transition of a society from a situation with high birth and death rates to one in which birth and death rates are low. During the transition, population usually increases rapidly.

Dendrochronology

Dating via tree rings.

Desertification

The spread of deserts.

Deuterium

Heavy hydrogen, an atom whose nucleus consists of one proton and one neutron. Ordinary hydrogen has only one proton.

Dione

A satellite of Saturn.

Diorite
A coarse-grained igenous rock of intermediate chemical composition.

Dipole magnetic field
A magnetic field that has two antipodal poles of opposite polarity.

Doldrums
A region of light winds in the intertropical convergence zone.

Dolomite
A calcium-magnesium carbonate, often similar in appearance to limestone.

Early Faint Sun Paradox
While the Sun has increased its output of energy by as much as 30% during the course of its evolution, worldwide climate has not changed so much as to prevent the continuance of life.

Echo sounder
An acoustic device for measuring the depth of the ocean floor.

Eckman-layer flow
The tendency of water in the upper layer of the ocean to flow at right angles to the direction of the wind that blows along the sea surface.

Ecliptic
The plane of Earth's orbit as projected on the sky. This is the apparent path of the Sun through the constellations of the Zodiac.

El Niño
A change in the normal interaction of prevailing winds and ocean currents in the equatorial Pacific Ocean, characterized by a failure of the normal upwelling currents off the coast of Peru and Chile and other effects.

Eolian
Applied to the erosive action of the wind and its ability to transport sediment.

Epicenter
The point on the surface of the Earth directly above the principal energy release during an earthquake.

Erosion
Any process that loosens or disintegrates rock or soil and allows it to be transported elsewhere.

Escape velocity
The velocity that must be imparted to a mass (neglecting air friction) in order for it to permanently leave the vicinity of a planet or star.

Eukaryote
An organism whose genetic material is organized into a well-defined nucleus which is surrounded by a nuclear wall membrane.

Europa
A satellite of Jupiter.

Eutrophication
An oversupply of nutrients in that can cause algal blooms and depletion of free oxygen in bodies of water.

Evaporite
Mineral deposits left behind after evaporation of the fluid in which they were dissolved.

Extratropical cyclone
The air circulation system around low-pressure cells; not necessarily associated with major storms.

Fault
A fracture in a rock mass across which there has been displacement of one side relative to the other.

Felsic
Rocks characterized by high silica content and low iron and magnesium content. Generally low in density and light in color.

Fetch
The extent of open ocean across which prevailing winds may blow unhindered.

Firestorm
A general conflagration characterized by high temperatures, fierce hot winds, and exhaustion of oxygen in the air.

Flares
See Solar flares.

Flood basalt
Huge outpouring of basaltic lava from fissures in the ground; not necessarily associated with volcanic mountains.

Fossil fuels
Fuels derived from organic matter in geologic deposits coal, oil, natural gas, tar sands, and oil shales.

Fossils
The remains or traces of animals or plants that have been preserved in rocks by natural processes.

Fractionation
Separation of one or more substances from a mixture.

Fraunhofer lines

Dark absorption lines in the solar spectrum that can be used to determine the composition of the outer layers of the Sun.

Fumarole

A vent in the ground emitting volcanic gases and sometimes hot water.

Fusilinids

Important microfossils in the Pennsylvanian and Permian systems. They became extinct in the great dying at the end of the Paleozoic.

Gabbro

A coarse-grained mafic igneous rock.

Galilean satellites

The four largest moons of Jupiter Io, Europa, Ganymede, and Callisto, that were observed first by Galileo.

Gamma ray

Extremely energetic electromagnetic radiation, with wavelengths less than $1 \times 10_{10}$ meter.

Ganymede

A satellite of Jupiter.

Geocentric model of the solar system

An early model of the solar system in which the Sun revolves about the Earth, which stands in the center.

Geoid

The shape of the Earth as defined by sea level.

Geophone

A sensitive microphone designed to detect low-frequency vibrations produced in seismic surveying.

Geosphere

The solid part of the Earth.

Geostrophic wind

A wind that moves parallel to isobars and perpendicular to the pressure gradient. It is a consequence of the Coriolis Force.

Geothermal energy

Energy derived from the Earth's crust, especially in volcanic regions.

Geyser

A hot spring in which water and steam are periodically expelled with some force.

Globigerina

A microscopic single-celled marine organism that secretes a hard calcium carbonate test, or outer shell, which makes it a useful fossil for stratigraphic studies.

Gondwana

A supercontinent consisting of South America, Africa, Madagascar, Antarctica, Australia, New Zealand, and other smaller landmasses, that existed throughout the Paleozoic.

Grab sampler

A scoop lowered to the ocean floor for sampling sediment and rock.

Granite

A coarse-grained felsic igneous rock.

Granulation

A mottling of the photosphere of the Sun due to the motion of convection cells below it.

Gravimeter

A device designed to measure the strength of the force of gravity.

Greenhouse Effect

Warming in the lower atmosphere due to the ability of air to transmit visible light from the Sun but to absorb infrared radiation attempting to escape into space.

Gyre

A large-scale rotating mass of fluid, such as those found in the ocean basins.

Hadley cell

Vertically-oriented circulating air patterns present in the tropical atmosphere. See Figure VII-9 on page 229.

Half-life

The time needed for the atoms of a parent radioisotope to decay to one-half their original number.

Heat islands

A region of somewhat higher temperatures produced by man-made activities around cities.

Heavy oil

Tar or very viscous oil that cannot be pumped from the ground like crude oil.

Heat capacity

The amount of heat that must be put into a gram of a substance in order to produce a one degree Celsius temperature rise in it. Similarly, the amount of heat that leaves a gram of the substance when it cools by one degree Celsius.

Heliocentric model of the solar system

The present model of the solar system in which Earth and the other planets revolve in orbits around the Sun, which stands in the center.

Herbivore
Plant eaters.

Holocene epoch
The relatively warm period since the last Ice Age. Approximately the last 10,000 years.

Horse latitudes
A region of light winds in the high pressure zone between the westerlies and the trade winds.

Hot spot
A localized source of heat within the mantle that produces volcanism above it. Hot spots often produce lines of volcanoes.

Humidity
See Relative humidity.

Hydrocarbons
Compounds consisting only of carbon and hydrogen.

Hydrologic cycle
The recirculating path that water takes in the natural world, evaporating from the oceans, falling as rain, traveling in rivers back to the oceans.

Hydrothermal
Hot or warm water, usually within porous or permeable rocks.

Hyperion
A satellite of Saturn.

Hypsometric diagram
A diagram showing the proportion of Earth's rocky surface that stands at different elevations.

Iapetus
A satellite of Saturn.

Igneous rocks
Rock produced by the freezing of molten rock.

Impact craters
Depressions blasted out of rock by infalling meteorites.

Inclination
The angle that the magnetic field makes with the horizontal.

Inner core
The innermost part of the Earth, probably solid iron.

Interglacial
A warm period separating Ice Ages.

Intertidal zone
The elevation zone between high and low tides.

Intertropical convergence zone
A region, generally near the equator, where the trade winds converge to replace air that rises into the Hadley cells.

Io
The innermost satellite of Jupiter.

Ion
An atom or molecule that has become electrically charged through the loss or gain of one or more electrons.

Ionization
The process of producing ions from neutral (uncharged) atoms or molecules.

Ionosphere
The lower portion of the thermosphere, characterized by the presence of ions. The aurora are produced within this region.

Isobar
A line on a map showing those points that all have the same atmospheric pressure.

Isoseismal map
A map depicting zones of equal intensity on the Modified Mercalli scale for a given earthquake.

Isostasy
The concept that explains large-scale surface elevations in terms of the buoyancy of the low-density crust as it "floats" on the higher-density mantle.

Isotopes
Forms of single element that have differing numbers of neutrons in their nuclei, but all with the same number of protons. Different isotopes of an element have different masses, but nearly the same chemical properties.

Jet stream
Fast-flowing rivers of air in the upper atmosphere.

Jovian planets
Jupiter, Saturn, Uranus, Neptune -- the large, gaseous, low-density outer planets with many satellites and rings.

Kazakhstania
A portion of central Asia east of the Ural Mountains that may have traveled as part of a separate microplate at some time in the Paleozoic.

Kinematics

The description of motion, as opposed to dynamics, which is a description of the forces that produce motion.

Land bridges

An early explanation for the appearance of similar land fossils on continents now separated by ocean basins, in which a narrow bridge of land connected the continents and has since sunk into the sea. Continental drift is now the favored explanation.

Landsat

A remote-sensing satellite designed to carry out multispectral studies of Earth's landmasses.

Latent heat of vaporization

The energy required to vaporize a unit mass of liquid. The same amount of heat is released when the vapor condenses back into liquid.

Laterite

A red soil characteristic of tropical regions, in which silica has been leached out, leaving behind a residual soil rich in aluminum and iron oxides.

Laurentia

A continental landmass consisting of present-day North America, Greenland, and a bit of northern Europe, including Scotland, that apparently existed during the Early Paleozoic era.

Laurussia

A continental landmass formed by the union of Laurentia and Baltica that existed during the Late Paleozoic.

Lava

Molten rock extruded onto the surface of the Earth, as by a volcano.

Light water reactor

The most common type of nuclear reactor in use today.

Lignite

A brownish black coal that is higher-grade than peat but lower-grade than sub-bituminous coal.

Limestones

Sedimentary rocks formed from calcium-rich sediments, often made up largely of the shells and skeletal remains of biota.

Lineaments

Straight lines that appear on maps of landscape features, often indicating the presence of faults.

Lithification

The process by which sediments are made into rock.

Lithosphere

The rigid outermost portion of the Earth, comprising the moving plates. It includes the crust and a portion of the upper mantle.

Little Ice Age

A period extending from roughly 1500-1850 AD in which global temperatures were significantly colder than more recent averages.

Love wave

A type of surface wave in which the particle motion is horizontal.

Low velocity zone

A region of low seismic velocity that is observed most clearly beneath oceanic crust at a depth of roughly 100 to 300 km. It is a portion of the asthenosphere.

Mafic

Rocks characterized by high iron and magnesium content and low silica content. Generally dense and dark in color.

Magma

Molten rock occurring at depth within the Earth.

Magmatic concentration

The process in which magma tends to segregate into different constituents as it cools and solidifies.

Magnetic polarity reversal

An event in which the direction of the Earth's magnetic field reverses its direction everywhere on Earth's surface. In the process, the North and South Magnetic Poles are interchanged.

Magnetic storm

Short-term changes in Earth's magnetic field excited by events on the Sun.

Magnetograms

Applied to the Sun's magnetic field, a photograph that allows the polarity of the Sun's local magnetic fields to be determined. Applied to Earth's magnetic field, a record of short-term magnetic field activity that generally reflects the level of solar activity.

Magnetometer

A device designed to measure the strength, and sometimes the direction, of a magnetic field.

Magnetopause

The boundary between the Earth's magnetic field and that of the solar wind.

Magnetosheath

The region of magnetic turbulence between the magnetopause and the bow shock.

Magnetosphere

The region of interaction between Earth's and the solar magnetic fields.

Magnetospheric substorm

Disturbances that originate on the night side of the magnetosphere that can produce auroral displays.

Magnetotail

A long tail-shaped region of the Earth's magnetic field that points away from the Sun due to the action of the solar wind.

Manganese nodules

Concretions of manganese and iron oxides with other metals found in certain places lying on the deep-sea floor.

Mantle

The largest part of the Earth's interior, below the crust and above the core, probably composed of ultramafic rocks.

Maria

Large basaltic plains on the Moon.

Maunder Minimum

A period from 1645 to 1715 AD during which sunspots were hardly ever observed. Solar activity was presumably uncommonly low during this period.

Mechanical concentration

A process of concentration for heavy metals that do not react readily with other elements. They are concentrated by stream action or other mechanical action due to their density.

Medieval Optimum

An abnormally warm period that extended from about 900 to 1200 AD.

Mercalli Intensity Scale

See Modified Mercalli Intensity Scale.

Mesoscale

Middle scale; that is, neither the largest nor smallest possible.

Mesoscale eddy

Rotating mass of water on a smaller scale than that of a gyre. Same as ring current.

Mesosphere

A region of the atmosphere above the stratosphere in which temperature falls with increasing elevation.

Metamorphic rocks

Any rocks that have heated and/or squeezed to such an extent that their chemical and physical makeup are altered.

Meteor

A piece of extraterrestrial rock, metal, or ice falling to Earth and glowing incandescent in its flight through the atmosphere.

Meteorite

A piece of extraterrestrial rock or metal that has survived its fall to Earth.

Meteorology

The study of the atmosphere and weather.

Microplate

A smaller than continent-sized plate.

Milankovitch hypothesis

A hypothesis that relates Pleistocene climate changes to slight changes in Earth's orbit around the Sun.

Mimas

A satellite of Saturn.

Mineral

A naturally-occurring compound or element of definite compostion.

Mitochondria

Organelles within cells that act to combine carbohydrates and fatty acids with oxygen in order to release useful energy.

Modified Mercalli Intensity Scale

A measure of the destruction caused by an earthquake.

Mohorovicic discontinuity (Moho)

The boundary between the crust and mantle.

Monsoons

Seasonal shifts in atmospheric circulation that brings alternating wet and dry seasons, especially marked in the eastern hemisphere.

Moraine

Piles of gravel and rocks left by retreating glaciers.

Multispectral scanner

A device that is capable of scanning (and often, constructing images) at several specific wavelengths.

Neap tide

A lower-than-average tidal amplitude due to opposing effects of the Sun and Moon.

Nebula

An interstellar dust cloud.

Nemesis

The name given to a hypothetical companion star to the Sun, which has been postulated to explain periodic bombardments of the inner solar system by large numbers of comets.

Neutrino

A subatomic particle with no electric charge and a vanishingly small rest mass that travels nearly at the speed of light and seldom interacts with other matter.

Nitrogen fixation

The production of water-soluble nitrogen compounds from free nitrogen gas.

Noble gas

One of the inert gases helium, neon, krypton, radon. They do not normally form chemical compounds and are found in nature as elements.

Normal fault

A fault that occurs in response to horizontal tensional forces.

Nova

The explosion of a star, often triggered by the exhaustion of the star's nuclear fuels. An enormous amount of energy is released, accompanied by extremely high temperatures and pressures that can cause the nucleosynthesis of heavy elements.

Nuclear Winter

Global lowering of temperature expected to take place in the aftermath of a substantial nuclear weapons exchange.

Nutrients

Elements essential to life.

Occluded front

Occurs when a cold front overtakes a warm front. The warm air mass is lifted above the cold air masses and does not contact the ground.

Ocean basin

The deep parts of an ocean.

Oil shale

Fine-grained sediments containing organic matter that can be distilled to yield oil.

Olivine

An ultramafic class of minerals, magnesium-iron silicates, that are presumed to be abundant in the mantle.

Omnivores

Creatures that eat both plants and meat.

Oort cloud

A cloud of comets that is presumed to be orbiting the Sun beyond the orbit of Pluto.

Ooze

Ocean sediments rich in biogenic material.

Ophiolite

A suite of mafic and ultramafic igneous rocks associated with deep-sea sediments found on a landmass; now believed to be portions of uplifted sea floor.

Orbital ellipticity

The extent to which the Earth's orbit around the Sun is elliptical rather than circular.

Ore

Rocks in which certain minerals are concentrated by natural processes to such an extent that it is profitable to mine them.

Organelles

Small specialized units within eukaryotic cells.

Outer core

The liquid portion of the Earth between the inner core and the mantle, probably composed largely of molten iron.

Ozone layer

A concentration of ozone (O_3) in the lower stratosphere that absorbs much of the ultraviolet light from the Sun.

P wave

The first-arriving seismic wave characterized by particle motion parallel to the direction of wave travel.

Paleoclimatology

The study of ancient climates.

Paleomagnetism

The study of the ancient Earth's magnetic field as recorded in rocks.

Paleontology

The study of fossils.

Pangea

The supercontinent that existed at the end of the Paleozoic Era and that included all the landmasses of the world.

Peat

Partially decomposed remains of marsh plants that may be used as a low-grade fuel.

Pegmatite dikes

Coarse-grained igneous rocks found in dikes or cracks in the surrounding rock; usually associated with igneous intrusions.

Penumbra
The mottled outer portion of sunspots.

Permeability
The ability of a rock to transmit a fluid through pores and cracks.

Permo-Carboniferous
At the boundary between the Carboniferous and Permian Eras, 290 million years ago.

Phobos
A satellite of Mars.

Phosphorites
Phosphorus-rich deposits that tend to form on the sea floor near continents.

Photosphere
The outer portion of the Sun that constitutes its visible surface.

Photovoltaic solar cells
Solid-state cells that are capable of using radiant energy from the Sun directly to generate a flow of electricity.

Phytoplankton
Microscopic marine plants.

Pillow basalts
Rounded pillow-shaped basalt tongues that form underwater as fluid lava breaks through a cooled outer crust. The result looks somewhat like a pile of sandbags.

Pinger
An acoustic device lowered on an instrument package that determines how far the package is from the sea floor.

Piston corer
A device for sampling sediment on the ocean floor.

Placer deposit
Mechanically concentrated deposits of metals such as gold found in the sand and gravel of streams.

Planetesimal
A body in the solar system of smaller size than a planet.

Plankton
Free-floating marine organisms that are not able to swim rapidly. Many are quite small.

Plasma
A gas consisting largely of ionized particles.

Pleistocene epoch
The time of the Ice Ages, spanning roughly the past two million years.

Polar wander
The apparent motion of the magnetic poles as seen from a particular landmass. In most cases, it turns out that the pole is fixed and it is the landmass that is moving.

Polarity reversals
See Magnetic polarity reversals.

Porosity
The proportion of a rock that is pore space (the space between grains). Fluids may migrate freely through porous rocks.

Porphyry copper deposits
Large and often relatively low-grade copper deposits associated with igneous intrusions.

Positron
A sub-atomic particle of the same mass as an electron but with a positive charge.

Precession of the Equinoxes
A gradual and periodic change in direction of the Earth's axis in space.

Pressure gradient
The change in pressure over a unit horizontal distance, where the distance is measured in the direction of maximum pressure change.

Prevailing winds
The average direction of winds in a particular place and season.

Primordial cloud
The dust cloud from which the Sun and the rest of the solar system formed.

Principle of Original Horizontality
States that sedimentary rocks generally form with each layer flat and approximately horizontal.

Principle of Superposition
States that in an undisturbed series of strata, the oldest are on the bottom and the youngest are on the top.

Prokaryote
Primitive single-celled organisms, represented today by blue-green algae and bacteria.

Prominences
See Solar prominences.

Pyroclastic surge
A mixture of molten or nearly molten rock and hot gases that may be explosively erupted from volcanoes.

Pyroxene
An ultramafic class of minerals.

Radiant energy
Energy that travels in the form of electromagnetic waves.

Radioisotope
Radioactive isotopes.

Radiolaria
Marine protozoans that possess complex internal siliceous skeletons. They may be used as an indicator of ocean temperatures at the time that they lived.

Radiosonde
A weather balloon carrying measuring instruments and a radio for transmitting information back to the ground.

Rain shadow
The lee side of a mountain chain with respect to the prevailing winds, where dry conditions often result.

Rayleigh wave
A surface seismic wave in which the particle motion is similar to that in a water wave.

Refraction
The bending of a wave as it travels across the boundary between two substances with different wave velocities.

Relative humidity
A measure of the amount of moisture (as vapor) in the air compared to its carrying capacity.

Replacement
The process in which rocks are altered when they come into contact with hot fluids, sometimes forming mineral ores.

Reserves
Known deposits from which minerals can be extracted profitably using existing technology under present economic and legal conditions.

Residual concentration
Concentration of a mineral because it is less easily weathered than surrounding rock, which is carried away.

Resolution of a model
The smallest increment, in space or time, that can be resolved or distinguished in a numerical model.

Resources
Known potential sources of extractable minerals that might be used in the future if changes in technology or economic and legal conditions allow.

Rhea
A satellite of Saturn.

Rhyolite
A fine-grained felsic igneous rock.

Richter magnitude scale
A measure of the strength of an earthquake, based on measurement of the amplitude of seismic waves that is corrected for distance to the earthquake.

Rift valley
A valley caused by extensional forces acting to produce normal faults.

Rilles
Sinuous valleys cut into the lunar maria.

Ring current
See mesoscale eddy.

Rock
A solid, naturally-occurring mass of mineral matter. Rocks can occur with a wide variety of mineral content.

Rock magnetism
The study of magnetic properties and behavior of rock materials.

S wave
A seismic wave characterized by particle motion perpendicular to the direction of wave motion. It travels slower than P waves.

Salinity
Measure of the quantity of total dissolved solids in water.

Sandstones
Sedimentary rocks formed from sandy deposits.

Scientific Method
The logical procedure by which scientists attempt to explain natural phenomena. It is based on an interplay between observations and explanations (hypotheses, theories, models).

Sea-floor spreading
The process by which plates move apart at oceanic ridges, creating new sea floor.

Seasat
The first remote-sensing satellite dedicated solely to study of the oceans.

Sedimentary basin

Downwarped basin filled with sedimentary rocks. The sediments are thickest in the middle and become thinner toward the edges of the basin.

Sedimentary rocks

Rocks formed by the deposition and cementation of sediment grains.

Seiche

A periodic wave set up in a lake or bay by a seismic disturbance.

Seismic ray

A convenient geometrical representation of the direction of travel of seismic waves.

Seismic sea wave

A sea wave caused by an earthquake. Popularly, but incorrectly, known as a tidal wave.

Seismic tomography

A technique for constructing a three-dimensional seismic velocity model of Earth's interior.

Seismicity

The tendency of a region to be prone to the occurrence of earthquakes.

Seismograph

A device designed to detect vibrations from earthquakes or man-made sources.

Seismology

The study of earthquakes and the travel of vibration waves through the Earth and other planets.

Sensible heat

The heat content of a mass of air.

Shales

Sedimentary rocks formed from mud.

Shepherd moons

Small moons associated with the ring system of Jupiter that may be responsible for maintaining the integrity of some of the rings.

Shot point

The point at which an artificially-induced seismic wave is introduced into the ground.

Solar flares

An extremely large release of energy originating in a localized region of the solar corona.

Solar prominences

Arches and streamers extending far beyond the Sun's disk, consisting of material ejected during solar flares and guided by the Sun's magnetic field.

Solar wind

A flux of charged particles streaming outward from the Sun.

Sonar

A device that measures distance to the hard surface of the Earth by measuring the time for a pulse of sound to travel that distance, be reflected off the surface, and return to the instrument.

Source rock

The rock in which a hydrocarbon deposit forms.

Southern oscillation

The shift of a major persistent low-pressure cell in the western Pacific Ocean, probably linked to the occurrence of the El Nino phenomenon.

Spectrograph

A device designed to display or record the spectrum of a light source.

Spectroheliograph

A telescopic device that allows photographs to be made in the light of a single emission line.

Spectroscopy

The study of emission and absorption lines in spectra.

Spectrum analysis

The separation of a complex waveform into its constituent sinusoidal components, similar to the separation of a musical chord into its constituent notes.

Spicules

Sharp protruberances in the upper boundary of the chromosphere that give a furry appearance to the edge of the Sun.

Spreading ridge

The manifestation in the ocean basins of a diverging plate boundary, where new ocean floor is being created in the process of sea floor spreading.

Spring tide

A larger-than-average tidal amplitude, caused by additive effects of the Sun and Moon.

Squall line

A series of storms organized ahead of an advancing storm front.

Stable eddy

A persistent cyclone in a planetary atmosphere. These are especially prevalent on Jupiter.

Strata

Plural of stratum.

Stratosphere

The layer of the atmosphere between roughly 10 and 50 km in which the temperature rises with elevation.

Stratum

A single sedimentary rock unit that may be distinguished in its features from other rock units above or below it.

Structural trap

A fold, fault, or other structural feature of strata that forms a trap for migrating oil and gas.

Subduction

The process by which plates move toward one another and overlap, with the lowermost bending down into the mantle and being consumed.

Substorms

See Magnetospheric substorms.

Suture zone

The zone that marks an ancient collision between continents.

Swell

Long-wavelength sea waves that may travel long distances in the open ocean.

Synoptic map

Large-scale maps giving a synopsis of weather conditions at a particular time.

Taconite ores

Low-grade iron ores.

Tar sands

Porous sandstones that contain heavy oil.

Tectonics

The study of the deformation and movement of the crust.

Terrane

A region characterized by a particular rock type or origin.

Terrestrial planets

Mercury, Venus, Earth, and Mars -- the innermost planets that are rocky, dense, relatively small, and deficient or lacking in satellites and ring systems.

Terrigenous sediments

Sediments that originate on the landmasses.

Tethys Sea

The ocean that separated India and Africa from Asia at the time that Pangea existed.

Thermocline
A region of rapid temperature change in the oceans that divides the surface waters from the deep waters.

Thermonuclear fusion
The process by which four hydrogen atoms are fused to form one helium atom. A large amount of energy is released in the process.

Thermosphere
A region of the atmosphere above the mesosphere in which temperature rises rapidly with elevation above about 80 km.

Tidal heating
Frictional heating of a planet's or satellite's interior due to the flexing caused by the passage of tides raised by the gravity field of its partner.

Tillite
Lithified glacial deposits.

Titan
The largest satellite of Saturn.

Trace elements
Elements present in very small amounts.

Trade winds
Tropical prevailing winds that blow toward the equator from the northeast in the northern hemisphere and from the southeast in the southern hemisphere.

Transition zone
A narrow region separating the chromosphere from the solar corona in which the temperature rises rapidly.

Transform fault
A fault characterized by side-to-side motion (no gap or overlap formed) that connects two or more active plate boundaries.

Travel-time curves
A timetable that describes how long it takes various seismic waves to travel from a source to a destination.

Trench
The deepest parts of the oceans, associated with subduction zones.

Triple junction
The place where three plate boundaries come together.

Triton
A satellite of Neptune.

Tropopause
The boundary between the troposphere and the stratosphere.

Troposphere
The lowest portion of the atmosphere, in which the temperature drops with increasing elevation. Weather is largely confined to the troposphere.

Tsunami
The Japanese term for a seismic sea wave.

Turbidity current
A mass of water and sediment that flows down slopes in ocean or lake bottoms, sometimes attaining considerable force and speed.

Typhoon
Hurricane.

Umbra
The dark inner portion of sunspots.

Uniformitarianism
The principle that states that no ancient geological process may be invoked that could not be observed to be in operation in historical times.

Ultramafic
Dark and dense rocks composed largely of mafic minerals.

Volcanic dust veil effect
Atmospheric cooling produced by decreased solar radiation arriving at the surface due to dust injected into the atmosphere by a volcanic eruption.

Volcanic island arc
An arc of generally andesitic island volcanoes situated behind an ocean trench and above the downgoing lithospheric slab in a subduction zone, when the uppermost slab is oceanic.

Wavelength
The distance measured between successive crests of a wave.

Westerlies
Prevailing winds from the west that are found in northern and southern mid-latitudes.

White smokers
Sea-floor vents emitting warm water that is cloudy due to the presence of chemosynthetic bacteria.

Zooplankton
Marine animals that live in a floating state and drift with the ocean currents.

INDEX

D

E